Stallcup's Master Electrician's Study Guide

2005

National Fire Protection Association
Quincy, Massachusetts

Copyright © 2005 by Grayboy, Inc.

Published by the National Fire Protection Association, Inc.
One Batterymarch Park
Quincy, Massachusetts 02169

All rights reserved. No part of the material protected by this copyright notice may be reproduced or utilized in any form without acknowledgment of the copyright owner nor may it be used in any form for resale without written permission from the copyright owner and publisher.

Notice Concerning Liability: Publication of this work is for the purpose of circulating information and opinion among those concerned for fire and electrical safety and related subjects. While every effort has been made to achieve a work of high quality, neither the NFPA nor the authors and contributors to this work guarantee the accuracy or completeness of or assume any liability in connection with the information and opinions contained in this work. The NFPA and the authors and contributors shall in no event be liable for any personal injury, property, or other damages of any nature whatsoever, whether special, indirect, consequential, or compensatory, directly or indirectly resulting from the publication, use of or reliance upon this work.
    This work is published with the understanding that the NFPA and the authors and contributors to this work are supplying information and opinion but are not attempting to render engineering or other professional services. If such services are required, the assistance of an appropriate professional should be sought.

National Electrical Code® and *NEC*® are registered trademarks of the National Fire Protection Association, Inc.

NFPA No.: SME05
ISBN: 0-87765-675-4
Library of Congress Card Catalog No.: 2004109241

Printed in the United States of America
05   06   07   08   09   5   4   3   2   1

Introduction

Master – A skilled person who knows all there is to know about his or her work. A skilled worker, qualified to teach apprentices; craftsman in business for himself or herself.
(World Book Dictionary)

Stallcup's Master Electrician's Study Guide, based on the 2005 *NEC*® is designed for journeyman electricians who are preparing for a Master's exam. Even though this workbook is comprehensive in its approach to a technical subject, it may be used by anyone wishing to learn more about the 2005 *National Electrical Code*®. This book begins with pertinent information and progresses through calculations with step-by-step procedures found in the illustrations and including text with a "show and tell" presentation of how to apply provisions of the 2005 *NEC* when answering electrical questions and solving problems. Included are example problems and guided exercises to help students have a better understanding of how to perform calculations and solve problems that may appear on the Master's exam. Frequent reference to and study of the 2005 *NEC* is required to obtain ultimate results.

Stallcup's Master Electrician's Study Guide is a result of the author's preparing thousands of electricians wishing to pass their exam the first time they take it. He has been doing this since 1974. Typically, a Master's exam will dedicate twenty-five percent to electrical questions and basic calculations, forty percent to the *National Electrical Code* and thirty-five percent to calculations. This book contains hundreds of explanations with illustrations, examples and over two thousand (2,000) practice questions and problems covering all of these subjects.

To get the most out of this book, students should answer the questions at the end of each chapter. The over two thousand questions and problems contained in this book are typical questions from Master's exams given in each state, country, and city.

The Instructor's Guide for this book contains the answers with *Code*® substantiation for all questions. Answers, solutions, and *Code* substantiations are given for all problems. Copies of the Instructor's Guide may be purchased from NFPA®.

Table of Contents

SERVICES
CHAPTER 1

CLEARANCES 1-7
    VERTICAL 1-8
CLEARANCE FROM BUILDING OPENINGS 1-5
CONDUCTOR SIZE AND RATING 1-7
DISCONNECTING MEANS FOR SERVICE EQUIPMENT 1-13
    LOCATION 1-13
    MAXIMUM NUMBER 1-14
    GROUPING 1-15
    RATING 1-15
    MORE THAN ONE BUILDING 1-16
        DISCONNECT REQUIRED FOR EACH 1-16
GROUND-FAULT PROTECTION 1-18
    SETTING 1-19
    FUSES 1-19
IDENTIFYING HIGHER VOLTAGE-TO GROUND 1-13
LATERAL -SIZE AND RATING 1-10
NUMBER PERMITTED TO A BUILDING 1-1
OVERCURRENT PROTECTION FOR SERVICE EQUIPMENT 1-16
OVERHEAD SERVICES 1-6
POINT OF ATTACHMENT 1-9
SERVICE DROPS OR LATERALS 1-1
SERVICE-ENTRANCE CONDUCTORS 1-10
SERVICE MASTS 1-9
SERVICE OUTSIDE OF BUILDING 1-5
SERVICE PASSING THROUGH
ONE BUILDING TO SUPPLY ANOTHER 1-5
TEST QUESTIONS 1-21
UNDERGROUND SERVICES 1-10
    NUMBER 1-10
    SIZE AND RATING 1-11

SWITCHBOARDS AND PANELBOARDS
CHAPTER 2

SWITCHBOARDS AND PANELBOARDS 2-1
SUPPORT AND ARRANGEMENT OF BUSBARS AND CONDUCTORS .. 2-1
    USED AS SERVICE EQUIPMENT 2-2
    HIGH-LEG MARKING 2-3
    PHASE ARRANGEMENT 2-3
MINIMUM WIRE BENDING SPACE 2-4
SWITCHBOARDS 2-6
    INSTALLATION 2-6
    CLEARANCES 2-6
TEST QUESTIONS 2-15
USED AS A SUBPANEL 2-7

GROUNDING AND BONDING
CHAPTER 3

AC CIRCUITS AND SYSTEMS 3-2
    LESS THAN 50 VOLTS 3-2
    50 TO 1000 VOLTS 3-2
ATTACHMENT. 3-32
BONDING 3-26
BONDING OTHER ENCLOSURES. 3-28
CIRCUIT AND SYSTEM GROUNDING. 3-1
CONDUCTOR GROUNDED FOR AC SYSTEM. 3-9
CORD-AND-PLUG CONNECTED COMPUTERS.. 3-36
EFFECTIVE GROUND-FAULT CURRENT PATH. 3-14
GROUNDED AND UNGROUNDED SYSTEMS. 3-12
GROUNDED CIRCUIT CONDUCTOR FOR GROUNDING EQUIPMENT..3-14
GROUNDED CONDUCTOR BROUGHT TO SERVICE EQUIPMENT 3-5
GROUNDING ELECTRODE.. 3-6
GROUNDING ELECTRODE SYSTEM........................3-20
GROUNDING SEPARATELY DERIVED AC SYSTEMS. 3-9
GROUNDING SERVICE SUPPLIED BY AC SYSTEM 3-3
GROUNDING SUBPANELS AND EQUIPMENT. 3-16
HOT TUBS, SPAS, AND HYDROMASSAGE TUBS. 3-33
    BONDING. 3-33
    GROUNDING 3-34
HYDROMASSAGE BATHTUBS. 3-35
MAIN AND EQUIPMENT BONDING JUMPERS. 3-29
OVER 250 VOLTS. 3-29
PERMANENT-WIRED COMPUTERS.. 3-37
RESISTANCE OF ROD ELECTRODES. 3-31
ROD, PIPE, AND PLATE ELECTRODES. 3-25
SERVICE EQUIPMENT BONDING. 3-26
    METHODS. 3-27
SIZING GROUNDING ELECTRODE CONDUCTOR. 3-18
SWIMMING POOLS 3-32
    BONDING. 3-32
    GROUNDING. 3-33
TEST QUESTIONS 3-39
TYPES OF EQUIPMENT GROUNDING CONDUCTORS. 3-17

OVERCURRENT PROTECTION DEVICES AND CONDUCTORS
CHAPTER 4

ABOVE THE AMPACITY OF THE CONDUCTORS 4-40
BELOW THE AMPACITY OF THE CONDUCTORS 4-41

v

CIRCUIT BREAKERS 4-28
   SIZING 4-28
   INDICATING 4-29
   MARKING 4-30
   APPLICATIONS 4-30
CONDUCTORS
   CALCULATING LOAD FOR CONDUCTORS 4-34
   DERATING AMPACITY OF CONDUCTORS 4-34
   APPLYING 50 PERCENT LOAD DIVERSITY
   FACTOR FOR CONDUCTORS 4-36
   LOADING MORE THAN HALF 4-36
   PROTECTION OF CIRCUIT CONDUCTORS 4-36
   SERVICE CONDUCTORS 4-38
   FEEDER-CIRCUIT CONDUCTORS 4-38
   BRANCH-CIRCUIT CONDUCTORS 4-38
CONDUCTORS IN PARALLEL 4-54
EQUIPMENT GROUNDING CONDUCTOR 4-52
FEEDER CIRCUIT PROTECTION - OVER 600 VOLTS 4-31
FUSE MARKINGS 4-26
GROUNDED CONDUCTOR 4-51
LUMINAIRES (LIGHTING FIXTURES)
   PROTECTING CONDUCTORS 4-50
   SIZING WIRE TO FLUORESCENT LAY-INS 4-50
   SUPPLY WIRES FOR OUTSIDE LIGHTING
   STANDARDS 4-51
OCPD'S
   SIZING FOR PROTECTION OF EQUIPMENT 4-2
   ROUNDING UP OR DOWN OF OCPD 4-2
   SIZING OCPD FOR TAPS 4-4
   USING HANDLE TIES FOR CIRCUIT BREAKERS 4-18
   READILY ACCESSIBLE OF OCPD'S 4-22
   ACCESS OF OCCUPANT TO OCPD'S 4-23
   FUSES 4-25
   SIZING 4-26
OCPD'S FOR SYSTEMS OVER 600 VOLTS 4-31
SIZING CONDUCTORS
   FOR TAPS 4-41
   FOR WELDER LOADS 4-47
   TO NONMOTOR ARC WELDERS 4-47
   TO MOTOR OR NONMOTOR
   GENERATOR ARC WELDERS 4-49
   TO RESISTANCE WELDERS 4-49
TEST QUESTIONS 4-57
TERMINAL RATINGS 4-31
UNGROUNDED CONDUCTORS 4-53

RACEWAYS, GUTTERS, WIREWAYS AND BOXES
CHAPTER 5

AUXILIARY GUTTERS 5-22
BOXES 5-6
   OCTAGON BOXES 5-7
   SQUARE BOXES 5-9
   DEVICE BOXES 5-11
   OTHER BOXES 5-12
      PLASTER RINGS AND EXTENSION RINGS 5-15
CLAMP FILL 5-4
CONDUCTOR FILL 5-2
CONDUIT BODIES 5-16
DEVICE OR EQUIPMENT FILL 5-4
ENCLOSING DIFFERENT SIZE CONDUCTORS 5-20
ENCLOSING THE SAME SIZE CONDUCTORS 5-20
EQUIPMENT GROUNDING CONDUCTOR FILL 5-5
GUTTER SPACE 5-21
JUNCTION BOXES 5-17
METHODS OF COUNTING CONDUCTORS 5-1
PANELBOARDS 5-22
SIZING CABLE TRAYS 5-23
SIZING CONDUITS OR TUBING 5-20
SIZING NIPPLES 5-21
SUPPORT FITTINGS FILL 5-4
TEST QUESTIONS 5-27

FEEDER-CIRCUITS AND BRANCH-CIRCUITS
CHAPTER 6

AIR-CONDITIONING LOADS 6-42
APPLYING DEMAND FACTORS 6-6
CALCULATING AMPS 6-2
CONDUCTORS 6-20
   DERATING BY TABLE 310.15(B)(2)(a) 6-23
   AMBIENT TEMPERATURES 6-23
   VOLTAGE LIMITATIONS 6-25
   DETERMINING AMPERAGE 6-27
COMMERCIAL COOKING EQUIPMENT 6-38
COMMERCIAL AND INDUSTRIAL 6-28
CONTINUOUS OPERATED LOADS 6-5
DEMAND FACTORS 6-5
ELECTRIC DISCHARGE LOADS 6-30
FOR RECEPTACLE LOADS 6-6
HEATING LOADS 6-41
INDIVIDUAL 6-37
LIGHTING LOADS 6-28
LOADS 6-1
MOTOR LOADS 6-44
MULTIWIRE BRANCH-CIRCUITS 6-38
NEUTRAL 6-6
NONCONTINUOUS OPERATED LOADS 6-4
PERMISSIBLE LOADS 6-15
POWER FACTOR 6-2
RATINGS 6-15
RECEPTACLE LOADS 6-36
RESIDENTIAL 6-10
   GENERAL-PURPOSE CIRCUITS 6-11
   SMALL APPLIANCE CIRCUITS 6-11
   LAUNDRY CIRCUIT 6-12
   INDIVIDUAL CIRCUITS 6-12
   RESIDENTIAL COOKING EQUIPMENT 6-13
   DRYER EQUIPMENT LOADS 6-14

TEST QUESTIONS	6-47
WATER HEATER LOADS	6-40
VOLTAGE DROP	6-8

GENERATORS AND TRANSFORMERS
CHAPTER 7

GENERATORS	
AMPACITY OF CONDUCTORS FROM GENERATORS	7-6
BUSHINGS	7-9
GUARDS FOR ATTENDANTS	7-8
LOCATION OF GENERATORS	7-2
NAMEPLATE MARKINGS	7-2
OVERCURRENT PROTECTION	7-2
PROTECTION OF LIVE PARTS	7-8
TRANSFORMERS	
CALCULATING PRI. AND SEC. CURRENTS	7-10
FINDING AMPERAGE	7-11
GROUNDING	7-22
GUARDING	7-21
INSTALLING TRANSFORMERS	7-11
LOCATION	7-12
OVERCURRENT PROTECTION	7-14
VENTILATION	7-22
LOCATION OF TRANSFORMER VAULTS	7-23
DOORWAYS	7-23
DRAINAGE	7-24
STORAGE IN VAULTS	7-23
VENTILATION OPENINGS	7-245
WALLS, ROOF, AND FLOOR	7-23
WATER PIPES AND ACCESSORIES	7-25
TEST QUESTIONS	7-27

MOTORS AND COMPRESSORS
CHAPTER 8

AMPACITY RATING	8-56
BRANCH-CIRCUIT AND FEEDER-CIRCUIT CONDUCTORS	8-2
BRANCH-CIRCUIT CONDUCTORS	8-63
BRANCH-CIRCUIT REQUIREMENTS	8-70
DISCONNECTING MEANS	8-70
CAPACITOR	8-51
CLASS 1 CIRCUITS	8-54
CLASS 2 AND 3 CIRCUITS	8-54
CODE LETTERS	8-16
COMBINATION LOAD	8-64
CONTROLLERS FOR MOTOR COMPRESSORS	8-65
CORD AND ATTACHMENT PLUG CONNECTED MOTOR COMPRESSORS AND EQUIPMENT ON 15 OR 20 AMP BRANCH-CIRCUITS	8-69
DESIGNING MOTOR CURRENTS NOT LISTED IN TABLES	8-55
LOCATION OF THE DISCONNECTING MEANS FOR THE CONTROLLER AND MOTOR WITHIN SIGHT	8-42
LOCKED-ROTOR CURRENT UTILIZING HP	8-17
MAGNETIC STARTER CONTACTOR AND ENCLOSURE	8-53
MOTOR COMPRESSORS AND EQUIPMENT ON 15 OR 20 AMP BRANCH-CIRCUIT - NOT CORD-AND-PLUG CONNECTED	8-68
MOTOR CONTROL AND MOTOR POWER CIRCUIT CONDUCTORS	8-50
NAMEPLATE LISTING	8-55
DISCONNECTING MEANS	8-57
OVERLOAD RELAY APPLICATION AND SELECTION	8-68
RACEWAYS	8-52
OCCUPYING THE SAME ENCLOSURE	8-52
ROOM AIR-CONDITIONERS	8-70
RUNNING OVERLOAD PROTECTION FOR THE MOTOR	8-32
SINGLE-BRANCH CIRCUIT TO SUPPLY TWO OR MORE MOTORS	8-28
RATING AND INTERRUPTING CAPACITY	8-57
SINGLE MOTOR-COMPRESSORS	8-64
SIZING CONDUCTORS	
SINGLE MOTORS	8-2
SINGLE-PHASE MOTORS	8-2
THREE-PHASE MOTORS	8-2
MULTISPEED MOTORS	8-3
WYE-START AND DELTA RUN MOTORS	8-4
DUTY CYCLE MOTORS	8-5
PART-WINDING MOTORS	8-7
SEVERAL MOTORS	8-8
SIZING THE BRANCH-CIRCUIT PROTECTIVE DEVICE	8-9
SIZING AND SELECTING OCPD'S	8-18
ALLOW MOTORS TO START AND RUN	8-20
TWO OR MORE MOTORS	8-27
SIZING THE CONTROLLER TO START AND STOP THE MOTOR	8-34
SIZING THE DISCONNECTING MEANS TO DISCONNECT BOTH THE CONTROLLER AND MOTOR	8-38
OTHER THAN HP RATED	8-38
SIZING CONDUCTORS FOR CONTROL CIRCUIT	8-45
SIZING OCPD FOR CONTROL CIRCUIT	8-47
TEST QUESTIONS	8-73

RESIDENTIAL CALCULATIONS - SINGLE-FAMILY DWELLINGS
CHAPTER 9

APPLYING THE STANDARD CALCULATION	9-1
APPLYING THE STANDARD CALCULATION	9-10
SERVICE WITH GAS HEAT	9-10
SERVICE WITH ELECTRIC HEATING	9-12
SERVICE WITH ELECTRIC HEATING AND HEAT PUMP	9-13
SERVICE WITH 120 VOLT, A/C WINDOW UNIT	9-14
DIVIDING VA ON LINE AND NEUTRAL	9-16
DIVIDING AMPS ON LINE AND NEUTRAL	9-28
APPLYING THE OPTIONAL CALCULATION	9-28
HEAT OR A/C LOADS	9-29
EXISTING UNITS	9-35
OTHER LOADS	
ADDED APPLIANCE LOAD	9-35
ADDED A/C LOAD	9-35
ADDED 120 VOLT, A/C WINDOW UNITS	9-36
COOKING EQUIPMENT LOADS	9-2
DEMAND FACTORS	9-2
DRYER LOAD	9-7
FEEDER TO MOBILE HOME - STANDARD CALCULATION	9-36
FIXED APPLIANCE LOAD	9-6
GENERAL LIGHTING AND RECEPTACLE LOADS AND SMALL APPLIANCE PLUS LAUNDRY LOADS	9-2
LARGEST LOAD BETWEEN HEAT AND A/C	9-9
LARGEST MOTOR LOAD	9-9
MINIMUM SIZE SERVICE	9-9
TEST QUESTIONS	9-41

RESIDENTIAL CALCULATIONS - MULTIFAMILY DWELLING
CHAPTER 10

- APPLYING THE STANDARD CALCULATION 10-1
  - GENERAL LIGHTING LOAD 10-2
  - SMALL APPLIANCE AND LAUNDRY LOAD 10-2
- APPLYING THE STANDARD CALCULATION
  - SAME SQUARE FOOTAGE 10-15
  - DIFFERENT SQUARE FOOTAGES 10-16
  - SERVICE WITH UNITS HAVING GAS HEAT 10-17
  - SERVICE WITH 120 VOLT, A/C WINDOW UNITS 10-18
  - SERVICE WITH HEAT PUMPS 10-19
  - DIVIDING VOLT-AMPS ON LINE AND NEUTRAL 10-21
  - DIVIDING AMPS ON LINE AND NEUTRAL 10-22
- APPLYING THE OPTIONAL CALCULATION 10-23
  - SAME SQUARE FOOTAGE 10-24
  - DIFFERENT SQUARE FOOTAGE'S 10-24
  - SERVICE WITH UNITS AND HOUSE LOADS 10-24
  - DIVIDING VOLT-AMPS ON LINE AND NEUTRAL 10-25
- COOKING EQUIPMENT LOADS 10-4
- DEMAND FACTORS 10-2
- DRYER LOAD 10-9
- FIXED APPLIANCE LOAD 10-8
- GENERAL LIGHTING, RECEPTACLE LOADS AND
  SMALL APPLIANCE PLUS LAUNDRY LOADS 10-2
- LARGEST LOAD BETWEEN HEAT AND A/C 10-13
- LARGEST MOTOR LOAD 10-13
- PROBLEMS AND EXERCISES 10-26
- TEST QUESTIONS 10-47

COMMERCIAL CALCULATIONS
CHAPTER 11

- APPLYING THE STANDARD CALCULATION 11-1
  - LIGHTING LOADS 11-2
  - GENERAL LIGHTING LOADS 11-3
  - LISTED OCCUPANCIES 11-3
  - UNLISTED OCCUPANCIES 11-4
  - SHOW WINDOW LIGHTING LOAD 11-4
  - TRACK LIGHTING LOAD 11-6
  - LOW-VOLTAGE LIGHTING LOAD 11-6
  - OUTSIDE LIGHTING LOAD 11-6
  - SIGN LIGHTING LOAD 11-6
  - RECEPTACLE LOADS 11-8
  - APPLYING DEMAND FACTORS 11-8
  - MULTIOUTLET ASSEMBLIES 11-9
  - SPECIAL APPLIANCE LOADS 11-10
  - CONTINUOUS AND
    NONCONTINUOUS OPERATION 11-10
  - APPLYING DEMAND FACTORS 11-11
  - COMPRESSOR LOADS 11-11
  - MOTOR LOADS 11-12
  - HEATING OR A/C LOADS 11-13
  - LARGEST MOTOR LOAD 11-13
- STANDARD CALCULATIONS 11-15
  - STORE BUILDING SUPPLIED BY 120/240 VOLT,
    SINGLE-PHASE POWER SOURCE 11-15
  - STORE BUILDING SUPPLIED BY 120/208,
    THREE-PHASE VOLT POWER SOURCE 11-16
  - STORE BUILDING SUPPLIED BY 277/480 VOLT,
    THREE-PHASE POWER SOURCE 11-17
  - STORE BUILDING SUPPLIED BY 120/240 VOLT,
    THREE-PHASE POWER SOURCE 11-18
  - OFFICE BUILDING SUPPLIED BY 277/480 VOLT,
    THREE-PHASE POWER SOURCE 11-19
  - SCHOOL BUILDING SUPPLIED BY 277/480 VOLT,
    THREE-PHASE POWER SOURCE 11-21
  - RESTAURANT SUPPLIED BY 120/208 VOLT,
    THREE-PHASE POWER SOURCE 11-22
  - HOSPITAL BLDG. SUPPLIED BY 277/480 VOLT,
    THREE-PHASE POWER SOURCE 11-23
  - HOTELS AND MOTELS SUPPLIED BY
    120/208 VOLT, THREE-PHASE POWER SOURCE 11-25
  - BANKS SUPPLIED BY 120/208 VOLT,
    THREE-PHASE POWER SOURCE 11-26
  - WELDING SHOPS SUPPLIED BY 120/208 VOLT
    POWER SOURCE (UNLISTED OCCUPANCY) 11-27
- APPLYING THE OPTIONAL CALCULATION 11-27
  - KITCHEN EQUIPMENT 11-28
  - SCHOOLS 11-29
  - RESTAURANTS 11-29
- OPTIONAL CALCULATIONS FOR ADDITIONAL LOADS TO EXISTING INSTALLATIONS 11-29
- TEST QUESTIONS 11-55

TEST QUESTIONS
CHAPTER 12

TEST PROBLEMS
CHAPTER 13

TOPIC INDEX

Services

Master electricians must understand that the service equipment consists of two main parts: the service conduit and service conductors, the panelboard, and the overcurrent protection devices. The wiring methods for the service consists of four main parts: the service, the feeders, the subfeeders, and the branch-circuits. The size of the service-entrance conductors can be computed by adding the total volt-amps of all the branch-circuit loads and dividing by the configuration of voltage utilized. Loads are then calculated at continuous, noncontinuous operation, or with demand factors. This chapter covers such rules and regulations.

Quick Reference

NUMBER PERMITTED TO A BUILDING	1-1
SERVICE DROPS OR LATERALS	1-1
SERVICE PASSING THROUGH	
ONE BUILDING TO SUPPLY ANOTHER	1-5
SERVICE OUTSIDE OF BUILDING	1-5
CLEARANCE FROM BUILDING OPENINGS	1-5
OVERHEAD SERVICES	1-6
CONDUCTOR SIZE AND RATING	1-7
CLEARANCES	1-7
VERTICAL	1-8
POINT OF ATTACHMENT	1-9
SERVICE MASTS	1-9
UNDERGROUND SERVICES	1-10
LATERAL -SIZE AND RATING	1-10
SERVICE-ENTRANCE CONDUCTORS	1-10
IDENTIFYING HIGHER VOLTAGE-TO GROUND	1-13
DISCONNECTING MEANS FOR SERVICE EQUIPMENT	1-13
DISCONNECT REQUIRED FOR EACH	1-16
OVERCURRENT PROTECTION	1-16
GROUND-FAULT PROTECTION	1-18
TEST QUESTIONS	1-21

NUMBER OF SERVICES PERMITTED TO A BUILDING
ARTICLE 230, PART I

Only one service drop or lateral is permitted to be installed to a building. However, a second service drop or lateral is permitted to supply large loads and larger buildings. High-rise buildings permit additional service(s) to be installed to supply large loads and equipment. Any additional service drops or laterals for a building or multiple-occupancy building requires a plaque to be installed denoting locations at each service drop.

NUMBER OF SERVICE DROPS OR LATERALS
230.2

To accommodate the load requirements for a facility either one of the following conditions permits more than one service to be installed:

(1) Fire pumps per **230.2(A)(1)**

(2) Emergency lighting and power systems per **230.2(A)(2)**

(3) Multiple-occupancy buildings per **230.2(B)(1)**

(4) Buildings covering a large area per **230.2(B)(2)**

(5) Large capacity requirements per **230.2(C)(2)**

(6) Different characteristics per **230.2(D)**

(7) Underground sets of conductors per **230.2**

To ensure against the **interruption of power to fire pumps** a separate service may be installed. Fire pump overcurrent protection is provided and set above the pump motor's locked rotor rating including other accessory loads per **230.90(A), Ex. 4**. Running overload protection can be eliminated per **240.4(A)**. For continuous operation, a disconnecting means rated 1000 amps or more supplied by a 277/480 volt service does not require ground-fault protection per **230.95, Ex. 2** and **240.13(3)**. **(See Figure 1.1)**

Figure 1-1. To ensure against the interruption of power to fire pumps a separate service may be installed.

Master Test Tip 1: OCPD's for fire pumps shall be sized at 600 percent (General Rule) of the motor's FLA per **695.4(B)(1)**.

Master Test Tip 2: Transformers supplying fire pumps and accessories must be sized at 125 percent of all loads served per **695.6(C)(1) and (C)(2)**.

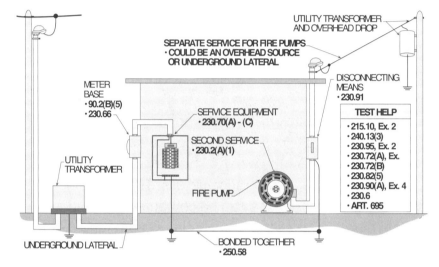

INTERRUPTION OF POWER TO FIRE PUMPS
NEC 230.2(A)(1)

Emergency lighting or power systems shall be permitted to be supplied by a separate service. Emergency lighting or power systems provide the needed power for lighting and equipment for the safety of occupants, if the main service is interrupted, for any reason such as loss of the normal power. For further information see **700.12(D)**. **(See Figure 1-2)**

Figure 1-2. Emergency lighting or power systems shall be permitted to be supplied by a separate service.

Master Test Tip 3: All wiring for emergency systems shall be kept separated from power wiring per **700.9(B)**.

Master Test Tip 4: Due to a power failure, generators shall restore power within 10 seconds per **700.12**.

Note that systems designed for connection to multiple sources of supply for the purpose of enhanced reliability per **230.2(A)(6)** are permitted.

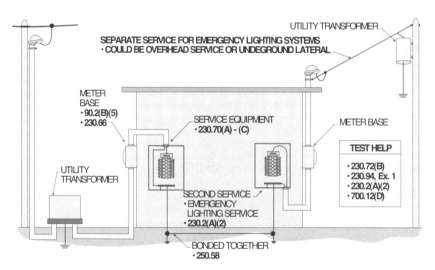

EMERGENCY LIGHTING OR POWER SYSTEMS
NEC 230.2(A)(2)

Note that by special permission, more than one service drop or lateral shall be permitted to be installed, provided there is space available for the service equipment to be mounted. Additional rules requires the OCPD's to be readily accessible for the user in each occupancy. For further information see **230.40, Ex. 1, 230.70(A) thru (C)**, and **230.71(A)**. For connection to multiple sources, per side bar.

When **multiple tenants** occupy individual units in an apartment complex or other types of buildings the service drop or underground lateral may be classified as a multiple occupancy and be served by more than one service. **(See Figure 1-3)**

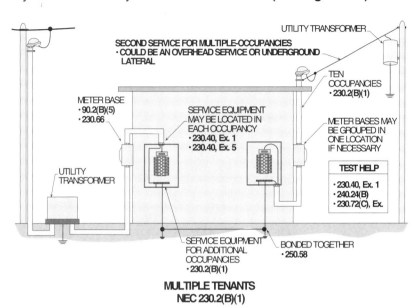

Figure 1-3. By special permission, more than one service drop or underground lateral shall be allowed to be installed.

Master Test Tip 5: A high rise building can have more than one service to supply loads and equipment per **230.2(B)(1)**.

Master Test Tip 6: An existing service of 1200 amp has a voltage of 277/480 V. A second service of 1200 amp can be added using the same voltage per **230.2(C)(1)**.

Load requirements in excess of 2000 amps shall be permitted to be supplied with more than one service to accommodate large loads that have been added. An additional service shall be permitted if the utility company requires more than one service due to the size of the load being added and the manner by which they must be served. **(See Figure 1-4)**

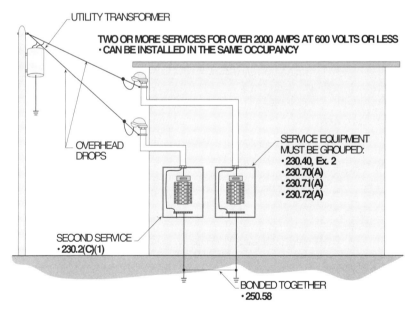

Figure 1-4. Load requirements in excess of 2000 amps shall be permitted to be supplied with more than one service per **230.2(C)(1)**.

Note: for connection to multiple sources, see **230.2(A)(6)** of the NEC.

Two or more services shall be permitted for **buildings that cover a large area.** Buildings that cover large areas usually have loads located a great distance from the service equipment. Two or more service drops or underground laterals are usually used to supply buildings with panelboards and switchboards located in different areas in the building because it is more practical than trying to use long routed feeder-circuits that will present voltage drop problems. **(See Figure 1-5)**

Master Note: High-rise buildings are not included in this requirement.

Figure 1-5. Two or more services shall be permitted for buildings that cover a large area.

Master Test Tip 7: A second service can be installed where the distance to the loads and equipment supplied are too far for feeder-circuits to be used per **215.2(A)(1), 215.9, 225.2, and 225.3**.

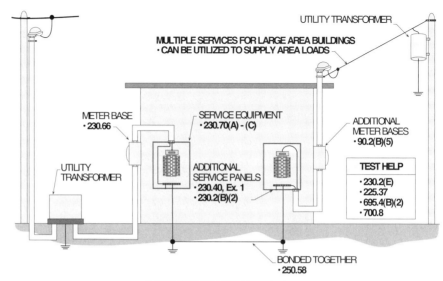

BUILDINGS THAT COVER A LARGE AREA
NEC 230.2(B)(2)

By special permission, a building of **different voltages (different characteristics)** is permitted to have two or more service drops or underground laterals. A building operating on a special rate schedule permits an additional service to supply specific loads or equipment such as computers, etc. **(See Figure 1-6)**

Figure 1-6. By special permission, a building of different voltages is permitted to have two or more service drops or underground laterals.

Master Test Tip 8: If a second service is installed, the service disconnects of each service must be grouped in a location acceptable to the AHJ per **230.72(A)** and **90.4**.

Master Test Tip 9: Only six service disconnects can be grouped in one location per **230.71(A)**.

Note that systems designed for connection to multiple sources of supply for the purpose of enhanced reliability per **230.2(A)(6)** are permitted.

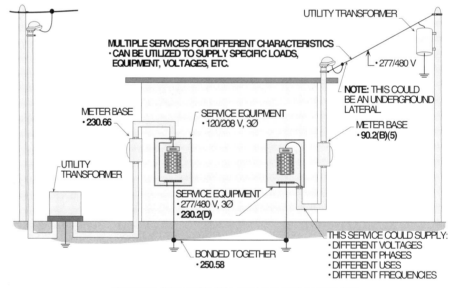

DIFFERENT VOLTAGES (DIFFERENT CHARACTERISTICS)
NEC 230.2(D)

For example, a service drop or underground lateral may consist of a 277/480 volt service and a 120/208 volt service supplying a building. The 120/208 volt service handles general-purpose outlets and equipment, while the 277/480 volt service takes care of special equipment.

A building shall be permitted to be supplied with **more than one set of underground laterals.** Section **230.2** requires all laterals to be grouped and located adjacent to one another. For this type of installation to be used, 1/0 AWG and larger conductors are required at the supply end, 1/0 AWG conductors must be in parallel and may be run and terminated to six or less main circuit breakers or disconnecting switches used exclusively for supplying individual services. **(See Figure 1-7)**

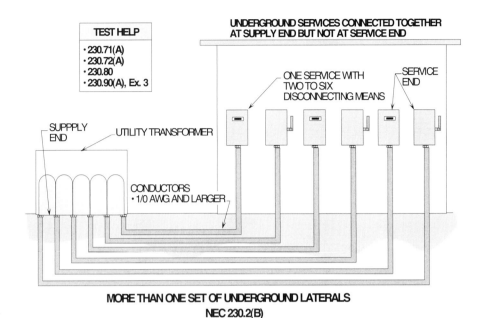

Figure 1-7. A building shall be permitted to be supplied with more than one set of underground laterals.

Master Test Tip 10: Either a master meter at the supply transformer or separate meters for each run must be provided per **90.2(B)(5)**.

SERVICE MUST NOT PASS THROUGH ONE BUILDING TO SUPPLY ANOTHER
230.3

Service conductors supplying a building or structure shall not pass through the interior of another building or structure, unless OCPD's are installed ahead of such conductors.

Service conductors installed in a duct or conduit and under at least 2 in. of concrete or encased with at least 2 in. of concrete or brick covering shall be considered outside of the building and may pass under, in, or on a building or structure to supply another building or structure. See **230.6** for the rules and regulations pertaining to this requirement.

Master Test Tip 11: A feeder-circuit with an OCPD ahead of it does not fall under this rule. Therefore, feeder-circuits can be routed through the building to another building without the need to apply **230.6**. (Also, see **230.3** and **408.36(A), Ex. 1**)

OUTSIDE OF THE BUILDING
230.6

Service conductors installed in a duct or conduit and under 2 in. of concrete or encased with at least 2 in. of concrete or brick covering shall be considered outside of the building. In other words, these conductors have never entered the building. Therefore, an OCPD does not have to be installed ahead of these conductors for them to be considered outside the building. **(See Figure 1-8)**

Master Test Tip 12: Service conductors and conduits routed outside of a building on the roof etc. are considered outside the building per **230.6**.

CLEARANCE FROM BUILDING OPENINGS
230.9

A clearance of 3 ft. shall be required from windows designed to be opened, including porches, platforms, etc. A clearance of 3 ft. shall not be required for service conductors attached above windows, for they are considered out of reach. Such clearance shall not be required for windows that do not open. **(See Figure 1-9)**

Figure 1-8. Service conductors installed in a duct or conduit and under 2 in. of concrete or encased with at least 2 in. of concrete or brick covering shall be considered outside of the building.

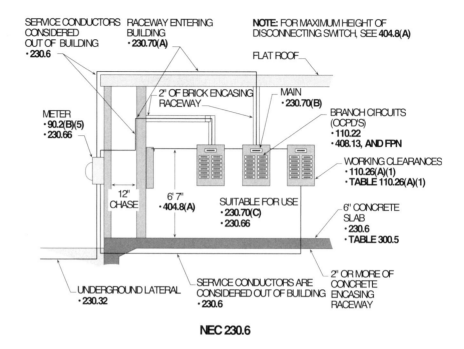

Figure 1-9. A clearance of 3 ft. shall be required from windows designed to be opened, including porches, platforms, etc.

Master Test Tip 13: Raceways or approved cables can be routed closer than 3 ft. to the window opening.

Master Test Tip 14: Service-drop grounded conductors shall not be smaller the minimum size calculated by **220.61** for normal currents and for ground-fault conditions per **250.24(C)(1);(C)(2)** and **310.15(B)(4)(A) thru (C)** to Table **310.16.**

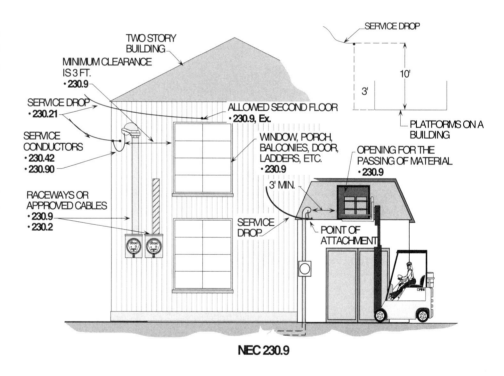

OVERHEAD SERVICES
ARTICLE 230, PART II

Service drops to a building can be supplied from an overhead supply from the utility pole to the attachment to the building. Such a supply is considered an overhead service drop. A minimum clearance from finished grade and the size and rating of the service drop conductors must be observed when installing these conductors. Before determining the proper height of a service drop, test takers taking the examination must consider who and what has access under these conductors.

SIZE AND RATING
230.23

Service-drop conductors shall not be smaller than 8 AWG copper or 6 AWG aluminum per **230.23(B)**. Service-drop conductors may be 12 AWG hard-drawn copper where installed to serve small loads of only one branch-circuit. Such installations of small loads are phone booths, small polyphase motors, etc. that do not draw large currents per **230.23(B), Ex.**

Master Test Tip 15: Service conductors must be capable of carrying the load as calculated per **220.5 thru 220.103**.

CLEARANCES
230.24

Service-drop conductors shall not be readily accessible for voltages of 600 volts or less. To ensure the protection of conductors and the safety of the general public, all clearances shall be provided underneath these conductors based on people travel or vehicle travel.

A clearance of 8 ft. shall be required for service-drop conductors passing over roofs. At least a 3 ft. clearance shall be maintained in all directions for vertical clearances. These clearances may be reduced by the following three exceptions:

Master Test Tip 16: The 8 ft. clearance rule of service conductors passing over flat roofs includes cooling towers and other such pieces of equipment per **230.24(A)**.

(1) Roofs subject to vehicular traffic per **230.24(A), Ex. 1**
(2) Voltage not exceeding 300 volts per **230.24(A), Ex. 2**
(3) Overhang portion of the roof per **230.24(A), Ex. 3**

Service-drop conductors shall have minimum clearances complying with **230.24(B)** where they are installed above **roofs subject to vehicular traffic.** Conductors shall have a vertical clearance of at least 8 ft. above flat roofs that are accessible to people per **230.24(A). (See Figure 1-10)**

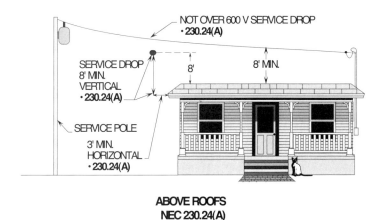

ABOVE ROOFS
NEC 230.24(A)

Figure 1-10. Service-drop conductors shall have minimum clearances, complying with **230.24(B)** where they are installed above roofs subject to accessibility.

Master Test Tip 17: Note that, if single service conductors are strung, they shall be insulated; this includes the neutral conductor.

If the roof has a slope of at least 4 in. by 12 in, service-drop conductors shall be permitted to pass above roofs where **voltage between conductors do not exceed 300 volts** and are installed at a height of 3 ft. Regardless of the slope of the roof, service-drop conductors shall be installed at a clearance of 8 ft. if the voltage between conductors exceeds 300 volts. **(See Figure 1-11)**

Figure 1-11. Service-drop conductors shall be permitted to pass above roofs where voltage between conductors does not exceed 300 volts and is installed at a height of 3 ft., and where the roof has a slope of at least 4 in. by 12 in.

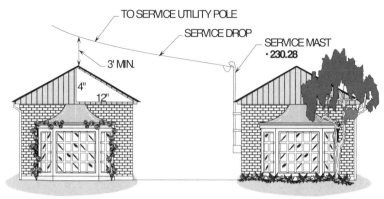

Figure 1-12. Service-drop conductors shall be permitted to be installed for roof overhangs provided no more than 6 ft. of such conductors does not pass over more than 4 ft. of the roof.

Master Test Tip 18: Note that the neutral conductor on the supply side can be used as a normal current carrying conductor and an EGC during a ground-fault condition per **250.142(A)(1)**.

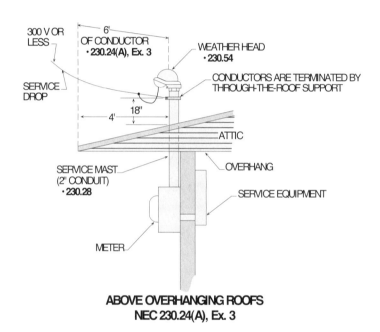

Service-drop conductors shall be permitted to be installed **above overhanging roofs** provided no more than 6 ft. of such conductors does not pass over more than 4 ft. of the roof. The attachment of the service-drop shall have a clearance of at least 18 in. from the conduit and roof line to prevent conductors from contacting the roof during a wind storm, etc. **(See Figure 1-12)**

VERTICAL CLEARANCE FROM GROUND
230.24(B)

Master Test Tip 19: The clearance from finished grade for overhead conductors is determined by "what" has access underneath. Having such access can be people, cars, trucks, etc. For voltage ratings, see **Figure 1-13**.

The vertical clearances from the finished grade to the service-drop conductors are determined by the voltage-to-ground. The following are the clearances required for each service voltage for service-drop conductors crossing or penetrating property or roof. Note that there are three levels of voltage (voltage-to-ground) on which these clearances are based.

Service-drop conductors **crossing property where people have access** shall have a clearance of at least 10 ft. from finished grade where the voltage is 150 volts or less to ground. This vertical clearance is considered a safe clearance, if only people have access under such conductors. **(See Figure 1-13)**

Service-drop conductors **crossing residential property or driveways** shall have a clearance of 12 ft. from finished grade where the **voltage is over 150 volts to 300 volts-to-ground.** Only a car or pickup truck, etc. should have access under these conditions, or greater heights shall be provided. **(See Figure 1-13)**

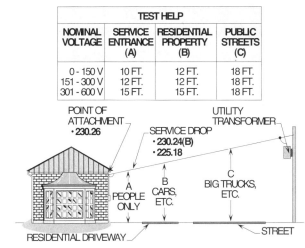

Figure 1-13. The vertical clearances from the finished grade to the service-drop conductors shall be determined by the voltage-to-ground.

Master Test Tip 20: When determining clearances for overhead conductors from finished grade, three voltages to ground shall be considered. See Column 1 of Test Help in **Figure 1-13**.

Service-drop conductors **crossing residential property or driveways** shall have a clearance of 15 ft. from finished grade where the **voltage exceeds 300 volts-to-ground.** Due to the higher voltage-to-ground, a greater height shall be provided to properly protect people and equipment from these higher voltages. **(See Figure 1-13)**

Service-drop conductors **crossing public streets, alleys, roads, parking areas** subject to traffic, or driveways on other than residential property, and other land transversed by vehicles such as cultivated, grazing, forest, and orchard shall have a clearance of 18 ft. from finished grade. This height is required to ensure safety for vehicles having greater heights and passing under such conductors. **(See Figure 1-13)**

POINT OF ATTACHMENT
230.26

The minimum point of attachment for service-drop conductors is 10 ft. from finished grade. The point of attachment shall be installed to provide the minimum clearances per **230.24**. The minimum point of attachment for service-drop conductors shall be installed based on the voltage level to ground. (For feeder-circuits, see **225.18**.)

Master Test Tip 21: The measurement for the point of attachment must be made to the attachment fitting itself or the drip loop, whichever is greater in height.

SERVICE MASTS
230.28

Service-drop cables shall be supported to a service mast of adequate strength to prevent damage to building structures. Service masts of adequate strength help prevent the strain on pipe where service-drop conductors are covered with ice or snow or subject to high winds. At least a 2 in. metal rigid or IMC conduit is normally required by most utility companies.

UNDERGROUND SERVICES
ARTICLE 230, PART III

An underground lateral is an underground service supplying electrical power to a building. Laterals may be terminated to a pad-mounted transformer or run underground and up a pole and terminated to an overhead transformer. An underground lateral may be installed inside or outside of a building with terminations in a terminal box or meter can. The conductors run between the terminal box or meter can and service equipment are considered service-entrance conductors and must be installed as such. **(See Figure 1-14)**

Figure 1-14. A lateral is an underground service supplying electrical power to a building.

Master Test Tip 22: The total rating of OCPD's in amps does not have to be equal to the ampacity of the conductors. Note that the conductors shall be sized with enough ampacity to supply the load per **230.90(A), Ex. 3.**

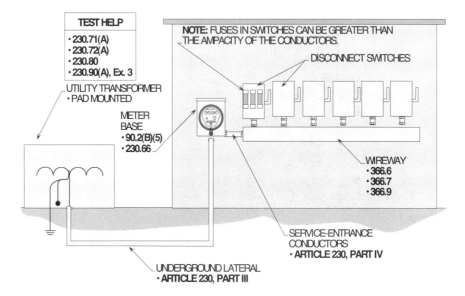

ARTICLE 230, PART III

SIZE AND RATING - LATERAL
230.31

Master Test Tip 23: Service-drop grounded conductors shall not be smaller than the minimum size calculated by **220.61**, used by **310.15(B)(4)** to Tables 0 - 2000 volts, and size d for fault-currents per **250.24(C)(1) and (C)(2)** as required by **230.31(C).**

Service lateral conductors shall not be smaller than 8 AWG copper or 6 AWG aluminum. However, service-drop conductors may be 12 AWG hard-drawn copper if they are installed to serve small loads of only one branch-circuit. Such installations of small loads are phone booths, small polyphase motors, etc. that require very small currents. **(See Figure 1-15)**

SERVICE-ENTRANCE CONDUCTORS
ARTICLE 230, PART IV

Service-entrance conductors must be sized and installed with enough capacity (amps) to supply the computed loads of the premises. Service-entrance conductors installed in a raceway or cable must be insulated to protect them from short-circuits and ground-faults, except the grounded neutral conductor can be bare under certain conditions of use.

NUMBER
230.40

Only one set of service-entrance conductors (to a building) can be supplied by a service-drop or service lateral. The following three exceptions allow service-entrance conductors to be installed with more than one set:

(1) More than one occupancy per **230.40, Ex. 1**
(2) Two to six services in separate enclosures per **230.40, Ex. 2**
(3) One or two-family dwelling units per **230.40, Ex. 3** and **Ex. 4**

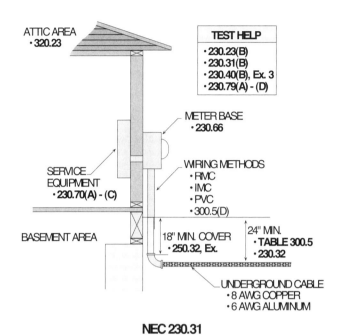

Figure 1-15. Service lateral conductors shall not be smaller than 8 AWG copper or 6 AWG aluminum per **230.31(B)**. (General Rule)

Each occupancy or group of occupancies in multiple occupancy buildings shall be permitted to have their service equipment supplied with one set of service-entrance conductors. Note that one to six service mains can be installed at such locations per **230.71(A)** and **230.72(A)**.

Two to six service disconnecting means in separate enclosures shall be permitted to be installed with one set of service-entrance conductors from a single service-drop or underground lateral.

Single-family dwelling units and separate structures shall be permitted to have one set of service-entrance conductors routed to their service equipment from a single service-drop or underground lateral. **Test Help:** Review **Ex. 5** to **230.40** very carefully.

Master Test Tip 24: Each one of the 2 to 6 service disconnects must identify the loads or equipment that they supply per **230.72(A)**.

SIZE AND RATING
230.42(A) and (B)

Service-entrance conductors shall be installed to carry the total calculated load of the premises per **Article 220**. Service-entrance conductors shall be sized and selected from the ampacities per **Tables 310.16 thru 310.19**, and all applicable Notes.

The ampacity for ungrounded conductors (hots) used for service-entrance shall be sized and selected by the minimum requirements as follows:

(1) A minimum 100 amp, three-wire service disconnecting means shall be installed for a one-family dwelling unit per **230.79(C)**. The minimum size service-entrance conductors allowed to be installed are 3 AWG THWN copper or 4 AWG THWN copper if the test taker is applying **310.15(B)(6)** and **Table 310.15(B)(6)**.

Master Test Tip 25: For other types of insulation permitted, see **310.15(B)(6)** and **Table 310.15(B)(6)** to Ampacity Tables 0-2000 volts.

Master Test Tip 26: Note that when applying **Table 310.15(B)(6) to Table 310.16**, a 4 AWG THWN copper conductor can be protected by a 100 amp OCPD.

(a) One or two two-wire branch-circuits shall have a 30 amp disconnecting means per **230.79(B)**.

(b) For limited loads, a 15 amp service disconnecting means may be installed per **230.79(A)**.

(c) Limited loads of a single branch-circuit shall be supplied with at least a 15 amp disconnecting means per **230.79(A).**

(d) Note; for all other installations, the service disconnecting means must have a rating of not less than 60 amps.

The ampacity for grounded neutral conductors for service-entrance must be sized and selected by the requirements of **220.61** and **250.24(C)(1)** and **(C)(2)**. **(See Figure 1-16)**

Figure 1-16. Service-entrance conductors shall be installed to carry the total calculated load of the premises per **230.42(A)(1)**. Service-entrance conductors shall be sized and selected from the ampacities per **Tables 310.16 thru 310.19**.

Master Test Tip 27: Service disconnects for dwelling units shall be sized and rated to comply with the rules of **230.79(A) thru (C)**.

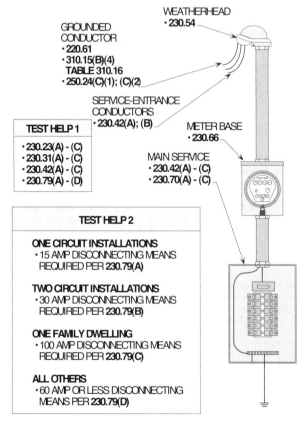

NEC 230.79(A) THRU (C)

Services

IDENTIFYING HIGHER VOLTAGE-TO-GROUND
230.56

When a master electrician is designing a three-phase, four-wire, delta-connected service and the midpoint of one phase winding is grounded, one phase leg will have a higher voltage-to-ground. An outer finish marked orange or identified by other effective means shall be used for the conductor at the higher voltage-to-ground. This identification will prevent an electrician from connecting 120 volt loads to 208 volt-to-ground (high-leg) conductor, for great damage can occur to 120 volt equipment if supplied by the high-leg. **(See Figure 1-17)**

Master Test Tip 28: The voltage to ground of the high leg can be found as follows:
- phase-to-phase = 240 V
- phase-to-ground = 120 V
- 240 V + 120 V ÷ 1.732 ($\sqrt{3}$) = 208 V

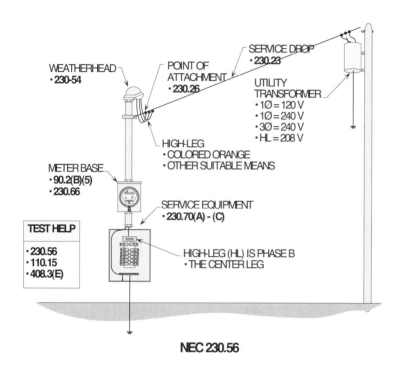

Figure 1-17. When installing a three-phase, four-wire, delta-connected service, where the midpoint of one phase winding is grounded, one phase leg will have a higher voltage-to-ground. An outer finish marked orange or identified by other effective means shall be used for the conductor with a higher voltage-to-ground.

DISCONNECTING MEANS FOR SERVICE EQUIPMENT
ARTICLE 230, PART IV

A means to disconnect the service shall be provided at each service. All ungrounded phase conductors must be disconnected from the power supply. The service disconnecting means (one or more) shall be permanently marked to identify the load it serves and then grouped in a common location where more than one service supplies a building.

Master Test Tip 29: Section **230.70(B)** requires a service main to be identified. Section **110.22** requires all disconnects to identify the loads that they serve. Section **408.4** requires the directory to be placed on the front or inside cover of the enclosure.

LOCATION
230.70(A)

The service disconnecting means shall be located in a readily accessible location for all premises. The service disconnecting means shall be located as close as possible to where the service-entrance conductors enter the premises. Service disconnecting means shall not be located in bathrooms. For OCPD rules, see Master Test Tip 30. **(See Figure 1-18)**

Master Test Tip 30: The service disconnecting means can be located on the outside or inside wall; such location complies with **230.70(A)**.

Note: OCPD's in panelboards must not be located in bathrooms of dwelling units and guest rooms in hotels or motels per **240.24(E)**.

Figure 1-18. The service disconnecting means shall be located in a readily accessible location for all premises.

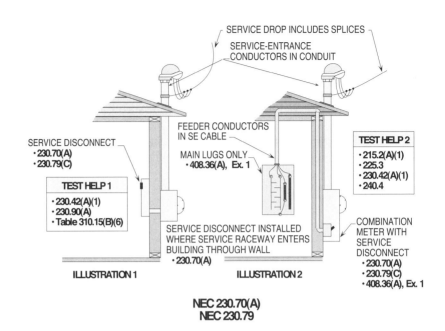

Master Test Tip 31: In Illustration 1, the service conduit must terminate to a service disconnecting means where it enters the building. In Illustration 2, there is an OCPD ahead of the conductors; therefore, there is no limit to the distance that they can be routed inside the building.

MAXIMUM NUMBER
230.71

No more than six fusible switches or six circuit breakers mounted in a single enclosure, a group of separate enclosures, or in or on a panelboard or switchboard can be installed for each set of service-entrance conductors. A power panel may have up to six circuit devices in a suitable enclosure. Six circuit breakers in a single enclosure may be used to supply air-conditioners, water heaters, heating units, ranges, dryers, and general-purpose circuits for lighting, receptacles, and appliances. **(See Figure 1-19)**

Figure 1-19. No more than six fusible switches or circuit breakers mounted in a single enclosure, a group of separate enclosures, or in or on a panelboard or switchboard can be installed for each set of service-entrance conductors.

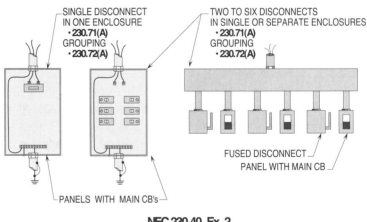

Master Test Tip 32: Equipment connected to the supply line of the service disconnecting means must comply with the provisions of **(1) thru (8)** to **230.82**.

A main circuit breaker may be installed ahead of other circuit breakers in a panel and one throw of the hand shuts the power OFF to all other circuit breakers and loads served. The disconnecting means for the service can also be six separate disconnecting means with fuses or separate circuit breakers, each in single enclosures.

GROUPING
230.72(A)

The service disconnecting means shall be grouped in a common location and be readily accessible so the entire service can be disconnected with six throws of the hand.

Dwelling units, apartment complexes, commercial facilities, or other types of premises shall be permitted to have their service designed with two to six disconnecting means. The disconnecting means shall be marked to identify the load they serve. Disconnects shall be required to be grouped in a common location where six or less are installed. A main disconnect shall be installed ahead of the disconnects if more than six disconnects are utilized. The following are two exceptions that allow switches to be ungrouped: **(See Figure 1-20)**

(1) Services installed per **230.2(A) thru (E)**
(2) Remote water pumps per **230.72(A), Ex.**

Master Test Tip 33: The service disconnecting means shall be grouped so the entire power supply of the service can be disconnected with no more than six throws of the hand per **230.71(A)** and **230.72(A)**.

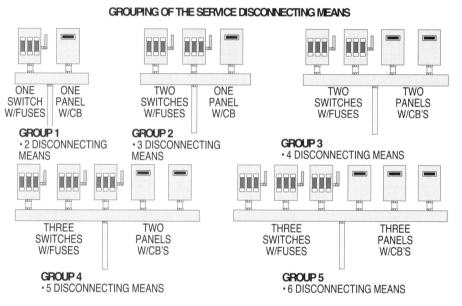

Figure 1-20. The service disconnecting means shall be grouped in a common location and readily accessible so the components of the service can be utilized, repaired, served, and maintained per **230.72(A)**.

RATING
230.79(A) THRU (D)

The main disconnecting means for the following installations shall have a rating with enough capacity (amps) to carry the calculated load per **Article 220**.

(1) One-circuit installation as permitted per **230.79(A)**
(2) Two-circuit installation as permitted per **230.79(B)**
(3) One family dwelling as permitted per **230.79(C)**

The service disconnecting means shall be required to be installed at a rating of 15 amps **for single branch-circuit loads** supplying limited loads such as phone booths, small polyphase motors, etc. that require very small amps to operate.

The service disconnecting means shall be required to be installed at a rating of 30 amps for **one or two, two-wire branch-circuit loads.** Note that these loads are small loads, which also require small amounts of current (amps) to operate.

Master Test Tip 34: The terminals of the service disconnecting means must be capable of interrupting the available fault-current per **110.9** and **110.10**.

MORE THAN ONE BUILDING
225.30 THRU 40

A separate disconnecting means shall be required if more than one building is on the same property and under the same management. In other words, each separate building shall be required to have a disconnecting means to completely disconnect all ungrounded phase conductors supplying such building or structure. **(See Figure 1-21)**

Figure 1-21. A separate disconnecting means is required if more than one building is on the same property and under the same management. However, disconnects can be located in the main building per **Ex. 1 to 225.32**.

Master Test Tip 35: There can be ("only") six or less disconnects grouped in any one location per **225.32, 225.33 and 225.34**.

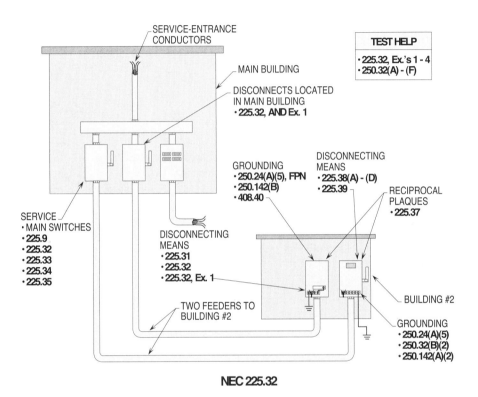

Note that a reciprocal plaque shall be placed at each feeder-circuit panelboard as required by **225.37**.

DISCONNECT REQUIRED FOR EACH
225.32

Each building shall be required to have the disconnecting means installed in a location, either outside or inside, in or on each building in a readily accessible location and as close as possible where the conductors enter in the building. This disconnect can be provided with OCPD or be of the nonfused or nonautomatic type.

OVERCURRENT PROTECTION
FOR SERVICE EQUIPMENT
ARTICLE 230, PART VII AND 230.90

The general rule requires each ungrounded conductor to be provided with an overcurrent protection device to protect the service equipment. The rating of service-entrance conductors shall not be exceeded when sizing overcurrent protection devices to protect the allowable ampacities of such conductors. However, there are exceptions that allow the OCPD to exceed the ampacity of the conductors, if a second stage of protection is provided to protect conductors and equipment from overloads.

REQUIRED
230.90(A)

The ampacity rating of service-entrance conductors must not be exceeded when installing overcurrent protection devices, unless one of the following exceptions is applied:

(1) Motors as permitted per **230.90(A), Ex. 1**
(2) Fuses and circuit breakers as permitted per **230.90(A), Ex. 2**
(3) Six circuit breakers or six sets of fuses as permitted per **230.90(A), Ex. 3**
(4) Fire pumps as permitted per **230.90(A), Ex. 4**

Motor conductors may be overfused for the start-up currents of motors per **430.52, 430.62,** or **430.63**. Overcurrent protection devices shall be sized to carry the high inrush current until the motor accelerates up to its running speed. **(See Figure 1-22)**

Master Test Tip 36: The OCPD for the service conductor can be sized greater than the ampacity of the conductor to allow large motors to start per 290.90(A), Ex. 1.

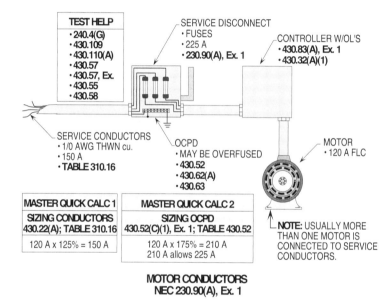

Figure 1-22. Overcurrent protection devices shall be sized to carry the high inrush current until the motor accelerates up to its running speed.

Note: For overload protection pertaining to service-conductors, see 230.90(A), Ex. 5, 310.15(B)(6), and Table 310.15(B)(6).

When the setting or rating does not correspond to the allowable capacity of service conductors, **the next higher standard fuse or circuit breaker is permitted to be installed.** This type of installation is permitted to be installed per **240.4(B)**. The standard ratings of fuses or circuit breakers are listed in **240.6(A)**. **(See Figure 1-23)**

The number of overcurrent protection devices allowed for a service can be **one to six circuit breakers or six sets of fuses** installed in a service-entrance panel. Note that there can be up to six separate enclosures with a single CB or single set of fuses in each. **(See Figure 1-24)**

For example, a 460 amp service-entrance (2 - 4/0 AWG) could be installed with four 100 amp and two 50 amp circuit breakers. The total selected rating of OCPD's would be 500 amps, which exceeds the 460 amp rating of the service-entrance. The service is required to be protected at 460 amps per **240.4(B)**, but the allowable ampacity of the service conductors is limited to the load per **230.90(A)** and it will never exceed the rating of the two 4/0 AWG copper conductors. Therefore, the six OCPD's rated at 500 amps are allowed.

The overcurrent protection device must be sized high enough to carry the locked-rotor current for a service supplying a **fire pump**. The overcurrent protection device must not clear until a short-circuit develops. This design technique allows the fire pump motor to pump water as long as possible to fight fire conditions. **(See Figure 1-25)**

Master Test Tip 37: Where the ampacities of the conductors do not correspond with a standard size OCPD, the next higher size may be used per 230.90(A), Ex 2.

Master Test Tip 38: If there are six or less disconnecting means, then the total rating in amps does not have to be equal to or less than the ampacity of the conductors per 230.90(A), Ex. 3 and 240.4(G).

Figure 1-23. When the setting or rating does not correspond to the allowable capacity of service conductors, the next higher standard fuse or circuit breaker shall be permitted to be installed per **230.90(A), Ex. 2** and **240.4(B)**.

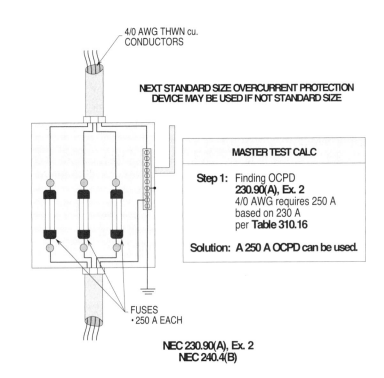

Figure 1-24. The number of overcurrent protection devices allowed for a service can be one to six circuit breakers or six sets of fuses installed in a service-entrance panel per **230.71(A)** and **230.72(A)**.

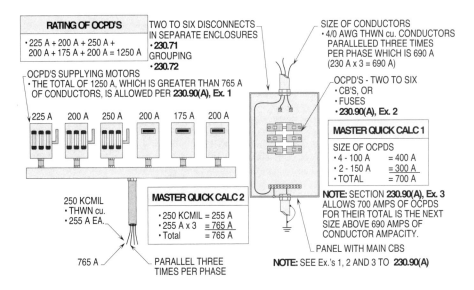

GROUND-FAULT PROTECTION
230.95

Master Test Tip 39: The ground-fault system shall be capable of clearing a fault of 3000 amps or more in one second.

Solidly grounded wye electrical services of more than 150 volts-to-ground having a service disconnecting means rated 1000 amps or more shall be protected by ground-fault protection. The service disconnecting means shall be rated for the largest fuse that can be installed or the highest continuous current trip setting for which the actual overcurrent device rating (circuit breaker) can be adjusted. **(See Figure 1-26)**

Services

SETTING
230.95(A)

The maximum setting permitted for ground-fault protection shall not exceed 1200 amps for any service disconnecting means exceeding 1000 amps or more. The ground-fault protection shall allow all ungrounded conductors of the faulted circuit to open, until such circuit(s) is clear of the low-magnitude fault.

To prevent damage to overcurrent protection devices, busbars, and service equipment the ground-fault system shall be required to clear a fault of 3000 amps or more within one second.

The system will be cleared once the low-magnitude fault-currents are detected and the overcurrent protection device is tripped open. Such trip will prevent low-magnitude faults from burning and damaging the electrical system beyond use.

Master Test Tip 40: Ground-fault protection shall not be required for six OCPD's that do not have a single OCPD equal to or greater than 1000 amps.

Master Test Tip 41: The maximum setting of the ground fault protection shall not exceed 1200 amps.

FUSES
230.95(B)

In a protection scheme consisting of a switch and fuse combination, the fuses shall be capable of interrupting any current higher than the interrupting capacity of the switch during a time when the ground-fault protective system will not cause the switch to open and clear the faulted circuit.

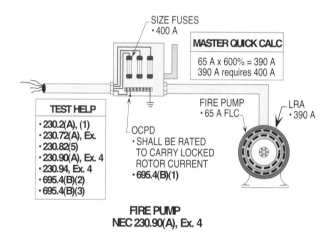

Figure 1-25. The overcurrent protection devices shall be sized high enough to carry the locked-rotor current for a service supplying a fire pump.

Note that on the Master's test, the OCPD for a fire pump is usually sized at 600 percent of the pump motor's FLA.

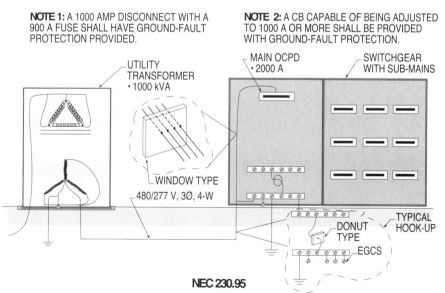

Figure 1-26. Solidly grounded wye electrical services of more than 150 volts-to-ground having a service disconnecting means rated 1000 amps or more shall be protected by ground-fault protection. These services are supplied by 277/480 V, three-phase, four-wire, wye systems.

Name _____   Date _____

Chapter 1
Services

                                                                Section   Answer

1. Service-drop conductors may be 14 AWG hard-drawn copper (based on 15 amp disconnecting means) where installed to serve small loads of only one branch-circuit.

      (a) True   **(b)** False

2. Service-drop conductors crossing property where people have access shall have a clearance of at least 8 ft. from finished grade.

      (a) True   **(b)** False

3. A minimum 100 amp service disconnecting means shall be installed for a single-family dwelling and may be supplied by 4 AWG THWN copper conductors per Table **310.15(B)(6)**.

      (a) True   **(b)** False

4. The minimum point of attachment for service-drop conductors shall be 10 ft. from finished grade.

      (a) True   **(b)** False

5. A connection ahead of the meter shall be permitted to be installed on the supply side of the main to serve a flat rate circuit for a water heater, heating equipment, or other such appliances.

      (a) True   **(b)** False

6. Service conductors installed in a duct or conduit and encased with at least _____ in. of concrete or brick covering shall be considered outside of the building.

      (a) 1   **(b)** 2   **(c)** 3   **(d)** 6

7. A clearance of _____ ft. shall be required from windows designed to be opened, including porches, platforms, etc.

      (a) 3   **(b)** 5   **(c)** 6   **(d)** 10

8. The service disconnecting means shall be grouped in a common location and readily accessible so the entire service can be disconnected with _____ throws of the hand.

      (a) 3   **(b)** 5   **(c)** 6   **(d)** 10

9. The service disconnecting means shall be required to be installed at a rating of _____ amps for single branch-circuit loads supplying limited loads.

      (a) 15   **(b)** 20   **(c)** 30   **(d)** 50

10. A circuit breaker rated over 600 volts can have a trip setting not exceeding _____ times the ampacity of the continuous rating of the conductors to be installed.

      (a) 3   **(b)** 5   **(c)** 6   **(d)** 10

Section	Answer

_____ _____ 11. Load requirements in excess of _____ amps shall be permitted to be supplied with more than one service.

        (a) 500   **(b)** 1000   **(c)** 1500   **(d)** 2000

_____ _____ 12. A clearance of _____ ft. shall be required for service-drop conductors passing over roofs. At least a _____ ft. clearance shall be maintained in all directions for vertical clearances.

        (a) 8 and 3   **(b)** 3 and 6   **(c)** 10 and 3   **(d)** 10 and 6

_____ _____ 13. Service-drop conductors shall have a vertical clearance of at least _____ ft. above flat roofs that are accessible to the public.

        (a) 6   **(b)** 7   **(c)** 8   **(d)** 10

_____ _____ 14. Service-drop conductors shall be permitted to be installed for overhanging roofs provided the attachment of the service-drop has a clearance of at least _____ in. from the conduit and roof line.

        (a) 6   **(b)** 12   **(c)** 18   **(d)** 24

_____ _____ 15. Service-drop conductors crossing residential property or driveways shall have a clearance of _____ ft. from finished grade.

        (a) 10   **(b)** 12   **(c)** 15   **(d)** 18

_____ _____ 16. Service-drop conductors crossing residential driveways shall have a clearance of _____ ft. from finished grade where the voltage exceeds 300 volts-to-ground.

        (a) 10   **(b)** 12   **(c)** 15   **(d)** 18

_____ _____ 17. Service-drop conductors crossing public streets, alleys, roads, and parking areas subject to traffic shall have a clearance of _____ ft. from finished grade.

        (a) 10   **(b)** 12   **(c)** 15   **(d)** 18

_____ _____ 18. A minimum _____ amp, service disconnecting means shall be installed for an installation having not more than two 2-wire branch-circuits.

        (a) 15   **(b)** 30   **(c)** 40   **(d)** 60

_____ _____ 19. Other cables shall be installed with a _____ in. set-off clearance from the wall.

        (a) 2   **(b)** 6   **(c)** 10   **(d)** 12

_____ _____ 20. No more than _____ fusible switches or circuit breakers mounted in a single enclosure, a group of separate enclosures, or in or on a panelboard or switchboard can be installed for each set of service-entrance conductors.

        (a) 2   **(b)** 4   **(c)** 6   **(d)** 10

Switchboards and Panelboards

Panelboards are intended to be installed and mounted in cabinets or cutout boxes placed "in or against a wall or partition." Panelboards are used for the control of small capacity circuits.

Switchboards are not intended to be installed and mounted in cabinets or cutout boxes. Switchboards usually are installed to stand on the floor. Switchboards are used for the control of high capacity circuits.

The space inside enclosures must have sufficient space to terminate conductors to lugs of overcurrent protection devices or busbars. This space is necessary to protect the conductors and equipment from physical damage when they are being terminated to the busbar lugs or to terminals of OCPD's.

Quick Reference

SWITCHBOARDS AND PANELBOARDS	2-1
SUPPORT AND ARRANGEMENT OF BUSBARS AND CONDUCTORS	2-1
MINIMUM WIRE BENDING SPACE	2-4
SWITCHBOARDS	2-6
PANELBOARDS	2-8
CONTINUOUS LOAD	2-11
SUPPLIED THROUGH A TRANSFORMER	2-12
DELTA BREAKERS	2-12
BACK-FED DEVICES	2-12
GROUNDING OF PANELBOARDS	2-13
ISOLATED EQUIPMENT GROUND	2-14
TEST QUESTIONS	2-15

SWITCHBOARDS AND PANELBOARDS
ARTICLE 408, PARTS II AND III

Switchboards are manufactured with busbars for fuses or circuit breakers to protect and supply feeders and branch-circuits. Panelboards are equipped with a permanently installed main circuit breaker or single set of fuses for disconnecting the number of circuit breakers or fuseholders installed to protect and supply feeders or branch-circuits.

SUPPORT AND ARRANGEMENT OF BUSBARS AND CONDUCTORS
408.3

Busbars shall be held firmly in place in the enclosure and shall be arranged in such a way to allow adequate space for conductors to be pulled in and terminated to lugs or OCPD's.

USED AS SERVICE EQUIPMENT
408.3(C)

Master Test Tip 1: Note that the grounded conductor must terminate to the same busbar as the neutrals, EGC's, and GEC per **250.24(A), (B) and (C)**.

Each switchboard or panelboard used as service equipment shall be provided with a grounding bar bonded to the switchboard frame or the casing of a panelboard. See **250.28(C)** and **408.40** for further information.

The main bonding jumper shall be sized per **250.28(D)** and installed within the panelboard or any section of a switchboard to which the grounded service conductor on the supply side of the switchboard or panelboard frame is connected. All sections of a switchboard shall be bonded together with an EGC that is sized per **Table 250.122**. Note that a screw or strap can be used instead of a main bonding jumper per **250.28(A)** and **(B)**. **(See Figure 2-1)**

Figure 2-1. Each switchboard, panelboard, or control board used as service equipment shall be provided with a grounding bar bonded to the switchboard frame or the casing of such boards.

Master Test Tip 2: For safety, enclosures housing panelboards must be of the dead front type per **408.38**.

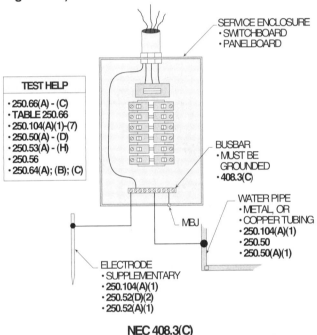

NEC 408.3(C)

HIGH-LEG MARKING
110.15; 408.3(E)

Master Test Tip 3: The center lug of utility meters is not required to be connected to the B phase of a three-phase, 240 volt, four-wire, delta connected system. For metering purposes only, this three-phase high-leg shall be connected to Phase C. A CT can with current transformers equipped with a remote meter must be connected to B phase (high-leg) in the CT can for meter conductors. However, it must be terminated to Phase C in the remote meter base. Check with testing agencies for verification of this rule. **(See Figure 2-3)**

For switchboards or panelboards supplied from a three-phase, four-wire, delta connected system, where the midpoint of one phase winding (neutral tap) is grounded, the voltage to the grounded neutral will have a higher voltage-to-ground than the other two phases-to-ground.

For example, the higher voltage-to-ground shall be identified by orange tape or tagged according to **110.15** and shall be connected to Phase B on a 240 volt, three-phase, four-wire delta connected system. The higher voltage-to-ground identification (high-leg, wild-leg, stinger-leg, or red-leg) prevents the 120 volt loads from being connected to 208 volt loads obtained from the high-leg. The 240 volt loads can be derived between the phases while the 120 volt loads must be derived from two of the phases-to-ground legs as follows: **(See Figure 2-2)**

Phase-to-Phase V

(1) A to B is 240 V
(2) A to C is 240 V
(3) B to C is 240 V

Phase-to-ground V

(4) A to N is 120 V
(5) B to N is 208 V
(6) C to N is 120 V

Switchboards and Panelboards

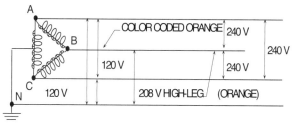

Figure 2-2. For switchboards or panelboards supplied from a three-phase, four-wire, delta-connected system, where the midpoint of one phase winding (neutral tap) is grounded, the voltage on one phase to the grounded neutral will have a higher voltage-to-ground than the other two phases-to-ground legs.

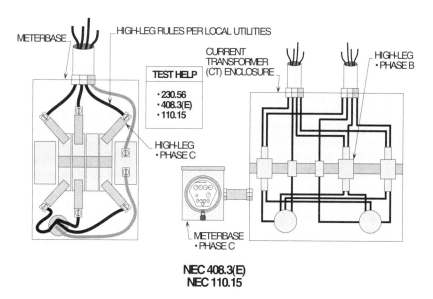

Figure 2-3. The center lug of utility meters are not required to be connected to the B phase of a three-phase, 240 volt, four-wire, delta system.

Master Test Tip 4: To derive a high-leg, transformer windings are connected open delta or closed delta. Note that two transformers are used for open delta and three transformers are used for closed delta system.

PHASE ARRANGEMENT
408.3(E)

The phase arrangement on three-phase buses shall be A, B, C from front to back, top to bottom, or left to right as viewed from the front of the panelboard, switchboard, or control board. The arrangement shall be viewed as Phases C, B, and A from the back of such boards. With this arrangement of the phases, the high-leg will always be placed as B and in the center. **(See Figure 2-4)**

Figure 2-4. The phase arrangement of three-phase buses shall be A, B, C from front to back, top to bottom, or left to right as viewed from the front of the panelboard, switchboard, or control board.

Master Test Tip 5: The color coding scheme that must be used to identify the high-leg is orange per **110.15** of the NEC.

Master Test Tip 6: Other power systems are usually color coded on the test as listed in the color code scheme. For application of this rule, check with testing authorities.

Master Test Tip 7: If there are different voltages available, each other system grounded conductor shall be identified white with a stripe (not green) running along its insulation.

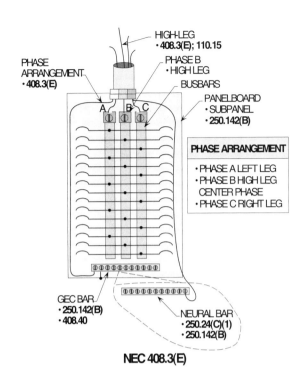

NEC 408.3(E)

Color Code Schemes sometimes on test

(1) Used for 120/208 volt or less systems
Phase A - black
Phase B - red
Phase C - blue

(2) Used for 277/480 volt systems
Phase A - brown
Phase B - orange
Phase C - yellow

(3) Used for 120/240 volt, three-phase systems
Phase A - black
Phase B - orange (high-leg)
Phase C - blue

(4) Used for 120/240 volt, single-phase systems
Phase A - black
Phase B - red or blue

MINIMUM WIRE BENDING SPACE
Table 312.6(A); (B); 408.55: 408.3(F)

Panelboards must comply with the provisions of **408.3(F)**, which requires the minimum gutter space in the board to meet the clearance rules listed in **Tables 312.6(A)** and **(B)** for L, S, or Z-bends.

If the lugs in the panelboard are removable, the clearance for S or Z-bends can be reduced from the dimensions in **Table 312.6(B)** based on the number and size of conductors connected to each lug. **(See Figure 2-5)**

Switchboards and Panelboards

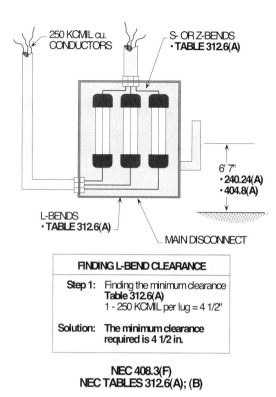

Figure 2-5. Panelboards must comply with the provisions of **408.3(F)**, which requires the minimum gutter space in the board to meet the clearance rules listed in **Tables 312.6(A)** and **(B)** for L, S, or Z-bends.

Master Test Tip 8: The space of the enclosure must not be filled to more than 40 percent of its cross-sectional area.

To prevent damaging conductors because of overbending, panelboards must comply with the provisions of **408.3(F)**, which requires the minimum gutter space in the board to meet the clearance rules in **Tables 312.6(A)** and **(B)** for L, S, or Z-bends.

If the lugs in the panelboard are removable, the clearance can be reduced from the dimensions in **Table 312.6(B)** based on the number and size of conductors connected to each lug. **(See Figure 2-6)**

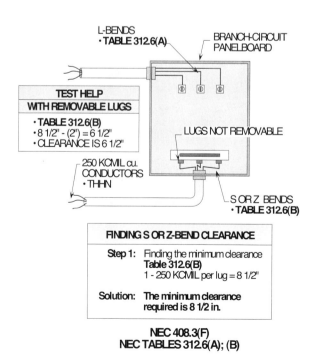

Figure 2-6. To prevent damaging conductors because of overbending, panelboards must comply with the provision of **408.3(F)**, which requires the minimum gutter space in the board to meet the clearance rules in Tables **312.6(A)** and **(B)** for L, S, or Z-bends.

Master Test Tip 9: Splices and taps must not fill the space of the enclosure to more than 75 percent of its cross-sectional area.

SWITCHBOARDS
ARTICLE 408, PART II

Switchboards that have any exposed live parts shall be installed in dry locations accessible only to qualified personnel. Total-enclosure type switchboards may be installed anywhere acceptable to the authority having jurisdiction. Switchboards can be used for service equipment with feeder-circuits supplying power to subpanels.

CLEARANCES
408.18(A); (B)

Switchboards that are not of the totally enclosed type shall have a clearance of 3 ft. from the top of the switchboard to a ceiling constructed of wood, ceiling paper, and any other type of material that will burn. This clearance is not required for totally enclosed type switchboards or for a fireproof shield provided above the switchboard. Clearances around and above switchboards shall comply with requirements of **110.26** and **408.18(B)**. **(See Figure 2-9)**

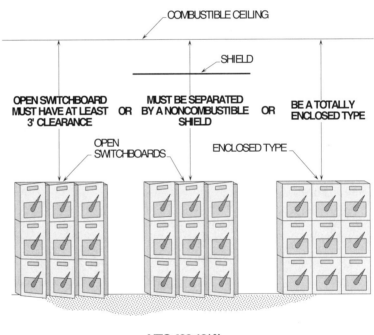

Figure 2-7. Switchboards that are not of the totally enclosed type shall have a clearance of 3 ft. from the top of the switchboard to a ceiling constructed of wood, ceiling paper, and any other type of material that will burn.

INSTALLATION
408.18(B); 110.26(A) THRU (F)

Master Test Tip 10: Control equipment is permitted in such panel room, if it is located within sight or adjacent to its operating machinery per **Ex.** to **110.26(F)** per **408.18(B)**.

Switchboards shall be installed in a dedicated space extending 6 ft. from the floor to the structural ceiling (roof). No piping, ducts, or equipment foreign to the electrical installation shall be installed in this dedicated space since leakage of water or condensation could damage the equipment. Any space extending above 6 ft. from the floor is not considered dedicated space. Sprinkler protection shall be permitted to be installed in this dedicated space to protect the equipment from fire hazards. See NFPA 13, 4-4.14 for further information.

Section **110.26(F)(1)(a), Ex.** allows equipment to be installed through industrial plants if isolated from foreign equipment by height or physical enclosures or covers that will afford adequate mechanical protection from vehicular traffic, accidental contact by unauthorized personnel, or accidental spillage or leakage from piping systems. However, such is only permitted in areas that do not have the dedicated space per **110.26(F)(1)(a)**. **(See Figure 2-7)**

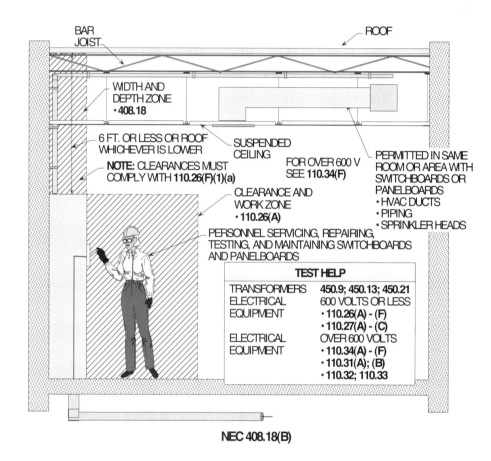

NEC 408.18(B)

Figure 2-8. Switchboards shall be installed with a dedicated space extending 6 ft. from its top, including width and depth, to the structural ceiling (roof). No piping, ducts, or equipment foreign to the electrical installation shall be installed in this dedicated space. (General Rule)

Master Test Tip 11: Note that when applying this rule, a dropped, suspended, or similar ceiling that does not add strength to the building structure is not considered a structural ceiling.

Master Test Tip 12: For the purpose of fire fighting, piping for sprinkler protection systems is allowed to be installed in this dedicated space.

Note: Section **110.26(F), (F)(1)(a) thru (d)** must be reviewed very carefully before taking the Master examination.

USED AS A SUBPANEL
408.34(A); 408.36(A), Ex. 1

Subpanels are supplied by feeders from the main service equipment. Subpanels are installed where the wiring of branch-circuits extends a length of great distance from the service equipment. Branch-circuit wiring is connected to the centralized subpanel, which reduces the amount of wiring and copper cost needed to wire electrical apparatus located in the area.

Master Test Tip 13: Subpanels must have their grounded busbars isolated from the metal enclosure and the grounding busbar bonded to such enclosures per **408.40**.

A subpanel that supplies the lighting and receptacle outlets plus small appliance loads can be supplied from one of the six single pole OCPD's in the power panel. Three double-pole OCPD's are usually utilized to supply larger fixed electrical appliances such as a heating unit, an A/C unit, a large computer, a piece of processing equipment, a water heater, etc. **(See Figure 2-8)**

Figure 2-9. A subpanel serving lighting and receptacle outlets plus special appliance loads can be supplied from one of the single-pole OCPD's in the power panel. Double-pole OCPD's are usually utilized to supply larger fixed electrical appliances such as a heating unit, an A/C unit, a large computer, a piece of cooking equipment, a water heater, etc.

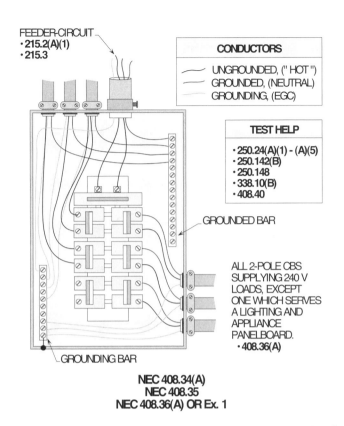

Master Test Tip 14: The **Ex.** to **408.40** allows an isolated EGC to be run through a subpanel without connecting it to the equipment grounding terminal bar.

GROUNDING USING SWITCHBOARD FRAMES
408.12

Master Test Tip 15: For rules and regulations pertaining to the grounding and bonding of instruments, relays, meters and instrument XFMR's, see **250.170 thru 250.178**.

Switchboard frames and structures supporting instruments, relays, meters, and instrument transformers shall be grounded. Proper grounding provides equipotential planes of electrical raceways, cables, and enclosures. For the definition of an equipotential plane, see **547.2 and 547.10**.

PANELBOARDS
ARTICLE 408, PART III

Panelboards shall have a rating equal to or higher than minimum capacity required for the load. When designing a panelboard with a computed load of 212 amps, the panelboard shall be sized at a rating of 212 amps. A panelboard of 225 amps is the next standard size above 212 amps and complies with these rules per **408.30**.

DIFFERENCE
408.34(A); (B)

Master Test Tip 16: Circuits in panelboards must be properly identified as to purpose and use per **408.30**.

There can be one to six mains in a service panel used as service equipment to supply and disconnect power to the various electrical loads. There are three types of panelboards used for the service equipment, and they are as follows:

(1) Power panels,
(2) Lighting and appliance, and
(3) Split-bus.

Power panels installed as service equipment usually consist of one to six mains, depending on the loads. Lighting and appliance panelboards have one to two mains. However, most of them are equipped with one in-line main to monitor the load.

Split-bus panelboards have one to six mains in the top section with one of the mains used to supply the bottom section, which is usually used for the lighting and receptacle outlet loads. Split-bus panelboards are required to have a main ahead of them since the 1981 NEC was published. This in-line main must be sized to protect the bus and loads of such panels.

Master Test Tip 17: Existing residential panelboards are allowed to have six or less OCPD's rated at 20 amps or less per **408.36(A), Ex 2**.

POWER PANEL
408.34(B); 408.36(B)

A power panel per **408.34(B)** can have one to six mains installed to supply electrical appliances and other electrical loads in residential, commercial, and industrial facilities. A power panel has 10 percent or less of its OCPD's rated at 30 amps or less with neutral branch-circuit connected loads. Power panels usually supply a great number of electrical appliances, in addition to their lighting and receptacle loads.

Master Test Tip 18: Snap switches rated 30 amps or less are allowed in panelboards equipped with 200 amp or less OCPD per **408.36(C)**.

For example, a panelboard with twenty-four, single-pole slots for installing overcurrent protection devices can have two 30 amp or less (24 OCPD's x 10% = 2.4) overcurrent protection devices installed with a neutral connection and still be classified as a power panel. Note that a single-pole circuit breaker usually counts as one circuit breaker. A double-pole circuit breaker counts as two circuit breakers. A three-pole circuit breaker counts as three circuit breakers in counting the number of single-pole slots in a panelboard per **408.35** to determine its classification. For power panel installations, also see **408.36(B)** and **Ex**.

Master Test Tip 19: Panelboards mounted in cabinets, cutout boxes, or enclosures must be of the dead front type per **408.38**.

LIGHTING AND APPLIANCE PANELBOARD
408.34(A); 408.36(A)

A lighting and appliance panelboard is equipped with one or two mains installed to supply power to a lighting section and power section per **408.36(A)**. A lighting and appliance panelboard has over 10 percent of its branch-circuit overcurrent protection devices rated at 30 amps or less with neutral connections. (Review **408.34(A)** and **408.36(A)**)

Master Test Tip 20: Fuses installed in panelboards must be installed on the load side of any switches per **408.39**.

For example, a panel with 42, single-pole slots has five 20 amp overcurrent protection devices with neutral loads. This is a lighting and appliance panelboard because it has five 20 amp OCPD's with neutral loads (42 x 10% = 4.2). If only four OCPD's are present, it is classified by **408.34(A)** and **408.35** as a power panel. Lighting and appliance panelboards with one in-line main cannot be overloaded. Therefore, they are safer to use and normally are installed because of this protection scheme. Note, see **Ex.'s** to **408.36(A)** for variations of this requirement. **(See Figure 2-10)**

Note that **408.34(A); 408.34(B)** and **408.36(B)** must be reviewed very carefully before classifying these panels.

SPLIT-BUS PANEL
408.36(A), Ex. 2

A split-bus panelboard (residential) has six mains in the top section with one main feeding the bottom section. The bottom section is used to supply power to the lighting and receptacle loads plus the small appliance loads. The five remaining mains in the top section are used to supply loads such as a heating unit, an A/C unit, a dryer, a cooktop, an oven, a range, a water heater, etc. **(See Figure 2-11)**

Master Test Tip 21: Exposed blades of knife switches must be deenergized when they are in the open position per **408.53**.

Figure 2-10. A lighting and appliance panelboard has over 10 percent of its branch-circuit OCPD's rated at 30 amps or less with neutral connections.

Master Test Tip 22: Instrument circuits protected by 15 amp OCPD or less are permitted in switchboards.

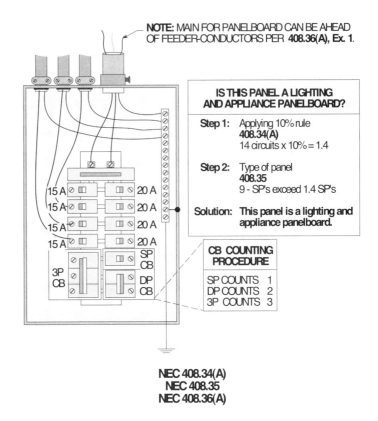

Figure 2-11. A split-bus panelboard has six mains in the top section with one main feeding the bottom section.

Master Test Tip 23: Instrument circuits greater than 15 amps are allowed in switchboards if interruption of such circuits creates hazard conditions.

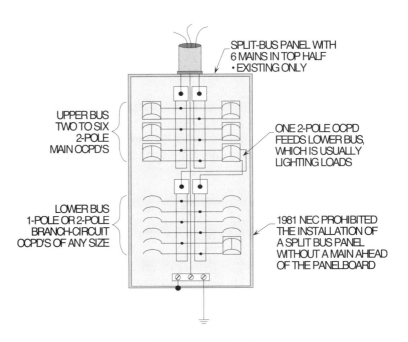

SNAP SWITCHES RATED AT 30 AMPS OR LESS
408.36(C)

Panelboards equipped with snap switches rated 30 amps or less shall be protected by overcurrent protection devices not exceeding 200 amps. Note that a 225 amp panelboard with a 225 amp in-line main cannot be used for this type of installation. **(See Figure 2-12)**

Switchboards and Panelboards

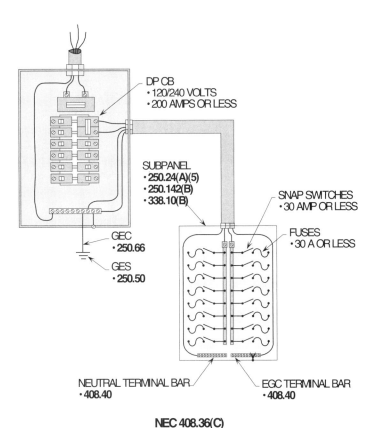

Figure 2-12. Panelboards equipped with snap switches rated 30 amps or less shall be protected by overcurrent protection devices not exceeding 200 amps.

Master Test Tip 24: Insulated or bare busbars installed in switchboards must be rigidly mounted per **408.40**.

CONTINUOUS LOAD
408.30; 215.2(A)(1); 230.42(A)(1)

The total load on any overcurrent device that serves continuous loads for feeders or branch-circuits shall not exceed 80 percent of its rating. For rules pertaining to computing the loads for branch-circuits, feeders, and services, see **210.19(A)(1), 210.20(A), 215.2(A)(1)** and **215.3** of the NEC. Note, for overcurrent protection devices rated over 600 volts, see **230.208(B)**. **(See Figure 2-13)**

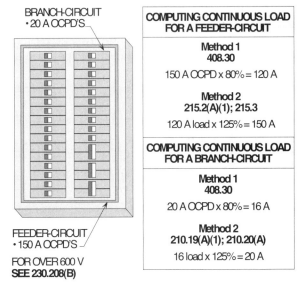

Figure 2-13. The total load on any overcurrent device that serves continuous loads for feeder or branch-circuits shall not exceed 80 percent of its rating.

Master Test Tip 25: Note that OCPD's (CB's) are not required to be derated 80 percent of their rating per **230.208(B)**.

2-11

SUPPLIED THROUGH A TRANSFORMER
408.36(D)

Where a lighting and appliance panelboard is supplied from the secondary of a transformer, the panelboard shall be provided with protection from the secondary side of the transformer.

However, under certain design techniques and conditions of use, the overcurrent protection device may be installed on the primary side for a two-wire primary to a two-wire secondary per **408.36(E)**. **(See Figure 2-14)**

Figure 2-14. Where a lighting and appliance panelboard is supplied from the secondary of a transformer, the panelboard shall be provided with protection from the secondary side of the transformer. (General rule)

Master Test Tip 26: Note that an OCPD on the primary of a 3-wire delta can protect the secondary of 3-wire delta XFMR per **240.4(F)**.

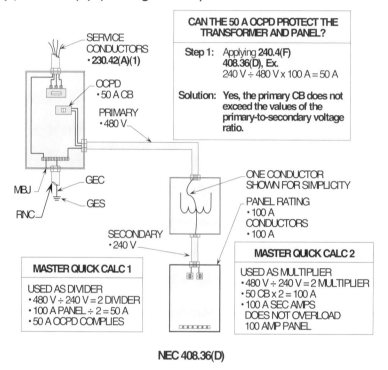

For example, if the load is supplied by a single-phase, two-wire transformer with a single voltage, the secondary may be protected by the overcurrent devices in the primary or supply side of the transformer, provided the primary is protected as covered in **Table 450.3(B)** and does not exceed the value determined by multiplying the transformer voltage ratio of the secondary-to-primary by the panelboard rating per **240.4(F)**.

DELTA BREAKERS
408.36(E)

Master Test Tip 27: Note that the distance between bare metal parts and busbars must comply with the spacing rules in **Table 408.56**.

Lighting and appliance panelboards shall not be installed with delta breakers. In other words, a three-phase disconnect shall never be terminated to the buses of panelboards having less than three buses. The reasoning behind this requirement was homeowners, in the past, were overloading the rating of the panel, due to the absence of an in-line CB for the high-leg.

BACK-FED DEVICES
408.36(F)

Plug-in type overcurrent protection devices or plug-in type main lug assemblies that are back fed shall be secured in place by an additional fastener other than the pull-to-release type. The past history of such installed devices has proved that plugging them in only was not enough support to secure them properly.

GROUNDING OF PANELBOARDS
408.40

An equipment grounding terminal bar shall be secured inside the cabinet for the termination of all equipment grounding conductors used to ground the metal of equipment, junction boxes, pull boxes, etc. The equipment grounding terminal bar ensures that the equipment grounding conductors are bonded to the cabinet and panelboard frame and, if done correctly, provide an effective path to ground.

Master Test Tip 28: Note that the bonding screw for the EGC terminal must be green in color and visible after installation per **250.28(B)**.

Panelboard enclosures or cases shall be grounded. Where installed with a metal conduit such as rigid metal conduit, electrical metallic tubing, intermediate metal conduit (any metal raceway), metal-clad cables, etc., the metal of these wiring methods may be used for grounding per **250.118**. An equipment grounding conductor shall be installed with a nonmetallic wiring method such as nonmetallic sheathed cable (romex or rope) or PVC. **(See Figure 2-15)**

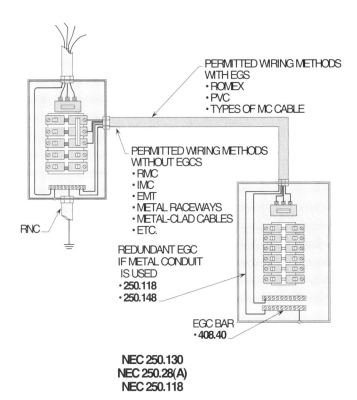

Figure 2-15. An equipment grounding terminal bar shall be secured inside the cabinet for the termination of all equipment grounding conductors used to ground the metal of equipment, junction boxes, pull boxes, etc.

The equipment grounding terminal bar and neutral bar in the service equipment shall be bonded to the panelboard case per **250.130(A)** and **250.66** and connected to the grounding electrode system per **250.52(A)(1) thru (A)(6)** by the grounding electrode conductor. The equipment grounding terminal bar and neutral bar in other than service equipment shall not be connected and bonded together in the panelboard enclosure per **250.142(B)**. Also see **250.24(A)(5)**. The neutral installed in the sub-panelboard enclosure shall be isolated back to the service equipment. Isolated equipment grounding conductors are permitted to pass through the panelboard enclosure per **250.146(D)**. However, all grounding conductors and neutrals in a subpanel must be terminated to separate terminal bars per **408.40**. **(See Figure 2-16)**

Master Test Tip 29: Equipment grounding conductors and neutrals can only terminate to a common bar at the service equipment, separately derived system, and separately fed buildings from a main building supply per **250.32(B)(2)**.

Figure 2-16. The equipment grounding terminal bar and the neutral bar in other than service equipment shall not be connected and bonded together in the panelboard enclosure.

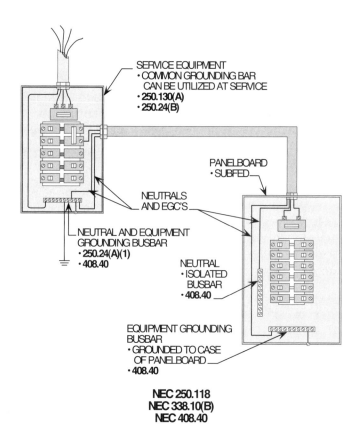

ISOLATED EQUIPMENT GROUNDING CONDUCTOR
408.40, Ex.; 250.96(B)

Master Test Tip 30: This rule is necessary for isolating computer grounding systems that are cord-and-plug connected per **250.146(D)**. Also, see **250.96(B)** for a similar rule when isolating the grounding system for hard-wired sensitivity electronic equipment.

Isolated equipment grounding conductors are permitted to pass through the panelboard with the power circuit conductors to reduce the threat of electromagnetic interference (electrical noise) if present. An isolated equipment grounding conductor shall be installed and connected directly to the equipment grounding terminal bar and neutral bar in the main service equipment or power source (separately derived system) to reduce the problem of electrical noise. Such noise, if not corrected, can disturb the operation of sensitivity electronic equipment. **(See Figure 2-17)**

Figure 2-17. Isolated equipment grounding conductors are permitted to pass through the panelboard with the power circuit conductors to reduce electromagnetic interference (electrical noise) if such noise is present.

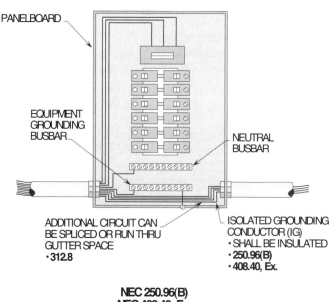

Name _____   Date _____

Chapter 2
Switchboards and Panelboards

Section   Answer

1. Piping, ducts, or equipment foreign to the switchboard installation shall be permitted to be installed in the dedicated space extending 6 ft. from the floor to the structural ceiling (roof) for switchboards.

    (a) True   **(b)** False

2. Switchboard frames supporting instrument equipment shall be grounded.

    (a) True   **(b)** False

3. Lighting and appliance panelboards are permitted to be installed with delta breakers.

    (a) True   **(b)** False

4. The equipment grounding terminal bar and the neutral terminal bar in a subfeed panelboard shall not be bonded to the panelboard case and connected to the grounding electrode system by the grounding electrode conductor.

    (a) True   **(b)** False

5. Isolated equipment grounding conductors shall not be permitted to pass through the panelboard with power circuit conductors.

    (a) True   **(b)** False

6. Switchboards shall be installed with a dedicated space extending _____ ft. from the floor to the structural ceiling (roof).

    (a) 10   **(b)** 20   **(c)** 6   **(d)** 50

7. The minimum clearance required between the bottom of the switchboard enclosure and busbars, with their supports, or other obstruction is _____ in. for noninsulated busbars.

    (a) 8   **(b)** 10   **(c)** 12   **(d)** 18

8. A lighting and appliance panelboard has over _____ percent of its branch-circuit overcurrent protection devices rated at _____ amps or less with neutral connections.

    (a) 10, 20   **(b)** 10, 30   **(c)** 20, 20   **(d)** 20, 30

9. The minimum wire bending space in panelboards can be reduced if the lugs in the panelboard are removable for _____ or _____ bends.

    (a) L, Z   **(b)** L, S   **(c)** S, Z   **(d)** All of the above

10. Switchboards that are not of the totally enclosed type shall be of a clearance of _____ from the top of the switchboard to a ceiling constructed of wood, ceiling paper, and any other type material that will burn.

    (a) 3   **(b)** 6   **(c)** 10   **(d)** 25

2-15

Grounding and Bonding

3

The master electrician's examination has many problems and questions on grounding and bonding electrical systems. This chapter covers the material that is necessary to prepare and answer such problems and questions correctly.

Equipment and circuits that are grounded and bonded properly are protected from lightning storms and excessive surges of voltage. Property is also protected, as well as safety for people provided. There are three grounding schemes in a well designed and installed grounding system, and they are as follows:

(1) Circuit and system grounding,
(2) Equipment grounding, and
(3) Bonding supply and load side.

In a well planned grounding scheme, there are grounding conductors, if sized and selected correctly that can be used as a network to ground and bond electrical systems for safety.

CIRCUIT AND SYSTEM GROUNDING ARTICLE 250

Circuit and system grounding is installed to:

(1) Limit excessive voltage from line surges, crossovers with higher voltage lines, and lightning strikes.

(2) To provide zero potential to ground for noncurrent-carrying enclosures and equipment.

(3) To help facilitate the opening of overcurrent protection devices, if they are sized to protect the circuit conductors from short-circuits and ground-faults.

Quick Reference

CIRCUIT AND SYSTEM GROUNDING.	3-1
AC CIRCUITS AND SYSTEMS	3-2
GROUNDING SERVICE SUPPLIED BY AC SYSTEM	3-3
GROUNDED CONDUCTOR BROUGHT TO SERVICE EQUIPMENT	3-5
GROUNDING ELECTRODE.	3-6
TWO OR MORE BUILDINGS/GROUNDED SYSTEMS	3-7
TWO OR MORE BUILDINGS/UNGROUNDED SYSTEMS	3-8
CONDUCTOR GROUNDED FOR AC SYSTEM.	3-9
GROUNDING SEPARATELY DERIVED AC SYSTEMS.	3-9
GROUNDED AND UNGROUNDED SYSTEMS.	3-12
EFFECTIVE GROUND-FAULT CURRENT PATH.	3-14
GROUNDED CIRCUIT CONDUCTOR FOR GROUNDING EQUIPMENT.	3-14
GROUNDING SUBPANELS AND EQUIPMENT.	3-16
TYPES OF EQUIPMENT GROUNDING CONDUCTORS.	3-17
SIZING GROUNDING ELECTRODE CONDUCTOR.	3-18
GROUNDING ELECTRODE SYSTEM	3-20
ROD, PIPE AND PLATE ELECTRODES.	3-25
BONDING	3-26
SERVICE EQUIPMENT BONDING.	3-26
BONDING OTHER ENCLOSURES.	3-28
OVER 250 VOLTS.	3-29
MAIN AND EQUIPMENT BONDING JUMPERS.	3-29
RESISTANCE OF ROD ELECTRODES.	3-31
ATTACHMENT.	3-32
SWIMMING POOLS	3-32
HOT TUBS, SPAS, AND HYDROMASSAGE TUBS.	3-33
HYDROMASSAGE BATHTUBS.	3-35
CORD-AND-PLUG CONNECTED COMPUTERS..	3-36
PERMANENT-WIRED COMPUTERS..	3-37
TEST QUESTIONS	3-39

AC CIRCUITS AND SYSTEMS
250.20

AC systems that may or may not have to be grounded:

(1) AC circuits of less than 50 volts
- See **Article 720**
- See **725.21(A)** and **725.41**

(2) AC circuits of 50 to 1000 volts
- See **250.26**
- See **250.24(A)**, **(B)** and **(C)**
- See **250.130**

AC CIRCUITS OF LESS THAN 50 VOLTS
250.20(A)

Master Test Tip 1: Low voltage electrical systems of less than 50 volts are usually Class 1, 2, or 3 circuits

Systems of less than 50 volts are found in **Article 720** of the NEC. Class 1, 2, and 3 circuits including rules pertaining to their installation procedures are listed in **725.21(A)** and **725.41** of the NEC.

AC circuits of less than 50 volts shall be grounded under the following conditions:

(1) A transformer installed to supply low-voltage and such transformer (primary) receives its supply from a circuit not exceeding 150 volts-to-ground.

For example, a circuit of 277 volts-to-ground supplying the primary side of such a transformer is considered a circuit of over 150 volts-to-ground and the secondary side must be grounded.

(2) A transformer is installed to receive its supply from a transformer with an ungrounded system. This condition includes a supply voltage and circuit obtained from a transformer which has an ungrounded secondary. Therefore, the secondary side of the transformer being supplied shall be grounded.

(3) Where low-voltage overhead conductors are installed outside and not inside.

For example, 50 volt AC conductors run outside (overhead) shall have one conductor grounded to ensure safety of wiring methods, equipment, and personnel.

AC CIRCUITS OF 50 TO 1000 VOLTS
250.20(B)

Master Test Tip 2: Transformers located outside and having a grounded output of 50 to 1000 volts must be earth grounded with additional earth ground at the service equipment. (See **250.24(A)(1)** and **(A)(2)**)

AC supply circuits of 50 to 1000 volts shall be grounded under any one of the following conditions:

(1) The maximum voltage-to-ground on the ungrounded conductors does not exceed 150 volts. This voltage-to-ground (circuit) is usually a 120 volt circuit derived from a 120/240 volt or 120/208 volt system.

(2) The system is nominally rated as 120/208 volt or 277/480 volt, three-phase, four-wire, wye-connected in which the neutral is used as a circuit conductor. The voltage-to-ground and voltage between phases can be any level for 50 to 1000 volt circuits. Wye systems of 2400/4160 volt are also used to supply circuits requiring a higher voltage.

(3) The system is nominally rated as 120/240 volt, three-phase, four-wire, delta-connected circuit in which the midpoint of one phase is used as a circuit conductor, with one phase conductor having a higher voltage-to-ground than the other two. The exceptions preceding the rules are as follows:

(a) Circuits installed for industrial electric furnaces or any other means of heating metals for refining, melting, or tampering.

(b) Separately derived systems used exclusively for rectifiers supplying only adjustable speed industrial drives. Such systems are used for speed control in industrial facilities and must not be utilized for anything else.

(c) Separately derived systems installed with primaries not exceeding 1000 volts used exclusively for secondary control circuits where conditions of maintenance and supervision ensure that only qualified persons will service the installation. However, the continuity of control power is required to be installed with ground detectors to sound an alarm if one phase should become grounded, causing a ground-fault condition to occur.

(d) Circuits supplying isolated systems such as anesthetizing locations per **517.61** and **517.160**. Note that **Article 517** prohibits a neutral to be grounded in operating rooms. Such supply circuits must be served by an isolation transformer, which uses a special metal barrier between the primary and secondary windings. **(See Figure 3-1 for grounded systems)**

Master Test Tip 3: A qualified person must know the operation, construction and hazards of the electrical equipment and wiring methods involved per **Article 100**.

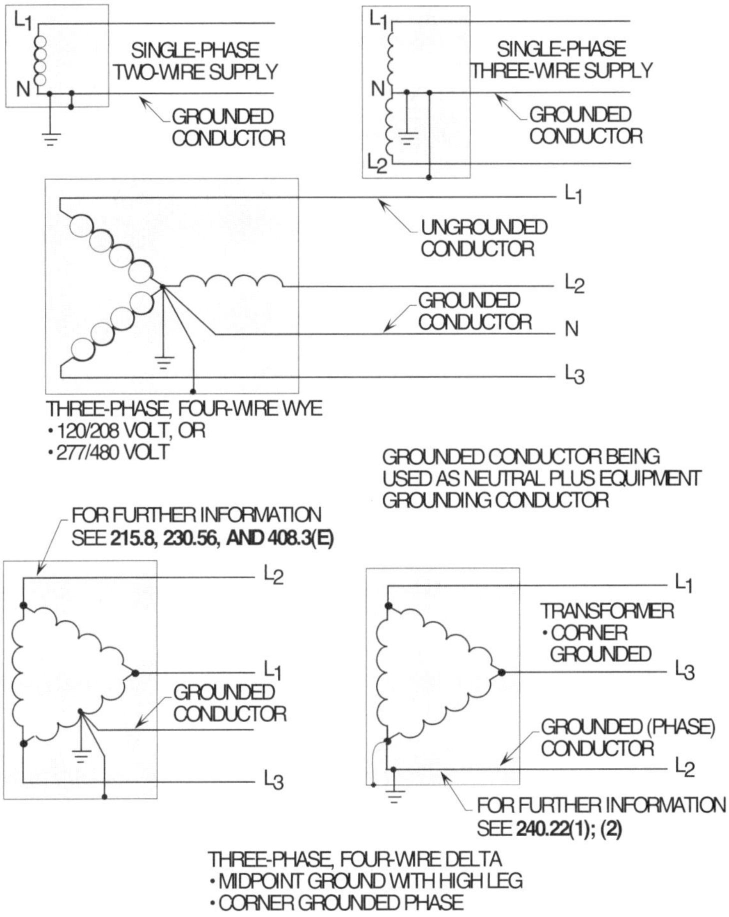

Figure 3-1. This illustration shows AC circuits of 50 to 1000 volts that must be grounded under certain conditions of use.

Master Test Tip 4: A corner grounded phase conductor is grounded in the same manner as a neutral conductor. They both measure zero to ground. (See **240.22(1)** and **(2)** and **250.26(4)**)

GROUNDING SERVICE SUPPLIED BY AC SYSTEM 250.24(A)(1); (A)(2)

AC grounded systems are required to be grounded at each service and the outside transformer. A grounding electrode conductor is usually installed from the neutral bus terminal to the grounding electrode system, which could be a cold water pipe, driven

Master Test Tip 5: The grounded circuit conductor must not have a grounding connection installed on the load side of the service disconnecting means per **250.142(B)** and **250.24(A)(5)**. (See Figure 3-3)

rod, or other electrodes, or any combination per **250.52(A)(1)** thru **(A)(7)**. At any accessible point on the load side of the service drop or service lateral the grounding electrode conductor can be installed and connected to the grounded service conductor. A terminal or bus is used to connect the grounded service conductor at the service disconnection means. **(See Figure 3-2)**

Figure 3-2. AC grounded systems are required to be grounded at each service and the outside transformer.

Master Test Tip 6: On the load side of the service, a range, cooktop, oven, or dryer can be grounded by the neutral conductor. Note that this rule applies only when there is an existing branch-circuit available. (See **250.140**)

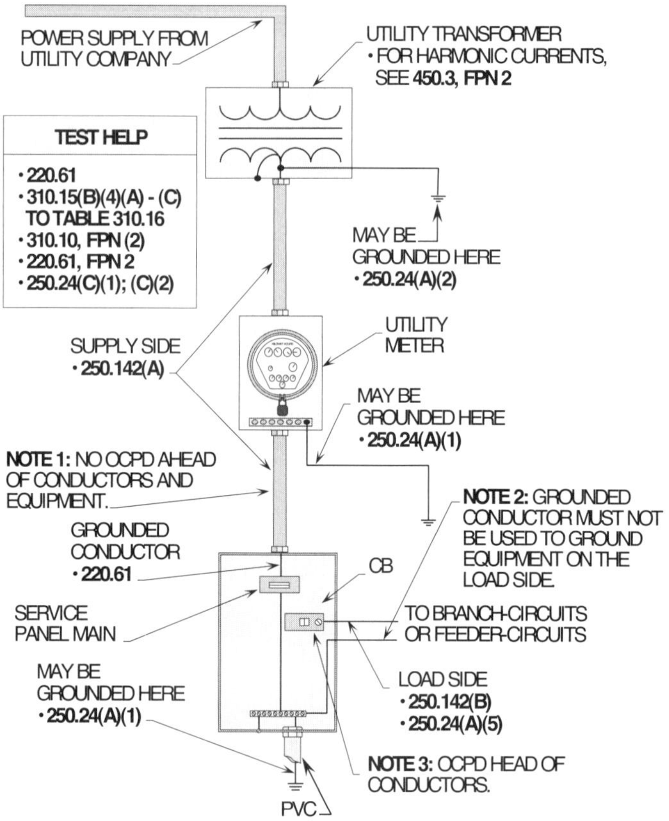

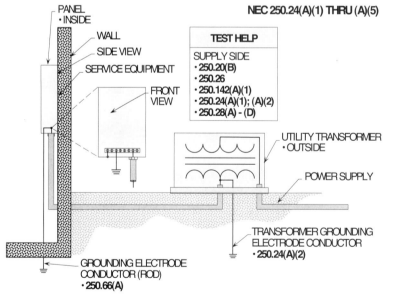

Figure 3-3. The grounded circuit conductor must not have a grounding connection installed on the load side of the service disconnecting means. (See **250.24(A)(5)** and **250.142(B)**)

Master Test Tip 7: The amount of fault-current is derived by the resistance of each conductor run added together and divided into the voltage-to-ground.

By dividing the length of wire between the supply and the load by 1000 and multiplying by the resistance, the amount of fault-current that will flow can be calculated. The resistance of each length is added together and divided into the voltage-to-ground to derive the fault-current at the location of such short. **(See Figure 3-4 and Quick Calc Tip 1 on page 3-5)**

Grounding

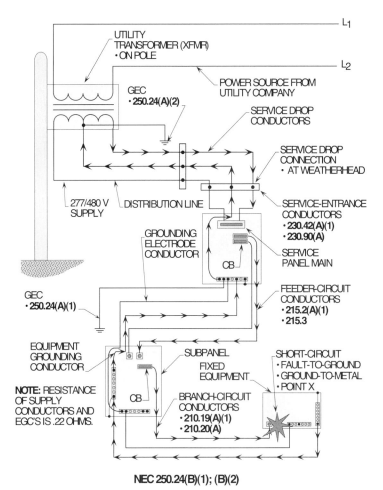

Figure 3-4. This illustration shows the flow of fault-current from the fault to the service panel and supply transformers outside.

QUICK CALC 1
With a total resistance of .22 for all conductors between the XFMR and the ground-fault location, what is the fault current (FC) in amps at point X ?
Step 1: Setting up formula $FC = \dfrac{V \text{ to } G}{\text{Total R}}$
Step 2: Applying Formula $FC = \dfrac{277 \text{ V}}{.22 \text{ R}}$ $FC = 1259 \text{ A}$
Solution: Fault current is equal to 1259 amps.

GROUNDED CONDUCTOR BROUGHT TO SERVICE EQUIPMENT
250.24(C)(1); (C)(2)

The grounded conductor is required to be installed with the phase conductors and run to each service if the secondary of the utility's power transformer is grounded. It is estimated that the amount of ground-fault current that will flow in the system is approximately 5 to 10 percent in the ground and 90 to 95 percent on the grounded conductor between the supply transformer and service equipment. The rating of the grounded conductor shall be required to be computed at least 12 1/2 percent of the largest ungrounding phase conductor. However, the equipment grounded conductor should be calculated at not less than 25 percent of the largest ungrounded phase conductor to ensure grounding conductors provide safe and dependable ground-fault paths. For an OCPD to clear a circuit safely, a fault-current of at least 6 to 10 times its rating should be available. **(See Figure 3-5)**

Master Test Tip 8: The grounded neutral conductor can be used to ground the service weatherhead and raceway, the meter enclosure, the service equipment panelboard, and all enclosures between the XFMR and service disconnecting means.

For example: What is the minimum and maximum fault-current required to clear a 150 amp OCPD?

  Step 1: Calculating percentage for 6 to 10 times OCPD rating
    250.24(C)(1)
    150 A x 600% = 900 A
    150 A x 1000% = 1500 A

 Solution: The minimum fault-current is 900 amps and the maximum fault-current is 1500 amps.

Note that when applying the 90 percent rule, the amount of fault-current that may flow over the grounded conductor, during a 10,000 amp fault is about 9000 amps. (10,000 A x 90% = 9000 A)

Figure 3-5. The grounded conductor shall be required to be installed with the phase conductors and run to each service if the secondary of the utility's power transformer is grounded.

QUICK CALC 2

What size grounded conductor is required for a service having three 250 KCMIL THWN copper conductors?

Step 1: Finding KCMIL
250.24(C)(1)
1-250 KCMIL per phase

Step 2: Finding GRD conductor
Table 250.66
250 KCMIL requires
2 AWG

Solution: The size grounded conductor required is 2 AWG copper.

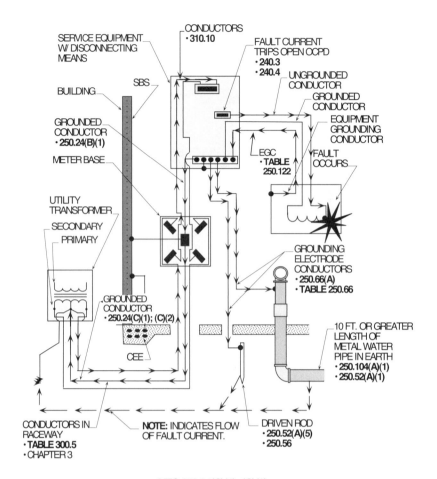

NEC 250.24(C)(1); (C)(2)

Fault-current will flow from the point of fault through the grounded phase or neutral conductor to the supply transformer and then will return through the phase that has faulted to ground and clear the overcurrent protection device. The overcurrent protection will trip quickly for a faulted phase that has a fault current of 6 to 10 times its rating in amperes.

Grounded conductors are installed to provide an effective path for fault-currents to travel over where phase-to-ground faults occur in the electrical system. The grounded neutral conductors shall always be sized as large as the grounding electrode conductor per **250.66(A) through (C)** and **Table 250.66**. The grounded conductor must be sized at least 12 1/2 (.125) percent of the area of the largest ungrounded conductor where the service conductors are installed larger than 1100 KCMIL copper or 1750 KCMIL aluminum.

Master Test Tip 9: Two-phase or three-phase conductors shall not be run to the service without installing a grounded conductor when the utility company's secondary system is grounded. An example of this rule is where all the service loads are three-phase, 480 volt and a step down separately derived system is installed to supply loads of 120/208 volt.

GROUNDING ELECTRODE
250.32(A), Ex.

A grounding electrode at a separate building or structure is not required to be installed where only one branch-circuit serves the building or structure and the branch-circuit includes an EGC for grounding noncurrent-carrying parts of all equipment.

For further information see **250.32(D)** for applying the requirements for installing the disconnecting means per **225.32** and **Ex.'s (See Figure 3-6)**

Grounding

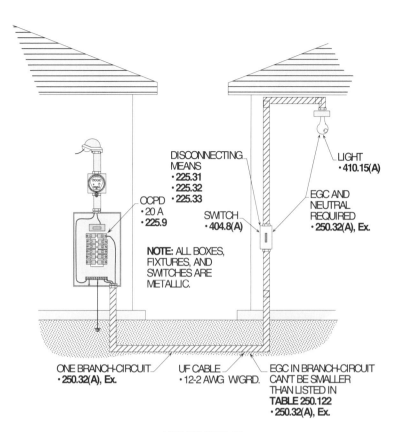

NEC 250.32(A), Ex.

Figure 3-6. A grounding electrode is not required to be installed at a separate building or structure where only one branch-circuit is run and the branch-circuit includes an EGC for grounding noncurrent-carrying parts of an equipment.

Master Test Tip 10: Test takers should take note that the grounded conductor in Figure 3-6 is being utilized as a neutral only and not as an EGC also.

Note: Bldg. 2 must not have any metal utility supply pipes run from Bldg. 1 that could create parallel paths on which current could flow from Bldg. 2 to Bldg. 1. (See **250.32(B)(2)** for this rule)

TWO OR MORE BUILDINGS GROUNDED SYSTEMS
250.32(B)(1); (B)(2)

Where one or more buildings or structures are supplied from a common AC grounded service, each panelboard at each building or structure must be separately grounded. The grounded or equipment grounding conductor run from the service panel of a building to a panel in a separate building or structure must at least be the size specified in **Table 250.122**. The supply to each building or structure must be disconnected by one to six disconnecting means per **225.32,** or one of the exceptions can be applied under certain conditions. OCPD's are required to comply with the rules of **Article 240**. See **240.4(A) thru (G)** for such rules and regulations.

EQUIPMENT GROUNDING CONDUCTOR
250.32(B)(1)

Where a building or structure is supplied from a service in another building by more than one branch-circuit, a grounding electrode shall be installed at the additional building(s) or structure(s) being served. Where livestock is housed, the equipment grounding conductor shall be insulated or covered where routed with the feeder or branch-circuit for bonding the metal case of the enclosure with the neutral bus and conductor isolated from the case. See **250.142(B)** and **408.40** for rules pertaining to this type of installation. **(See Figure 3-7(a))**

Design Tip: The equipment grounding conductor is installed to provide an effective return path for the fault current to travel over instead of using a grounded conductor

Master Test Tip 11: If the panelboard in a separate building is used as a subpanel, there must be an isolated neutral bar and grounded equipment grounding bar provided per **408.40**.

Master Test Tip 12: The grounded neutral conductor is installed to provide an effective return path for the fault current to travel over instead of using an equipment grounding conductor for such use. This particular installation is used to prevent the grounded neutral conductor and equipment grounding conductor from joining together at both ends, which if done, provides a parallel path for stray currents to travel over; this is not desirable. This type of installation can create problems for certain types of equipment.

Figures 3-7(a). Equipment grounding conductors that are routed in circuits to separate buildings must be sized per **Table 250.122**. **Note:** Grounded conductors are sized per **Table 250.122** and can be used as a neutral and EGC under certain conditions of use.

Master Test Tip 13: Note that the switch box, single-pole switch, light box, and lighting fixture are made out of nonmetallic or metallic material.

Master Test Tip 14: Test takers should note that the grounded conductor in Figure 3-7(b) is isolated from the metal enclosure and the EGC is bonded to the enclosure.

for such use. This particular installation is used to prevent the grounded neutral conductor and equipment grounding conductor from joining together at both ends, which if done, provides a parallel path for stray currents to travel over; this is not desirable for such an installation and can cause problems for certain types of equipment.

GROUNDED CONDUCTOR 250.32(B)(2)

If a grounded conductor, without an EGC, is routed in the circuit to a separate building, it must be equal to the size per **Table 250.122** to ensure the capacity to clear a ground-fault condition. See **250.142(A)(2)** for permission to use the grounded conductor as a current-carrying conductor and equipment grounding conductor in a feeder-circuit. **(See Figure 3-7(b))**

UNGROUNDED SYSTEMS 250.32(C)

A grounding electrode for an ungrounded system is required to be connected only to the service equipment enclosure where installed at one or more buildings. The feeder or branch-circuit must be grounded to an electrode at the other building being served from the service of the main building. All equipment grounding conductors at the separate building are connected to a grounding bus, which is bonded to the enclosure. Such enclosure is bonded and grounded to the grounding electrode conductor and grounding electrode system.

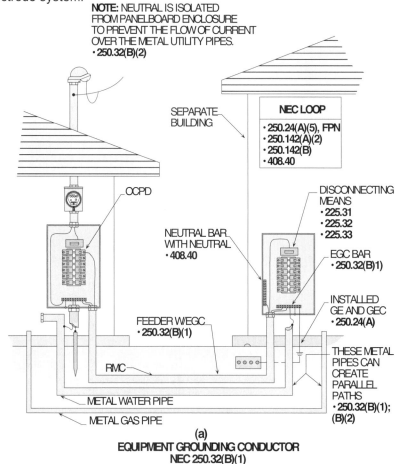

(a)
EQUIPMENT GROUNDING CONDUCTOR
NEC 250.32(B)(1)

Grounding

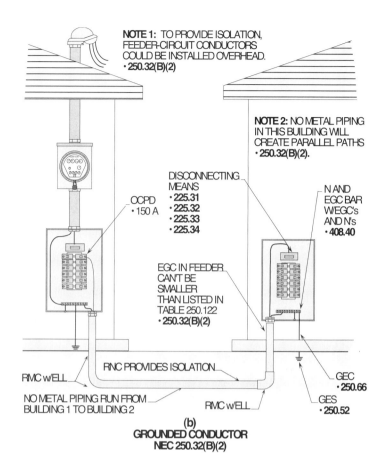

(b) GROUNDED CONDUCTOR NEC 250.32(B)(2)

Figures 3-7(B). Equipment grounding conductors that are routed in circuits to separate buildings shall be sized per **Table 250.122**. **Note:** Grounded conductors are sized per **Table 250.122** and can be used as a neutral and EGC under certain conditions of use.

Master Test Tip 15: Test takers beware that only 6 or less disconnecting means are allowed to be installed at a second building per **225.33** and **225.34**.

CONDUCTOR GROUNDED FOR AC SYSTEM
250.26

The grounded conductor whether a grounded phase or neutral is usually installed with a white or gray insulation or otherwise identified. The following systems are grounded and shall have a grounded phase (hot) or grounded neutral conductor. **(See Figure 3-8)**

   (1) Single-phase, two-wire (one hot and one neutral)
   (2) Single-phase, three-wire (two hots and one neutral)
   (3) Three-phase, four-wire, wye (three hots and one neutral)
   (4) Three-phase, four-wire, delta (two hots and one neutral)
   (5) Three-phase, with one phase grounded (three hots with one hot grounded)

Master Test Tip 16: Test takers take note that such disconnects shall be grouped per **225.33** and **225.34**.

GROUNDING SEPARATELY DERIVED AC SYSTEMS
250.30

Low-voltage and high-voltage feeder-circuits are sometimes installed from floor-to-floor in a high rise building with transformers installed on each floor to reduce the voltage to 120/240 or 120/208 volts for general use lighting and receptacle loads in large building applications. The secondary of a separately derived system is divided into three parts when designing and installing the bonding and grounding of a transformer system. Such grounding since the 1978 NEC can be installed either at the transformer or at the load served, which is tapped and supplied from the secondary side per **240.21(B)** and **(C)**. The following three parts shall be designed and installed for a transformer system: **(See Figure 3-9)**

   (1) Bonding jumper
   (2) Grounding electrode conductor
   (3) Grounding electrode

Master Test Tip 17: Note that the number of hot conductors that can be pulled with a neutral is the number measured to ground. For example, three hot conductors can be pulled with one neutral conductor derived from 120/208 volt, four-wire wye.

Figure 3-8. The grounded conductor, whether a grounded phase or neutral, is usually installed with a white or gray insulation or otherwise identified.

Master Test Tip 18: The grounded phase leg of a corner grounded delta system must not terminate in fuse unless it is providing overload protection for a motor per **240.22(2)**.

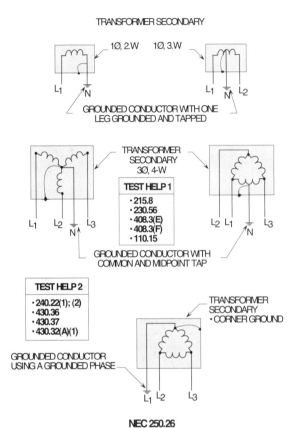

Figure 3-9. The equipment bonding jumper (BJ) and grounding electrode conductor (GEC) is designed and installed based on the derived phase conductors supplying the panel, switch, or other equipment tapped from the secondary side of the transformer.

Master Test Tip 19: Note that there is no OCPD ahead of the XFMR's secondary conductors in **Figure 3-9**, therefore **Table 250.66** must be used to size the grounded conductor, bonding jumper and GEC.

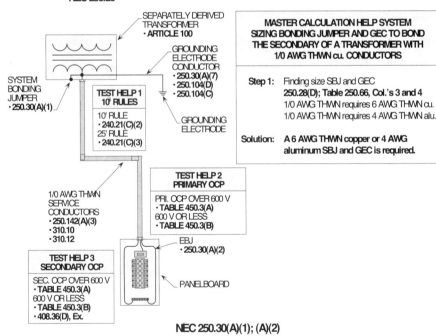

SYSTEM BONDING JUMPER
250.30(A)(1)

Master Test Tip 20: Bonding jumpers are used to ensure that the EGC's and metal conduits are permanently connected together.

The system bonding jumper is designed and installed based on the derived phase conductors supplying the panel, switch, or other system equipment tapped from the secondary side of the transformer and sized per **250.28(D)**. The system bonding jumper is required to be sized per **250.66** and **Table 250.66** based on the ungrounded phase conductors up to 1000 KCMIL for copper and 1750 KCMIL for aluminum. The system bonding jumper must sized at least 12 1/2 (.125) percent of the area of the largest phase conductors where the service conductors are installed larger than 1100 KCMIL copper or 1750 KCMIL aluminum. The bonding jumper shall be installed and connected at any point on the separately derived system from the source to the first system disconnecting means or overcurrent protection device. If the ungrounded phase conductors are larger than 1100 KCMIL for copper and 1750 KCMIL for aluminum, the bonding jumper will be larger than the grounding electrode conductor.

GROUNDING ELECTRODE CONDUCTOR 250.30(A)(3)

The grounding electrode conductor is designed and installed based on the derived phase conductors supplying the panel, switch, or other equipment connections from the secondary side of the transformer and sized per **Table 250.66**. The grounding electrode conductor shall be installed and connected at any point on the separately derived system from the source to the first system disconnecting means or overcurrent protection device. When the KCMIL rating is greater than 1100 for copper and 1750 for aluminum, the grounding electrode conductor will usually be smaller than the bonding jumper.

GROUNDING ELECTRODE 250.30(A)(7)

The grounding electrode conductor shall be as near as possible and preferably in the same area as the grounding electrode conductor connection to the system. The following three choices shall be installed in the order they appear in the NEC: **(See Figure 3-10)**

(1) Nearest building steel
(2) Nearest metal water pipe system
(3) Other electrodes specified in **250.52(A)(4) through (A)(7)**
(4) Metal water pipe located in the area must be bonded to the grounded conductor per **250.104(A), (A)(1) and (D)**

The grounding electrode conductor does not have to be installed larger than 3/0 AWG for copper or 250 KCMIL for aluminum when connecting to the nearest building steel or nearest metal water pipe system. The grounding electrode conductor does not have to be larger than 6 AWG copper or 4 AWG aluminum when connecting to a driven rod or other made electrodes. **Note:** See **250.66(B)** and **(C)** for the minimum size grounding electrode conductor to be connected to a concrete-encased electrode or ground ring.

QUICK CALC 3

If the service conductors are made up of 1800 KCMIL, what size THWN copper grounded conductor is required?

Step 1: Finding KCMIL
250.24(C)(1); (C)(2)
1800 KCMIL x .125
= 225 KCMIL

Step 2: Selecting KCMIL
Table 8, Ch 9
225 KCMIL
Requires 250 KCMIL copper

Solution: The size grounded conductor is 250 KCMIL copper.

Note, for selecting the size GEC for multiple separatley derived systems, see 250.30(A)(4).

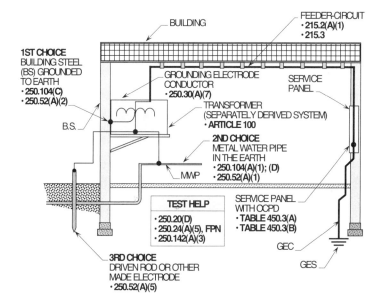

Figure 3-10. The grounding electrode conductor shall be as near as possible and preferably in the same area as the grounding electrode conductor connection to the system.

Master Test Tip 21: The connection to ground of the separately derived system can be made at the XFMR source or at the first OCPD, but not both places.

GROUNDED AND UNGROUNDED SYSTEMS
250.130

Equipment grounding conductor connections for a separately derived system and for service equipment shall be made in accordance with **250.30(A)** and **(B)**. The equipment grounding conductor connection shall be made at the service equipment and separately derived systems as follows:

(1) Grounded systems
(2) Ungrounded systems

GROUNDED SYSTEM
250.130(A)

Master Test Tip 22: Test takers should note that if a system is grounded, there will be a grounded neutral or phase present in the electrical system.

The equipment grounding conductor shall be bonded to the grounded (neutral) conductor and the grounding electrode conductor at the service equipment neutral busbar terminal. The grounded (neutral) conductor shall be installed and connected to the grounding electrode system per **250.52(A)(1)** thru **(A)(6)**. Note that grounded conductor must be connected to the busbar where the grounding electrode conductor is terminated per **250.24(A)**.

UNGROUNDED SYSTEM
250.130(B)

Master Test Tip 23: If the electrical power source is ungrounded, there will not be a grounded neutral or phase conductor present in the electrical system.

The equipment grounding conductor shall be bonded to the grounding electrode conductor at the service equipment neutral busbar terminal. An ungrounded (neutral) conductor is not available to be connected to ground. Equipment grounding conductors installed per **250.118** can be used to bond and connect all metal noncurrent-carrying parts of the wiring system to the grounding electrode conductor at the service equipment. **(See Figure 3-11)**

REPLACEMENT OF NONGROUNDING-TYPE RECEPTACLES
250.130(C)

The basic rule for replacement of receptacles is that grounding-type receptacles shall be used to replace existing receptacles where there is an EGC routed with the branch-circuit. A nongrounding-type receptacle shall be used with a branch-circuit that has no EGC or grounding means. The EGC shall connect the green grounding terminal of the receptacle to the grounding electrode conductor in the service equipment panelboard or to the closest metal water pipe per **250.104(A)(1)** and **250.52(A)(1)**. **(See Figure 3-12)**

Master Test Tip 24: Non-grounding receptacles can be installed on branch-circuits that do not have an EGC such as existing nonmetallic sheathed cable or knob-and-tube wiring methods.

The NEC allows five methods by which a nongrounding receptacle may be replaced with a circuit without an EGC and they are as follows:

(1) Install a nongrounding receptacle at each receptacle location
(2) A GFCI receptacle protecting each outlet at receptacle location
(3) A GFCI receptacle protecting a single outlet and additional outlets downstream
(4) An EGC routed from the receptacle to a metal water pipe per **250.52(A)(1)**
(5) A GFCI CB protecting all outlets

Grounding

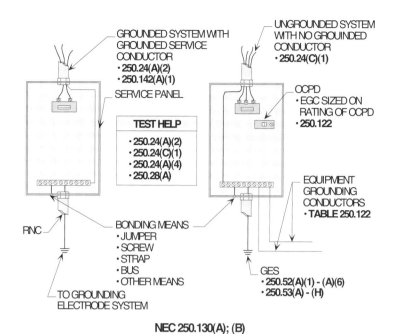

Figure 3-11. The equipment grounding conductor shall be bonded to the grounded (neutral) conductor and the grounding electrode conductor at the service equipment neutral busbar terminal for grounded systems. The equipment grounding conductor shall be bonded to the grounding electrode conductor at the service equipment neutral busbar terminal for ungrounded systems.

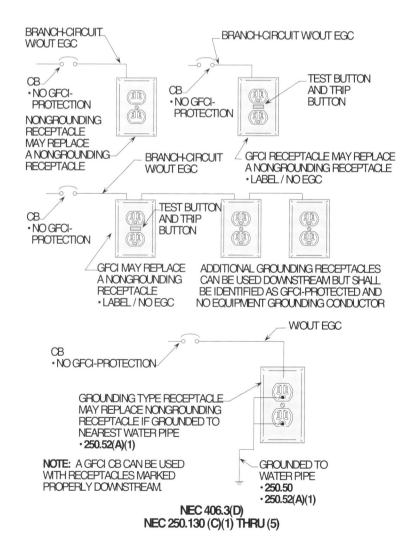

Figure 3-12. The basic rule for replacement of receptacles is that grounding-type receptacles shall be used to replace existing receptacles where there is an EGC routed with the branch-circuit.

Master Test Tip 25: When connecting the grounding terminal of a non-grounded receptacle to a metal water pipe, such connection must be made on the piping, within 5 ft. of its entry into the building per **250.52(A)(1)**.

Master Test Tip 26: To accomplish this effective grounding path, the installation must comply with **300.10, 300.12, 300.13,** and **300.15**.

Master Test Tip 27: To accomplish this capacity rule, all grounding conductors must comply with **Table 250.66** where there is not an OCPD ahead of the conductors and the equipment supplied. **Table 250.122** is used when there is an OCPD between the conductors and the load supplied.

Master Test Tip 28: To accomplish this low-impedance rule, the EGC, neutral, and phase conductors must be run together per **300.3(B), 300.5(I); 300.20(A)** and **(B)** and **250.134(B)**.

Figure 3-13. To provide a continuous path to ground, all connections shall be installed in a wiring system with metal-to-metal contact of parts using listed fittings and bonding jumpers.

EFFECTIVE GROUND-FAULT CURRENT PATH 250.4(A)(5)

To provide an effective continuous path to ground, all connections shall be installed in a wiring system with metal-to-metal contact of parts. This ground path from circuit, equipment, and metal enclosures housing conductors shall comply with the following:

(1) Be permanent and electrically continuous

(2) Have capacity to conduct safely any fault current likely to be imposed on it

(3) Have sufficiently low-impedance to limit the voltage-to-ground and to facilitate the operation of the overcurrent protection devices of the system. **(See Figure 3-13)**

In other words, the EGC and neutral must be routed with the phase conductors of the circuit.

The earth shall not be used as the sole equipment grounding conductor. To fully understand this rule, review **250.54** for supplementary electrodes and **250.136(A)** for grounding to the metal frame of the building.

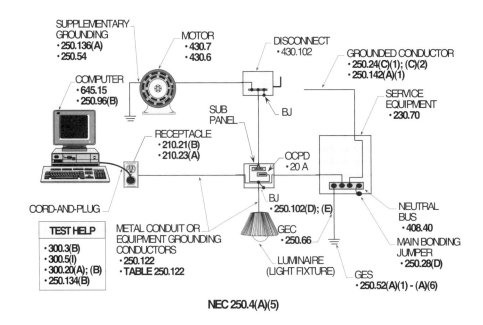

GROUNDED CIRCUIT CONDUCTOR FOR GROUNDING EQUIPMENT 250.142

Master Test Tip 29: There are three types of installations where the grounded conductor can be used as a neutral and EGC and they are at (1) services, (2) for separate buildings and at (3) separately derived systems. **(See Figure 3-14)**

Under certain conditions, all metal parts of enclosures used to install the service equipment may be grounded by the grounded neutral conductor on the supply side of the system. The service weatherhead, service raceway, service meter base, and the service equipment enclosure are included for this type of grounded system.

The grounded neutral conductor shall be installed and isolated from the other system conductors and from metal enclosures or metal conduits when installed on the load side of the system.

Grounding

USING GROUNDED CONDUCTOR - SUPPLY SIDE
250.142(A)

The grounded phase or neutral conductor can be used as a current-carrying conductor and grounding means on the supply side of the service disconnecting means and secondary side of a separately derived system as follows:

(1) On the supply side of service equipment per **250.142(A)(1)**

(2) On the supply side of the main service disconnecting means for separate buildings and structures per **250.142(A)(2)**

(3) On the supply of the disconnect or overcurrent protection device of a separately derived system per **250.142(A)(3) (See Figure 3-14)**

Master Test Tip 30: Note that there is no OCPD ahead of these conductors and equipment supplied per **250.142(A)(1) thru 250.142(A)(3)**.

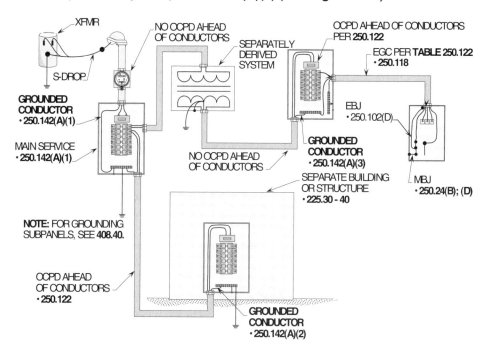

NEC 250.142(A)(1); (A)(2); (A)(3)

Figure 3-14. The grounded phase or neutral conductor can be used as a current-carrying conductor and grounding means on the supply side of the service disconnecting means and secondary side of a separately derived system.

Master Test Tip 31: Note that the neutral (General Rule) on the load side (OCPD ahead of conductors and equipment) must not connect or touch the metal of enclosures per **250.24(A)(5)** and **250.142(B)**, also, see **408.40**.

Note: Test takers must verify if there is an OCPD ahead of the conductors or there is not an OCPD ahead of the conductors, before answering questions on the Master examination pertaining to sizing grounding conductors.

USING GROUNDED CONDUCTOR - LOAD SIDE
250.142(B)

The grounded neutral conductor must not be used as an equipment grounding conductor on the load side except as follows:

(1) Frames of ranges, wall-mounted ovens, counter-mounted cooking units, and clothes dryers per **250.140**. (Only for existing branch-circuits without an EGC)

(2) Where one or more buildings or structures are supplied from a common AC grounded service, each grounded service at each individual building or structure shall be separately grounded per **250.32**.

(3) If no service ground-fault protection is provided, all meter enclosures located near the service disconnecting means shall be permitted to be grounded by the grounded neutral conductor on the load side of the service disconnect. The grounded neutral conductor shall not be smaller than the size specified in **Table 250.122**.

(4) DC systems shall be permitted to be grounded on the load side of the disconnecting means or overcurrent protection per **250.164(B)**.

Master Test Tip 32: Note that these installations all have OCPD's ahead of the conductors to the load supplied per **250.142(B)**.

Master Test Tip 33: Test takers use **Table 250.122** to size EGC's and BJ's, when there is an OCPD ahead of the conductors and equipment served.

QUICK CALC 4

What size EGC is required for 175 amp panel that was originally fed by a 2/0 AWG conductor but was increased to 4/0 AWG for voltage drop (175 A OCPD in panel)?

Step 1: Setting up formula
250.122, T8, Ch. 9
$M = \dfrac{4/0\ AWG = 211,600}{2/0\ AWG = 133,100}$
M = 1.5897821

Step 2: Finding EGC for 175 A OCPD
Table 250.122
175 A = 6 AWG cu.

Step 3: Finding size EGC
250.122(B); T8, Ch. 9
6 AWG = 26240 CM
1.5897821 = 41,716 CM

Step 4: Sizing EGC
250.122(B); T8, Ch. 9
41,716 CM = 4 AWG cu.

Solution: The size EGC is 4 AWG copper.

Figure 3-15. The equipment grounding conductor is the conductor used to ground the noncurrent-carrying metal parts of equipment per Article 100 of the NEC.

GROUNDING SUBPANELS AND EQUIPMENT TABLE 250.122

The equipment grounding conductor is the conductor used to ground the noncurrent-carrying metal parts of equipment. It keeps the equipment elevated above ground, at zero potential, and provides a path for the ground-fault current. Equipment grounding conductors protect elements of circuits and equipment and also ensure the safety of personnel from electrical shock. The size of the equipment grounding conductor is based on the size of the overcurrent protection device protecting the circuit conductors. **(See Figure 3-15)**

The equipment grounding conductor also has the job of providing a continuous path for the ground-fault current to flow over and trip open the circuit and protect the conductors and equipment from being damaged per **240.3** and **240.4**. The equipment grounding conductor shall be sized not less than the listed sizes in **Table 250.122** based on the size OCPD protecting the elements of the circuit.

Where current-carrying conductors have to be larger to compensate for voltage drop, the equipment grounding conductors shall be adjusted proportionately to the circular mil area.

Circuit conductors that are connected in parallel and routed through separate conduits must have a separate equipment grounding conductor in each conduit. The equipment grounding conductor must be sized from **Table 250.122**, based on the rating of the overcurrent protection device protecting the circuit per **250.122(D)**.

Where a single equipment grounding conductor is run with multiple circuits in the same raceway, it shall be sized for the largest overcurrent device protecting the conductors in the raceway or cable per **250.122(C)**.

If the overcurrent device is a motor short-circuit protector or an instantaneous trip breaker, see **430.52** and **Table 430.252**. The size of the equipment grounding conductor must be based upon the rating of the protective device to ensure a safe trip per **250.122(D)**.

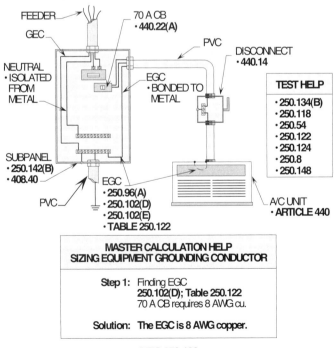

TYPES OF EQUIPMENT GROUNDING CONDUCTORS
250.118

The equipment grounding conductor run with or enclosing the circuit conductors shall be one or more or a combination of the following:

(1) A copper or other corrosion-resistant conductor
(2) Rigid metal conduit
(3) Intermediate metal conduit
(4) Electrical metallic tubing
(5) Flexible metal conduit
(6) Type AC cable
(7) Mineral-insulated sheathed cable
(8) Metal-sheathed cable
(9) Type MC cable
(10) Cable trays
(11) Other electrically continuous metal raceways

Sections **250.118(5), (6), and (7)** allow FMC, LTFMC, and FMT to be used as follows:

If the maximum length that can be used in any combination of flexible metal conduit, flexible metallic tubing, and liquidtight flexible metal conduit in any grounding run is 6 ft. or less in length. (See **250.118(5)c, (6)d** and **(7)b**)

The circuit shall be protected by overcurrent devices not to exceed 20 amps, but may be protected less than 20 amps. The flexible metal conduit, LTFMC, or FMT shall be terminated only in fittings listed for grounding per **250.118(5)a, (6)a** and **(7)**. **(See Figure 3-16)**

Master Test Tip 34: Flexible conduit (FMC and LTFMC) longer than 6 ft. length must be equipped with an EGC.

Master Test Tip 35: If a piece of FMC or LTFMC is used for flexibility, such as a connection to a motor, an EGC must be used per **348.60** and **350.60**.

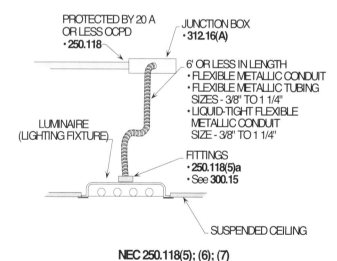

Figure 3-16. The circuit shall be protected by overcurrent devices not to exceed 20 amps, but may also be less than 20 amps.

Master Test Tip 36: Note that only one 6 ft. piece of FMC or LTFMC can be used as a grounding means in any one run between junction box and load.

Liquidtight flexible metal conduit in sizes up to 1 1/4 in. that have a total length not to exceed 6 ft. is allowed to be used as the ground return path. Fittings for terminating the flexible conduit must be listed for grounding. Liquidtight flexible metal conduit in 3/8 and 1/2 in. trade sizes shall be protected by overcurrent devices at 20 amps or less, and sizes of 3/4 in. through 1 1/4 in. trade size must be protected by overcurrent devices of 60 amps or less. For these rules, see **250.118(6)a thru e. (See Figure 3-17)**

Figure 3-17. Liquidtight flexible metal conduit 3/8 and 1/2 in. trade size shall be protected by overcurrent devices at 20 amps or less, and sizes of 3/4 in. through 1 1/4 in. trade size, protected by overcurrent devices of 60 amps or less.

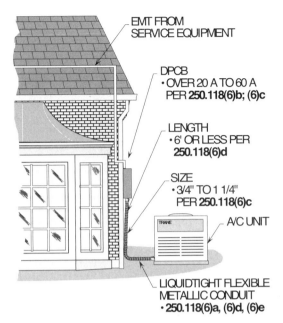

NEC 250.118(6)a THRU e

SIZING GROUNDING ELECTRODE CONDUCTOR TABLE 250.66

Master Test Tip 37: If the service conductors do not exceed 1100 KCMIL for cu. and 1750 KCMIL for alu., the grounded conductor, GEC, and the SBJ must be at least the same size per **250.24(C), 250.28(D),** and **Table 250.66.**

The following grounding electrode conductors must be grounded by conductors sized and selected as listed in **Table 250.66** and **250.66(A)** thru **(C)**.

(1) Metal water pipe
(2) Made electrodes
(3) Concrete-encased electrodes
(4) Ground rings
(5) Building structural steel

METAL WATER PIPE
250.104(A); 250.52(A)(1); 250.66; TABLE 250.66

Master Test Tip 38: For an MWP to be considered a grounding electrode, at least 10 ft. of metal piping must be in the earth (ground).

The procedure for selecting the grounding electrode conductor (GEC) to ground the service to a metal water pipe (MWP) is determined by the size of the service-entrance conductors. **(See Figure 3-18)**

For example: What size copper GEC is required to ground a service to a metal water pipe supplied by 250 KCMIL copper conductors?

   Step 1: Finding size GEC
   Table 250.66
   250 KCMIL = 2 AWG cu.

Solution: The size GEC is 2 AWG copper.

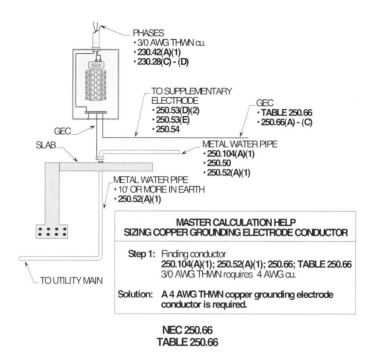

Figure 3-18. The procedure for selecting the grounding electrode conductor (GEC) to ground the service to a metal water pipe is determined by the size of the service-entrance conductors per **Table 250.66**.

Master Test Tip 39: Table 250.66 is used to size the grounded conductor, the GEC, and all SBJ's, when there is not an OCPD ahead of the conductors and equipment served.

ROD ELECTRODES
250.66(A)

The service equipment can be grounded with a driven rod or plate in cases where there are no other electrodes available in **250.50**. A driven rod with a resistance of 25 ohms or less is considered low enough to allow the grounded system to operate safely and function properly. If the driven rod is used as a supplementary electrode for the water pipe system, it should be connected to the grounded bar in the service equipment panelboard. For further information see **250.52(A)(1) thru (A)(6)** for the application of this rule when using one of the made electrodes listed there. The grounding electrode conductor (GEC) shall not be required to be larger than 6 AWG cu. or 4 AWG alu. where connected to made electrodes such as driven rods. Note that for other electrodes, the size GEC is selected per **250.66(B) and (C)**.

Master Test Tip 40: Test takers should note that the GEC used to ground the service to a rod electrode does not have to be sized larger than 6 AWG cu. per **250.66(A)**.

For example: What is the current flow in a 6 AWG copper grounding electrode conductor connecting the common grounded terminal bar in the service equipment to a driven rod? (The supply voltage is 120/240 volt, single-phase)

    Step 1: Finding amperage
        250.56
        I = 120 volts ÷ 25R
        I = 4.8 A

Solution: The normal current flow is about 4.8 amps.

CONCRETE-ENCASED ELECTRODE
250.52(A)(3); 250.66(B)

Lengths of rebar 1/2 in. in diameter and at least 20 ft. long can be used as a grounding electrode to ground the service equipment. The size of the grounding electrode conductor must be a 4 AWG or larger copper conductor per **250.66(B)** and **250.52(A)(3)**. The rebar system can be one length that is 1/2 in. x 20 ft. long or a number of such that are spliced together to form a 20 ft. length. This grounding method is known in the electrical industry as the UFER ground. **Watch for this terminology on the test.**

A concrete-encased electrode can also be a 4 AWG copper conductor instead of 1/2 in. x 20 ft. rebar system. The 4 AWG copper conductor must be located within 2 in. of concrete located within or near the bottom of the foundation.

The size of the grounding electrode must be at least a 4 AWG copper per **250.66(B)**. The 4 AWG copper grounding electrode is selected based on the 4 AWG copper grounding electrode installed in the foundation. **(See Figure 3-19)**

Figure 3-19. The size of the grounding electrode must be at least a 4 AWG copper per **250.66(B)**. The 4 AWG copper grounding electrode conductor is selected based on the 4 AWG copper grounding electrode installed in the foundation.

Master Test Tip 41: Note, for the steel rebar to be considered a grounding electrode there must be at least 20 ft. in length (one piece) or a number of short pieces spliced together to create a continuous 20 ft. length.

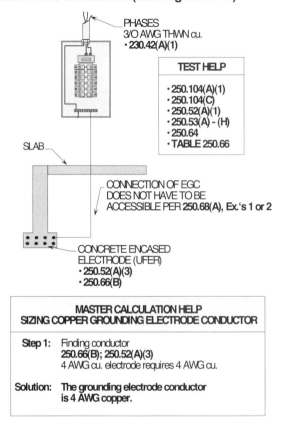

GROUND RING
250.52(A)(4); 250.66(C)

Master Test Tip 42: When a ground ring is used as a grounding electrode, it must be buried at least 2 ft. in the ground and connected into a complete circle around the building.

A bare copper conductor not smaller than 2 AWG and at least 20 ft. long shall be installed to create a ground ring for encircling the building or structure. This ring can be utilized as the main grounding electrode, if necessary, or just bonded in as a GE per **250.52(A)(4)**.

For example: What size grounding electrode conductor is required to ground 400 KCMIL copper service-entrance conductors to a ground ring?

Step 1: Finding conductor
250.52(A)(4); 250.66(C)
2 AWG cu. electrode requires 2 AWG cu.

Solution: The grounding electrode is 2 AWG copper.

GROUNDING ELECTRODE SYSTEM
250.52(A)(1) THRU (A)(6)

The grounding electrode system may consist of more than one electrode. If there is a 4 AWG copper conductor or 1/2 in. metal rebars in lengths of 20 ft. or longer (can be spliced to form 20 ft.) installed in the foundation, they shall be bonded to the service equipment to create the grounding electrode system. Metal water piping with at least 10 ft. installed in the earth or installed above grade and not in contact with the earth must be bonded to the grounding electrode system per **250.104(A)(1)** and **250.52(A)(1)** thru **(A)(6)**.

Interior metal water piping further than 5 ft. from its entrance, and intended to be used as splicing point for a concrete-encased electrode, structural steel, ground ring, or driven rod, is prohibited. The interior metal water piping, located more than 5 ft. from the point of entrance to the building, is not to be used as part of the grounding electrode system or as a conductor to interconnect electrodes that are a part of the grounding electrode system. The following five types of electrodes must, if available, be bonded together:

(1) Metal water pipe
(2) Metal frame of building
(3) Concrete-encase electrode
(4) Ground ring
(5) Driven rods if used
(6) Plates if used

Master Test Tip 43: If any of the electrodes that are listed in **250.52(A)(1)** thru **(A)(6)** are present, they must be bonded together to complete the grounding electrode system. The method in which the grounding electrode system is connected (bonded) depends upon whether the installation is new, existing, or has been remodeled. **(See Figure 3-20)**

Note that metal water pipe passing concrete floors, walls and ceilings are considered exposed per **250.52(A)(1), Ex.**

INDUSTRIAL AND COMMERCIAL BUILDINGS
250.52(A)(1), Ex.

The grounding electrode conductor shall be permitted to be installed further than 5 ft. in industrial and commercial buildings where conditions of maintenance and supervision ensure that only qualified personnel will service the installation and the entire length of the interior metal water pipe that is being used for the conductor is exposed. Under these conditions, such metal water pipes are allowed to be used as grounding electrode conductors.

METAL WATER PIPE
250.104(A)(1); 250.52(A)(1)

The copper or metal water pipe in a new installation may be installed in the earth or above grade. Either installation requires the copper or metal water pipe to be supplemented by an additional electrode. The additional electrode used to supplement the metal water pipe can be any of the electrodes listed in **250.52(A)(1) thru (A)(6)**.

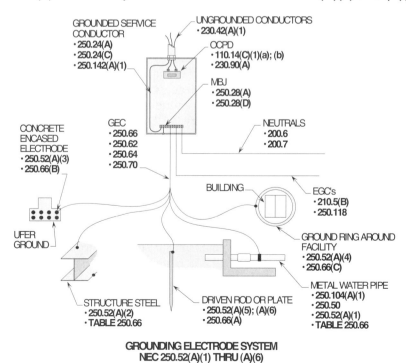

Figure 3-20. If any of the electrodes that are listed in **250.52(A)(1)** thru **(A)(6)** are present, they must be bonded together to complete the grounding electrode system.

Master Test Tip 44: In existing installations, where the service-entrance conductors or elements of the service equipment have been upgraded to supply additional loads, the copper or metal water pipe lines must be supplemented by an additional electrode if the existing grounding system does not already provide such supplementary grounding electrode.

Master Test Tip 45: *Above grade* is where the copper or metal water pipes emerge from the earth and connect to the plumbing fixtures. *Below grade* is considered to be in the earth or foundation.

Metal water pipes with lengths of at least 10 ft. long in the earth must be connected to the grounded neutral bar in the service equipment enclosure. The grounded (neutral) bar must be bonded to the service equipment enclosure, the grounded conductor, the neutral conductor, and the equipment grounding conductors by a grounding electrode sized from **Table 250.66**.

In most all installations, (existing) the grounding electrode conductor connects the service equipment grounded conductor (neutral) to a copper tubing or metal water pipe system or a driven rod. All other electrodes must be connected, if they are present. **(See Figure 3-21)**

If the metal water piping from the utility is nonmetallic (PVC) and is converted to metal or copper tubing above ground, it still has to be bonded into the grounding electrode system.

Figure 3-21. In most all installations, (existing) the grounding electrode conductor connects the service equipment grounded conductor (neutral) to a copper or metal water pipe system or a driven rod. All other electrodes if present must be bonded together to form an electrode system.

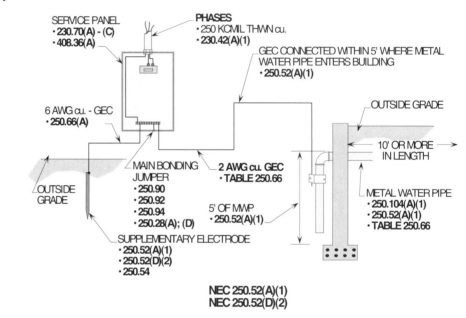

Master Test Tip 46: The grounding connection to the metal water pipe must be made at a point no more than 5 ft. where the piping enters per **250.52(A)(1)**. It is recommended that the grounding conductor from the supplementary driven ground rod be terminated to the common grounded bar in the service equipment panel.

METAL WATER PIPING
250.104(A)(1)

Regardless of whether the water piping is supplied by nonmetallic pipe to the building or structure, proper bonding of the interior metal water piping to the service equipment enclosure is required. This bonding jumper is sized using **Table 250.66**. It may be run to the service-entrance equipment if the service grounding conductor is of sufficient size. Note that the bonding jumper must be installed by the rules and regulations of **250.92(B)(1) thru (4). (See Figure 3-22)**

Master Test Tip 47: Metal water piping shall be electrically continuous or made electrically continuous by bonding around insulated joints or sections.

Beware that in multiple occupancies if the interior piping is metal and the piping system of all occupancies are tied together, it is required that metal piping be isolated from all other occupancies when nonmetallic pipe is used as a supply line. Each individual metal water pipe system is required to be bonded separately to the panelboard or switchboard enclosure, but not the service-entrance equipment. Bonding jumpers and points of attachment are required to be accessible and sized according to **Table 250.12**2. This ensures that the metal piping system in each occupancy is at ground potential and the threat of electrical shock and fires is eliminated.

Master Test Tip 48: With circuit conductors (hots) that could energize metal piping, it is permitted to use the equipment grounding conductor if it is routed with the circuit conductors to bond and ground such piping systems.

Note that where there is a separately derived system using a grounding electrode near an available point of metal water piping, which is in the area supplied by such derived systems, such metal water pipe systems must be bonded to the derived systems grounded conductor. See **250.104(D)** for such rules explained in greater detail.

REVIEW
- Metal water pipe
  — 250.104(A)(1); (A)(2)
  — 250.104(B)
- Gas metal pipe
  — 250.104(B)
  — 250.52(B)

Grounding

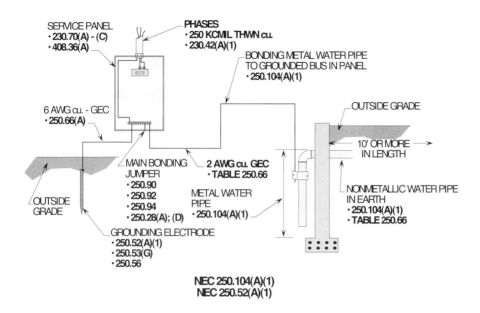

Figure 3-22. Regardless of whether the water piping is supplied by nonmetallic pipe to the building or structure, proper bonding of the interior metal water piping to the service equipment enclosure is required.

Master Test Tip 49: If the rebar in the foundation is coated to protect against corrosion, such rebar is not considered a grounding electrode.

METAL FRAME OF BUILDING
250.104(C)

In most all installations, (existing) the grounding electrode conductor connects the service equipment grounded conductor (neutral) to a copper tubing or metal water pipe system or a driven rod is used. Where all other electrodes are available, they must be bonded together. **(See Figure 3-23)**

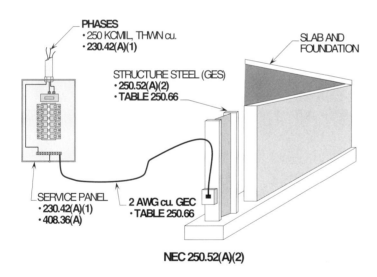

Figure 3-23. In most all installations, (existing) the grounding electrode conductor connects the service equipment grounded conductor (neutral) to a copper or metal water pipe system or a driven rod or all other electrodes, if they are present.

Master Test Tip 50: Note that the MWP can be bonded to the structural steel and then connected to the grounding electrode system.

STRUCTURAL METAL
250.104(C)

Exposed interior building steel frames that are not intentionally grounded must be bonded to either the service enclosure, the neutral bus, the grounding electrode conductor, or to a grounding electrode (except a metal water pipe). The bonding conductor must be size according to **Table 250.66**, and installed according to **250.64(A)** and **(B)**. **(See Figure 3-24)**

Figure 3-24. Exposed interior building steel frames that are not intentionally grounded must be bonded to either the service enclosure, the neutral bus, the grounding electrode conductor, or to a grounding electrode.

Master Test Tip 51: For additional safety, it is recommended to bond all metal air ducts and metal piping systems on the premises together and connect them to the grounding electrode system.

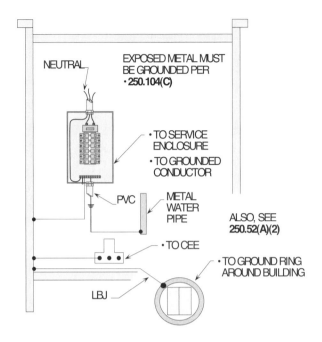

CONCRETE-ENCASED ELECTRODE 250.52(A)(3)

Master Test Tip 52: For the rules pertaining to an equipotential plane, see **547.2** and **547.10**.

Concrete-encased electrodes consist of 1/2 in. x 20 ft. lengths of reinforcing rebar located in the foundation. The 20 ft. length of rebar can be in one continuous piece or several pieces spliced together to form a 20 ft. or greater continuous length. The metal reinforcing rebars are required to be of the conductive type. A concrete-encased electrode may also be a 4 AWG bare copper conductor at least 20 ft. long that is installed in the footing of the foundation. The 20 ft. long 4 AWG conductor should be located in at least 2 in. of concrete near the bottom of the foundation or footing. The reinforcing rebars or 4 AWG bare copper conductor, if installed properly, usually provides a resistance of a little above zero ohms resistance. **(See Figure 3-25)**

Figure 3-25. Concrete-encased electrodes consist of 1/2 in. x 20 ft. lengths or longer lengths of reinforcing rebar located in the foundation. A concrete-encased electrode can also be a 4 AWG bare copper conductor at least 20 ft. long that is installed in the footing of the foundation.

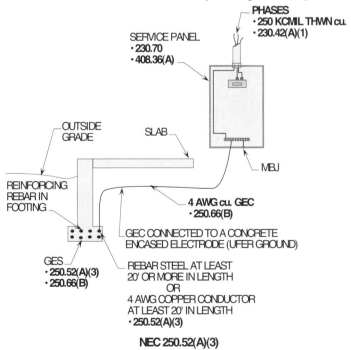

Grounding

GROUND RING
250.52(A)(4)

To create a ground ring grounding system, a bare copper conductor not smaller than 2 AWG and at least 20 ft. long in length shall be installed at least 2 1/2 ft. below the ground and must completely encircle the building or structure. **(See Figure 3-26)**

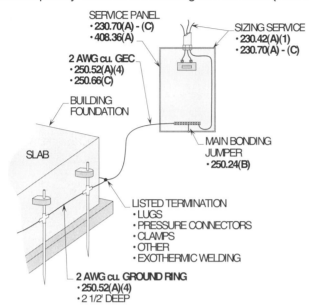

Figure 3-26. A bare copper conductor not smaller than 2 AWG and at least 20 ft. long in length shall be installed at least 2 1/2 ft. below the earth to create a ground ring grounding system. A ground ring must circle the building or structure.

Master Test Tip 53: A more reliable and dependable ground (counter-poise ground) may be obtained by burying the conductor about 3 ft. from the building and periodically driving ground rods and attaching them to the bare conductor by approved connecting methods.

ROD, PIPE, AND PLATE ELECTRODES
250.52(A)(5) AND (A)(6)

Where none of the electrodes are present in **250.52(A)(1)** thru **(A)(6)**, one or more of the electrodes specified below shall be used: **(See Figure 3-27)**

(1) Local metal underground systems or structures
(2) Rod and pipe electrodes
(3) Plate electrodes

Note: Underground metal gas piping systems and aluminum electrodes shall not be permitted to be installed as a grounding electrode per **250.52(B)(1)** and **(B)(2)**.

ROD AND PIPE ELECTRODES
250.52(A)(5) and (A)(6)

A driven electrode used to ground the service can be a 1/2 in. x 8 ft. copper (brass) rod or a 3/4 in. x 10 ft. galvanized pipe. A copper rod is usually the type of driven electrode that is used. Driven rods are utilized to connect the neutral bar terminal to earth ground. Where the metal water pipe system in the ground is PVC and converts to metal piping or copper tubing above ground, the driven rod can be used to bond and ground the metal water pipe system and connect the electrical system to ground.

(1) 1/2 in. x 8 ft. copper rod
(2) 3/4 in. x 10 ft. pipe or conduit
(3) 5/8 in. x 8 ft. rebar
(4) 5/8 in. or larger stainless steel rod or 1/2 in. 8 ft., if listed for such use

Master Test Tip 54: The metal water pipe system must be bonded into the grounding electrode system, even if the piping in the ground is PVC or metal. The following are acceptable made electrodes which can be used to supplement the metal water pipe:
- Structural metal
- Concrete encased electrodes
- Rod electrodes

Figure 3-27. Local metal underground systems or structures, rod and pipe electrodes, and plate electrodes must be installed when none of the electrodes are present in **250.52(A)(1) thru (A)(6).**

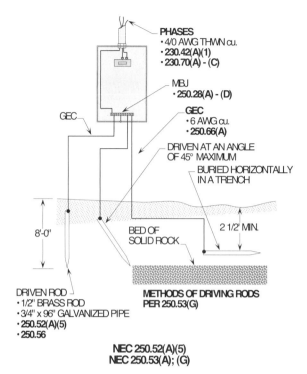

PLATE ELECTRODES
250.52(A)(6)

Plate electrodes shall be a minimum of 2 sq. ft., and if made of iron or steel, they shall be at least 1/4 in. thick. If made of a nonferrous metal such as copper, they may be .06 in. in thickness.

BONDING
250.90

Master Test Tip 55: A bonding jumper on the supply side has to be as large as the grounded conductor and GEC where the phase conductors do not exceed 1100 KCMIL for cu. and 1750 KCMIL for alu. per **250.24(C)(1).**

Bonding is the permanent joining of metallic parts to form an electrically conductive path that will assure electrical continuity and the capacity to conduct safely any current likely to be imposed on it. Such bonding jumpers must be sized to handle these larger available fault-currents, when necessary, to prevent damaging components and equipment.

SERVICE EQUIPMENT BONDING
250.92(A)(1) THRU (A)(3)

The following noncurrent-carrying metal parts of equipment shall be bonded together:
  (1) Service-entrance raceway, cable, and weatherhead
  (2) Meter base enclosure including CT cans
  (3) Service equipment enclosure
  (4) All other metal enclosures between the service equipment and supply transformer

Service-entrance raceways, meter base enclosures, service equipment enclosures, and all other metal enclosures of the service equipment must be bonded together to ensure a continuous metal-to-metal system.

Section **250.94, FPN 2** deals with the bonding of other systems together. For further information on these bonding and grounding requirements such as communication circuits, radio and television equipment, and CATV circuits, see **800.100, 810.21,** and **820.100. (See Figure 3-28)**

Grounding

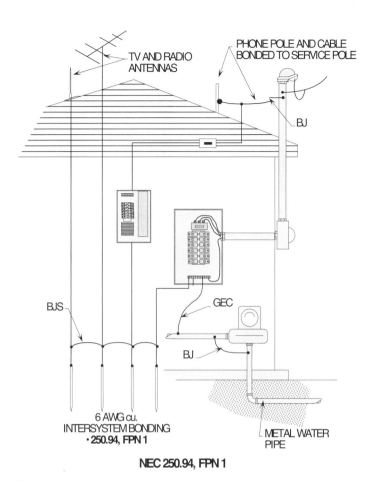

Figure 3-28. Bonding and grounding requirements for communication circuits, radio and television equipment, and CATV circuits.

Master Test Tip 56: Electrodes of other systems, when bonded together and connected to the service's grounding electrode, become one electrode system per **250.58**.

Note: For bonding around devices that could interrupt the continuity of the metal water pipe, see **250.52(A)(1)** and **250.68(B)**.

METHODS
250.92(B)

Any one of the following five methods can be utilized for the bonding of service equipment, enclosures, etc.: **(See Figure 3-29)**

- **(1)** Grounded service conductor
- **(2)** Threaded connections
- **(3)** Threadless couplings and connectors
- **(4)** Bonding jumpers
- **(5)** Other devices such as bonding type locknuts and bushings

GROUNDED SERVICE CONDUCTOR
250.92(B)(1)

The grounded service conductor shall be permitted for bonding equipment if connected by one of the following methods per **250.8**:

- **(1)** Exothermic welding
- **(2)** Listed pressure connectors
- **(3)** Listed clamps
- **(4)** Other listed means

THREADED CONNECTIONS
250.92(B)(2)

Threaded connections such as threaded couplings or threaded bosses may be installed where the service equipment is threaded into the meter base or the service equipment enclosure. Such threaded connections must be made up wrench-tight.

Master Test Tip 57: If standard locknuts and bushings are used for a service-entrance raceway, a bonding jumper must be used to ground the raceway to the service enclosure per **250.92(B)**. However, fittings listed for grounding, such as an extra thick locknut with a set screw, can be used.

Figure 3-29. Service equipment can be bonded by the grounded service conductor, threaded connections, threaded couplings and connectors, bonding jumpers, and other devices such as bonding type locknuts and bushings.

Master Test Tip 58: The grounded conductor and main bonding jumper of the service must be sized at 12 1/2 percent of the largest phase conductor(s), if such conductors exceed 1100 KCMIL for cu. and 1750 KCMIL for alu. per **250.24(C)** and **250.28(D)**.

Master Test Tip 59: Note that the sq. in. area of a grounded conductor is less for a bare conductor than for an insulated conductor per **Table 5, Ch. 9, Col. 4** and **Table 8 to Ch. 9, Col. 6**.

- 6 AWG THWN = .0507 sq. in.
- 6 AWG Bare = .0270 sq. in.

Note that under certain conditions, a metal raceway supplying a piece of computer equipment can be isolated from the computer's enclosure to help prevent electromagnetic interference. (See **250.96(B)** for this rule)

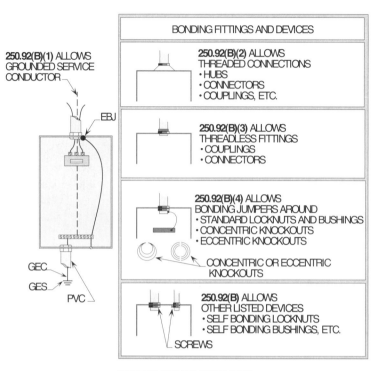

NEC 250.92(B)(1) THRU (B)(4)

THREADLESS COUPLINGS AND CONNECTORS
250.92(B)(3)

Threadless couplings and connectors installed for rigid metal conduit (RMC), intermediate metal conduit (IMC), and electrical metallic tubing (EMT) shall be made up tight. Standard locknuts and bushings shall not be installed for bonding the raceway to the enclosure. Only locknuts approved for bonding, such as extra thick locknuts and and bushings with a set screw, can be used for this purpose.

BONDING JUMPERS
250.92(B)

A bonding conductor shall be used to connect and bond the raceway to the enclosure where smaller concentric or eccentric knockouts are removed, leaving larger ones. A metal bushing with a lug on the threaded raceway is used to terminate the bonding jumper to a lug connected to the enclosure or grounded bus terminal inside such enclosure.

OTHER DEVICES
250.92(B)(4)

A wedge fitting is an application of other approved devices that can be used. A grounding wedge is equipped with a set screw that tightens the metal conduit and ensures the proper bonding of service equipment enclosures.

BONDING OTHER ENCLOSURES
250.96

The rules in this section require all metal enclosures, raceways, and cables to be bonded together to provide a continuous electrical path for ground-faults to travel over, to clear the OCPD in the supply panel.

OVER 250 VOLTS
250.97 and Ex.

One of the following wiring methods shall be installed for bonding circuits over 250 volts-to-ground, not including services:

(1) Install the following wiring methods between sections of metal conduit, intermediate metal conduit, electrical metallic tubing, or other type of metal raceways with:

   (a) Threaded couplings
   (b) Threadless couplings

(2) Install the following wiring methods between metallic raceways, metallic boxes, cabinet enclosures, etc. with:

   (a) Jumpers
   (b) Two locknuts and bushing
   (c) Threaded (bosses) hubs
   (d) Other approved devices

For further information on other bonding techniques, see **250.92(B)(1)** thru **(B)(4)**.

Master Test Tip 60: If the voltage to ground measures 250 volts or less to ground from all phase conductors, bonding jumpers are not required from raceways to enclosures per **250.97** with exception. Note that bonding jumpers are not required even if punched-out rings are present.

MAIN AND EQUIPMENT BONDING JUMPERS
250.28(D); 250.102(C)

The size of equipment bonding jumper and service equipment main bonding jumper is based on the size of the service-entrance conductors and sized per **Table 250.66**. The grounded conductor shall be sized at least 12 1/2 (.125) percent of the area of the largest conductor where the service conductors are installed larger than 1100 KCMIL copper or 1750 KCMIL aluminum. **(See Figure 3-30)**

Note that the equipment bonding jumper shall be sized based on the largest ungrounded phase conductor in each raceway. Such a case is where three raceways enter a switchgear and each raceway is individually bonded to the enclosure, cabinet, etc. **(See Figure 3-31)**

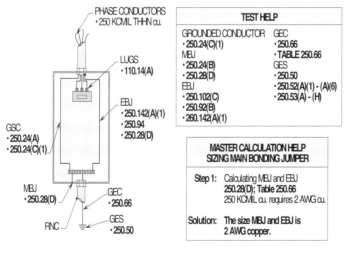

Figure 3-30. The size of equipment bonding jumpers and the main bonding jumper is based on the size of the service-entrance conductors per **Table 250.66**.

Figure 3-31. Where three raceways enter a switchgear with each raceway individually bonded to the enclosure, cabinet, etc., the equipment bonding jumper shall be sized based on the largest ungrounded phase conductor of the raceway.

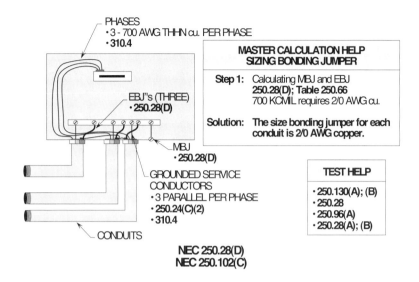

The grounded conductor and MBJ must be sized at least 12 1/2 (.125) percent of the total KCMIL rating per phase. Note that the main bonding jumper shall be installed as large as or larger than grounding electrode conductor per **250.28(D)**. **(See Figure 3-32)**

Figure 3-32. The main bonding jumper and equipment bonding jumper must be sized at least 12 1/2 percent of the total KCMIL rating per phase, if only one bonding jumper is used.

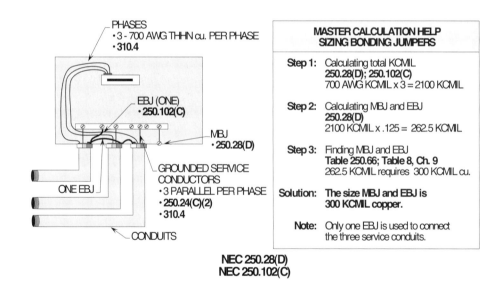

EQUIPMENT BONDING JUMPER ON LOAD SIDE OF SERVICE 250.102(D)

Table 250.122 lists the size of the equipment bonding jumper on the load side of the service OCPD's. The bonding jumper must be a single conductor sized according to **Table 250.122** for the largest OCPD that is provided to supply the electrical circuits. If the bonding jumper supplies two or more raceways or cables, the same rules apply and proper size conductors must be selected. Note that it is not necessary to size the equipment bonding jumper larger than the phase conductors, but in no case shall it be smaller than 14 AWG per **250.102(D)**.

Master Test Tip 61: Note that raceways and conductors are kept continuous by using listed fittings and connectors per **300.10** and **300.15**.

Grounding

INSTALLATION OF EQUIPMENT BONDING JUMPER
250.102(E)

It is permitted to install the equipment bonding jumper (EBJ) either inside or outside of the raceway or enclosure. When it is installed outside of the raceway or enclosure, it shall not be over 6 ft. in length and shall be routed with the raceway or enclosure. If the equipment bonding jumper is inside the raceway, refer to **310.12(B)** and **250.119** for the conductor identification requirements. Note that if such EBJ is routed on the outside of the raceway, listed grounding fittings must be used to terminate raceway and EBJ.

Master Test Tip 62: Note that the EBJ, if run inside of raceways, must be green, with one or more yellow stripes, or bare per **310.12(B)** and **250.119**.

RESISTANCE OF ROD ELECTRODES
250.56

Rod electrodes shall have a resistance to ground of 25 ohms or less, wherever practicable. When the resistance is greater than 25 ohms, two or more rod electrodes may be connected in parallel or extended to a greater length. Note that a made electrode that measures more than 25 ohms is required to be augmented by one additional electrode of a type permitted by **250.53(D)(2)**.

Continuous metal water piping systems usually have a ground resistance of less than 3 ohms. Metal frames of buildings normally have a good ground and usually have a resistance of less than 25 ohms. As pointed out in **250.52(A)(2)** and **(D)(2)**, the metal frame of a building, if effectively grounded, may be used as the ground. Local metallic water systems and well casings make good grounding electrodes in most all types of installations. **(See Figure 3-33)**

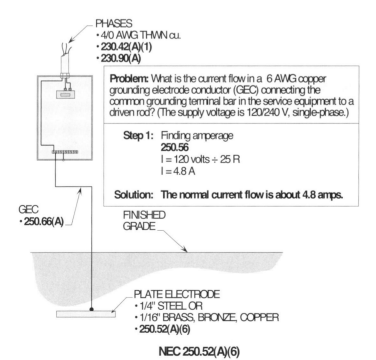

Figure 3-33. Rod electrodes shall have a resistance to ground of 25 ohms or less, wherever practicable. When the resistance is greater than 25 ohms, two or more electrodes may be connected in parallel or extended to a greater length.

Master Test Tip 63: If two driven rods are used, they must be separated at least 6 ft. apart per **250.56**.

ATTACHMENT
250.148

Master Test Tip 64: Test takers should take note that an isolated ground for a computer does not have to be bonded to the metal of junction boxes or panelboards per **250.96(B)** and **408.40, Ex.**

Where more than one equipment grounding conductor enters a box, all such conductors shall be spliced within the box with devices identified for the purpose. Equipment grounding conductors shall be connected and bonded to metal boxes to ensure grounding continuity. In other words, all grounding conductors entering the box are to be made electrically and mechanically secure, and the pigtail from them serves as the grounding connection to the device being installed. If the device is removed, the continuity of the EGC's will not be disrupted. **(See Figure 3-34)**

SWIMMING POOLS
ARTICLE 680, PART I

Swimming pools and decorative fountains are required to be grounded and bonded to prevent electrical shock to people wading and swimming in the pool water. Properly sized bonding and grounding conductors are essential to ensure the safety of persons in or around the wet areas of the swimming pool.

Master Test Tip 65: Note that the 8 AWG solid copper conductor is used to bond metal parts within 5 ft. of the walls of the pool. The BJ must not be used as an EGC to clear a ground fault or short circuit. Such BJ is used to help provide an equipotential plane.

BONDING
680.26(A); (B); (C)

The bonding of all metal parts and grounding of the elements keeps the interconnected system and steel at the same potential, which creates an equipotential plane. Stray currents moving in the water, metal, and steel of the pool that are not cleared by the overcurrent protection device are kept at zero or at earth potential. This prevents electrical shock hazards to persons in the pool water or the area surrounding the pool, such as the deck, where there may be stray currents in the steel or mesh in the concrete or gunite.

Note: When the grid steel is coated, see **680.26(C)(3)** for the installation of an alternate means (grid structure).

An 8 AWG or larger solid copper conductor shall be used to bond together all metal parts to the common bonding grid that is the foundation steel of the pool. The reinforcing bars in the pool and metal in the walls of the pool are used to form the common grid system. The bonding conductor should not be run to the service equipment and connected to the grounded terminal bar in the service equipment enclosure per **680.26(A), FPN.** Note that this connection could allow more current from the service neutral to flow to the pool.

Note that for clearances of overhead conductors crossing the pool, see **680.8** and **Table 680.8**.

The equipment and metal parts of a swimming pool that are required to be bonded are as follows:

(1) All fixed parts located within 5 ft. of the walls of the pool, or they must be separated by a permanent barrier.

(2) All metal parts and reinforcing steel in the pool or deck.

(3) All forming shells housing wet-niche or dry-niche luminaires (fixtures).

(4) All metal parts and fittings of recirculating motor, etc.

(5) All fixed metal parts, such as conduits, pipes, cables, equipment, etc. located within 5 ft. of the pool and not separated by a permanent barrier, and within 12 ft. vertically of the maximum water level. **(See Figure 3-37)**

Grounding

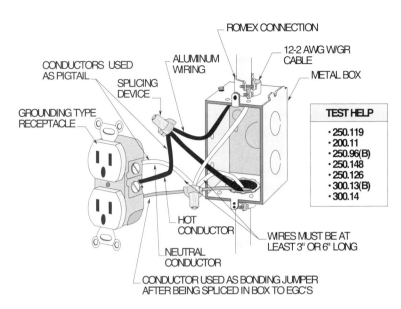

NEC 250.148

Figure 3-34. Where more than one equipment grounding conductor enters a box, all such conductors shall be spliced within the box with devices identified for the purpose and bonded to the box.

Note that there are lengths of 3 in. and 6 in. now required based on the size of the box used. See **300.14** for this requirement.

HOT TUBS, SPAS, AND HYDROMASSAGE TUBS
ARTICLE 680, PARTS IV and VII

Hot tubs, spas, and hydromassage tubs are required to be bonded and grounded for the same reasons as swimming pools for the safety of those using them. Such tubs can be located inside or outside of the dwelling unit or building.

Master Test Tip 66: Note that a hot tub or spas located outside must comply with the requirements for installing a swimming pool outdoors per **680.40**.

All metal parts around the spa or hot tub are required to be bonded and grounded to ground potential and thus reduce the threat of shock hazard. The equipment and parts that must be grounded includes the following items:

(1) Wet-niche luminaire (lighting fixture)
(2) Dry-niche luminaire (lighting fixture)
(3) All electrical equipment located within 5 ft. of the inside walls
(4) Circulating pump
(5) All other electrical equipment

BONDING
680.43(D)

Section **680.42(D)** lists the items that must be bonded together. Section **680.42(D)** lists the methods by which the bonding can be accomplished. The following items must bonded together per **680.43(D)**.

(1) All metal fittings of the spa or hot tub.
(2) All metal parts of electrical equipment associated with the spa or hot tub.
(3) All metal conduit, piping, and metal surfaces located within 5 ft. of the inside walls of the spa.

Any of the following bonding methods may be used per **680.42(E)** to provide a safe and reliable bonding system.

(1) Interconnected threaded metal piping and fittings.
(2) Metal-to-metal bonding on a common frame or base.
(3) Size 8 AWG or larger copper bonding jumper. The bonding jumper may be insulated, covered, or a bare wire. **(See Figure 3-35)**

Master Test Tip 67: The 8 AWG bonding jumper used to bond the metal of the equipment and steel in and around pools, spas, and hot tubs is used for bonding and not as an equipment grounding conductor.

Figure 3-35. An 8 AWG solid copper conductor shall be used to ground all noncurrent-carrying parts associated with the hot tub or spa, including metal piping, etc. located within 5 ft. of the tub.

Master Test Tip 68: Switches must be located at least 5 ft. measured horizontally from the inside walls of the spa or hot tub.

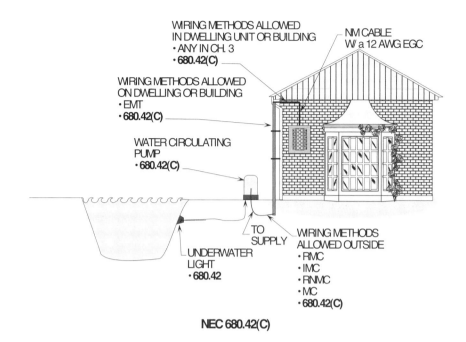

GROUNDING
680.42; 680.43

Master Test Tip 69: The EGC is routed from the panelboard to a junction box usually located under the diving board and bonded to the metal of the box per **680.23(B)(2)** and **680.24(A)**.

All metal parts around the pool are required to be bonded and grounded to ground potential and thus reduce the threat of shock hazard. The equipment and parts that must be grounded include the following items:

(1) Wet-niche luminaire (lighting fixture)
(2) Dry-niche luminaire (lighting fixture)
(3) All electrical equipment located within 5 ft. of the inside walls
(4) Circulating pump
(5) All other electrical equipment

GROUNDING
680.43(F)

Grounding may be accomplished according to the provisions of **Article 250**. The equipment grounding conductor can be done by using metal conduits or equipment grounding conductors per **250.118**. The equipment grounding conductor within a flexible cord, flexible cable, or flexible conduit must be connected to the fixed metal part of the assembly.

The equipment grounding conductor used to ground the metal parts of equipment, junction boxes, etc. must be sized from **Table 250.122**, based on the rating of the overcurrent protection device protecting the branch-circuit conductors. **(See Figure 3-36)**

Master Test Tip 70: The 12 AWG equipment grounding conductor is used to clear a fault, while the 8 AWG bonding jumper is utilized to bond the metal parts of the pool to the grid system.

Grounding of the junction box and wet-niche fixtures must be accomplished by a 12 AWG copper insulated equipment grounding conductor. The grounding conductor is sized according to the rating of the overcurrent protection device per **Table 250.122**. It must not be smaller than 12 AWG and must be unspliced unless the exception is applied.

The grounding conductor must be installed and run with the circuit conductors in rigid metal conduit, intermediate metal conduit, or rigid nonmetallic conduit from the panelboard to the deck box. The main function of the 12 AWG equipment grounding conductor is to clear the overcurrent protection device in case of a ground-fault.

Grounding

The rules in **680.23(F)(1)** permit conductors on or within buildings to be installed in electrical metallic tubing instead of rigid metal conduit, intermediate metal conduit, PVC (Schedule 80), or electrical nonmetallic tubing. See **680.23(F)(1)** for a complete list of these wiring methods. **(See Figure 3-37)**

Flexible cords used to ground wet-niche luminaires (fixtures) shall be provided with a 16 AWG or larger insulated copper equipment grounding conductor connected to a terminal in the supply deck box. This type of connection provides continuity when the fixture is removed for servicing.

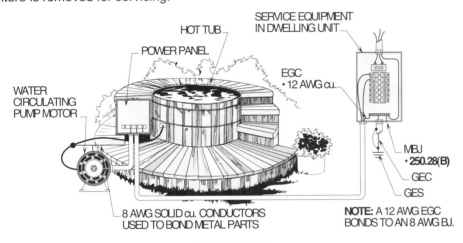

Figure 3-36. The bonding of all metal parts and the grounding of the elements keep the interconnected system and steel at the same potential, which creates an equipotential plane.

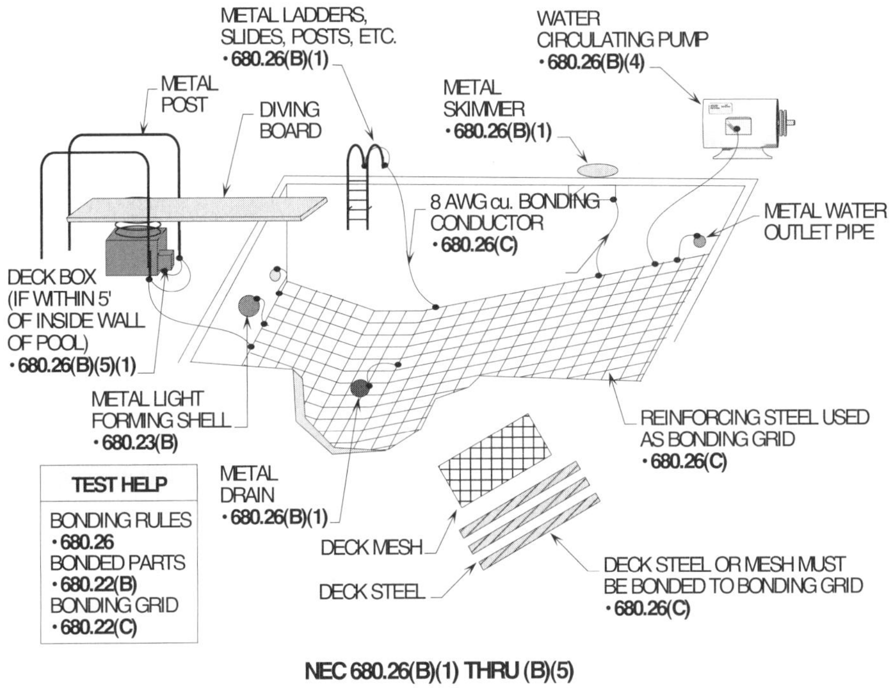

Figure 3-37. The rules in **680.23(F)(1)** permit conductors on or within buildings to be installed in electrical metallic tubing instead of rigid metal conduit, intermediate metal conduit, PVC (Schedule 80), or electrical nonmetallic tubing. The letter wiring methods can be used inside the building.

Note: When a new pool is constructed, review **680.26(C)** very carefully.

HYDROMASSAGE BATHTUBS
680.70; 680.71; 680.72; 680.73

Hydromassage bathtubs are treated in the same manner as conventional bathtubs. The wiring methods for hydromassage bathtubs shall comply with **Chapters 1-4** of the NEC. All elements for hydromassage bathtubs shall be supplied by GFCI-protected circuits. See **410.4(D)** for hanging fixtures over and around the tub. Location of switches for lighting fixtures and receptacle outlets are treated in the same manner as a regular bathtub. **(See Figure 3-38)**

Master Test Tip 71: See **404.4** and **406.8(C)** for locating and installing switches not associated with the hydromassage tub. Access to the circulating motor for servicing is required per **680.73**.

3-35

Figure 3-38. Hydromassage bathtubs are treated in the same manner as regular bathtubs except they must be supplied with a GFCI-protected circuit to protect the user from electric shock. Lighting fixtures shall comply with **410.4(D)**. Switches and receptacles shall meet the provisions of **404.4** and **406.8(C)**.

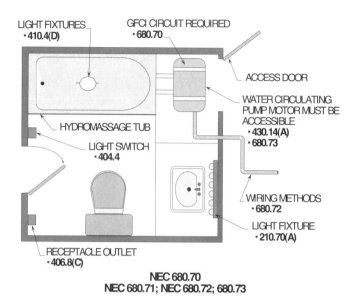

CORD-AND-PLUG CONNECTED COMPUTERS 250.146(D)

An isolated equipment grounding conductor is used to connect the receptacle isolated ground terminal back to the grounded connection at the service equipment or separately derived system to reduce electromagnetic interference.

This rule applies when electrical noise known as electromagnetic interference occurs in the grounding circuit. To help prevent such noise, an insulated equipment grounding conductor may be run to the isolated terminal with the circuit conductors. This insulated grounding conductor may pass through one or more panelboards in the same building without being connected thereto as allowed in **408.40, Ex**. Note that it still has to terminate at the equipment grounding terminal at a separately derived system or service panel. **(See Figure 3-39)**

Master Test Tip 72: The isolation ground can be routed and connected to the service equipment grounding bus or the grounding bus of a separately derived system.

Figure 3-39. An isolated equipment grounding conductor is used to connect the receptacle isolated ground terminal back to the grounded connection at the service equipment or separately derived system (SDS) to reduce electromagnetic interference.

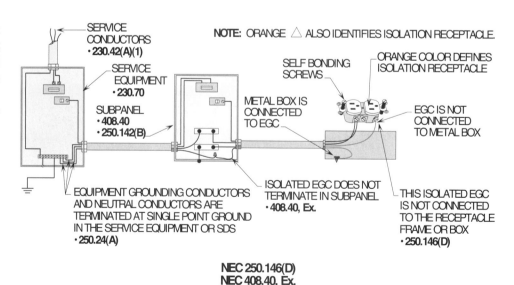

Grounding

PERMANENT-WIRED COMPUTERS
250.96(B)

When it is necessary for the reduction of electrical noise in a grounding circuit, a listed nonmetallic fitting is allowed to be installed between the enclosure and metal raceways. In these cases, a separate equipment grounding conductor must be used to properly ground the enclosure. Note that electromagnetic interference that creates noise is reduced in sensitive electronic equipment when this type of installation is applied properly. **(See Figure 3-40)**

Master Test Tip 73: When a nonmetallic spacer is used to isolate a sensitive electronic piece of equipment, the EGC must not be disconnected and the metal of the unit grounded to an isolation ground such as a driven rod.

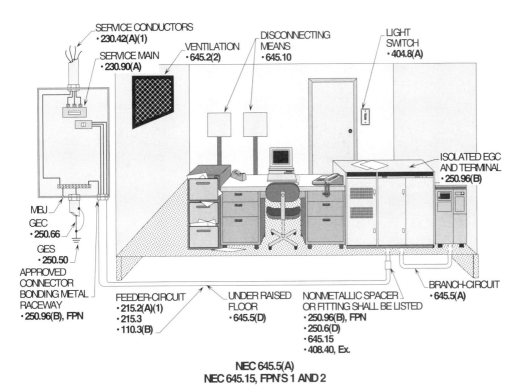

Figure 3-40. A nonmetallic spacer is allowed to be installed between sensitive electronic equipment and the service panel or source. However, an EGC must be run to such equipment.

Name _____   Date _____

Chapter 3
Grounding

Section   Answer

1. The frame of a portable generator shall be grounded if it supplies only the equipment on the generator or cord-and-plug connected equipment connected receptacles mounted on the generator.

      (a) True   **(b)** False

2. Where one or more buildings or structures are supplied from a common AC grounded service, each service at each individual building or structure shall be separately grounded. (More than one branch-circuit is provided.)

      (a) True   **(b)** False

3. The equipment grounding conductor shall be bonded to the grounding electrode conductor at the service equipment neutral busbar terminal for a grounded system.

      (a) True   **(b)** False

4. A GFCI receptacle may be used to protect a single outlet and additional outlets downstream when replacing nongrounding-type receptacles.

      (a) True   **(b)** False

5. In 6 ft. or less lengths, it shall permitted to install the equipment bonding jumper either inside or outside of the raceway or enclosure.

      (a) True   **(b)** False

6. The GEC, when connected to a metal water pipe, shall be installed within 5 ft. where the MWP enters an industrial or commercial building.

      (a) True   **(b)** False

7. Lightning conductors that are connected to lightning rods for earth grounding shall be bonded to the grounding electrode system of the service equipment.

      (a) True   **(b)** False

8. Grounding electrode conductors shall be only permitted to be installed as copper.

      (a) True   **(b)** False

9. All electrical equipment located within 5 ft. of the inside wall of the pool shall be bonded and grounded.

      (a) True   **(b)** False

10. When the copper service conductors are 1100 KCMIL or less, the grounded conductor, equipment bonding conductor and main bonding jumper shall be equal in size to the grounding electrode conductor per **Table 250.66**.

      (a) True   **(b)** False

Section	Answer

_____ _____ **11.** Existing branch-circuit installations for frames of ranges, cooktops, ovens, clothes dryers, and junction boxes or outlet boxes which are part of the circuit shall be permitted to be bonded and grounded with a grounded (neutral) conductor provided it is installed no smaller than _____ AWG copper.

      (a) 14   **(b)** 12   **(c)** 10   **(d)** 8

_____ _____ **12.** When a screw is used as the main bonding jumper, it can be identified by being colored _____.

      (a) White   **(b)** Brass   **(c)** Silver   **(d)** Green

_____ _____ **13.** When an equipment bonding jumper is installed outside of the raceway or enclosure, it shall not be over _____ ft. in length.

      (a) 5   **(b)** 6   **(c)** 10   **(d)** 12

_____ _____ **14.** Which of the following methods can be utilized for the bonding of service equipment?

      (a) Grounded service conductor   **(b)** Threaded connections
      (c) Bonding jumpers   **(d)** All of the above

_____ _____ **15.** The interior metal water piping, located more than _____ ft. from the point of entrance to the building, shall not to be used as part of the grounding electrode system to connect grounding electrode conductors. (General Rule)

      (a) 5   **(b)** 6   **(c)** 10   **(d)** 20

_____ _____ **16.** Metal water pipe lengths of at least _____ ft. long in earth (grounding electrode) shall be connected to the grounded neutral bar in the service equipment enclosure.

      (a) 5   **(b)** 10   **(c)** 15   **(d)** 20

_____ _____ **17.** A concrete-encased electrode no smaller than _____ AWG bare copper conductor at least 20 ft. long can be installed in the footing of the foundation and shall be considered a grounding electrode.

      (a) 8   **(b)** 4   **(c)** 2   **(d)** 1

_____ _____ **18.** A bare copper conductor not smaller than 2 AWG and at least 20 ft. in length shall be installed at least _____ ft. below the earth to create a ground ring (grounding system). Note that such ring shall encircle the building or structure.

      (a) 2 1/2   **(b)** 4 1/2   **(c)** 5 1/2   **(d)** 6 1/2

_____ _____ **19.** Rod and pipe electrodes used in forming the grounding electrode system shall have a resistance to ground of _____ ohms or less. (General Rule)

      (a) 10   **(b)** 20   **(c)** 25   **(d)** 50

_____ _____ **20.** Liquidtight flexible metal conduit, in sizes up to 1 1/4 in. that has a total length not to exceed _____ ft. (from the ground return path) and equipped with listed fittings for terminating the flexible conduit shall be permitted to be used as a grounding means.

      (a) 3   **(b)** 5   **(c)** 6   **(d)** 10

Overcurrent Protection Devices and Conductors

4

The master electrician examination is loaded with questions and problems about overcurrent protection of conductors and equipment. The material in this chapter covers most of the questions and problems found on a test.

The purposes of overcurrent protection devices are to monitor the current in a circuit and keep it at a level that will prevent overheating of conductors, elements, and equipment. Excessive current flowing in an electrical circuit generates heat, which raises the conductor temperature. Such temperature depends entirely upon the amount of current flowing through the electrical circuit and equipment.

In selecting the proper size conductors and overcurrent protection devices for supplying power in a circuit from the source to the load, it is most important that the appropriate rules of the *National Electrical Code* (*NEC*) be applied. Rules require the overcurrent protection device (OCPD) for conductors and equipment to be sized in such a manner to open the circuit if currents reach a value that causes an excessive or dangerous temperature in conductors or insulation. Such OCPD must be selected, sized, and wired to protect both conductors and equipment or a second stage of protection must be provided. The ampacities of conductors can vary depending upon their conditions of use which are mainly based upon number and exposure to surrounding ambient temperatures.

Quick Reference

OCPD'S	4-2
FUSE MARKINGS	4-26
CIRCUIT BREAKERS	4-28
OCPD'S FOR SYSTEMS OVER 600 VOLTS	4-31
FEEDER CIRCUIT PROTECTION - OVER 600 VOLTS	4-31
TERMINAL RATINGS CONDUCTORS	4-31
SIZING OCPD'S ABOVE THE AMPACITY OF THE CONDUCTORS	4-40
SIZING OCPD'S BELOW THE AMPACITY OF THE CONDUCTORS	4-41
SIZING CONDUCTORS	4-41
LUMINAIRES (LIGHTING FIXTURES)	4-50
GROUNDED CONDUCTOR	4-51
EQUIPMENT GROUNDING CONDUCTOR	4-52
UNGROUNDED CONDUCTORS	4-53
CONDUCTORS IN PARALLEL	4-54
TEST QUESTIONS	4-57

SIZING OCPD FOR PROTECTION OF EQUIPMENT
240.3

Master Test Tip 1: OCPD's shall protect conductors and equipment from short-circuits (phase to phase), and ground-faults (phase to ground). Overloads (slow heat build-up) shall be protected by such OCPD's if possible.

For overcurrent protection of appliances, motors, generators, etc. it is necessary to refer to the different Articles listed in this section. In all installations, there are three main parts to be protected and they are as follows:

(1) The circuit conductors,
(2) Circuit elements, and
(3) The equipment.

The fuse or circuit breaker protecting an installation must be of a rating small enough for protection, according to the rules and regulations of **240.3** and **240.4** of the NEC.

ROUNDING UP OR DOWN OF OCPD
240.4

The general rule is that conductors shall be protected against overcurrent by a fuse or circuit breaker setting rated no higher than the ampacity of the conductor. Conductors have specific current-carrying ampacities for different sizes of conductors, for different insulations, and for different ambient-temperature conditions. The overcurrent protection device shall be sized to protect the insulation of the conductors from damage caused by the current reaching an excessive value. **(See Figure 4-1)**

Figure 4-1. The OCPD must be sized and selected to properly protect the electrical equipment served.

Master Test Tip 2: If the OCPD does not protect the conductors and equipment from overload, a second stage of protection such as overloads in a controller that protects the conductors and motor windings from overload must be provided.

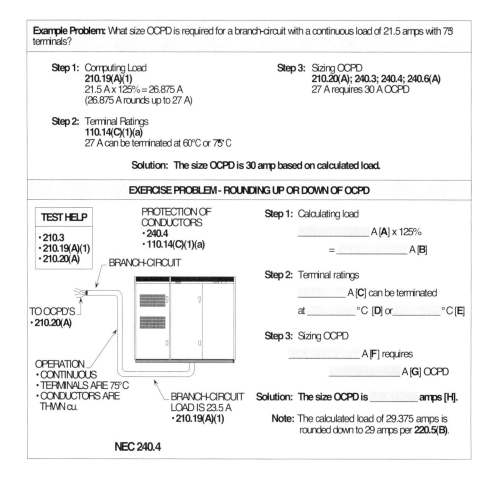

Overcurrent Protection Devices and Conductors

SIZING OCPD'S RATED 800 AMPS OR LESS
240.4(B)

If the standard current ratings of fuses or nonadjustable circuit breakers do not conform to the ampacity of the conductors being used, it is permissible to use the next larger standard rating when such OCPD is rated 800 amps or less. **(See Figure 4-2)**

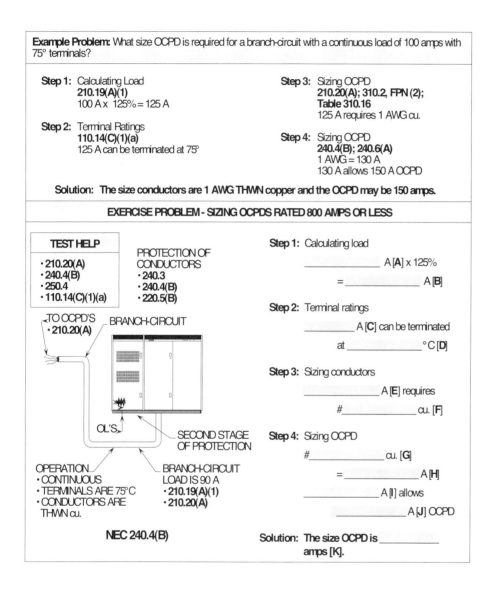

Figure 4-2. Where the ampacity of conductor(s) does not correspond to a standard device per **240.6(A)**, the next higher rating can be used, if 800 amps or less in rating.

Master Test Tip 3: OCPD's installed on grounded systems are usually marked with slash markings, such as 277/480, etc. OCPD's installed on ungrounded systems must be of the straight markings such as 600 volts, 480 volts, etc. per **240.85**.

Master Test Tip 4: CB's that are not marked with an IC rating are 5000 amps. Fuses without such markings are 10,000 amps per **240.60(C)(3)** and **240.83(C)**.

Note, for rounding up or rounding down rule, see **220.5(B)**.

SIZING OCPD'S RATED OVER 800 AMPS
240.4(C)

If the overcurrent protection device is greater than 800 amps, the ampacity of the conductors shall be equal to or greater than the ampacities of the overcurrent protection device per **240.6(A)**. Note that the requirements in **240.6(A)** allow an OCPD not listed in **240.6(A)** to be used, if listed for such use. **(See Figure 4-3)**

Figure 4-3. Where the ampacity of conductor(s) does not correspond with a standard device per **240.6(A)**, the next lower size must be used if such OCPD is rated over 800 amps. (General rule, see **240.4(B)**)

Master Test Tip 5: When applying the 10 ft. tap rule, such tap conductors shall be capable of supplying whatever they terminate to.

Master Test Tip 6: Tapped conductors utilizing the 10 ft. tap rule shall be at least equivalent to 10 divided into the feeder-circuit's OCPD.

Note that **240.6(A)** permits an OCPD to be used that is not listed as a standard size per **240.6(A)**.

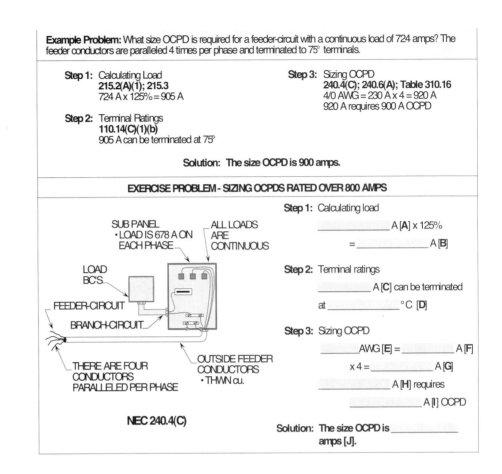

SIZING OCPD FOR TAPS
240.4(E)

Small conductors can be tapped from larger conductors under certain conditions. Sections **240.5, 240.21,** and **210.19(A)(4)** of the NEC have tap rules that allow a 18 in., 10 ft., 25 ft., and 100 ft. tap respectively. Before taking the master examination, review these requirements carefully on how to calculate and size such taps properly.

SIZING OCPD FOR TAPS 10 FT. OR LESS IN LENGTH
240.21(B)(1)

An example of applying this rule is a tap feeding a lighting and appliance branch-circuit panelboard. The tap cannot be over 10 ft. long and shall terminate at such panel. Mechanical protection by conduit, tubing, or metal gutter is required, and conductors shall be sized to carry the total load. If these requirements are followed, no overcurrent protection device shall be required at the point of such tap. **(See Figure 4-4)**

SIZING OCPD FOR TAPS OVER 10 FT. TO 25 FT. IN LENGTH
240.21(B)(2)

Master Test Tip 7: When making a tap over 10 ft. and up to 25 ft. in length, the tapped conductors must be equal to 1/3 of the feeder's OCPD rating.

A tap from a larger conductor to a smaller conductor may extend over 10 ft. up to a distance of 25 ft., provided the current-carrying capacity of the tap is at least 1/3 that of the feeder.

Overcurrent Protection Devices and Conductors

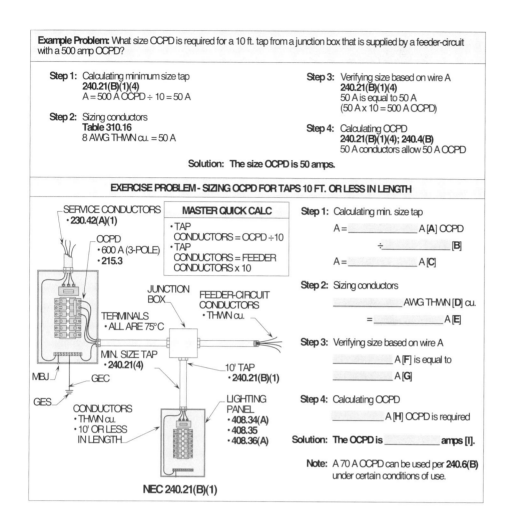

Figure 4-4. This illustration shows a test problem for sizing and selecting the OCPD for taps 10 ft. or less in length. **Note:** For sizing conductors, see **Figure 4-42**.

Master Test Tip 8: The 10 ft. tapped conductors shall terminate to a CB or fused disconnect if they terminate to a lighting and appliance panelboard.

Master Test Tip 9: 10 ft. tapped conductors terminating to a power panelboard are permitted to be connected to the lugs of the panel under certain conditions of use.

OCPD. Overcurrent protection shall not be required at the point of the tap, but the conductors shall have overcurrent protection at the end of the tap. Such tap shall be properly sized and protected from physical damage and enclosed in a raceway. **(See Figure 4-5)**

SIZING OCPD FOR TAPS INCLUDING TRANSFORMER
240.21(B)(3); (C)(5); (C)(6)

Transformer primary tap with primary conductors plus secondary conductors not over 25 ft. in length can be made without overcurrent protection at the point of such tap. The following requirements must be applied when applying this tap rule:

(1) Tap conductor ampacity is at least 1/3 that of the feeder's OCPD,
(2) Secondary conductor ampacity is at least 1/3 that of the feeder's OCPD based on the primary-to-secondary transformer ratio,
(3) Total length of conductors is not over 25 ft., primary plus secondary,
(4) All conductors are protected from physical damage, and
(5) Secondary conductors terminate at a fuse or circuit breaker sized to protect the secondaries.

See Figure 4-6 for a detailed illustration of applying the requirements of this tap rule.

Master Test Tip 10: The secondary connected conductors and OCPD shall be sized at least 1/3 of the ratio of the secondary-to-primary voltage based upon the feeder's OCPD.

Figure 4-5. This illustration shows an test problem for sizing and selecting the OCPD for taps over 10 ft. to 25 ft. in length. **Note:** For sizing conductors, see **Figure 4-43**.

Master Test Tip 11: A 25 ft. tap must have its conductors terminating to a main CB or fused disconnect switch.

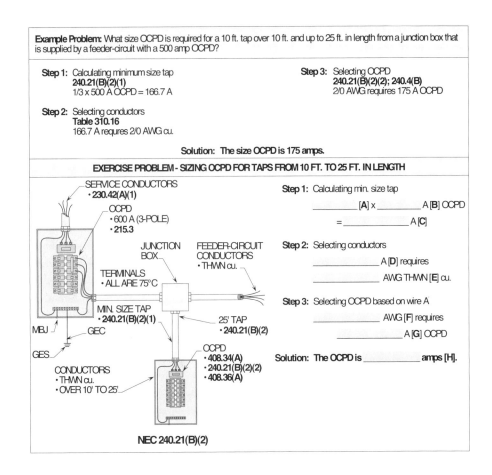

SIZING OCPD FOR TAPS OVER 25 FT. UP TO 100 FT. IN LENGTH
240.21(B)(4)

Master Test Tip 12: Tapped conductors must be equal to 1/3 of the feeder's OCPD.

Master Test Tip 13: The junction box for tapping conductors must be at least 30 ft. from the finished floor.

Master Test Tip 14: The smallest tap conductor must be at least 6 AWG cu. or 4 AWG alu.

For taps in high bay manufacturing buildings, the overcurrent protection for the tap may be at the end of the tapped conductors if the tap is not over 25 ft. long horizontally and not over 75 ft. long vertically. Note that the total run both horizontally and vertically is limited to 100 ft. or less in length. The following requirements must be complied with when applying this tap rule:

(1) The ampacity of the tap conductors is at least 1/3 of the rating of the overcurrent device protecting the feeder conductors,

(2) The tap conductors terminate at a single circuit breaker or a single set of fuses that will limit the load to the ampacity of the tap conductors. This single overcurrent device shall be permitted to supply any number of additional overcurrent devices on its load side,

(3) The tap conductors are suitably protected from physical damage or are enclosed in a raceway,

(4) The tap conductors are continuous from end-to-end without splices,

Overcurrent Protection Devices and Conductors

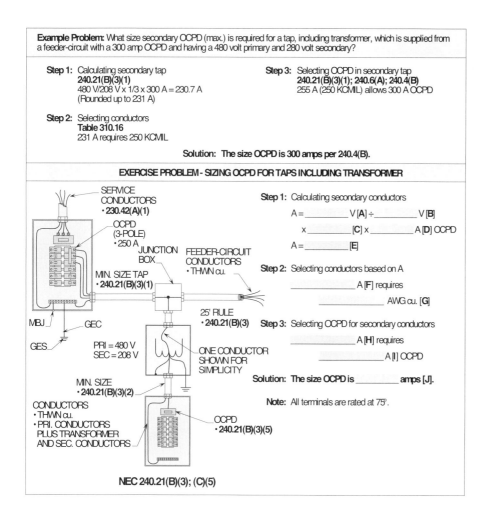

Figure 4-6. This illustration shows an test problem for sizing and selecting the OCPD for taps, including transformer. **Note:** For sizing conductors, see **Figure 4-44**.

Master Test Tip 15: The primary tap in **Figure 4-6** is determined by dividing the feeder-circuit OCPD by 1/3 (250 A x 1/3 = 83 A) after the calculation is performed, the size tap must be at least 4 AWG THWN cu. conductors per **240.21(B)(3)(1)**.

Master Test Tip 16: The primary tap conductors, plus the XFMR, plus the secondary conductors, must not exceed 25 ft. in total length from the point of the tap.

Master Test Tip 17: The largest branch-circuit for lighting loads is 20 amps or less in dwelling units.

(5) The tap conductors are sized 6 AWG copper or 4 AWG aluminum or larger,

(6) The tap conductors do not penetrate walls, floors, or ceilings, and

(7) The tap is at least 30 ft. from the floor.

See Figure 4-7 for a detailed illustration of applying the requirements of this tap rule.

BRANCH-CIRCUIT TAPS
240.21(A)

Taps to individual outlets and circuit conductors are considered as protected by the branch-circuit overcurrent protection devices when they comply with the requirements of **210.19(A)(1)** and **210.20(A)**.

Section **210.19(A)(1)** allows taps of smaller conductors to be made to larger branch-circuit conductors for certain purposes, including taps for a small range.

> For example, a 14 AWG conductor may be tapped to a 20, 25, or 30 amp circuit. A 12 AWG conductor may be tapped to a 40 or 50 amp branch-circuit, and overcurrent protection is not required at the point of the tap. **(See Figure 4-8)**

Figure 4-7. This illustration shows a test problem for sizing and selecting taps over 25 ft. and up to 100 ft. in length. **Note:** For sizing conductors, see **Figure 4-45**.

Master Test Tip 18: When using the 100 ft. tap rule, the horizontal run is limited to 25 ft. in length and must not be spliced or routed thru walls, floors, ceilings, etc. per **240.21(B)(4)(2)**.

Note, to use the 100 ft. tap rule, the bay of the facility must be at least 35 ft. in height per **240.21(B)(4)**.

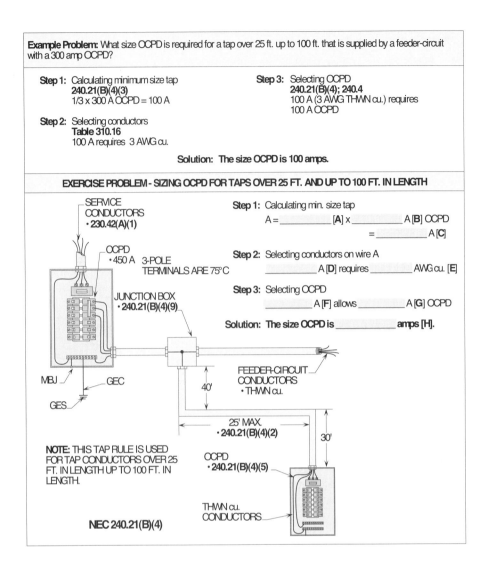

Figure 4-8. This illustration shows 12 AWG THWN copper fixture wire being tapped from a branch-circuit having 8 AWG THWN copper supply conductors.

Master Test Tip 19: Note that mogul base lampholders must be rated 750 watts or larger to be installed on a branch-circuit rated at 30, 40, or 50 amps. respectively per **210.23(B)** and **(C)** and **210.21(A)**.

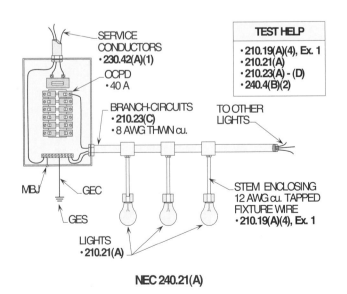

SIZING OCPD FOR MOTOR CIRCUIT TAPS
240.21(F)

Where more than one motor is on a feeder-circuit, each motor circuit is tapped from the feeder conductors. The tap is protected by the motor branch-circuit protection. The motor branch-circuit protection is located where the tap conductors terminate, and the following requirements must be applied:

(1) If the length of the tap is 25 ft. or less but more than 10 ft., conductor ampacity must be at least 1/3 that of the feeder, provided that the conductors are protected from physical damage. Conduit, EMT, flexible metal conduit, or AC or MC cable, etc. provide suitable protection for the conductors.

(2) If the length of the tap is 10 ft. or less, conductors with ampacity less than 1/3 that of the feeder may be used, provided that the conductors are in a raceway or entirely within a controller. If field installed, the OCPD on the line side of the tap conductors must not exceed 1000 percent of the tapped conductors ampacity.

(3) In high bay manufacturing facilities, a tap to a feeder may be up to 100 ft. in length. For a horizontal run, conductors are limited to 25 ft. or less in length; in addition, the tap conductors must:

(a) Be unspliced,
(b) Be installed in raceways,
(c) Have an ampacity 1/3 that of the feeder,
(d) Be at least 6 AWG cu. or 4 AWG alu., and
(e) Be unspliced along their entire length.

Master Test Tip 20: Test takers must be aware that the tap rules for motor circuits are not found in **240.21** but are listed in **430.28**.

Master Test Tip 21: Note that the general rule for determining taps is to divide the OCPD by 10 when applying the 10 ft. tap and by 1/3 for the 25 ft. tap rule. See **Figure 4-9** for the application of such tap rules.

Master Test Tip 22: If the conductors from the generator are sized for the load served and they are smaller than 115 percent of the generator's output, OCPD must be provided at the generator's output.

See Figure 4-9 for a detailed illustration of applying the requirements of these tap rules.

CONDUCTORS FROM GENERATOR TERMINALS
240.21(G)

Generator conductors must have an ampacity equal to at least 115 percent of the nameplate current rating of the generator. The following are exceptions to the general rule:

(1) If the design or operation of the generator is such as to prevent overloading, an ampacity of 100 percent may be utilized.

(2) Where an integral overcurrent protection device is provided by the manufacturer, and conductors are terminated to the device.

See Figure 4-10 for a detailed illustration of applying the requirements of this tap rule.

Figure 4-9. This illustration shows a test problem for sizing and selecting a circuit tap for a motor. **Note:** For sizing conductors, see **Figure 4-47**.

Master Test Tip 23: If the motor in Figure 4-9 was a 25 ft. tap, the size tapped conductors would be 400 KCMIL THWN copper (800 A x 1/3 = 266.7 A.)

Master Test Tip 24: Note that the conductors for each 25 ft. tap must be equal to the output of the XFMR and the OCPD's in each panelboard must not exceed the output of the XFMR.

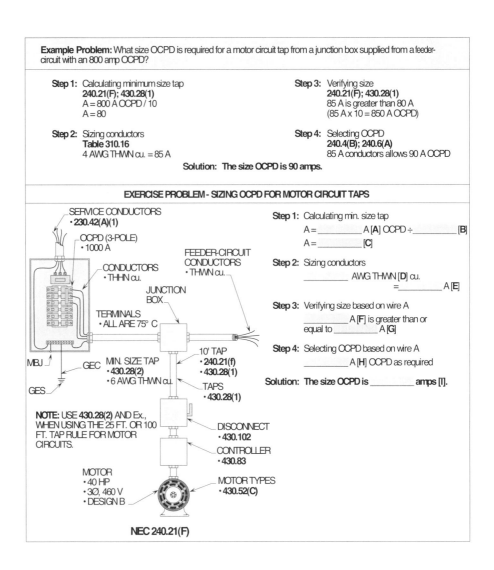

SIZING OCPD FOR CONDUCTORS OF 25 FT. FROM SECONDARY OF TRANSFORMER
240.21(C)(3)

Conductors are permitted to be connected to the secondary side of a separately derived system for industrial installations, without overcurrent protection at the connection, where all the following conditions are complied with:

(1) The length of the secondary conductors does not exceed 25 ft. in length.

(2) The ampacity of the secondary conductors is not less than the secondary current rating of the transformer, and the sum of the ratings of the overcurrent devices does not exceed the ampacity of the secondary conductors.

(3) All overcurrent protection devices are grouped.

(4) The secondary conductors are properly protected from physical damage.

See Figure 4-11 for applying the requirements pertaining to secondary conductors connected to transformers.

Overcurrent Protection Devices and Conductors

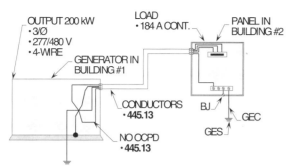

Figure 4-10. This illustration shows a test problem for tapping conductors from the secondary of a generator. **Note:** For another example, see Figure 4-51.

Master Test Tip 25: Generators are classified by the NEC as temporary per **250.34, 702.6,Ex.** and **702.10**, emergency per **700.1, FPN 3**, legally required standby per **701.2, FPN**, optional standby per **702.2, FPN**, and interconnected type per **705.1 and 2**.

Master Test Tip 26: The OCPD in the panelboard or switchgear must protect the conductors and output of the XFMR from short-circuits, ground-faults, and overloads. For example, a XFMR with a 421 amp output must be protected by a 400 amp CB or 400 amp fuses.

SIZING OCPD FOR OUTSIDE SECONDARY CONNECTIONS 240.21(C)(4)

Outside conductors are allowed to be tapped to a feeder or to be connected at the transformer secondary, without overcurrent protection at the point of connection. However, the following conditions must be complied with:

(1) The conductors are properly protected from physical damage.

(2) The conductors terminate at a single circuit breaker or a single set of fuses that limits the load to the ampacity of the conductors. This single overcurrent protection device can supply any number of overcurrent protection devices on its load side.

(3) The tap conductors are installed outdoors, except at the point of termination.

(4) The overcurrent device for the conductors is an integral part of a disconnecting means or is located immediately adjacent thereto.

(5) The disconnecting means for such conductors are installed in a readily accessible location either outside of a building or structure, or inside nearest the point of entrance of the conductors.

See Figure 4-12 for applying the requirements pertaining to secondary conductors connected to transformers.

4-11

Figure 4-11. This illustration shows an test problem for sizing and selecting connections of 25 ft. from the secondary of a transformer. **Note 1:** For sizing conductors for a 10 ft. connection or a 25 ft. connection, see **Figures 4-48** and **4-49**. **Note 2:** The OCPD for a 10 ft. connection is determined by applying the same procedure.

Master Test Tip 27: Note that the secondary tapped conductors do not have an OCPD ahead of them, therefore **Table 250.122** is used to size the minimum grounded conductor and bonding jumper per **250.24(C)** and **250.28(D)**.

Master Test Tip 28: When used for test questions and problems, an appliance is considered utilization equipment generally other than industrial types. Appliances are units such as clothes dryers, air conditioners, food fixing, deep frying, etc.

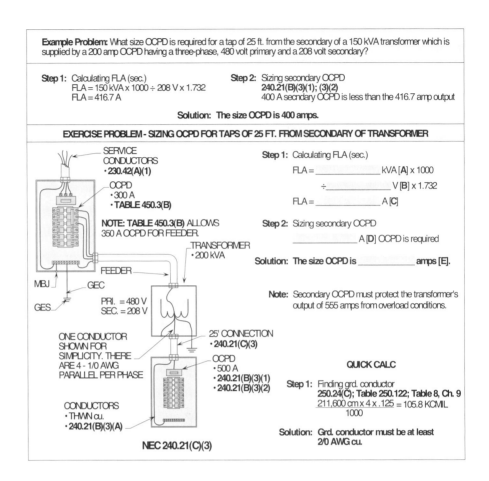

SIZING OCPD FOR MOTOR-OPERATED APPLIANCE CIRCUIT CONDUCTORS
240.4(G); TABLE 240.4(G)

Motor-operated appliance circuit conductors are normally protected against overcurrent by the provisions listed in **Parts II** and **IV** of **Article 422**. **(See Figure 4-13)**

SIZING OCPD FOR MOTOR AND MOTOR CONTROL CIRCUIT CONDUCTORS
240.4(G); Table 240.4(G)

Motor circuits are another exception listed in **240.4(G)** (to the general rule), where the OCPD can be sized above the conductor's ampacity. Motor circuits are sized according to the requirements of **Article 430**. A study of these requirements will reveal that for motor circuits, a fuse size or circuit breaker setting in excess of the ampacity of the conductor is permitted by the NEC. This exception is intended to provide fuse or circuit breaker protection large enough to hold the high momentary inrush current required for starting and running the driven load.

(See Figure 4-14) Also, review **422.10** and **422.11**.

Overcurrent Protection Devices and Conductors

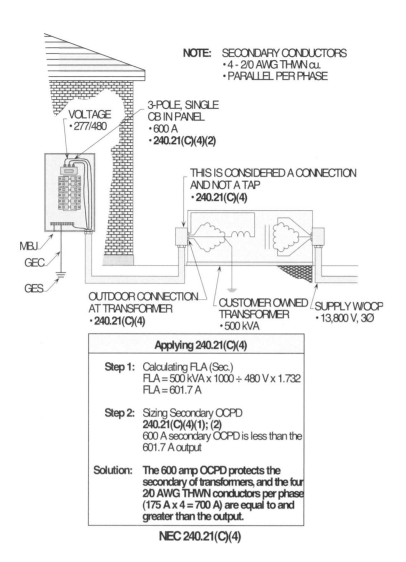

Figure 4-12. This illustration shows a test problem for connecting conductors from the secondary of a transformer using the outside connection rule. **Note:** For sizing conductors, see **Figure 4-50**.

Master Test Tip 29: Note that the 600A CB in the panel must protect the output of the XFMR and the conductors routed between the panel and XFMR.

Master Test Tip 30: To select the percentage to size the OCPD for the primary side of the XFMR in Figure 4-12, see **Table 450.3(A)** based upon supervised or nonsupervised locations.

Master Test Tip 31: The OCPD for compressors can be sized at 175 percent or 225 percent of the FLC rating in amps. However, conductors can be sized at 125 percent of the compressor's FLA. See **440.22** and **440.32**.

SIZING OCPD FOR PHASE CONVERTERS
240.4(G); Table 240.4(G)

Phase converters supply conductors for motor-related loads, and nonmotor-related loads can be protected against overcurrent by the rules and regulations of **455.7**. Before sizing the OCPD for such loads, review **Article 455**, which contains different rules than the requirements of **Article 430** for sizing the elements of motor circuits. **(See Figure 4-15)**

SIZING OCPD FOR AC AND REFRIGERATION EQUIPMENT CIRCUIT CONDUCTORS
240.4(G); Table 240.4(G)

Circuit conductors supplying air conditioning and refrigeration equipment shall be protected against overcurrent by the provisions of **Parts III** and **VI** of **Article 440**. Note that these rules are used only for hermetically sealed motors and not individual motors per **Article 430**. **(See Figure 4-16)**

Figure 4-13. This illustration shows a test problem for sizing and selecting the OCPD for a motor-operated appliance.

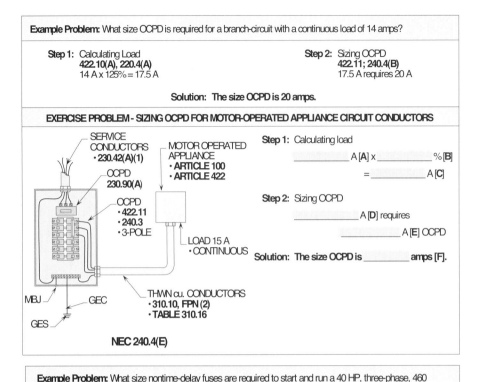

Figure 4-14. This illustration is a test problem for sizing and selecting the minimum and next size nontime-delay fuses to start and run a motor.

Master Test Tip 32: The OCPD and conductors must be sized by the FLA of **Table 430.248** for single-phase motors and **Table 430.250** for three-phase motors per **430.6(A)(1)**. Note that the FLA on the motor's nameplate is used to size the overload protection.

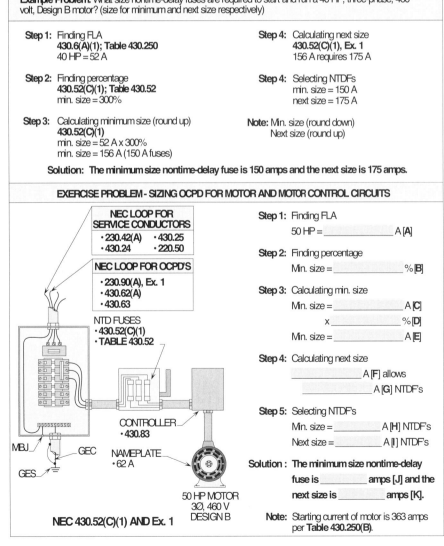

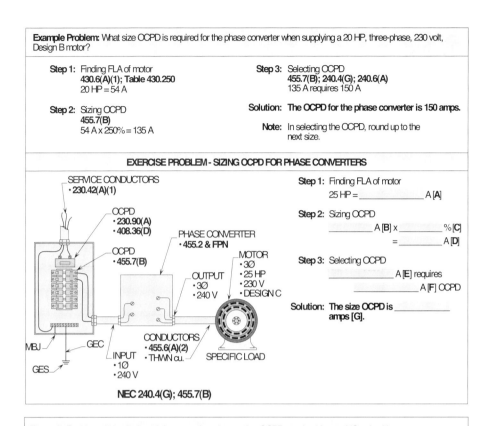

Figure 4-15. This illustration is a test problem for sizing and selecting the OCPD for phase converters.

Master Test Tip 33: Note that on the test, the next size OCPD (round up) (**Ex. 1 to 430.52(C)(1)**) is usually selected when sizing the OCPD to allow a motor to start and run. However, when sizing the OCPD to allow a compressor motor to start and run, the next lower size is selected if it will allow the compressor to start and run.

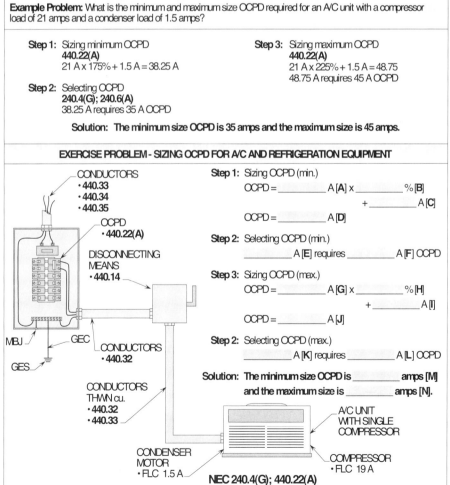

Figure 4-16. This illustration shows a test problem for sizing and selecting the minimum and maximum size OCPD for A/C and refrigeration equipment circuits.

Master Test Tip 34: When sizing the OCPD for an A/C unit be sure to read the question carefully; if the minimum size is asked for, use 175 percent x the compressor's FLA or 225 percent of such FLA for the maximum size.

Note that **440.14** requires a disconnecting means to be provided, if one does not come with the unit.

SIZING OCPD FOR TRANSFORMER SECONDARY CONDUCTORS
240.4(F); 240.21(C)(1)

Conductors supplied by the secondary side of a single-phase transformer having a two-wire (single voltage) secondary or a three-wire delta secondary shall be considered as protected by overcurrent protection provided on the primary (supply) side of the transformer, providing this protection is in accordance with **450.3** and does not exceed the value determined by multiplying the secondary conductor ampacity by the secondary-to-primary transformer voltage ratio. **(See Figures 4-17(a) and (b))**

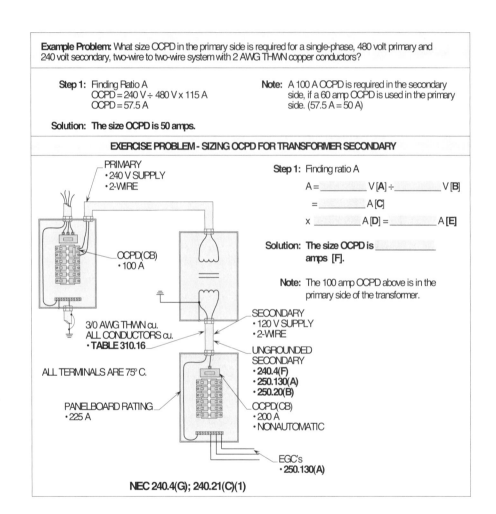

Figure 4-17(a). This illustration shows a test problem for sizing and selecting the OCPD for a two-wire to two-wire system.

Master Test Tip 35: Note that the **Ex.** to **408.36(D)** allows the primary OCPD to protect the primary windings and conductors, and the secondary windings and conductors, plus the equipment served. For this rule to apply, the primary and secondary must be a two-wire to two-wire system.

Note that **240.21(C)(1)** must also be reviewed when protecting the transformer and conductors.

SIZING OCPD FOR ELECTRIC WELDER CIRCUIT CONDUCTORS
240.4(G); 630.12(A) and (B); 630.32(A) and (B)

Circuit conductors supplying welders shall be protected against overcurrent by the provisions of **Parts II, III,** and **IV** of **Article 630**. Nonmotor generator arc welders and conductors are protected per **630.12(A)** and **(B)**. Motor-generator arc welders and conductors are protected per **630.12(A)** and **(B)**. Resistance welders and conductors are protected per **630.32(A)** and **(B)**. **(See Figures 4-18(a); (b); and (c))**

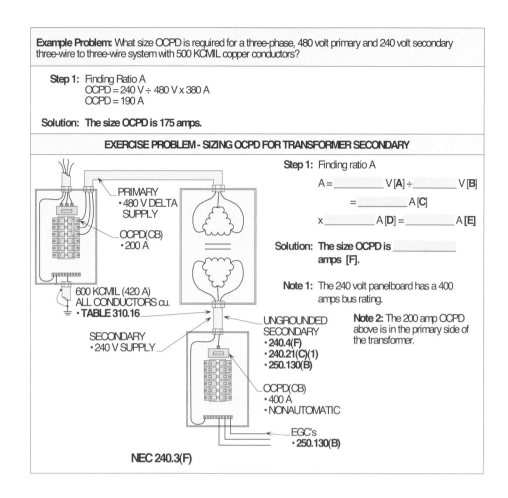

Figure 4-17(b). This illustration shows a test problem for sizing and selecting the OCPD for a three-wire to three-wire system.

Master Test Tip 36: For the primary OCPD in **Figure 4-17(B)** to be allowed to protect the secondary conductors and equipment, there must not be any phase to neutral loads involved.

Master Test Tip 37: Note that 18 AWG control wire must be protected at 7 amps or less and 16 AWG can be protected at 10 amps or less per **725.23**.

SIZING OCPD FOR REMOTE-CONTROL, SIGNALING, AND POWER-LIMITED CIRCUIT CONDUCTORS
240.4(G); 725.23, Ex.

Remote-control circuits can have overcurrent protection, sized up to three times the ampacity of the conductors per **725.23, Ex 3**. Motor-control circuits may be fused above their ampacity. A good example of a motor-control circuit is a circuit to a pushbutton station (start-stop) derived from a magnetic motor controller.

For motor control circuits, the NEC allows the protection of the control circuit to be sized up to 300 or 400 percent of the ampacity of the control circuit conductors. Such protection is based upon control circuits remaining in the control enclosure or leaving the enclosure to supply a remote feed stop and start station, etc. **(See Figures 4-19(a); (b); (c); (d))**

Note that control circuits can be designed to remain in the controller's enclosure or leave the enclosure and routed to remote control stations. Control-circuits remaining in the controller's enclosure are protected per **Column B** of **Table 430.72(B)**. For those leaving the enclosure, **Column C** of **Table 430.72(B)** is utilized.

SIZING OCPD FOR FIRE ALARM CIRCUIT CONDUCTORS
240.4(G); 760.23, Ex.

Circuit conductors used in fire alarm systems shall be protected against overcurrent conditions by the provisions of **Parts II** and **III** of **Article 760**. For further information on sizing and selecting OCPD's for fire alarm circuits, see **760.23, 760.24,** and **760.41** of the NEC. These circuits are special, and such rules and regulations must be studied and well understood. **(See Figure 4-20)**

Figure 4-18(a). This illustration is a test problem for sizing and selecting the size OCPD required for a nonmotor generator arc welder. **Note:** For sizing the conductors, See **Figure 4-52**.

Master Test Tip 38: In the example problem, the OCPD for protecting the welder and conductors of the branch-circuit must be sized based on the ampacity of such conductors times 200 percent. The disconnecting means at the welder must be sized based on the welder's FLC rating in amps times 200 percent per **630.12(A)** and **(B)**. Note that if nuisance tripping occurs, the next size OCPD above 200 percent can be used per **630.12**.

Note that all terminals in exercise problem are rated at 75°C.

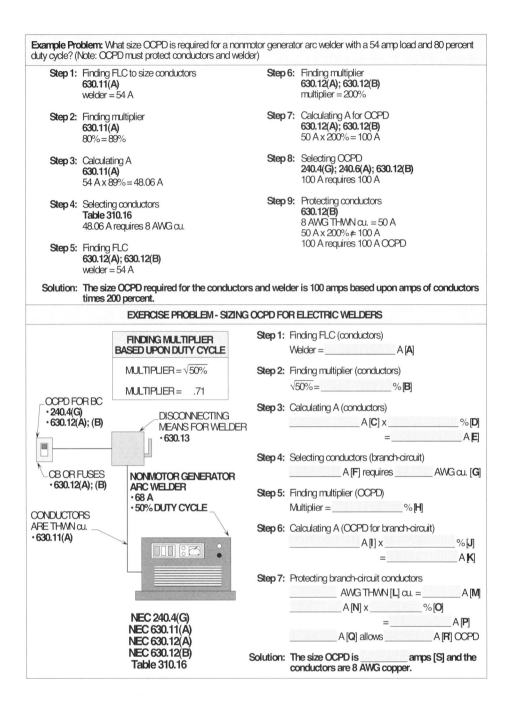

USING HANDLE TIES FOR CIRCUIT BREAKERS 240.20(B)(1); (2); (3)

Master Test Tip 39: For ungrounded two-wire circuits, two single-pole circuit breakers with approved handle ties may be used in place of a two-pole circuit breaker.

Circuit breakers must open all ungrounded conductors of the circuit.

For example, when one of the trip elements operates, it trips the circuit breaker, which opens all poles of the circuit. A circuit breaker used for short-circuit protection is required to have a trip unit for each ungrounded conductor, and a pole for each ungrounded conductor that opens the circuit under short-circuit and ground-fault conditions.

Overcurrent Protection Devices and Conductors

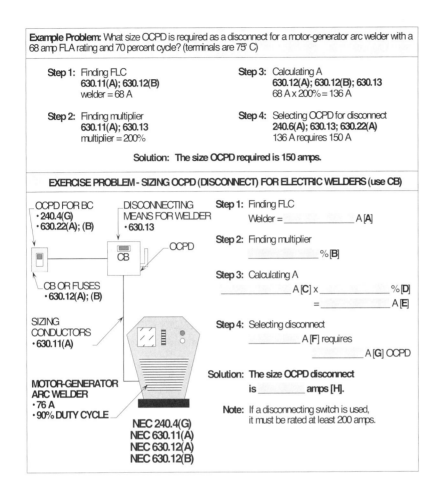

Figure 4-18(b). This illustration is a test problem for sizing and selecting the size disconnect required for a motor-generator arc welder. **Note:** For sizing the conductors, see **Figure 4-53**.

Master Test Tip 40: In **Figure 14-18(b)**, the multiplier used to size the conductors and OCPD is selected from the chart in **Table 630.11(A)** based on the duty cycle of the welder.

Master Test Tip 41: Handle ties used to tie the handles of two single-pole CB's together must be approved for such use per **240.20(B)(3)**.

Note that circuit breakers for three-phase circuits must be three-pole breakers with a simultaneously common trip. Circuit breakers with approved handle ties can be used as follows:

(1) Except where limited by the provisions of **210.4(B)** and **240.20(B)(1)** through **(B)(3)**, individual single-pole CB's, with or without approved handle ties, are allowed to be utilized as the protection for each ungrounded conductor of multiwire branch-circuits that serve only single-phase, line-to-neutral loads.

(2) In grounded systems, individual single-pole circuit breakers with approved handle ties are allowed to be utilized as the protection for each ungrounded conductor for line-to-line connected loads for single-phase circuits or three-wire DC circuits.

(3) For line-to-line loads in four-wire, three-phase systems or five-wire, two-phase systems having a grounded neutral and no conductor operating at a voltage greater than allowed in **210.4**, individual single-pole circuit breakers with approved handle ties are allowed to be utilized as the protection for each ungrounded conductor.

See Figure 4-21 for a detailed illustration of using handle ties with single-pole circuit breakers to connect and disconnect circuits.

Note that a disconnecting means must be sized and installed, if one is not provided with the unit.

Stallcup's Master Electrician's Study Guide

Figure 4-18(c). This illustration is a test problem for sizing and selecting the size OCPD required for a resistance welder. **Note:** For sizing conductors, see **Figure 4-54**.

Master Test Tip 42: To size the OCPD in the example problem, multiply the ampacity of the conductor by 300 percent. To select the proper size disconnect for the welder, multiply the FLA of the welder by 300 percent.

Master Test Tip 43: When selecting the size OCPD and disconnect for a resistance welder, the general rule requires that the 300 percent must not be exceeded. In other words, always round the size of the OCPD down and not up to the next size.

Note that test takers must always check to verify if terminals are 60°C or 75°C respectively per **110.14(C)(1)(a)** and **(b)**.

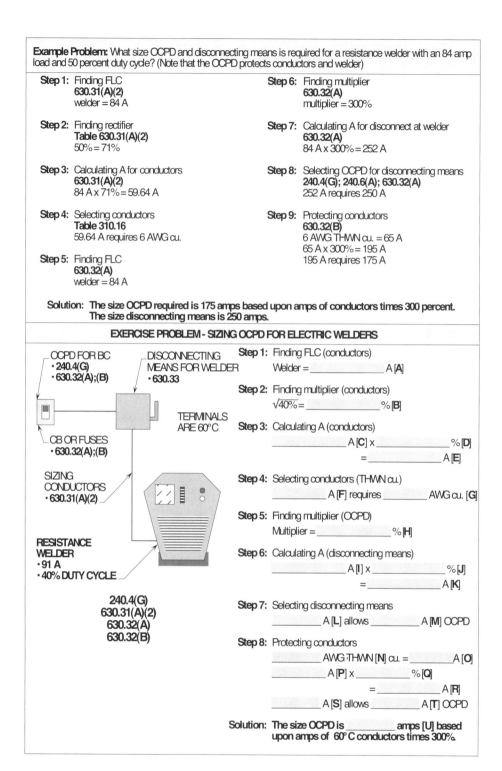

Example Problem: What size OCPD and disconnecting means is required for a resistance welder with an 84 amp load and 50 percent duty cycle? (Note that the OCPD protects conductors and welder)

Step 1: Finding FLC
630.31(A)(2)
welder = 84 A

Step 2: Finding rectifier
Table 630.31(A)(2)
50% = 71%

Step 3: Calculating A for conductors
630.31(A)(2)
84 A x 71% = 59.64 A

Step 4: Selecting conductors
Table 310.16
59.64 A requires 6 AWG cu.

Step 5: Finding FLC
630.32(A)
welder = 84 A

Step 6: Finding multiplier
630.32(A)
multiplier = 300%

Step 7: Calculating A for disconnect at welder
630.32(A)
84 A x 300% = 252 A

Step 8: Selecting OCPD for disconnecting means
240.4(G); 240.6(A); 630.32(A)
252 A requires 250 A

Step 9: Protecting conductors
630.32(B)
6 AWG THWN cu. = 65 A
65 A x 300% = 195 A
195 A requires 175 A

Solution: The size OCPD required is 175 amps based upon amps of conductors times 300 percent. The size disconnecting means is 250 amps.

EXERCISE PROBLEM - SIZING OCPD FOR ELECTRIC WELDERS

Step 1: Finding FLC (conductors)
Welder = _____ A [A]

Step 2: Finding multiplier (conductors)
√40% = _____ % [B]

Step 3: Calculating A (conductors)
_____ A [C] x _____ % [D]
= _____ A [E]

Step 4: Selecting conductors (THWN cu.)
_____ A [F] requires _____ AWG cu. [G]

Step 5: Finding multiplier (OCPD)
Multiplier = _____ % [H]

Step 6: Calculating A (disconnecting means)
_____ A [I] x _____ % [J]
= _____ A [K]

Step 7: Selecting disconnecting means
_____ A [L] allows _____ A [M] OCPD

Step 8: Protecting conductors
_____ AWG THWN [N] cu. = _____ A [O]
_____ A [P] x _____ % [Q]
= _____ A [R]
_____ A [S] allows _____ A [T] OCPD

Solution: The size OCPD is _____ amps [U] based upon amps of 60°C conductors times 300%.

Overcurrent Protection Devices and Conductors

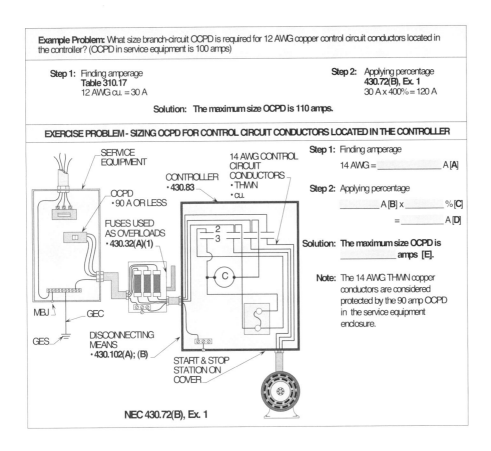

Figure 4-19(a). This illustration is a test problem for sizing and selecting the size OCPD for circuits in the enclosure.

Master Test Tip 44: In **Figure 4-19(a)**, the size of the OCPD for the branch-circuit can be increased in size by increasing the size of the control-circuit conductors. To select the ampacity of the control-circuit conductors, use **Table 310.17**. Note that the control circuit conductors, in some cases, are considered protected without providing individual OCPD's.

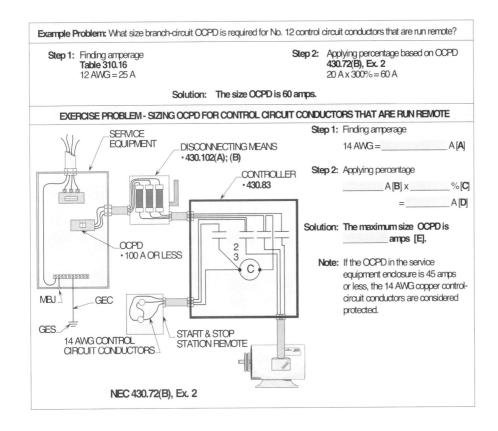

Figure 4-19(b). This illustration is a test problem for sizing and selecting the size OCPD for remote control circuits.

Master Test Tip 45: In the example problem, the remote control conductors are considered protected by the 60 amp OCPD in the service equipment panel. Therefore, individual OCPD's are not required in the control enclosure to protect the control-circuit conductors.

4-21

Figure 4-19(c). This illustration is a test problem for sizing and selecting the size OCPD for the primary circuits of a control transformer.

Master Test Tip 46: The control XFMR is considered protected by the 6 A fuses per **Table 450.3(B)** and **430.72(C)(4)**. Note that the primary windings and conductors plus secondary windings and conductors are protected by such fuses.

Master Test Tip 47: The maximum height of handles on switches and CB's is limited to 6 ft. 7 in. However, there is not a minimum height requirement in the NEC.

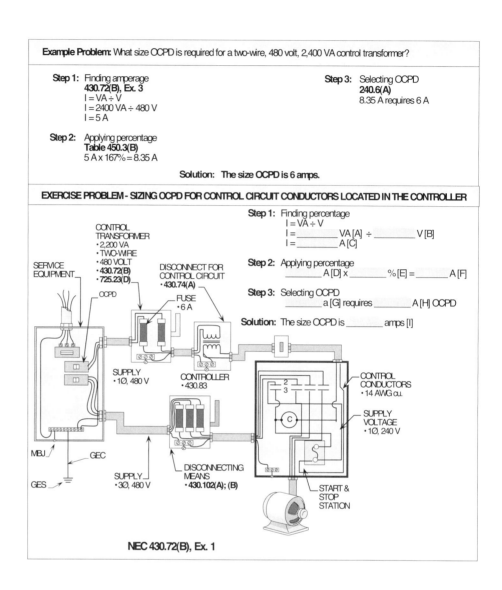

READILY ACCESSIBLE OF OCPD'S 240.24(A)

For safety, the requirement that overcurrent protection devices be readily accessible is mandatory. They shall be located so that they may be readily reached in case of emergencies or for servicing, without reaching over objects, climbing on chairs, ladders, etc.

See Figure 4-22 for a detailed illustration for OCPD's being readily accessible. However, there are four exceptions to the accessibility rule, and they are as follows:

(1) For busways, **368.12** permits overcurrent protection for a plug-in to be out of reach from the finished grade.

(2) For lighting fixtures and appliances that are manufactured with built-in fuses, the fuses are not required to be readily accessible per **240.10**.

(3) For services, the overcurrent protection device can be located at the beginning of the run, at the point where the service-entrance tap to the service drop or lateral is made. In such cases, the main overcurrent protection device is not required to be readily accessible inside the facility. See **230.40** and **230.92** for further details concerning this type of installation.

Overcurrent Protection Devices and Conductors

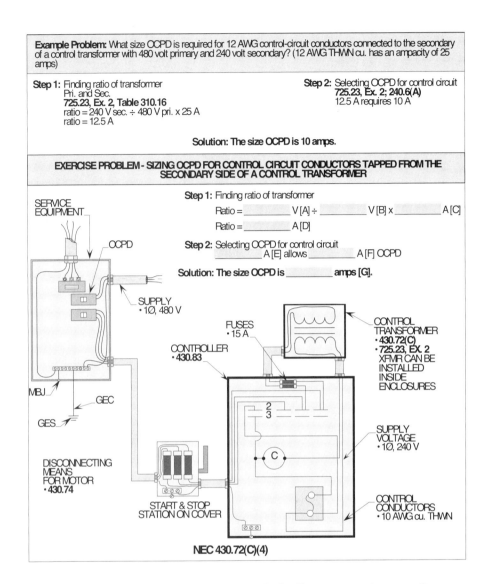

Figure 4-19(d). This illustration is a test problem for sizing and selecting the size OCPD necessary for protecting the secondary side of a control transformer.

Master Test Tip 48: In the exercise problem of Figure 4-19(c), the 6 amp fuses on the primary side of the control XFMR protects the secondary control conductors because they do not exceed the primary to secondary ratio of voltage and current $\frac{(240\text{ V} \times 20\text{ A} = 10\text{ A})}{(480\text{ V}\qquad)}$
The 6 amp set of fuses are less the computed 10 amp ratio per **240.21(C)(1)**.

Master Test Tip 49: Note that storage boxes, mop buckets, etc. must not be placed under or in front of panelboards having OCPD's for feeder or branch-circuits.

(4) For overcurrent protection devices installed adjacent to motors, appliances, or other equipment that are out of reach, such OCPD'S do have to be readily accessible. However, such devices can be capable of being reached by a portable means. For further information refer to **408.8(A), Ex. 2** and **240.24(A)(4)**.

See Figure 4-23 for a detailed illustration for applying the requirements of these rules.

ACCESS OF OCCUPANT TO OCPD'S
240.24(B)

The general rule does not allow overcurrent protection devices to be located where they are exposed to physical damage. Furthermore, they are not to be installed where ignitable materials are near enough to catch fire in case of accidental sparking or overheating of such devices.

Section **240.24(B)** requires each occupant of a building or apartment to have ready access to the overcurrent devices for his or her occupancy. Exceptions to this rule are made in the case of apartment buildings and guest rooms in hotels or motels where a building employee is constantly in attendance. Remember, such OCPD's must be readily accessible to whom. Basically, they must be readily accessible to occupants or maintenance personnel. **(See Figure 4-24)**

Note that OCPD's (CB's) that are not marked with an IC rating are 5000 amps and if fuses are not marked, they are rated at 10,000 amps per **240.83(C)**.

Ex. 1 and **Ex. 2** to **240.24(B)** reads as follows:

(1) In a multiple-occupancy building where electric service and electrical maintenance is provided by the building management and where these are under continuous building management supervision, the service overcurrent device and feeder overcurrent devices supplying more than one occupancy may be accessible to authorized management personnel only.

(2) Ex. 2 to **240.24(B)** allows for guest rooms of hotels and motels, which are intended for transient occupancy, to have the overcurrent protection devices for such occupancies to be accessible to authorized management personnel only.

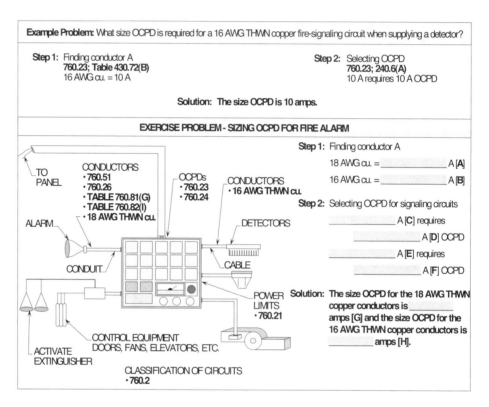

Figure 4-20. This illustration is a test problem for sizing and selecting the size OCPD for fire alarm circuits.

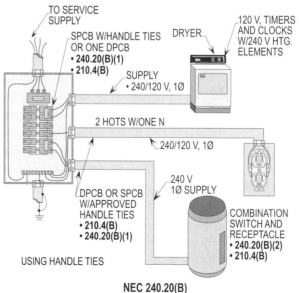

Figure 4-21. In certain installations and types of equipment, single-pole CB's can be used with approved handle ties.

Overcurrent Protection Devices and Conductors

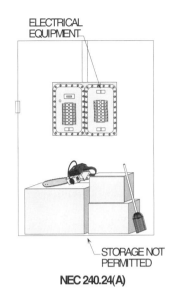

Figure 4-22. OCPD's must be readily accessible without having to remove or climb over objects such as chairs, desks, etc.

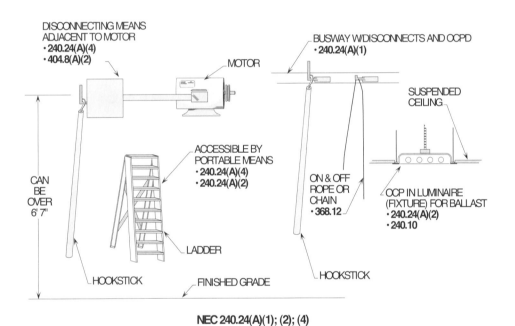

Figure 4-23. This illustration shows installations where the OCPD's are not required to be readily accessible.

Master Test Tip 50: Fuses not marked with an IC rating are considered having a rating of 10,000 amps per **240.60(C)(3)**.

FUSES
ARTICLE 240, PARTS E AND F

Fuses for conductors with specific types of insulations are available to supply power to equipment. Conductors are sized to carry the load of the equipment in amps without deteriorating the insulation per **310.10**. The overcurrent protection devices are sized to protect the conductors and equipment from short-circuits, ground-faults, and overloads.

The calculated load for fuses is determined by calculating the loads at noncontinuous duty (100 percent) or continuous duty (125 percent) or both and applying demand factors where applicable. Depending on the type of fuse used, a fuse will hold five times its rating for different periods of time.

Note that a 100 amp TDF will allow a motor having a LRC of 450 amps to start and run (100 A TDF x 5 = 500 A). For further information, see text on page 4-26.

Figure 4-24. OCPD's must be readily accessible to each occupant in a premises. However, they can be located behind locked doors if qualified personnel are present.

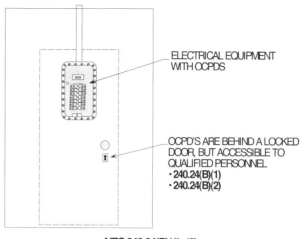

NEC 240.24(B)(1); (2)

SIZING
210.20(A); 215.3; 408.13

Master Test Tip 51: An exception to the general rule allows the fuse to carry 100 percent of the continuous load current if the fuse and enclosure has been rated by specific design for 100 percent rated current operation. See the exceptions to **210.20(A)** and **215.3**.

Fuses are sized based upon the operation characteristics of the load served. The NEC defines a continuous load as a load that operates continuously for three hours or more. Such continuous current must not exceed 80 percent of the fuse rating to prevent opening the circuit.

For example, a 200 amp fuse can supply a continuous load current of 160 amps (200 A x 80% = 160 A).

The NEC recommends that load currents in amps be calculated by multiplying the 160 amp continuous load current by 125 percent (160 A x 125% = 200 A) to obtain the size, which is 200 amps. By calculating the load served by 125 percent of 160 amps (1/80% or .80 = 125%) the 80 percent loading of the circuit breaker for continuous operation has been accomplished. **(See Figure 4-25)**

Master Test Tip 52: Time delay fuses will clear a short-circuit or ground-fault in a 1/2 cycle, which is about .0084 second in time.

Fuses are equipped with a time-delay feature that is designed to allow loads requiring high inrush currents to start and run. Such loads are motors, compressors, welders, etc. For further information pertaining to these loads, see **240.4(A) thru (G)**.

Note that the 5 times fuse method in **Figure 4-26(a)** can be used to estimate the size fuse needed to start a motor, A/C unit etc.

For example, a 200 amp time-delay fuse will hold about 5 times its rating (200 A x 5 = 1000 A), which is 1000 amps. Such fuse will hold this value of 1000 amps for about 10 seconds based upon its time characteristics. A 200 amp nontime-delay fuse will hold 1000 amps for about 2 seconds or less without blowing and opening the circuit due to inrush current. Therefore, a piece of equipment with an inrush current that is less than 1000 amps will start and run without blowing the fuse under normal starting and running conditions. **(See Figures 4-26(a) and(b))**

FUSE MARKINGS
240.60(C)

Fuses are required to be plainly marked, either by printing on the fuse barrel or by a label attached to the barrel. So that fuses can be sized, selected, and installed correctly, the following information shall be provided:

(1) Ampere rating
(2) Voltage rating
(3) Interrupting rating where other than 10,000 amps
(4) "Current limiting" where applicable
(5) The name or trademark of the manufacturer

Overcurrent Protection Devices and Conductors

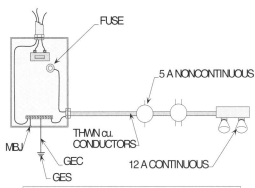

Figure 4-25. This illustration shows the procedure for calculating the size fuse and conductors based upon continuous operation. Notice that multiplying the load by 125 percent is the same as derating such fuse by 80 percent and limiting the load to this value.

Master Test Tip 53: Loads on the Master Electricians' examination must be computed at continuous operation or noncontinuous operation, or demand factors can be applied for certain loads. For examples of some of these rules, See **210.19(A)(1), 215.2(A)(1), and 230.42(A)(1).**

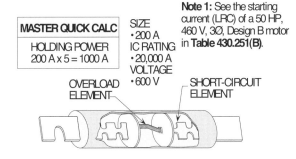

Figure 4-26(a). This illustration shows that certain amperage markings must be placed on fuses. Note that a fuse will hold 5 times its rating for 2 to 10 seconds based upon type.

Master Test Tip 54: Fuses rated at 600 volts can be used on electrical circuits with voltage ratings less than 600 volts.

Also, review the starting current of the motor in the exercise problem of **Figure 4-14**.

4-27

Figure 4-26(b). This illustration is detailed on how fuses must be marked.

Master Test Tip 55: If two 20,000 amp IC rated fuses are in a switch with one 10,000 amp fuse, the 10,000 amp fuse must be replaced with a 20,000 amp IC fuse per **110.9** and **110.10**.

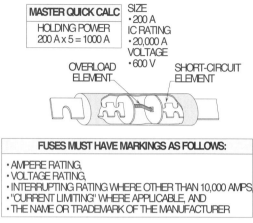

CIRCUIT BREAKERS
ARTICLE 240, PART VII

Circuit breakers and conductors with insulations are available to supply power to equipment. Conductors are sized to carry the load of the equipment in amps without deteriorating the insulation per **310.10**. The overcurrent protection devices are sized to protect the conductors and equipment from short-circuits, ground-faults, and overloads.

All circuit breakers and conductors shall be designed and installed according to the latest provisions of the NEC to ensure protection and proper installation.

The calculated load for circuit breakers is determined by calculating the loads at non-continuous operation (100 percent) and continuous operation (125 percent) and applying demand factors where applicable. Depending on the frame size of the unit, a circuit breaker can hold approximately three times their rating for different periods of time.

SIZING
210.20(A); 215.3

Master Test Tip 56: An exception to the general rule allows circuit breakers to carry 100 percent of the continuous load current if the circuit breaker and enclosure have been rated by specific design for 100 percent rated current operation. See the exceptions to **210.20(A)** and **215.3**.

Circuit breakers are sized based upon the operation characteristics of the load served. The NEC defines a continuous load as one that operates continuously for three hours or more. Such continuous current shall not exceed 80 percent of the circuit breaker's rating.

For example, a 200 amp circuit breaker can supply a continuous load current of 160 amps (200 A x 80% = 160 A). The NEC recommends that the load current in amps be computed by multiplying the 160 amp continuous load current by 125 percent (160 A x 125% = 200 A) to obtain the size circuit breaker, which is 200 amps. By computing the load by 125 percent of 160 amps (1/80% or .80 = 125%) the 80 percent loading of the circuit breaker for continuous operation has been accomplished.

Circuit breakers are equipped with a time-delay feature that is designed to allow loads requiring high inrush currents to start and run. Such loads are motors, compressors, welders, etc. For further detailed information pertaining to these loads see **240.4(A)** thru **(G)**.

For example, a 200 amp circuit breaker will hold about 3 times its rating (200 A x 3 = 600 A), which is 600 amps. Such circuit breaker will hold this value (600 amp) for a number of seconds based upon its frame size. A 200 amp circuit breaker will hold 600 amps for about 35 seconds without tripping open the circuit. Therefore, a piece of equipment with an inrush current that is less than 600 amps will start and run without tripping the circuit breaker. **(See Figure 4-27)**

Note that a 200 A CB will allow the motor in the exercise problem of Figure 4-14 to start and run.

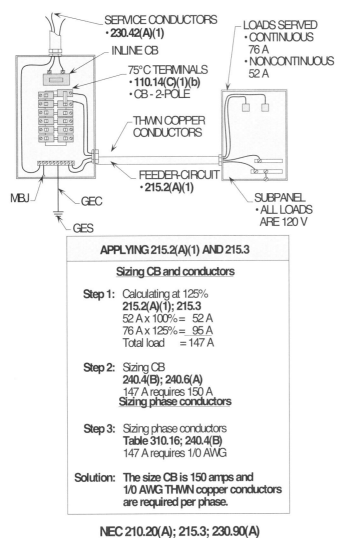

Figure 4-27. This illustration shows the procedure for sizing the size CB and conductors based upon continuous and noncontinuous operation. Notice that by multiplying the continuous load by 125 percent is the same as derating such CB by 80 percent and limiting the load to this value.

Master Test Tip 57: CB's used for switching 120 or 277 volt fluorescent lighting circuits are required to be marked for switching duty (SWD) per **240.83(D)**. Note that CB/s used to switch HID lighting units must be listed for such use.

Master Test Tip 58: CB's must protect electrical circuits operating within their voltage range per **240.83(E)**.

INDICATING
240.81

When the circuit breaker is turned ON, the handle shall be installed so that it is in the UP position. Circuit breakers shall be clearly marked in the open "OFF" or closed "ON" position. Circuit breaker handles that operate vertically instead of rotationally or horizontally shall be installed with the ON position of the handle in the UP position.

MARKING
240.83

Master Test Tip 59: Circuit breakers used for supplementary protection shall not require an interrupting rating marked on the device per **240.83(C)**.

The following markings shall be applied before installing circuit breakers:

(A) **Durable and visible.** The ampere rating shall be durable and visible after installation.

(B) **Location.** The ampere rating shall be molded, stamped, etched, or similarly marked into the handles of circuit breakers rated 100 amps or less.

(C) **Interrupting rating.** Interrupting current (IC) ratings other than 5000 amps shall be marked with IC rating.

(D) **Used as switches.** Circuit breakers used to switch 120 or 277 volt fluorescent or HID lighting circuits are required to be marked SWD, which means switching duty.

(E) **Voltage marking.** Voltage ratings no less than the nominal system voltage shall be marked on circuit breakers for the capability to interrupt fault-currents between phases or phase-to-ground.

See Figure 4-28 for a detailed illustration pertaining to the rules for marking circuit breakers.

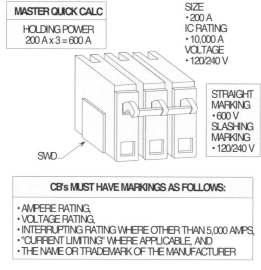

Figure 4-28. This illustration shows the markings that must be placed on CB's. Note that a CB will hold 3 times its rating for a period based upon its frame size and rating. A 200 amp circuit breaker will start a load with a starting current of 600 amps or less.

Master Test Tip 60: A slash marked CB must not be installed on any ungrounded electrical system per **240.85**.

APPLICATIONS
240.85

Circuit breakers can be used with grounded, ungrounded, or grounded neutral systems when identified with a straight voltage marking.

> For example, a circuit breaker can be used with grounded, ungrounded or grounded neutral systems when identified with a straight voltage marking of 240 volts, 480 volts, etc.

Circuit breakers can be used with grounded neutral systems when identified with a slash voltage marking. Slash markings identify phase-to-phase and phase-to-ground voltages.

> For example, a circuit breaker can only be used with grounded neutral systems when identified with a slash voltage marking of 208/120 volts, 480/277 volts, etc.

Overcurrent Protection Devices and Conductors

OCPD'S FOR SYSTEMS OVER 600 VOLTS

OCPD's may be sized properly to protect the elements of electrical systems rated over 600 volts. High-voltage electrical systems behave differently from systems of 600 volts or less. When short-circuits or ground-faults develop on high-voltage circuits, great damage to conductors, elements, and equipment can occur. Therefore, it is imperative to protect such systems from these hazards.

Master Test Tip 61: The operating time of the protective device, the available short-circuit current, and the conductor used need to be coordinated to prevent damaging or dangerous temperatures in conductors or conductor insulation under short-circuit conditions.

FEEDER CIRCUIT PROTECTION - OVER 600 VOLTS
240.101(A)

For short-circuit protection of high-voltage feeders, the fuse rating may be increased up to three times the conductor ampacity. Circuit breaker setting may be increased up to six times the conductor ampacity. **(See Figures 4-29(a) and (b))**

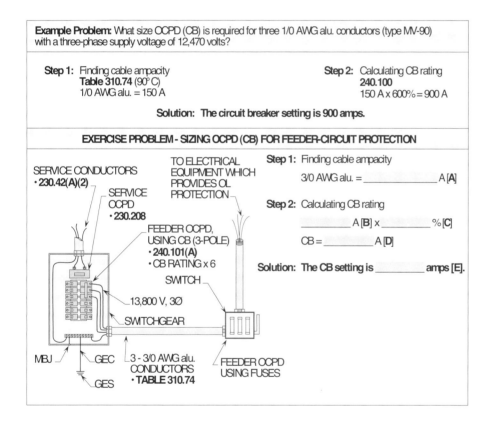

Figure 4-29(a). This illustration is a test problem for sizing and selecting the circuit breaker setting.

Master Test Tip 62: In **Figure 4-29(a)**, overload protection for conductors must be provided per **240.100(A)(1)** and **490.20(A)**.

Master Test Tip 63: A 6 AWG cu. THHN insulated conductor connected to a 60°C terminal on a CB has an ampacity of 55 amps and not 70 amps per **110.14(C)(1)(a)** and **(b)** and **Table 310.16**.

TERMINAL RATINGS
110.14(C); 310.10, FPN (1) THRU (4)

The procedure used in verifying a terminal's overcurrent protection device rating is to check the rating listed on the overcurrent protection device to see if it is 60°C, 60°C/75°C, or 75°C. Where the rating of the overcurrent protection device is 100 amps or less, the terminal rating is 60°C if it is not marked as mentioned above. Overcurrent protection devices rated over 100 amps are rated at 75°C and can be loaded to the 75°C ampacities of conductors that are found in **Tables 310.16** thru **310.19** of the NEC.

Note that motor loads or terminals rated 100 amps or less are considered 75°C for selecting ampacities per **110.14(C)(1)(a)(4)** and **Table 310.16**.

Figure 4-29(b). This illustration is a test problem for sizing and selecting the size fuse ratings.

Master Test Tip 64: In **Figure 4-29(b)**, overload protection for conductors must be provided per **240.100(A)**, **(B)**, and **(C)**. Also, see **490.21(A)** and **(B)**.

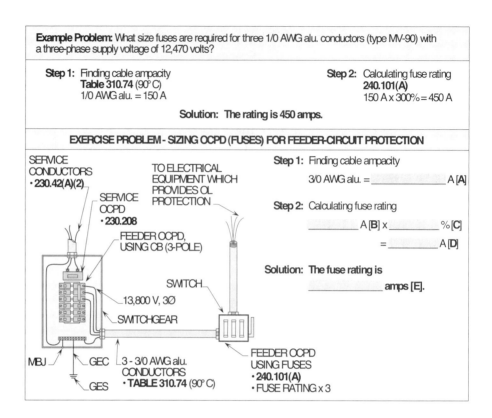

OCPD'S RATED 100 AMPS OR LESS
110.14(C)(1)(a); 310.10, FPN 3

Terminations for circuits that are rated 100 amps or less, and that use conductors sizes 14 AWG through 1 AWG, are limited to 60°C ampacities. Conductors that have higher temperature ratings, such as THHN conductors, can be used for these circuits. However, the ampacity of such conductors must be determined by the 60°C columns of **Tables 310.16** thru **310.19** if not marked 60°C/75°C or 75°C. In cases where the termination devices for such circuits are listed for operation at higher temperatures, conductors can have their ampacity selected at the higher temperatures per **110.14(C)(1)(a)** and **(b)**.

Master Test Tip 65: OCPD's rated 100 A or less are 60°C terminals if they are not marked 60/75°C. Motors can be cabled with 75°C ampacities.

See **Figure 4-30** for a test problem pertaining to selecting the terminal ratings to determine the conductor's ampacity based upon 100 amps or less devices.

OCPD'S RATED OVER 100 AMPS
110.14(C)(1)(b); 310.10, FPN 3

Master Test Tip 66: Continuous load is any load that can operate for three hours or more per **Article 100** of the NEC.

Terminations for circuits that are rated over 100 amps and that use conductors larger than No. 1 are limited to 75°C. Conductors that have higher temperature ratings, such as THHN conductors, can be used for these circuits. However, the ampacity of such conductors must be determined by the 75°C columns of **Tables 310.16** thru **310.19**. In cases where the termination devices for the circuit are listed for operation at higher temperatures, the conductors can have their ampacity calculated at the higher temperatures per **110.14(C)(1)(a)** and **(b)** of the NEC.

See **Figure 4-31** for a test problem pertaining to selecting the terminal ratings to determine the conductor's ampacity based upon devices rated over 100 amps.

Overcurrent Protection Devices and Conductors

Example Problem: What is the ampacity of 4 AWG THHN copper conductors connected to an overcurrent protection device supplying power to equipment where all terminals are rated 75°C?

Step 1: Finding ampacity
110.14(C)(1)(a) and (b)
75°C terminals per Table 310.16 permits 4 AWG THHN cu. to have an ampacity of 85 A

Solution: A 4 AWG THHN copper conductor can be loaded to 85 amps with 75°C terminals.

EXERCISE PROBLEM - SIZING OCPD RATED 100 AMPS OR LESS

- 60°C TERMINALS
  - TABLE 310.16
  - 110.14(C)(1)(a)
- NEWLY ADDED EQUIPMENT 75°C
- LOAD • 100 A
- 100 A OCPD OR LESS • 100 A
- EXISTING SERVICE PANEL

Step 1: Finding terminals
Panel terminals = _____ °C [A]
Equip. terminals = _____ °C [B]

Step 2: Finding ampacity
Load = _____ A [C]
Panel terminals = _____ °C [D]
Equip. terminals = _____ °C [E]

Step 3: Selecting conductors
OCPD rating = _____ A [F]
Wire ampacity = _____ A [G]
Wire size = _____ AWG cu. [H]

Solution: The ampacity of the THHN copper conductors is _____ amps [I] based upon _____ °C [J] terminals.

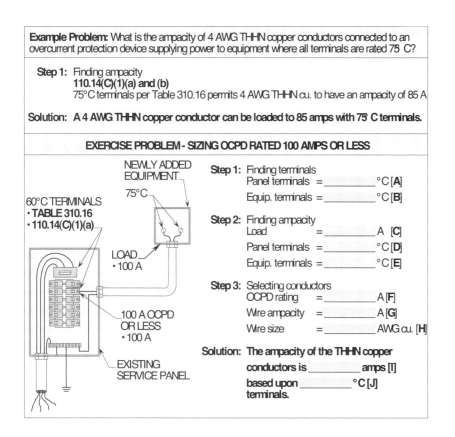

Figure 4-30. Test problem for selecting the ampacity of conductors, which are determined by the markings on the terminals of the OCPD's and equipment supplied. **Note:** OCPD in panel is rated 100 amps or less.

Master Test Tip 67: In **Figure 4-30**, the ampacity of the conductors is limited to 60°C ampacities because the terminals of the CB's in the panelboard are only rated at 60°C.

For motors,
Review 110.14(C)(1)(a)(4)

Example Problem: What is the ampacity of a 1/0 AWG THHN copper conductor connected to an overcurrent protection device supplying power to equipment where all terminals are rated 75°C?

Step 1: Finding ampacity
110.14(C)(2)(a) and (b)
75°C terminals per Table 310.16 permits 1/0 AWG THHN cu. to have an ampacity of 150 A

Solution: A 1/0 AWG THHN copper conductor can be loaded to 150 amps with 75°C terminals.

EXERCISE PROBLEM - SIZING OCPD RATED OVER 100 AMPS

- NEWLY ADDED EQUIPMENT 75°C
- 75°C TERMINALS
  - TABLE 310.16
  - 110.14(C)(1)(b)
- LOAD • 200 A
- OCPD RATED OVER 100A
  - 2-POLE CB
  - 200 A
- EXISTING SERVICE PANEL

Step 1: Finding terminals
Panel terminals = _____ °C [A]
Equip. terminals = _____ °C [B]

Step 2: Finding ampacity
Load = _____ A [C]
Panel terminals = _____ °C [D]
Equip. terminals = _____ °C [E]

Step 3: Selecting conductors
OCPD rating = _____ A [F]
Wire ampacity = _____ A [G]
Wire size = _____ AWG cu. [H]

Solution: The ampacity of the THHN copper conductors is _____ amps [I] based upon _____ °C [J] terminals.

Figure 4-31. Test problem for selecting the ampacity of conductors, which are determined by the markings on the terminals of the OCPD's and equipment supplied. **Note:** OCPD in the panel is rated over 100 amps.

Master Test Tip 68: Note that the No. 1/0 THHW cu. conductors in the example problem must not be connected to 75°C terminals on a 175 A OCPD and used at 170 amps (90°C Col. per **Table 310.16**). The 75°C terminals limits the ampacity to 150 amps per 75°C Col. in **Table 310.16**.

CALCULATING LOAD FOR CONDUCTORS
210.19(A)(1); 210.20(A)

Loads that are continuous shall have their full-load currents increased by 125 percent, and this value shall be added to noncontinuous loads at 100 percent or demand factors, if any are present.

The above concept is utilized to compute the load for branch-circuits, feeders and services. Loads are calculated in VA and amps to size and select the elements of such systems.

DERATING AMPACITY OF CONDUCTORS
310.10, FPN's (1) THRU (4)

Master Test Tip 69: Correction factors are based upon the temperature surrounding the cable or conduit.

The ampacity of a conductor is the current rating (amps) that a conductor can carry continuously without exceeding its temperature rating, which is determined by its condition of use. Such condition of use is too many current-carrying conductors in a raceway or cable or a surrounding ambient temperature that exceeds 86°F. Note that both of these conditions may exist in a circuit.

APPLYING ADJUSTMENT FACTORS FOR CONDUCTORS
310.10, FPN (4); TABLE 310.15(B)(2)(a)

Master Test Tip 70: Where there are four or more current-carrying conductors in a cable or raceway (adjustment factors), a derating factor must be applied and the ampacity of the conductors reduced.

If four or more current-carrying conductors are pulled through a raceway or bundled in a cable for a distance greater than 24 in., the allowable ampacity of such conductors shall be reduced (derated) by the factors (percentages) listed in **310.15(B)(2)(a)** and **Table 310.15(B)(2)(a)**.

For example, nine current-carrying conductors in a raceway shall have their ampacities listed in **Table 310.16** reduced by 70 percent so they will not overheat their insulation ratings if they should become fully loaded per **Table 310.15(B)(2)(a)**. **(See Figure 4-32)**

Figure 4-32. Test problem for adjustment factors per **310.15(B)(2)(a)** and **Table 310.15(B)(2)(a)** where there are four or more current-carrying conductors in a raceway or cable.

Master Test Tip 71: Note that EGC's are not considered current-carrying conductors per **310.15(B)(5)** to **Table 310.16**. (See Master Test Tips 75, 76, and 77 of this chapter.)

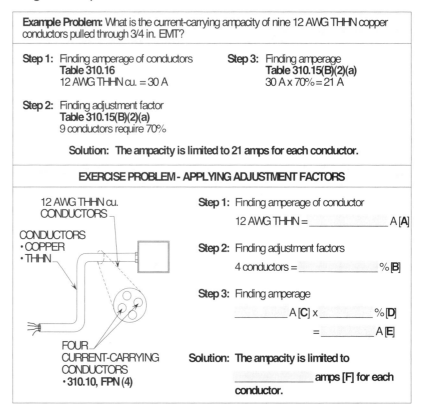

APPLYING CORRECTION FACTORS FOR CONDUCTORS
310.10, FPN (1); TABLE 310.16

Conductors routed through ambient temperatures exceeding 86°F are required to be derated according to the correction factors of **Table 310.16**. The derating factors are listed in the ampacity correction factor chart below **Table 310.16**, and they are selected based on the ambient temperature that the conductors are exposed to. The ampacity correction factors are based on the material, insulation, and the size of the conductors utilized. **(See Figure 4-33)**

Master Test Tip 72: After derating a conductor for condition of use, the reduced ampacity must be capable of supplying the computed load.

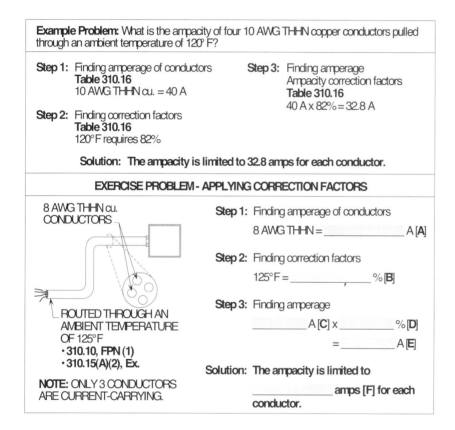

Figure 4-33. Test problem for calculating the ampacity of conductors where they are routed through ambient temperatures above 86°F.

Master Test Tip 73: Current-carrying conductors are considered conductors that have to have derating factors applied to their allowable ampacities per **Table 310.15(B)(2)(a)**. (See Master Test Tips 71, 74, and 75 of this chapter.)

Master Test Tip 74: Neutrals in a 4-wire, 3-phase wye system, the neutral is considered to be current-carrying if the major portion of such current contains harmonics per **310.15(B)(4)(c)** to **Table 310.16**. (See Master Test Tips 71, 73, and 75 of this chapter.)

APPLYING BOTH ADJUSTMENT AND CORRECTION FACTORS FOR CONDUCTORS
TABLE 310.15(B)(2)(a) AND
CORRECTION FACTORS TO TABLES 0-2000 VOLTS

Four or more current-carrying conductors enclosed in cables and raceways and routed through an ambient temperature above 86°F are required to have their ampacities derated twice.

Too many current-carrying conductors in a cable or raceway have problems dissipating heat into the surrounding ambient medium. The load on the conductors must heat the cable or raceway above the surrounding ambient temperature before harmful heat can be dissipated. **(See Figure 4-34)**

Figure 4-34. Test problem for determining the ampacity for conductors routed through an ambient temperature above 86°F with more than three current-carrying conductors in a cable or raceway.

Master Test Tip 75: If the amps on control-circuit conductors exceed 10% of their ampacity rating, they are considered current-carrying for the purpose of applying derating factors. See **725.28(B)(1)** and **(B)(2)** (See Master Test Tips 71, 73, and 74 of this chapter.)

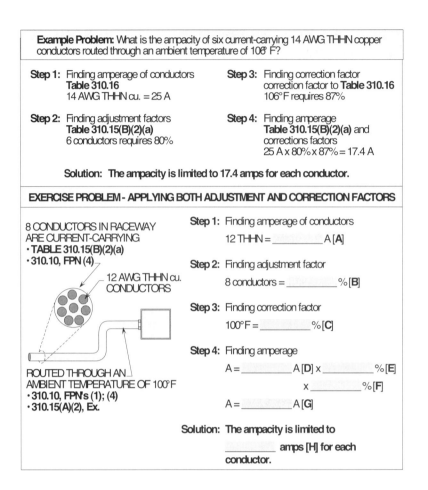

APPLYING 50 PERCENT LOAD DIVERSITY FACTOR FOR CONDUCTORS
TABLE B.310.11

A 50 percent load diversity can be applied to a conductor (cond.) ampacity when all the conductors are not loaded at the same time, if allowed by the AHJ. **(See Figure 4-35)**

LOADING MORE THAN HALF OF THE CONDUCTORS FORMULA TO TABLE B.310.11

Diversity can be applied to more than half of the conductors (cond.'s) in a raceway or cable if they are limited to a certain calculated ampacity. **(See Figure 4-36)**

PROTECTION OF CIRCUIT CONDUCTORS
240.4(A) THRU (G)

Master Test Tip 76: For some types of loads, the OCPD must be sized larger than the ampacity of the conductors to allow starting and running of the equipment such as motors, A/C units, welders, etc.

The general rule is that conductors shall be protected against overcurrent by a fuse or circuit setting rated no higher than the ampacity of the conductor. Conductors have specific current-carrying ampacities for different sizes of conductors, for different insulations, and for different ambient temperature conditions. The overcurrent protection device shall be sized to protect the insulation of the conductors from damage, caused by the current reaching an excessive value.

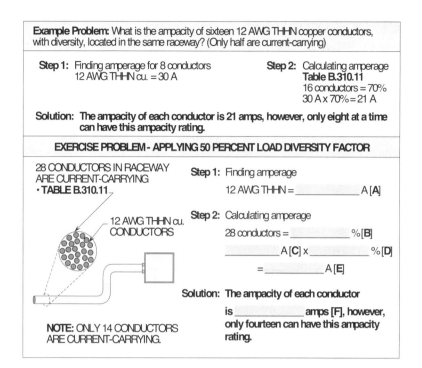

Figure 4-35. Test problem for applying derating factors to ampacities of conductors when there is a load diversity. Note that only half of the conductors carry current.

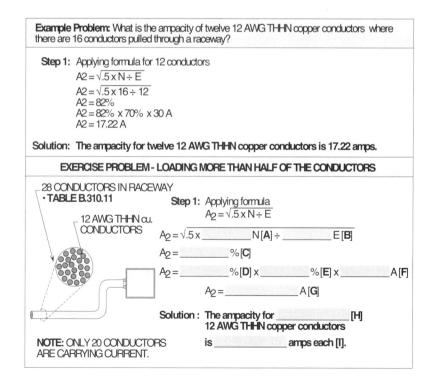

Figure 4-36. Test problem for applying derating factors to ampacities of conductors when there is a load diversity. Note that more than half of the conductors carry current.

SERVICE CONDUCTORS
215.2(A)(1); 230.42(A)(1)

Service-entrance conductors shall be of sufficient size to carry the loads as computed in **Article 220**. Ampacity is usually determined from **Table 310.16** of the NEC. In other words, the ampacity of the service conductors must not be smaller than required to serve the load, and in no case be smaller than the sum of the loads of the branch-circuits it serves, as computed in **220.12** thru **220.61**. This rule is subject to applicable demand factors that may apply in **220.80** thru **220.103**. **(See Figure 4-37)**

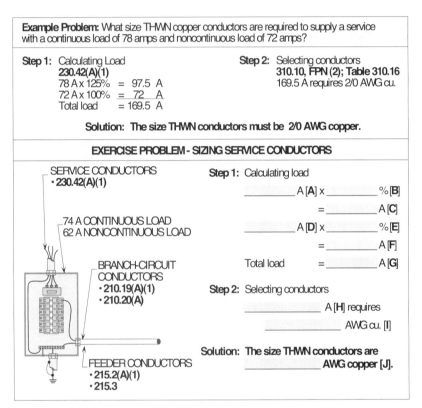

Figure 4-37. Test problem for determining the service load in amps and selecting conductors based upon continuous and noncontinuous operation.

Master Test Tip 77: Note that 10 pieces of electric cooking equipment under certain conditions of use can have a 65 percent demand factor applied per **Table 220.56**. In other words, such loads are not calculated at 125 percent or 100 percent of the total connected load.

FEEDER-CIRCUIT CONDUCTORS
215.2(A)(1); 215.3

If all of the feeder load is a continuous load, the feeder must be sized to carry 125 percent of the load. However, in cases where part of the feeder load is continuous and part noncontinuous, only the continuous part of the load is subjected to the 125 percent rule. **(See Figure 4-38)** Note: Demand factors can be applied to certain types of loads.

Master Test Tip 78: For a circuit to be classified as a feeder, it must be rated at least 30 amps and have capacity of at least 10 AWG copper conductors per **215.2(A)(2)**.

BRANCH-CIRCUIT CONDUCTORS
210.19(A)(1); 210.20(A)

Where a branch-circuit supplies continuous loads or any combination of continuous or noncontinuous loads, the ampacity of such conductors shall be the ampacity of the noncontinuous load plus 125 percent of all continuous loads. In other words, the computed continuous load (at 100% of amps or VA) shall not exceed 80 percent of the branch-circuit rating per **210.19(A)** and **210.20(A)**. **(See Figure 4-39)**

Overcurrent Protection Devices and Conductors

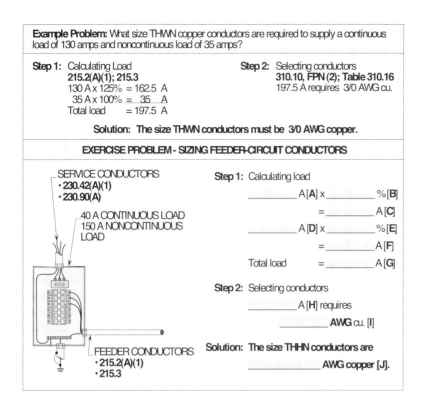

Figure 4-38. Test problem for determining the feeder load in amps and selecting conductors based upon continuous and noncontinuous operation.

Master Test Tip 79: A 50,000 VA general-purpose receptacle load can be computed at 100 percent of the first 10,000 VA and 50 percent of the remaining 40,000 VA per **Table 220.44**. Note that a demand factor is applied per **220.44**.

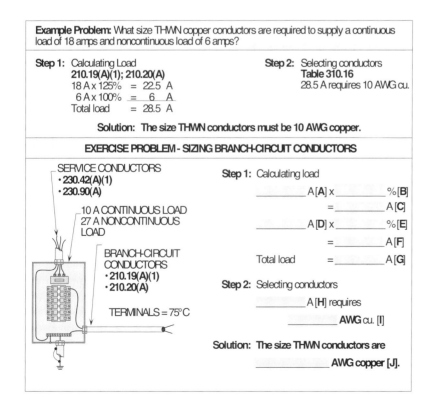

Figure 4-39. Test problem for determining the branch-circuit load in amps and selecting conductors based upon continuous and noncontinuous operation.

Master Test Tip 80: Note that a branch-circuit is the last OCPD between the panelboard and load served.

Master Test Tip 81: In the Example problem, a 30 amp OCPD can be used, because it is the next size above 28.5 amp per **240.4(B)(3)**.

SIZING OCPD'S ABOVE THE AMPACITY OF THE CONDUCTORS 240.4(B)(1) THRU (B)(3)

Master Test Tip 82: Note that the next size OCPD above the ampacity of conductor that does not correspond with a standard OCPD can only be selected at 800 amp or less in rating per **240.4(B)(3)**.

The general rule is that conductors shall be protected against overcurrent by a fuse or circuit breaker setting rated no higher than the ampacity of the conductor. However, in some installations, the NEC allows the fuse or circuit breaker setting to be above the ampacity of the conductor.

Such an installation, where the ampacity of the conductor does not correspond with a standard size fuse or circuit breaker setting, the next higher size may be used.

For example, a 4 AWG THHN copper conductor connected to 75°C terminals has an ampacity of 85 amps. A 80 amp fuse or circuit breaker setting would be largest size permitted for protection of this conductor per **240.4(B)(1)** thru **(B)(3)**.

Note that **240.4(B)(2)** permits the next size OCPD, which is a 90 amp fuse or circuit breaker, to be connected ahead of the circuit. **(See Figure 4-40)**

Figure 4-40. Test problem for sizing the OCPD above the ampacity of the conductors or load in amps.

Master Test Tip 83: In the Example problem, to select the OCPD at 125 amps, the load of the circuit must not exceed the calculated load of 112.5 amps. Note that the ampacity of the conductor at 115 amps supplies the 90 amp connected load plus the 112.5 amp calculated load.

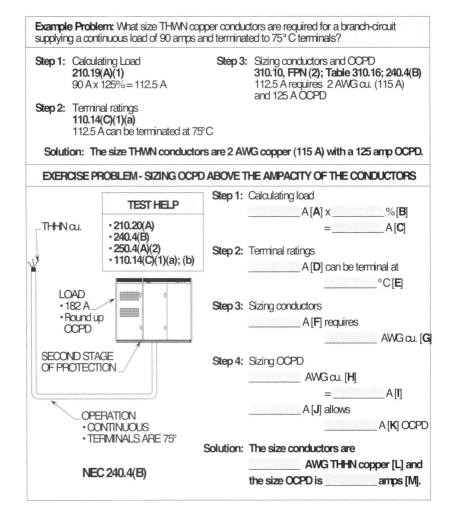

Example Problem: What size THWN copper conductors are required for a branch-circuit supplying a continuous load of 90 amps and terminated to 75°C terminals?

Step 1: Calculating Load
210.19(A)(1)
90 A x 125% = 112.5 A

Step 2: Terminal ratings
110.14(C)(1)(a)
112.5 A can be terminated at 75°C

Step 3: Sizing conductors and OCPD
310.10, FPN (2); Table 310.16; 240.4(B)
112.5 A requires 2 AWG cu. (115 A) and 125 A OCPD

Solution: The size THWN conductors are 2 AWG copper (115 A) with a 125 amp OCPD.

EXERCISE PROBLEM - SIZING OCPD ABOVE THE AMPACITY OF THE CONDUCTORS

- TEST HELP
  - 210.20(A)
  - 240.4(B)
  - 250.4(A)(2)
  - 110.14(C)(1)(a); (b)

THHN cu.

LOAD
- 182 A
- Round up OCPD

SECOND STAGE OF PROTECTION

OPERATION
- CONTINUOUS
- TERMINALS ARE 75°

NEC 240.4(B)

Step 1: Calculating load
_____ A [A] x _____ % [B]
= _____ A [C]

Step 2: Terminal ratings
_____ A [D] can be terminal at
_____ °C [E]

Step 3: Sizing conductors
_____ A [F] requires
_____ AWG cu. [G]

Step 4: Sizing OCPD
_____ AWG cu. [H]
= _____ A [I]
_____ A [J] allows
_____ A [K] OCPD

Solution: The size conductors are _____ AWG THHN copper [L] and the size OCPD is _____ amps [M].

Overcurrent Protection Devices and Conductors

SIZING OCPD'S BELOW THE AMPACITY OF THE CONDUCTORS
240.4(C)

If the overcurrent protection device is greater than 800 amps, the ampacity of the conductors shall be equal to or greater than the ampacities of the overcurrent protection device per **240.6(A)**. Note that **240.6(A)** allows an OCPD that is not listed in **240.6(A)** to be used, if listed for such use.

For example, a paralleled hookup of five 250 KCMIL THWN copper conductors per phase have an ampacity of 1275 amps (255 A x 5 = 1275 A). By applying **240.4(C)** of the NEC, a 1200 amp OCPD must be used. **(See Figure 4-41)**

Master Test Tip 84: A calculated load of 1560 amps can be protected with 1500 amp OCPD instead of 1200 amp per **240.6(A)**.

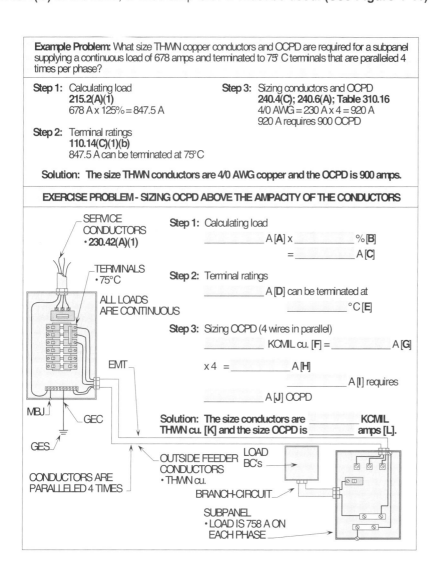

Figure 4-41. Test problem for sizing the OCPD below the ampacity of the conductors.

Master Test Tip 85: The general rule for paralleling conductors is that all phase conductors and, if used, all grounded conductors and EGC's be routed with each circuit per **300.3(B)** and **310.4**. For application of these rules, See page 4-54 and **Figure 4-59** of this chapter.

SIZING CONDUCTORS FOR TAPS
240.21

The general rule for installing fuses or circuit breakers in a circuit is that they must be installed at the source of the circuit. However, there are exceptions to this rule.

Master Test Tip 86: A tapped conductor is usually from a feeder-circuit having a larger conductor sufficient to supply the tapped loads.

For example, when a tap is made from a feeder, the fuses or circuit breaker protecting the tap conductors can be installed in the equipment enclosure, where such conductors are terminated.

SIZING CONDUCTORS FOR TAPS 10 FT. OR LESS IN LENGTH 240.21(B)(1) THRU (4)

An example of applying this rule is a tap feeding a lighting and appliance branch-circuit panelboard. The tap conductors cannot be over 10 ft. long and must terminate at such panel. Mechanical protection by conduit, tubing, or metal gutter is required, and conductors must be sized to carry the total load. If these requirements are followed, no overcurrent protection device is required at the point of such tap. **(See Figure 4-42)**

Figure 4-42. Test problem for sizing the conductors for a tap that is 10 ft. or less in length. **Note:** For sizing OCPD, see **Figure 4-4**.

Master Test Tip 87: The conductors of the 10 ft. tap must have enough capacity in amps to supply whatever they are terminated to. For example, 2/0 AWG THWN cu. conductors(175 A) can terminate to a 175 amp OCPD, but 4/0 AWG THWN cu. conductors (230 A) must be used if connected to the lugs of a 225 amp panelboard.

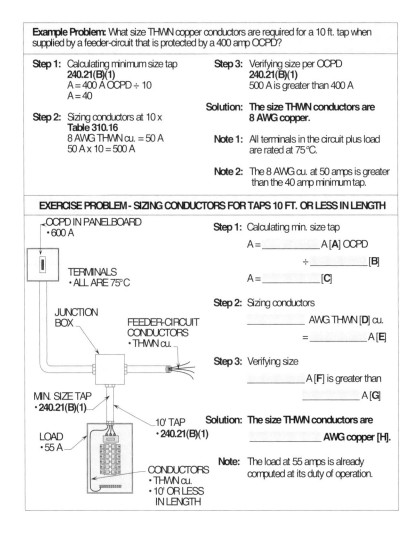

Overcurrent Protection Devices and Conductors

SIZING CONDUCTORS FOR TAPS OVER 10 FT. TO 25 FT. IN LENGTH 240.21(B)(2)(1) THRU (3)

A tap from a larger conductor to a smaller conductor may extend over 10 ft. up to a distance of 25 ft., provided the current-carrying capacity of the tap is at least 1/3 that of the OCPD. Overcurrent protection is not required at the point of the tap, but the tap must have overcurrent protection at the end of the run. Such tap must be properly sized and protected from physical damage and enclosed in a raceway system. **(See Figure 4-43)**

Master Test Tip 88: When applying the 25 ft. tap rule, conductors must terminate in an OCPD.

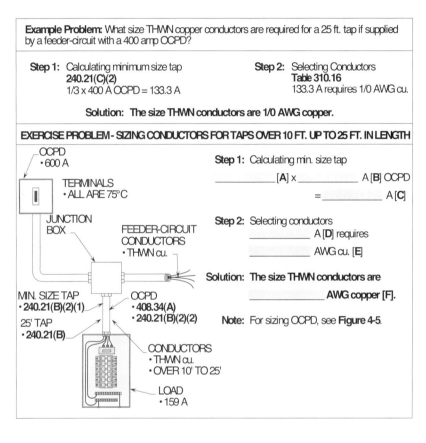

Figure 4-43. Test problem for sizing the conductors for a tap over 10 ft. up to 25 ft. in length.

Master Test Tip 89: Note that the conductor of 25 tap must terminate to a main CB or set of fuses. They are never allowed to terminate to the lugs of the equipment.

In the exercise problem, a 2/0 AWG THWN cu. conductor will supply the load of 159 amps, but the 1/3 rule per **240.21(B)(2)(1)** requires at least 3/0 AWG THWN cu. conductors.

SIZING CONDUCTORS FOR TAPS AND CONNECTIONS INCLUDING TRANSFORMER 240.21(B)(3)(1) THRU (5)

Transformer with primary taps plus secondary conductors that are not over 25 ft. in length can be made without overcurrent protection at the point of such taps. Conductors must be sized and based on the primary and secondary side of the transformer. **(See Figure 4-44)**

SIZING CONDUCTORS FOR TAPS OVER 25 FT. UP TO 100 FT. IN LENGTH 240.21(B)(4)(1) THRU (9)

For taps in high bay manufacturing buildings, the protection for the tap may be at the end of the tapped conductors, if the tap is not over 25 ft. long horizontally, and not over 75 ft. long vertically. Note that the total run, both horizontally and vertically, is limited to 100 ft. or less in length. **(See Figure 4-45)**

Master Test Tip 90: When applying the 100 ft. tap rule, the tapped box (junction) must be at least 30 ft. above the finished grade.

Figure 4-44. Test problem for sizing the conductors for a 25 ft. tap including the primary tap, transformer, and secondary connected conductors. **Note:** For sizing OCPD, see **Figure 4-6**.

Master Test Tip 91: In the exercise problem, the 300 amp OCPD of the feeder-circuit protects the primary conductors, XFMR windings, and secondary conductors from ground-faults and short-circuits. The OCPD in the panelboard (to be sized) will protect the primary XFMR and secondary from overloads.

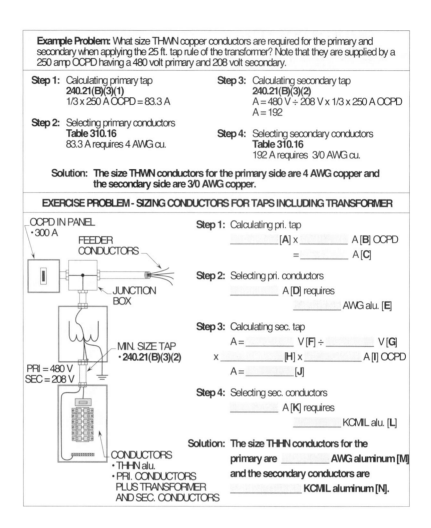

SIZING BUSWAY TAPS
240.21(E)

Master Test Tip 92: Note that OCPD's do not have to be readily accessible for busway taps per **240.21(A)(1), 368.17(C)** and **404.8(A), Ex. 1**.

Section **368.17(B)** permits the reduction in the size of a busway. In such cases, an additional overcurrent protection device is not required for smaller busways at the point of the tap. **(See Figure 4-46)**

SIZING CONDUCTORS FOR MOTOR CIRCUIT TAPS
240.21(F)

Where more than one motor is on a feeder-circuit, each motor circuit is tapped from the feeder conductors. The tap is protected by the motor's branch-circuit protection. The motor's branch-circuit protection is located where the tap conductors terminate in a circuit breaker or set of fuses. **(See Figure 4-47)**

Overcurrent Protection Devices and Conductors

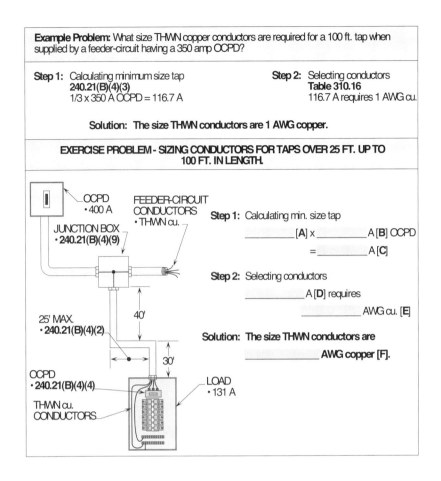

Figure 4-45. Test problem for sizing the conductors for a tap over 25 ft. up to 100 ft. in length. **Note:** For sizing OCPD, See **Figure 4-7**.

Master Test Tip 93: For taps up to 100 ft. in length, the smallest tap permitted is 6 AWG cu. or 4 AWG alu. per **240.21(B)(4)(7)**.

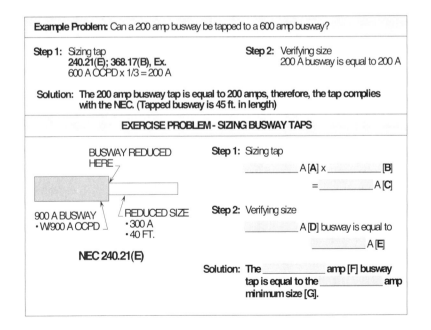

Figure 4-46. Test problem for sizing a smaller busway that is tapped from a larger busway.

Master Test Tip 94: Note that the smaller busway rated at 200 amps in the example problem is not required to have an OCPD installed at the point of such tap.

4-45

Figure 4-47. Test problem for sizing the conductors for a 10 ft. tap to supply a motor. **Note:** For sizing OCPD, see **Figure 4-9**.

Note that the 65 A conductors at 65 amp (65 A x 10 = 650 A) are greater than the 600 amp OCPD protecting the feeder-circuit conductors.

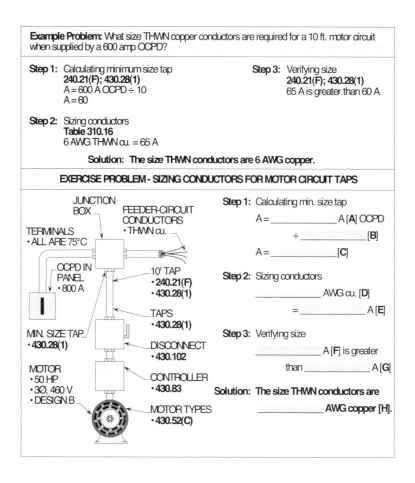

CONNECTIONS OF 10 FT. FROM SECONDARY OF XFMR 240.21(C)(2)(1) THRU (3)

Master Test Tip 95: The OCPD at the end of the connected conductors must protect both the output of the transformer and the tapped conductors from overloads.

Conductors of 10 ft. or less may be connected from the secondary of a separately derived system. **(See Figure 4-48)**

CONNECTIONS OF 25 FT. FROM SECONDARY OF XFMR 240.21(C)(3)(1) THRU (3)

Conductors are permitted to be connected to the secondary of a separately derived system for industrial locations, without overcurrent protection at the point of connection, where the OCPD in the panel protects the conductors and output of the transformer from overloads. **(See Figure 4-49)**

OUTSIDE FEEDER TAPS 240.21(C)(4)(1) THRU (4)

Outside conductors are allowed to be connected as a feeder from the secondary of a transformer, without overcurrent protection at the point of connection. However, to apply this rule, the OCPD must protect the conductors and output of the transformer from overloads. **(See Figure 4-50)**

Overcurrent Protection Devices and Conductors

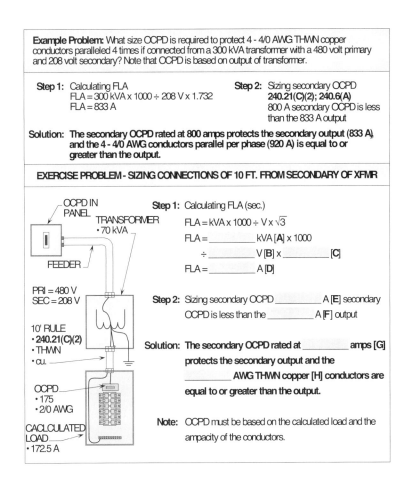

Figure 4-48. Test problem for connection of conductors from the secondary of a transformer using the 10 ft. connecting rule. **Note:** For sizing OCPD, see Figure 4-11.

Master Test Tip 96: Parallel conductors of each phase circuit must be of the same type insulation, material, terminated in the same way, and have the same length per **310.4** and **250.24(C)(2)**.

In Exercise Problem 4-48, 360 V is used instead of the full square root of 3 times 208 V, which is equal to 360.256 V.

SIZING CONDUCTORS FROM GENERATOR TERMINALS 240.21(G)

According to the general rule, generator conductors must have an ampacity equal to at least 115 percent of the nameplate current rating of the generator. **(See Figure 4-51)**

SIZING CONDUCTORS FOR WELDER LOADS ARTICLE 630, PARTS II AND IV

There are three types of welders used in modern day welding shops. The type used determines how the circuit elements must be designed and selected.

The procedure for calculating the full-load amps for welders is to obtain the duty-cycle factor and select the multiplier. The primary amps of the welder is multiplied by the multiplier to derive the full-load amps to size the elements of the branch-circuit.

SIZING CONDUCTORS TO NONMOTOR ARC WELDERS 240.4(G); 630.11(A)

When sizing the branch-circuit conductors for AC/DC arc welders, the current-carrying capacity shall not be less than the primary current of the welder times a duty cycle factor listed in **Table 630.11(A)**. The OCPD can be sized larger than the ampacity of the conductors to allow the welder to operate. The ON and OFF welding cycles of the welder will protect the conductors from being overloaded. **(See Figure 4-52)**

Master Test Tip 97: If the conductors tapped from a generator are sized at least 115 percent of the generator's output, no OCPD at the point of tap is required per **445.13** and **445.12**.

Stallcup's Master Electrician's Study Guide

Figure 4-49. Test problem for connecting conductors from the secondary of a transformer using the 25 ft. connection rule. **Note:** For sizing OCPD, see **Figure 4-11**.

Master Test Tip 98: Note that all 25 ft. connections made from the secondary side of XFMR's must be in amps equal to the output of such XFMR's.

Example Problem: What size OCPD is required to protect 3 - 300 KCMIL THWN copper conductors paralleled 3 times and connected to the secondary of a 300 kVA transformer having a 208 secondary? Note that OCPD is based on output of the transformer.

Step 1: Calculating FLA (secondary)
FLA = 300 kVA x 1000 ÷ 208 V x 1.732
FLA = 833 A

Step 2: Sizing secondary OCPD
240.21(C)(3); 240.6(A)
800 A secondary OCPD is less than the 833 A output

Solution: The secondary OCPD rated at 800 amps protects the secondary output (833 A), and the 3 - 300 KCMIL parallel per phase (285 A x 3 = 855 A) is equal to or greater than the output.

EXERCISE PROBLEM - SIZING CONNECTIONS OF 25 FT. FROM SECONDARY OF XFMR

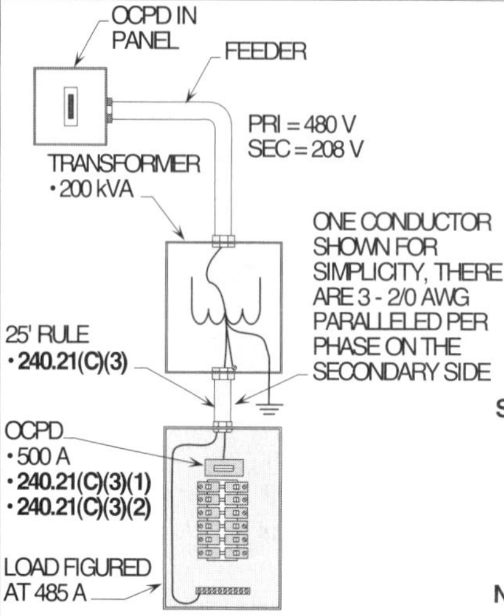

- OCPD IN PANEL
- FEEDER
- PRI = 480 V
- SEC = 208 V
- TRANSFORMER • 200 kVA
- ONE CONDUCTOR SHOWN FOR SIMPLICITY, THERE ARE 3 - 2/0 AWG PARALLELED PER PHASE ON THE SECONDARY SIDE
- 25' RULE • 240.21(C)(3)
- OCPD • 500 A • 240.21(C)(3)(1) • 240.21(C)(3)(2)
- LOAD FIGURED AT 485 A

Step 1: Calculating FLA (sec.)
FLA = kVA x 1000 ÷ V x √3
FLA = _____ kVA [A] x 1000
÷ _____ V [B] x _____ [C]
FLA = _____ A [D]

Step 2: Sizing secondary OCPD
_____ A [E] secondary OCPD is less than the _____ A [F] output

Solution: The secondary OCPD rated at _____ amps [G] protects the secondary output and the 3 - _____ AWG THWN copper [H] parallel per phase is equal to or greater than the output.

Note: The parallel hook-ups of 3 - 2/0 AWG THWN copper conductors are equal to 525 amps (3 x 175 A = 525 A).

Figure 4-50. Test problem for connecting conductors from the secondary of a transformer using the outside connection rule. **Note:** For sizing OCPD, see **Figure 4-12**.

Note that the 13,800 volt, three-phase supply in the Exercise Problem can be routed from inside building No. 1 or from another building or location.

Example Problem: What size OCPD is required to protect a 250 KCMIL THWN copper conductors paralleled 3 times that are connected from a 600 kVA customer owned transformer with a 13,800 volt primary and 480 secondary?

Step 1: Calculating FLA (secondary)
FLA = 600 kVA x 1000 ÷ 480 V x 1.732
FLA = 722 A

Step 2: Sizing secondary OCPD
240.21(C)(4); 240.6(A)
700 A secondary OCPD is less than the 722 A output

Solution: The 700 amp OCPD protects the secondary of a transformer (722 A) and the 3 - 250 KCMIL THWN copper conductors per phase (255 A x 3 = 765 A) are equal to and greater than the output.

EXERCISE PROBLEM - SIZING OUTSIDE FEEDER CONNECTIONS

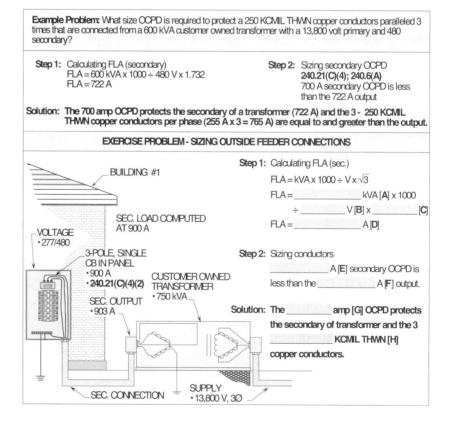

Step 1: Calculating FLA (sec.)
FLA = kVA x 1000 ÷ V x √3
FLA = _____ kVA [A] x 1000
÷ _____ V [B] x _____ [C]
FLA = _____ A [D]

Step 2: Sizing conductors
_____ A [E] secondary OCPD is less than the _____ A [F] output.

Solution: The _____ amp [G] OCPD protects the secondary of transformer and the 3 _____ KCMIL THWN [H] copper conductors.

Overcurrent Protection Devices and Conductors

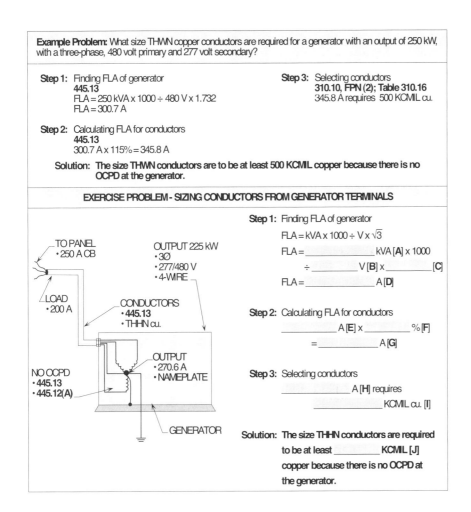

Figure 4-51. Test problem for tapping conductors from the secondary of a generator. **Note:** For another example, see Figure 4-10.

Master Test Tip 99: If the tapped conductors from the output side of the generator are not equal to 115 percent of the generator's output in amps, an OCPD is required at the point of such tap.

SIZING CONDUCTORS TO MOTOR OR NONMOTOR GENERATOR ARC WELDERS
240.4(G); 630.11(A)

When sizing the branch-circuit conductors for motor or nonmotor generator arc welders, the current-carrying capacity shall not be less than the rated primary current of the welder times a duty cycle factor listed in **630.11(A)**. **(See Figure 4-53)**

Master Test Tip 100: The multiplier for a nonmotor generator arc welder can be determined by taking the square root of the welder's duty cycle.

For example:
$$\sqrt{80\%} = .89$$

SIZING CONDUCTORS TO RESISTANCE WELDERS
240.4(G); 630.31(A)

When sizing the branch-circuit conductors for resistance welders, the current-carrying capacity shall not be less than the primary current of the welder times a duty cycle factor listed in **630.31(A)**. **(See Figure 4-54)**

Figure 4-52. Test problem for calculating the amps to size the conductors to a nonmotor generator arc welder. **Note:** For sizing OCPD, see Figure 4-18(a).

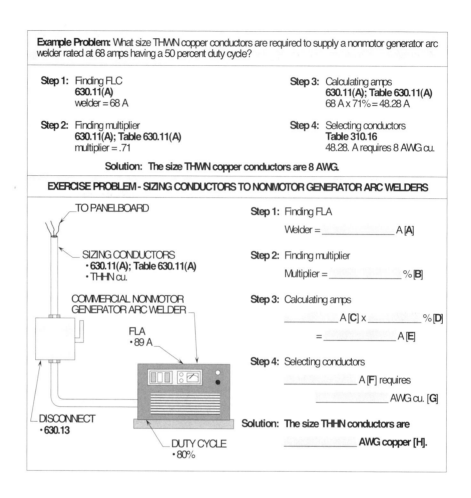

PROTECTING FIXTURE CONDUCTORS
240.5(B)(2); 410.67(C); 410.15(B)

Master Test Tip 101: Heavy duty lampholders can be installed on a 50 amp branch-circuit per **210.21(A)** and **210.23(C)**.

A 20 amp overcurrent protection device is considered as being adequate protection for fixture wires or flexible cords or for tinsel cord, in sizes 16 AWG and 18 AWG, respectively. Fixture wire taps that comply with **210.19(A)(4), Ex. 1** are permitted to be protected if they are tapped from 30, 40, and 50 amp branch-circuits per **Article 210**.

SIZING FIXTURE WIRE TO FLUORESCENT LAY-INS
240.5(B)(2); 410.67(C)

Master Test Tip 102: Fixture wires in fixture whips (trade term) are allowed in lengths of 18 in. to 6 ft.

Fixture wires shall be considered as protected by the overcurrent protection device of the branch-circuit if sized and selected per **Article 210** and comply as follows:

20 amp circuits, 14 AWG and greater
30 amp circuits, 14 AWG and greater
40 amp circuits, 12 AWG and greater
50 amp circuits, 12 AWG and greater

Note that 18 AWG fixture wire can run up to 50 ft. if protected by a 20 amp OCPD or less. 16 AWG fixture wire can be run up to 100 ft. if protected by a 20 amp OCPD or less. **(See Figure 4-55)**

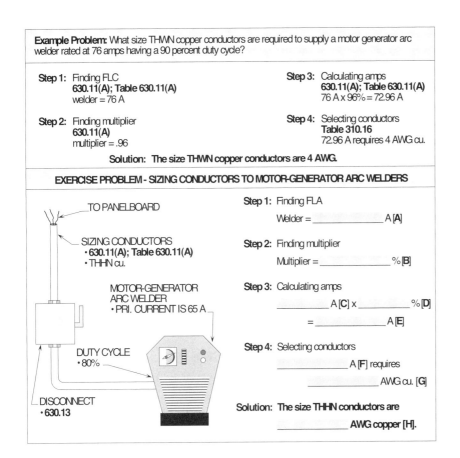

Figure 4-53. Test problem for calculating the amps when sizing the conductors to a motor generator arc welder. **Note:** For sizing OCPD, See **Figure 4-18(b)**.

SUPPLY FIXTURE WIRES FOR OUTSIDE LIGHTING STANDARDS
240.5(B)(2); 410.15(B)

For 20 amp circuits, 18 AWG, up to a tapped length of 50 ft. and 16 AWG, up to a tapped length of 100 ft. can be used as fixture whips per **410.67(C)** and outside lighting standards per **410.15(B)**. Note that the fixture wires can be routed from the fixtures to the base of the pole and terminated in the hand hole to the branch-circuit conductors. **(See Figure 4-56)**

GROUNDED CONDUCTOR
310.12(A); 200.6(A)

Insulated conductors of 6 AWG and smaller wire, when used as grounded conductors, shall have a white or natural gray colored insulation. Where the insulated conductors are larger than 6 AWG, they shall be identified by a white or natural gray colored insulation, or by a distinctive marking (white) at the terminals at the time they are being installed.

The grounded conductor used in service-entrance raceways and cables can be uninsulated per **230.41, Ex. (1)** and **230.22, Ex.**.

See **200.6(A)** and **(B)** and **210.5(A)** for further information considering identification of grounded conductors.

Master Test Tip 103: Note that the grounded conductor must be terminated to an isolated busbar, if they are routed to subpanels on the same premise.

Figure 4-54. Test problem for calculating the amps to size the conductors to a resistance welder. **Note:** For sizing OCPD, see Figure 4-18(c).

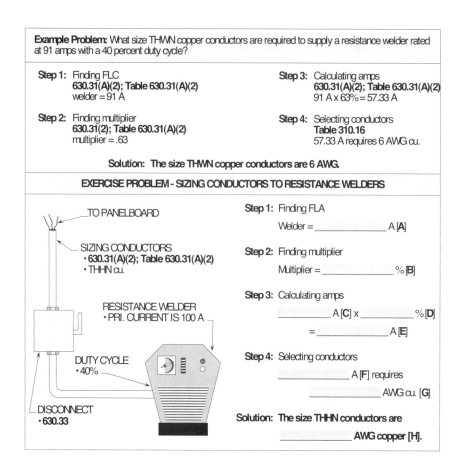

EQUIPMENT GROUNDING CONDUCTOR
310.12(B); 250.119

Master Test Tip 104: Note that EGC's must be terminated to a grounding busbar bonded to the metal enclosure enclosure of a subpanel that it serves.

An equipment grounding conductor must be contained in the same raceway, cord, or cable, or may otherwise be run with circuit conductors. It shall be a part of the cable or cord, or run in the same raceway with the circuit conductors to keep the impedance to the lowest value. When using NM, NMC, or UF cables, these cables should contain the equipment grounding conductor as part of the cable to ground noncurrent-carrying parts of the circuit.

The equipment grounding conductor may be bare, covered, or insulated. Where individual covered or insulated equipment grounding conductors are run, it is required that they shall have a continuous green outer finish or a continuous green outer finish with one or more yellow stripes. Bare conductors can always be used for such grounding techniques.

When the equipment grounding conductor is larger than 6 AWG copper or aluminum, this conductor shall be permanently marked at the time of installation, at each end or any other accessible point. It shall be identified by any one of the following means:

(1) Stripping the insulation from the entire exposed length.

(2) Coloring the exposed insulation green.

(3) Marking the exposed insulation with green colored tape or green colored adhesive labels.

See **210.5(B)** and **250.119** for a detailed description of identifying the equipment grounding conductor.

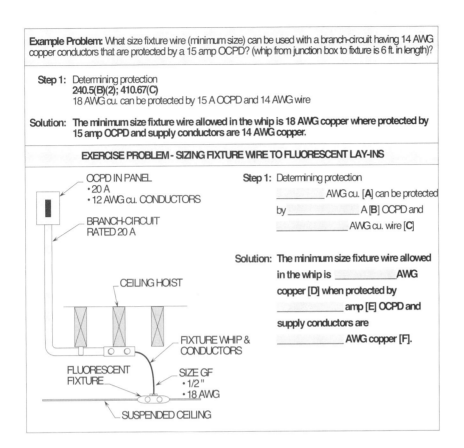

Figure 4-55. Test problem for determining if the branch-circuit OCPD and conductors protect the smaller tapped fixture wires.

UNGROUNDED CONDUCTORS
310.12(C); 110.15

The ungrounded phase conductors can be identified with any color of insulation or tagging except white, natural gray, or green. The following are recommended colors:

(1) 120/240 volt, single-phase system
 • Phase A - Black
 • Phase B - Blue or red

(2) 120/240 volt, three-phase, four-wire delta system
 • Phase A - Black
 • Phase B - Orange
 • Phase C - Blue

(3) 120/208 volt, three-phase, four-wire wye system
 • Phase A - Black
 • Phase B - Red
 • Phase C - Blue

(4) 277/480 volt, three-phase, four-wire wye system
 • Phase A - Brown
 • Phase B - Orange
 • Phase C - Yellow
 • Known as BOY system

See Figure 4-57 for a detailed illustration of color coding ungrounded conductors by recommended colors based on voltage levels of 600 volts or less.

Master Test Tip 105: Until the 1975 NEC, the color coding scheme for ungrounded phase conductors of an electrical system provided only the recommended colors.

Master Test Tip 106: This is not a color code requirement; it is a means of identification so each system's voltage can be easily determined.

Figure 4-56. Test problem for determining if the branch-circuit OCPD and conductors protect the smaller tapped fixture wires.

Master Test Tip 107: At the base of the metal pole in the exercise problem, a 2 in. x 4 in. handhole is required and a grounding terminal must be accessible for connecting the EGC's.

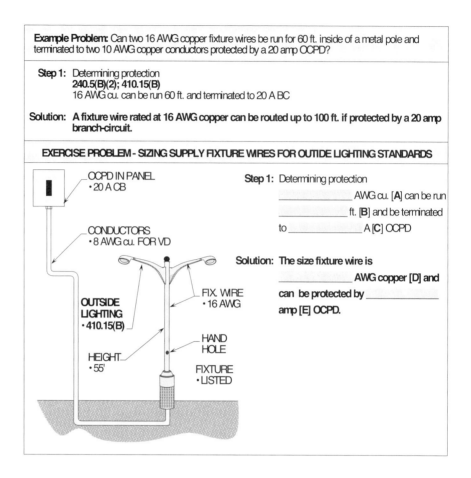

Section **210.5(C)** requires the ungrounded conductors of multiwire circuits where more than one nominal voltage system exists in a building's wiring system to be identified. The identification of each conductor, based on the voltage, is required to be marked with tape, tagging, spray paint where permitted, or other effective means accepted by the authority having jurisdiction.

The means of identification is required to be permanently posted at each panelboard housing the OCPD's of such branch-circuits. **(See Figure 4-58)**

CONDUCTORS IN PARALLEL
310.4

Master Test Tip 108: When necessary to run equipment grounding conductors with paralleled conductors, these equipment grounding conductors are to be treated the same as the other conductors and sized by the provisions of **Table 250.122**.

These requirements apply to copper and aluminum conductors connected in parallel so as to form a common conductor. Size 1/0 AWG or larger, comprising each phase or neutral, will be permitted to be paralleled, that is, both ends of the paralleled conductors are connected together so as to form a single conductor. In paralleling conductors, the following conditions shall be complied with:

(1) They shall be the same length.
(2) Of the same conductor material.
(3) Same circular-mil area.
(4) Same type of insulation.
(5) Terminated in the same manner.
(6) Where run in separate raceways or cables, the raceways or cables shall have the same physical characteristics.

See **Figure 4-59** for a detailed illustration pertaining to the rules of paralleling conductors 1/0 AWG and larger. Exceptions that permit conductors of any size to be run in parallel are as follows:

(1) For elevator lighting (**620.12(A)(1)**)
(2) Conductors supplying control power to indicating instruments, contactors, relays, and the like may be run in parallel in all sizes, provided that the following conditions are complied with:

(a) They are joined electrically at both ends so that they form a single conductor.
(b) They are in the same raceway or cable.
(c) The ampacity of each conductor is sufficient to carry the entire load shared by the parallel conductors.
(d) If one parallel conductor should happen to become disconnected, the ampacity of each conductor is not exceeded.

(3) Conductors supplying power for frequencies of 360 hertz and higher can be connected in parallel if they comply with the four rules listed above in item (2).

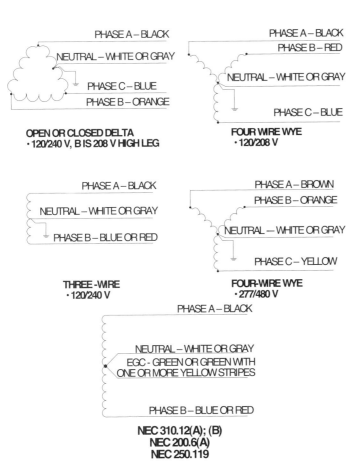

Figure 4-57. Until the 1975 NEC was published, ungrounded phase conductors were recommended to be color coded.

Master Test Tip 109: Note that only grounded conductors and EGC's are required to be color coded per **310.12(A)** and **(B)**. See **Figure 4-57** for the color code rules for grounded conductors and EGC's.

Note that the color of the high-leg (orange), grounded conductor (white or natural gray), and equipment grounding conductor (green or green with one or more yellow stripes) are the conductors on the test that are usually asked to be color coded.

Figure 4-58. Where there is more than one system voltage supplying a premises, proper identification of circuits in panelboards must be provided.

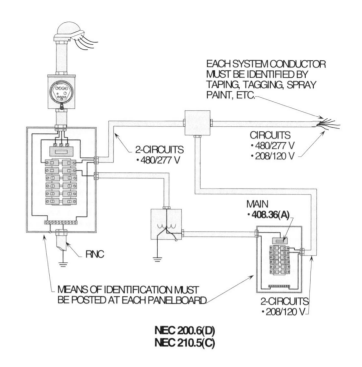

Figure 4-59. Conductors of one phase does not have the same characteristics as other phase, grounded phase, or grounded neutral conductor.

Master Test Tip 110: In a completely nonmetallic paralleled installation, Phase A conductors, Phase B conductors, Phase C conductors and neutrals can be routed in separate conduits per **300.5(I), Ex. 2**.

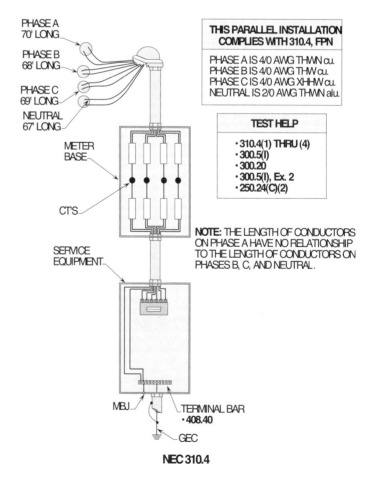

Name _____   Date _____

Chapter 4
Overcurrent Protection Devices and Conductors

Section   Answer

1. Conductors routed through an ambient temperature greater than 87°F shall be derated per correction factors based upon surrounding temperatures.

    (a) True   **(b)** False

2. If a conductor with 90°C insulation is connected to a terminal of 60°C, the ampacity of the 90°C conductor shall be selected at 60V per **Table 310.16**.

    (a) True   **(b)** False

3. A neutral of a four-wire, three-phase wye system shall be considered current-carrying when it carries the major portion of the load involving nonlinear related loads.

    (a) True   **(b)** False

4. Four or more current-carrying conductors enclosed in a cable or raceway routed through an ambient temperature above 86°F shall have their ampacities derated twice.

    (a) True   **(b)** False

5. Conductors listed as suitable for dry locations shall be permitted to be used in damp or wet locations.

    (a) True   **(b)** False

6. Overcurrent protection devices are never required to be readily accessible.

    (a) True   **(b)** False

7. Continuous current (three hours or more) shall not exceed 80 percent of the circuit breaker's rating or be calculated at 125% of the load.

    (a) True   **(b)** False

8. When the circuit breaker is turned ON, the handle shall be installed so that it is in the DOWN position.

    (a) True   **(b)** False

9. Circuit breakers shall be permitted to be used with grounded, ungrounded, or grounded neutral systems when identified with a slash voltage marking.

    (a) True   **(b)** False

10. Performance testing shall be performed at time of installation when installing service equipment with a ground-fault protection system.

    (a) True   **(b)** False

Section	Answer

11. A continuous load is where the current in amps can operate for a period of time of _____ hours or more.

    (a) One   **(b)** Two   **(c)** Three   **(d)** Five

12. The _____ connects all the metal parts of enclosures and equipment together.

    (a) Grounding electrode conductors   **(b)** Equipment grounding conductors
    (c) Bonding jumpers   **(d)** Neutrals

13. Copper and aluminum conductors shall be permitted to be connected in parallel in sizes _____ AWG or larger.

    (a) 3   **(b)** 1/0   **(c)** 3/0   **(d)** 250

14. Control circuits are current-carrying when they carry continuously _____ percent of their ampacity.

    (a) 10   **(b)** 20   **(c)** 30   **(d)** 50

15. Insulated conductors of _____ AWG and smaller wire, when used as grounded conductors, shall have a white or natural gray insulation.

    (a) 10   **(b)** 8   **(c)** 6   **(d)** 4

16. A tap feeding a lighting and appliance panelboard cannot be over _____ ft. long.

    (a) 25   **(b)** 50   **(c)** 75   **(d)** 125

17. No overcurrent protection device shall be permitted to be connected in _____ with any conductor that is intentionally grounded. (General Rule)

    (a) Series   **(b)** Parallel   **(c)** All of the above   **(d)** None of the above

18. To prevent damage to overcurrent protection devices, busbars, and service equipment, the ground-fault system shall be required to clear a fault of _____ amps or more within one second. (GFPE)

    (a) 1000   **(b)** 3000   **(c)** 5000   **(d)** 10,000

19. If the standard current ratings of fuses or nonadjustable circuit breakers do not conform to the ampacity of the conductors being used, it shall be permissible to use the next larger standard rating when below _____ amps.

    (a) 100   **(b)** 400   **(c)** 800   **(d)** 1000

20. A tap from a larger conductor to a smaller conductor may extend up to a distance of _____ ft., provided the current-carrying capacity of the tap is at least 1/3 that of the OCPD protecting the larger conductor.

    (a) 10   **(b)** 20   **(c)** 25   **(d)** 50

	Section	Answer

21. What is the ampacity rating for 4 - 10 AWG THHN copper conductors that are current-carrying?

      (a) 28 amps   **(b)** 32 amps   **(c)** 35 amps   **(d)** 38 amps

22. What is the ampacity rating for 8 - 10 THHN copper conductors routed through an ambient temperature of 120°F?

      (a) 20.84 amps   **(b)** 21.32 amps
      (c) 22.96 amps   **(d)** 24.72 amps

23. What size THWN copper conductors and OCPD (75°C) are required for a tap using the 10 ft. tap rule with the feeder overcurrent protection device rated at 400 amps?

      (a) 8 AWG conductors and 45 amp OCPD
      (b) 8 AWG conductors and 50 amp OCPD
      (c) 6 AWG conductors and 50 amp OCPD
      (d) 6 AWG conductors and 60 amp OCPD

24. What size THWN copper conductors and OCPD (75°C) are required for a tap using the 25 ft. tap rule with the overcurrent protection device rated at 400 amps?

      (a) 1 AWG conductors and 100 amp OCPD
      (b) 1 AWG conductors and 125 amp OCPD
      (c) 1/0 AWG conductors and 125 amp OCPD
      (d) 1/0 AWG conductors and 175 amp OCPD

Raceways, Gutters, Wireways, and Boxes

5

Journeyman and other persons interested in taking the Master electrician's examination must learn to size and select boxes and conduits that will accommodate the number of conductors required to supply power to various pieces of electrical equipment in residential, commercial, and industrial locations. Boxes, in addition to housing conductors, shall have enough fill area to provide sufficient space for receptacles, switches, dimmers, and combination devices without damaging the conductor's insulation. When boxes and conduits are not sized properly, conductors are sometimes packed in undersized boxes and are pinched by large size devices that intrude into the box's fill area. Therefore, conductors, devices, fittings, and clamps are required to be counted and the appropriate box selected. Boxes and conduits are sized on the number of conductors either being the same size and the same characteristics or being different sizes with different characteristics.

Quick Reference

METHODS OF COUNTING CONDUCTORS	5-1
BOXES	5-6
CONDUIT BODIES	5-16
JUNCTION BOXES	5-17
SIZING CONDUITS OR TUBING	5-20
SIZING NIPPLES	5-21
GUTTER SPACE	5-21
AUXILIARY GUTTERS	5-22
PANELBOARDS	5-22
SIZING CABLE TRAYS	5-23
TEST QUESTIONS	5-27

METHODS OF COUNTING CONDUCTORS 314.16(A)(1)

The size box selected is based on the number and size of conductors, devices, and fittings that are contained within the box. The test taker must count the number of items that are in the box and select the size box from **Table 314.16(A)** based upon the maximum number of conductors or the minimum cubic inch capacity rating. The column for the maximum number of conductors is used when the conductors are the same size. The minimum cubic inch capacity column is applied for combination conductors in the same box.

CONDUCTOR FILL
314.16(B)(1)

Conductors passing through the box unbroken and not pulled into a loop or spliced together with scotchlocks are counted as one conductor. Such conductors are passing through boxes supplying power to boxes supporting lighting fixtures or receptacles, etc. **Note:** Spliced conductors in a box count as one each.

See Figures 5-1(A) and (b) for counting the number of conductors passing through or spliced in a box.

Figure 5-1(a). Conductors passing straight through the box are counted as one each.

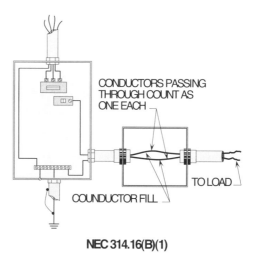

NEC 314.16(B)(1)

Figure 5-1(b). Spliced conductors count as one for each conductor spliced in the box.

Master Test Tip 1: Conductors passing through the box in Figure 5-1(b) count as one each. However, spliced conductors are counted as one each.

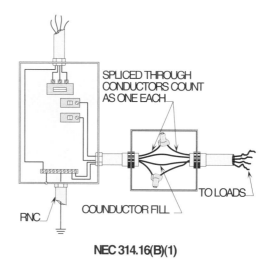

NEC 314.16(B)(1)

CONDUCTOR FILL
314.16(B)(1)

A conductor that is not entirely within the box is not required to be counted as a conductor that is used toward the box fill. A pigtail from a splice within the box to a receptacle or a switch mounted to the box or a bonding jumper from the receptacle yoke bonded to the box is an example of such a conductor. **(See Figure 5-2)**

Raceways, Gutters, Wireways and Boxes

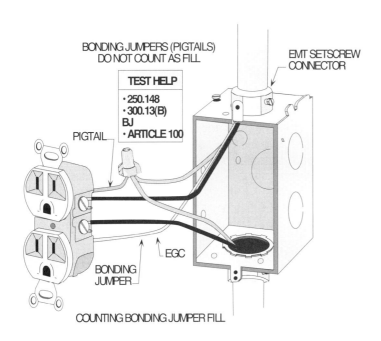

Figure 5-2. A pigtail or bonding jumper does not have to be counted toward the fill space of the box.

Master Test Tip 2: In Figure 5-2, the pigtail and bonding jumper do not count as fill in the box.

Note that conductors looped through box shall be counted twice the minimum length required for free conductors per **300.14** and be counted twice per requirements of **314.16(B)(1)**.

CONDUCTOR FILL
314.16(B)(1), Ex.

Wires from fixtures shall be counted in determining the size of the box. Fixture wire sizes from 14 AWG to 6 AWG shall be counted to determine the size box used to support the fixture. The Exception to **314.16(B)(1)** permits the canopy of a fixture to be used to house four or fewer wires in sizes 18 AWG through 16 AWG, plus an EGC. It is not necessary to count these fixture wires toward the fill of the ceiling box supporting the fixture or ceiling fan where there's a canopy. (See side bar)

Review both the fixture box and fixture canopy and verify if they can be used for fill area. If so, a smaller ceiling box can usually be installed to support the fixture.

See Figure 5-3 for the procedures used to count fixture wires.

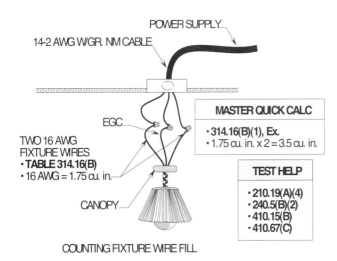

Figure 5-3. Fixture wires must be counted in determining the size box to support a fixture. **Note:** Fixture wires smaller than 14 AWG that are inside a fixture canopy are not required to be counted toward the fill space for sizing a box.

Master Test Tip 3: Four or fewer fixture wires 18 AWG through 16 AWG plus EGC installed in a fixture canopy do not count toward the fill in the box.

CLAMP FILL
314.16(B)(2)

Master Test Tip 4: All cable clamps count as one based upon the largest conductor entering the box.

Boxes with one or more cable clamps that are installed to support cables in the box shall have one conductor added toward the fill computation to determine the size box. The conductor used to represent the one or more cable clamps is selected on the largest conductor in any one cable entering the box. The outer covering of the cable that is used to protect the insulation of the conductors should be extended 1/4 in. past the clamp in the box to prevent the clamp from damaging the insulation of the conductor. When the clamp is tightened to support the cable, this extended covering beyond the clamp will prevent the insulation of the conductor from being pinched or nicked. Arcing or sparking can occur from bare conductors to metal if it becomes grounded and can cause a fire hazard. **(See Figure 5-4)**

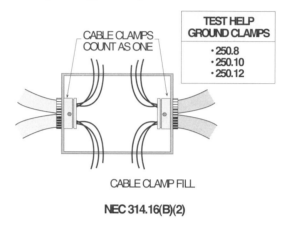

Figure 5-4. Boxes with one or more cable clamps shall have one conductor counted toward the fill space of the box.

SUPPORT FITTINGS FILL
314.16(B)(3)

Master Test Tip 5: Fixture studs and hickeys count as one each based upon the largest conductor entering the box.

Boxes that contain fixture studs or hickeys shall have one conductor added for each fitting. The conductor for each fitting is selected and based on the largest conductor entering the box. By adding a conductor for each fitting in the box, a larger box is required to be selected and helps prevent an undersized box. Conductors in boxes shall have sufficient room in the box per **110.7** and **314.16** to breathe and dissipate heat to prevent the overheating of its insulation per **310.10, FPN's (1) thru (4)**.

For example, a box contains cables of 14-2 AWG and 12-2 AWG w/ground and is equipped with a fixture stud and hickey. The size conductors used for the count of the stud and hickey are determined by each one being counted as a 12 AWG. In other words, the two fittings count as 12 AWG conductors based on the largest conductor entering the box, which is the 12-2 AWG w/ground cable. **(See Figure 5-5)**

DEVICE OR EQUIPMENT FILL
314.16(B)(4)

Master Test Tip 6: A yoke or strap supporting a receptacle, switch, or pilot light counts as two based upon the sized conductors terminating to its terminal.

A receptacle or switch that is mounted on a strap or yoke is counted as two conductors in determining the fill of the box. The count of two is based on the size conductor that is connected to the receptacle or switch. One strap or yoke could have one, two or three receptacles mounted on it and still counted as two conductors. **Article 100** defines a receptacle as a single receptacle having a contact device with no other contact device on the same strap. A multiple receptacle is a single device with two or more receptacles mounted to it. A duplex device has two receptacles, and a despar device may have three or four receptacles mounted on it. The same rules apply to switches mounted to a yoke (strap). **(See Figure 5-6)**

Raceways, Gutters, Wireways and Boxes

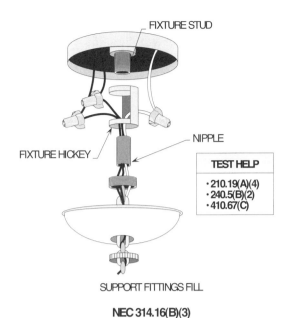

Figure 5-5. Boxes containing fixture studs or hickeys shall have one conductor added for each fitting.

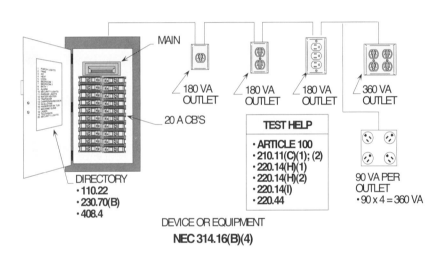

Figure 5-6. The yoke or strap with mounted devices counts as two conductors based on the size conductors connecting to its terminals.

Note that a yoke or strap with four or more receptacles are counted as 90 VA for each device per **220.14(I)**.

EQUIPMENT GROUNDING CONDUCTOR FILL
314.16(B)(5)

Equipment grounding conductors passing through or spliced together in the box count as one. It doesn't matter how many equipment grounding conductors are installed in a box. All the equipment grounding conductors added together count as one. It is a good recommendation to place the equipment grounding conductors into the box first. The equipment grounding conductors may be insulated or bare but are required to be bonded to the box if the box is the metal type. Should the bare copper of one of the equipment grounding conductors come in contact with the metal of the box it would not create a problem because it is bonded to the box anyway per **250.148(A)** and **110.14(B)**.

The only time that equipment grounding conductors in a box count more than one is when an isolation equipment grounding conductor (IEGC) is run to an isolation receptacle and used to ground the metal enclosure of a PC. The isolation equipment grounding conductor is routed with the equipment grounding conductor in the same conduit and bypasses the metal yoke (strap) of the receptacle and connects directly to the

Master Test Tip 7: An isolation equipment grounding conductor may be used to ground the metal of any type of cord-and-plug connected sensitive electronic equipment to reduce electromagnetic noise. Note that an IEGC counts as one conductor toward the fill of the box. **(See Figure 5-7(a))**

metal enclosure of the PC. All the regular equipment grounding conductors count as one plus one for the IEGC for a total of two. See **250.146(D)** and **480.40, Ex.** for the requirements of installing the isolation equipment grounding conductor.

See **Figures 5-7(A) and (b)** for a detailed illustration of installing EGC's and isolation grounding conductors in boxes.

Figure 5-7(a). Equipment grounding conductors passing through or spliced together in the box count as one.

Master Test Tip 8: EGC's count as one except where an isolated EGC is involved, then two conductors must be counted if other EGC's are present.

Figure 5-7(b). One additional conductor shall be added to the count where an isolation EGC is contained in the box.

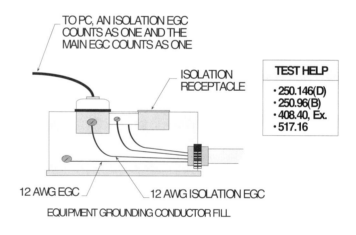

BOXES
314.16(A), Table 314.16(A)

Master Test Tip 9: A yoke is a metal strap used to support a switch, receptacle, or a pilot light or combination of such.

Boxes are usually required to be mounted to support fixtures, devices, fans, and etc. Boxes are available in four main types from **Table 314.16(A)**. The four types most used in electrical wiring systems are octagonal, round, square, or device. The octagonal or round box is used to mount lighting fixtures, ceiling fans, or for splicing conductors. Square boxes are used with plaster rings to mount and support lighting fixtures, ceiling fans, or devices. They are also used for the splicing of conductors.

Device boxes are used to support the yoke (strap) of receptacles, switches, or combinations of receptacles, switches, pilot lights, etc. They are also used to contain splice conductors. **Table 314.16(A)** lists the sizes of these boxes in cubic inches (cu. in.), which can be used for conductors of the same size in ratings from 18 AWG through 6 AWG. See **314.28(A)(1) and (A)(2)** to size boxes for conductors sized 4 AWG and larger.

Raceways, Gutters, Wireways and Boxes

Based upon the number of the same size conductors or cubic inch ratings of combination conductors (18 AWG through 6 AWG), the size box can be determined from **Table 314.16(A)**.

The requirements **of 314.16(A)(1) and (B)(1) thru (B)(5)** shall be applied to compute the number of conductors or cubic inch value of each conductor used to select the size box from **Table 314.16(A)**.

Master Test Tip 10: Pigtails and bonding jumpers do not count as fill in the box.

OCTAGON BOXES
TABLE 314.16(A)

Octagon boxes are used mainly to mount and properly support lighting fixtures. They are also utilized for splicing the conductors of branch-circuits or feeder-circuits supplying power to various pieces of electrical apparatus such as ranges, cooktops, ovens, heating units, AC units, etc. Special octagon boxes that are listed may be used to support ceiling fans. Octagon boxes that are installed to support lighting fixtures shall not support lighting fixtures that weigh more than 50 lb per **410.16(A)**, unless listed to support a greater weight. Ceiling fans that are supported to listed octagon boxes shall weigh 35 lb or less per **422.18**, **314.23(D)** and **314.27(D)**, unless listed to listed to support a greater weight. **(See Figures 5-8(A) and (b))**

Master Test Tip 11: Reference **314.23(D)** and **314.27(D)** for the rules pertaining to the proper methods of supporting boxes to framing members, etc.

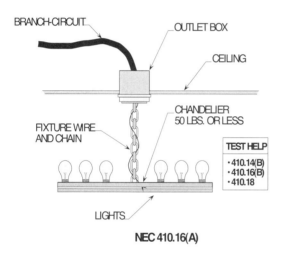

Figure 5-8(a). Outlet boxes shall not support fixtures weighing more than 50 lb unless identified for such use.

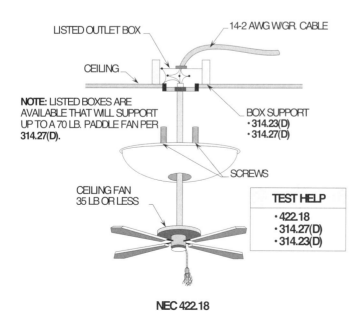

Figure 5-8(b). Outlet boxes shall not support ceiling fans weighing more than 35 lb.

Master Test Tip 12: Boxes that are used to support ceiling fans must be identified for such use per **422.18; (B)**, **314.23(D)**, and **314.27(D)**.

SAME SIZE CONDUCTORS
314.16(B)(1) THRU (5), Table 314.16(A)

Boxes containing conductors that are the same size shall have their total fill space determined by adding the number of conductors plus additional conductors for each fitting or device. The cubic inch value listed in **Table 314.16(B)** for the largest conductor in the box is used to determine the fill capacity for the fittings. For devices, use the size conductors connecting to its terminals.

See **Figure 5-9** for a step-by-step procedure on sizing and selecting octagon boxes where the conductors are the same size.

Figure 5-9. Sections **314.16(B)(1)** thru **(B)(5)** and **Table 314.16(A)** are utilized to size and select an octagon box based on the box containing the same size conductors. **Note:** The canopy is used for the fill of the fixture wires.

Master Test Tip 13: Boxes housing conductors of the same size with the same characteristics can be selected based on the number only per **Table 314.16(A)**. (See **314.16(B)(1), Ex.**)

Example Problem: What size octagon box is required to support a lighting fixture supplied with two 14-2 AWG with ground nonmetallic sheathed cables (romex) with a fixture stud, hickey, and two 14 AWG fixture wires? Neutrals are spliced with a pigtail and romex connections are used instead of romex clamps.

Step 1: Same size conductors
314.16(B)(1) thru (5)
2 - 14 AWG hots = 2
2 - 14 AWG neutrals = 2
2 - 14 AWG grounds = 1
1 fixture stud = 1
1 hickey = 1
1 pigtail = 0
2 romex connectors = 0
2 - 14 AWG fixture wires = 2
Total = 9

Step 2: Selecting box
Table 314.16(A)
9 - 14 AWG conductors require 4" x 2 1/8" box

Solution: A 4 in. x 2 1/8 in. octagon box is required.

EXERCISE PROBLEM - SAME SIZE CONDUCTORS (OCTAGON BOX)

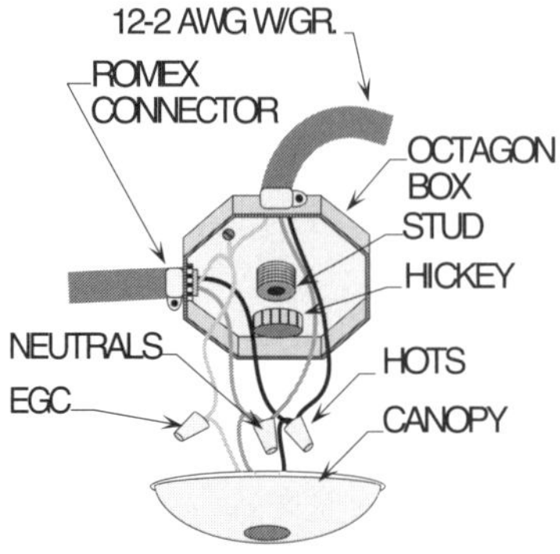

Step 1: Same size conductors
2 - 12 AWG hots = _____ [A]
2 - 12 AWG neutrals = _____ [B]
2 - 12 AWG EGC's = _____ [C]
1 fixture stud = _____ [D]
1 hickey = _____ [E]
2 - 18 AWG fixture wires = _____ [F]
plus EGC = _____ [G]
Total = _____ [H]

Step 2: Selecting box
_____ in. [I] x _____ in. [J]

Solution: The size box is _____ in. [K] x _____ in. [L].

CONDUCTORS ARE THE SAME SIZE
NEC 314.16(B)(1) THRU (B)(5)
NEC 314.16(A)
NEC 314.16(B)(1), Ex.

DIFFERENT SIZE CONDUCTORS
314.16(B)(1) THRU (5); Table 314.16(B)

Master Test Tip 14: Conductors of different sizes or with different types of insulation require the calculation to be based on cu. in. ratings per **Table 314.16(B)**.

Boxes containing conductors that are not the same size are computed on the cubic inch rating of each size conductor listed in **Table 314.16(B)**. The cubic inch rating of each conductor is added together plus the number of fittings or devices. The cubic inch rating used for each fitting is based on the largest conductor entering the box. The cubic inch rating for yokes or straps is based on the size conductor that is connected to the switch or receptacle.

Raceways, Gutters, Wireways and Boxes

Many times the fixture wires are 18 AWG or 16 AWG, which allows more flexibility for terminating such wires to the branch-circuit conductors in the box. Where the fixture wire is a different size than the branch-circuit conductors, the fill of the box is determined by applying the combination calculation per **Table 314.16(B)**.

There are installations where different size conductors are spliced in boxes and fed out of the boxes to supply power to loads. The boxes are used as junction points for the branch-circuits between the panelboard and loads. **See Figure 5-10** for a step-by-step procedure on sizing and selecting octagon boxes where the conductors are different sizes.

Example Problem: What size octagon box is required to support a lighting fixture supplied with two 12-2 AWG with ground nonmetallic sheathed cables (romex) with a fixture stud, two cable clamps, two pigtails, and two 18 AWG fixture wires? (No fixture canopy)

Step 1: Combination conductors
314.16(B)(1) thru (B)(5); Table 314.16(B)
2 - 12 AWG hots
2.25 cu. in. x 2 = 4.5 cu. in.
2 - 12 AWG neutrals
2.25 cu. in. x 2 = 4.5 cu. in.
2 - 12 AWG EGC's
2.25 cu. in. x 1 = 2.25 cu. in.
1 fixture stud
2.25 cu. in. x 1 = 2.25 cu. in.
2 cable clamps
2.25 cu. in. x 1 = 2.25 cu. in.
2 pigtails = 0 cu. in.
2 - 18 AWG fixture wires
1.5 cu. in. x 2 = 3.0 cu. in.
Total = 18.75 cu. in.

Step 2: Selecting box
Table 314.16(A)
18.75 cu. in. requires 4 in. x 2 1/8 in.

Solution: A 4 in. x 2 1/8 in. octagon box is required.

EXERCISE PROBLEM - DIFFERENT SIZE CONDUCTORS (OCTAGON BOX)

12-2 AWG W/GR. OCTAGON BOX WITH ROMEX CONNECTOR
CEILING
EGC — HOT
NEUTRAL
NO CANOPY
FIXTURE WIRES 16 AWG

NEC 314.16(B)(1) THRU (5)
NEC TABLE 314.16(B)

Step 1: Different size conductors
1 - 12 AWG hot = _____ [A]
1 - 12 AWG neutral = _____ [B]
1 - 12 AWG EGC = _____ [C]
2 - 16 AWG fixture wires = _____ [D]
1.75 cu. in. x 2 = 3.5
1 - 16 AWG EGC = _____ [E]
Total cu. in. = _____ [F]

Step 2: Selecting box
_____ in. [G] x _____ in. [H]

Solution: The size box is _____ in. [I] x _____ in. [J].

Figure 5-10. Sections **314.16(B)(1)** thru **(B)(5)** and **Table 314.16(B)** are utilized to size and select the box based on using an octagon box containing different size conductors. The cubic inch rating of each conductor times the number derives the size box.

Master Test Tip 15: Sections **314.16(B)(1)** thru **(B)(5)** and **Table 314.16(B)** are utilized to size and select a square box based on the box containing different size conductors. The box is sized based upon the number of cu. in. rating of each conductor.

SQUARE BOXES
Table 370.16(A)

Square boxes have greater cubic inch ratings than octagon boxes. Square boxes are usually installed where more than four romex cables enter and leave the box. Square boxes may be used with plaster rings, which are designed to accommodate the mounting of a receptacle, switch, light, etc.

The rise on the plaster ring may be 1/4, 1/2, or 3/4 in. to provide more space inside to mount receptacles and switches. This extra space in the box due to the rise of the plaster ring permits a receptacle or switch with a greater depth to be installed. Plaster rings are also used to extend the surface of the box inside the wall or ceiling to the face of the wall or ceiling where a receptacle (device) or luminaire (fixture) may be mounted. Square boxes are utilized as a junction to splice circuit conductors to supply power to various electrical loads in the dwelling unit or building.

SAME SIZE CONDUCTORS
314.16(B)(1) THRU (B)(5); Table 314.16(A)

Section **314.16(A)** lists the items in or out of the box and the number of conductors that shall be added for each item based on the condition of use. Where all conductors are the same size, each item in or out of the box is counted based on these same size conductors.

Figure 5-11. This illustration shows the proper size required to contain conductors of the same size.

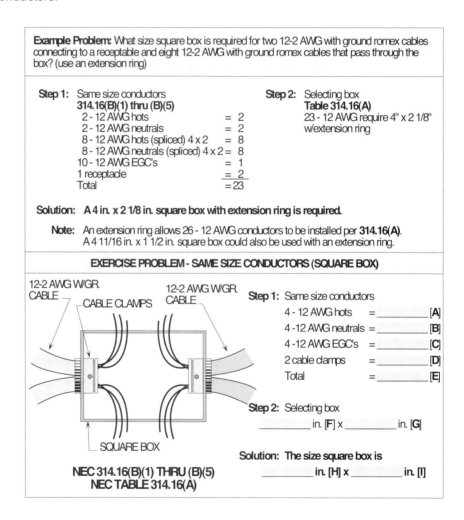

Master Test Tip 16: Sections **314.16(B)(1)** thru **(B)(5)** and **Table 314.16(A)** are utilized to size and select a square box based on the box containing the same size conductors. Note that no extension ring is needed.

DIFFERENT SIZE CONDUCTORS
314.16(B)(1) THRU (5); Table 314.16(B)

Master Test Tip 17: When conductors of different sizes are installed in a box, the cu. in. rating of each conductor must be added together to determine the size box per **Tables 314.16(B)** and **314.16(A)**.

Section **314.16(B)** refers to **Table 314.16(B)**, which requires the cubic inch rating of each conductor in the box to be used for determining the fill space in cubic inches. The total rating is then used for selecting the box size. Section **314.16(B)** refers to **314.16(A)** for selecting the cubic inch rating of each fitting, cable clamp, etc. in the box based on the larger conductor entering the box. The cubic inch rating for each yoke is based on the size conductor connecting to the device terminal.

5-10

See Figure 5-12 for a step-by-step procedure on sizing and selecting square boxes where the conductors are different sizes.

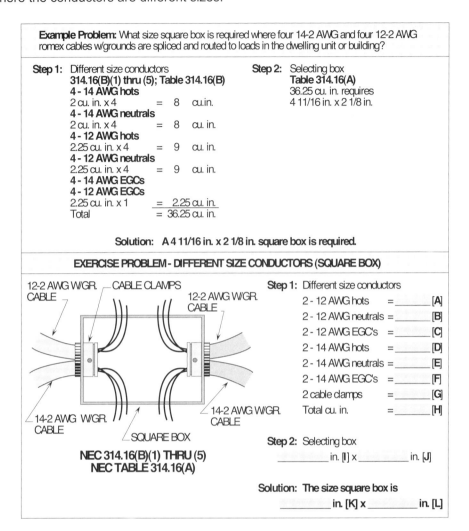

Figure 5-12. Sections **314.16(B)(1)** thru **(B)(5)** and **Table 314.16(B)** are utilized to size and select a square box based on the box containing different-size conductors. The box is sized based upon the number of cu. in. ratings of each conductor.

Note that the use of an extension ring doubles the fill in the box. For example, a 4 in. x 2 1/2 in. square box with an extension ring has a total fill area of 42 cu. in. (21 cu. in. x 2 = 42 cu. in.) per **Table 314.16(A)** and **314.16(A)**.

DEVICE BOXES
Table 314.16(A)

Device boxes can be used to mount switches to control luminaires (lighting fixtures) and other electrical loads. They also are used to mount receptacles for the cord-and-plug connection of electrical appliances. The depth space (fill) in the box must be sized to accommodate the size device installed. A dimmer switch (device) will take up more depth space in the box than a regular-sized toggle switch. A GFCI receptacle in most cases takes up more depth (space) for installations than other size devices. Testing agencies require test takers to be capable of selecting the proper size device box to accommodate not only conductors and fittings but also the depth of the device.

Master Test Tip 18: Device boxes can be used to support toggle switches, receptacles, dimmers, etc., with the aid of a plaster ring mounted to a round or square box.

SAME SIZE CONDUCTORS
314.16(B)(1) THRU (5); Table 314.16(A)

This section contains specific rules that shall be applied where the device box has the same size conductors with the various types of fittings, devices, etc. all present in the box. Devices are required to be counted as two conductors to allow more fill (space) to accommodate the different-size devices that are available in the industry today. The count of two conductors is based upon the size conductors that are connected to the terminals of the device that is mounted on the yoke (strap).

See Figure 5-13 for a step-by-step procedure for sizing and selecting device boxes where the conductors are the same size in the box.

Figure 5-13. Sections **314.16(B)(1)** thru **(B)(5)** and **Table 314.16(A)** are utilized to size and select a device box based on the box containing the same size conductors.

Master Test Tip 19: Note that set-screw or compression connectors are not required to be counted as fill in the calculation of the box.

Note that bonding jumpers (BJ's) and pigtails in the box do not count toward the fill of the box.

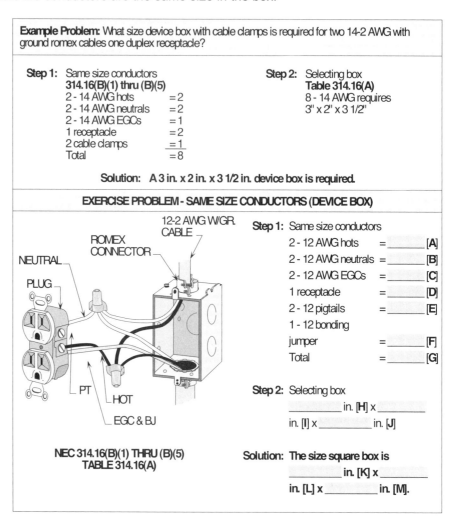

DIFFERENT SIZE CONDUCTORS
314.16(B)(1) THRU (B)(5); Table 314.16(B)

Section **314.16(B)** requires the cubic inch rating of each conductor from **Table 314.16(B)** to be utilized where conductors are not the same size in the box. The cubic inch rating of each conductor is multiplied by the number in the box, and the total cubic inch value is used to determine the size device box from **Table 314.16(A)**.

See Figure 5-14 for a step-by-step procedure on sizing and selecting device boxes where the conductors in the box are different sizes.

OTHER BOXES
314.16(A); 314.16(B)

Master Test Tip 20: If the cu. in. ratings of conductors exceed **Table 314.16(A)** with the use of an extension ring, a junction box must be used.

Other boxes are boxes that are not listed in **Table 314.16(A)**. They are junction boxes with greater fill (space) area that allows more conductors to be routed straight through or spliced in the box. Junction boxes come in many sizes and are built to house the different-size conductors in **Table 314.16(A)**. If a box size selected from **Table 314.16(A)**

Raceways, Gutters, Wireways and Boxes

with an extension ring will not accommodate the number of conductors ranging from 18 AWG to 6 AWG installed in the box, then a junction box must be used. See chart on the back inside cover of this book for the standard-size junction boxes used to contain spliced conductors.

Example Problem: What size device box is required to contain a 14-2 AWG nonmetallic sheathed cable, two 12-2 AWG nonmetallic sheathed cables passing through with one receptacle present? There is also one pigtail and one bonding jumper in the box (romex connectors are used)

Step 1: Different size conductors
314.16(B)(1) thru (B)(5); Table 370.16(B)
1 - 14 AWG hot
2 cu. in. x 1 = 2 cu.in.
1 - 14 AWG neutral
2 cu. in. x 1 = 2 cu. in.
2 - 12 AWG hots
2.25 cu. in. x 2 = 4.5 cu. in.
2 - 12 AWG neutrals
2.25 cu. in. x 2 = 4.5 cu. in.
1 receptacle
2 cu. in. x 2 = 4 cu. in.
2 - 12 EGC
2.25 x 1 = 2.25 cu. in.
1 pigtail = 0
1 bonding jumper = 0
Total = 19.25 cu. in.

Step 2: Selecting box
Table 314.16(A)
19.25 cu. in. requires 3" x 2" x 3 1/2" with extension ring.

Note: An extension ring is used due to the number of conductors in the box.

Solution: A 3 in. x 2 in. x 3 1/2 in. device box is required with extension ring.

EXERCISE PROBLEM - DIFFERENT SIZE CONDUCTORS (DEVICE BOX)

12 AWG NEUTRAL
PLUG
METAL BOX
ALUMINUM WIRING
METAL CONDUIT W/2 EGC'S
14 AWG EGC
12 AWG HOT
12 AWG EGC AND PT
3 - 14 AWG PASSING THROUGH

NEC 314.16(B)(1) THRU (B)(5)
TABLE 314.16(A)

Step 1: Different size conductors
1 - 14 AWG hot = _____ [A]
1 - 14 AWG neutral = _____ [B]
1 - 12 AWG hot = _____ [C]
1 - 12 AWG neutral = _____ [D]
1 - 12 AWG EGC = _____ [E]
1 receptacle = _____ [F]
1 - 14 AWG EGC = _____ [G]
Total cu. in. = _____ [H]

Step 2: Selecting box
_____ in. [I] x _____ in.
[J] x _____ in. [K]

Solution: The size device box is _____ in. [L] x _____ in. [M] x _____ in. [N]

Figure 5-14. Sections 314.16(B)(1) thru (B)(5) and Table 314.16(B) are utilized to size and select a device box based on the box containing different-size conductors.

Master Test Tip 21: Bonding jumpers and pigtails are not counted as fill when computing the fill space of boxes.

Note that extension rings mounted to boxes can be used to add extra fill area to the box.

Master Test Tip 22: If the number of conductors in a box exceed those permitted in **Table 314.16(A)**, a junction box with a greater rating must be used.

SAME SIZE CONDUCTORS
314.16(B)(1) THRU (B)(5); 314.16(A)

Because there are not any tables available with the number of conductors permitted in junction boxes for conductors that are not the same size, a calculation applying the cubic inch value of each conductor must be done. By multiplying the cubic inch rating of each conductor by the number in the box and using this total cubic inch rating, the proper size junction box may be selected. The cubic inch rating of each conductor is found in **Table 314.16(B)** for conductors 18 AWG through 6 AWG. The cubic inch fill space of a junction box is found by multiplying the dimensions of the box.

For example, a 4 in. x 4 in. x 4 in. junction box has a cubic inch (cu. in.) rating of 64 cu. in. (4" x 4" x 4" = 64 cu. in.). This box will contain any arrangement of conductors not exceeding 64 cu. in. in total rating.

Note that boxes housing conductors 4 AWG and larger must be sized by the rules listed in **314.28(A)(1) and (A)(2)** or junction boxes must be used per chart on the back inside cover of this book.

When an existing panelboard of a remodeling job is used for a junction box, it must comply with **240.24(D)** (also see **240.24(E)**), which prohibits a panelboard with OCPD's in a clothes closet around combustible material. Section **230.70(A)(2)** prohibits service disconnecting means to be located in a bathroom of dwelling units, commercial, or industrial locations. **See Figure 5-15** for a step-by-step procedure for sizing and selecting junction boxes with the same size conductors.

Figure 5-15. For sizing and selecting a junction box to contain the same size conductors, see chart for pull and junction boxes on inside back cover.

Example Problem: What size junction box is required to contain twelve 12-2 AWG w/ground, eight 12-2 AWG w/ground, and four 12-2 AWG w/ground cables where all the conductors in the box are spliced?

Step 1: Different size conductors
314.16(B)(1) thru (B)(5); Table 314.16(B)
12 - 12 AWG hots (spliced)
2.25 cu. in. x 12 = 27.0 cu. in.
12 - 12 AWG neutral (spliced)
2.25 cu. in. x 12 = 27.0 cu. in.
12 - 12 AWG EGCs (spliced)
2.25 cu. in. x 1 = 2.25 cu. in.
8 - 12 AWG hots (spliced)
2.25 cu. in. x 8 = 18.0 cu. in
8 - 12 AWG neutrals (spliced)
2.25 cu. in. x 8 = 18.0 cu. in.
8 - 12 AWG EGs
2.25 cu. in. x 0 = 0 cu. in.
4 - 12 AWG hots (spliced)
2.25 cu. in. x 4 = 9.0 cu. in.
4 - 12 AWG neutrals (spliced)
2.25 cu. in. x 4 = 9.0 cu. in.
4 - 12 AWG EGCs (spliced)
2.25 cu. in. x 0 = 0 cu. in.
Total = 110.25 cu. in.

Step 2: Selecting box
314.16(A)
6" x 6" x 4" box = 144 cu. in.
144 cu. in. will contain 110.25 cu. in.

Solution: A 6 in. x 6 in. x 4 in. junction box is required.

EXERCISE PROBLEM - SAME SIZE CONDUCTORS (OTHER BOXES)

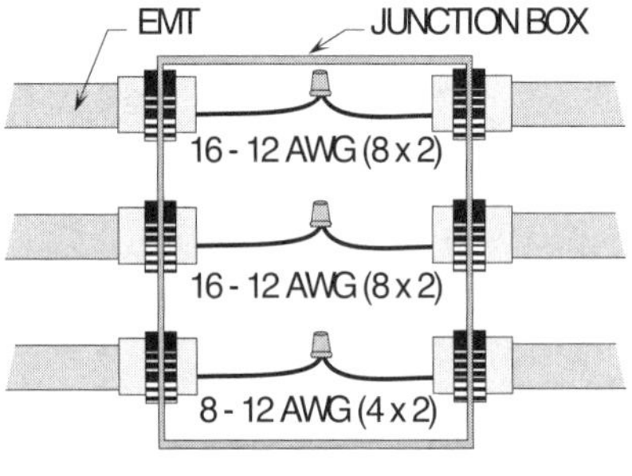

Step 1: Same size conductors
8 - 12 AWG conductors = _____ [A]
8 - 12 AWG conductors = _____ [B]
8 - 12 AWG conductors = _____ [C]
8 - 12 AWG conductors = _____ [D]
4 - 12 AWG conductors = _____ [E]
4 - 12 AWG conductors = _____ [F]
Total cu. in. = _____ [G]

Step 2: Selecting box
_____ in [H] x _____ in. [I]
x _____ in [J]

Solution: The size junction box is _____ in. [K]
x _____ in. [L] x _____ in. [M]

EMT — JUNCTION BOX
16 - 12 AWG (8 x 2)
16 - 12 AWG (8 x 2)
8 - 12 AWG (4 x 2)
NEC 314.16(B)(1) THRU (B)(5)
TABLE 314.16(B)

Master Test Tip 23: Note that all openings in boxes such as knockouts must be sealed per **314.17(A)**.

DIFFERENT-SIZE CONDUCTORS
314.16(B)(1) THRU (B)(5); 314.16(B); 314.16(A)

Master Test Tip 24: Openings for conduits containing conductors used for entry into boxes must be adequately closed per **314.17(A)**.

To select the proper size junction box housing different-size conductors, the cubic inch rating of each conductor from **Table 314.16(B)** must be selected and multiplied by the number of conductors for each cubic inch rating. The total calculation of cubic inch ratings is used to select the proper size junction box.

See Figure 5-16 for a step-by-step procedure for sizing and selecting junction boxes with different-size conductors.

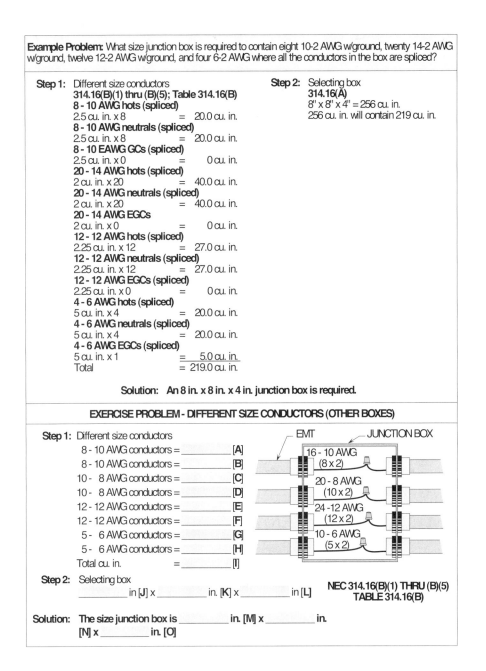

Figure 5-16. Sizing and selecting a junction box to contain different-size conductors (see chart on inside back cover).

Master Test Tip 25: Boxes installed in noncombustible material are allowed to be set back from the finished surface no more than 1/4 in. per **314.20**.

PLASTER RINGS AND EXTENSION RINGS
314.16(A)

Plaster rings and extension rings are accessories that are used in conjunction with octagon, round, square, or device boxes to extend the face of the box to mount luminaires (fixtures), ceiling fans, or devices, or to provide more fill space in the box.

For example, the face of an octagon box needs to be extended 1/2 in. to the surface of a sheet-rock ceiling. A 1/2 in. plaster ring with provisions for mounting a luminaire (lighting fixture) may be used to extend the face of the octagon box to the surface edge of the sheet-rock.

Extension rings can be used to extend the face of boxes a greater distance than plaster rings, because plaster rings only come with limited extension height to extend the face of a box to a given height.

Master Test Tip 26: In most cases, when extra fill space is needed, an extension ring can be used to double the fill area of a box.

For example, one or two extension rings may be used to extend the face of a box to the surface edge of a wall or ceiling where a plaster ring is mounted and a device, luminaire (fixture), etc. is mounted to the plaster ring. Most testing agencies interpret **314.22** in the NEC to allow one extension ring to be used to provide additional fill space in the box. Check with the local testing authority for an interpretation of how to apply this rule.

The reason for this requirement of limiting the number of extension rings for fill space is to allow access to the inside box area for servicing the spliced conductors. **(See Figure 5-17)**

Figure 5-17. Increasing the fill space in a box using an extension ring.

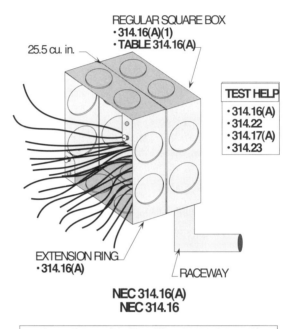

Master Test Tip 27: In complete installations, each box must be provided with a cover, faceplate, or luminaire (fixture) canopy per **314.25**.

CONDUIT BODIES
314.16(C)

Master Test Tip 28: Conduit bodies shall not contain splices, taps, or devices unless they are identified for such use.

Conduit bodies containing conductors 6 AWG and smaller shall not be less than twice the cross-sectional area of the largest conduit to which it is connected. The number of conductors that are permitted in conduit bodies is determined by the square inch area per **Table 5** and **Table 4** of **Chapter 9**.

Conduit bodies with fewer conduit entries shall not contain splices, taps, or devices unless they comply with **314.16(B)** and are supported properly. **(See Figure 5-18)**

Raceways, Gutters, Wireways and Boxes

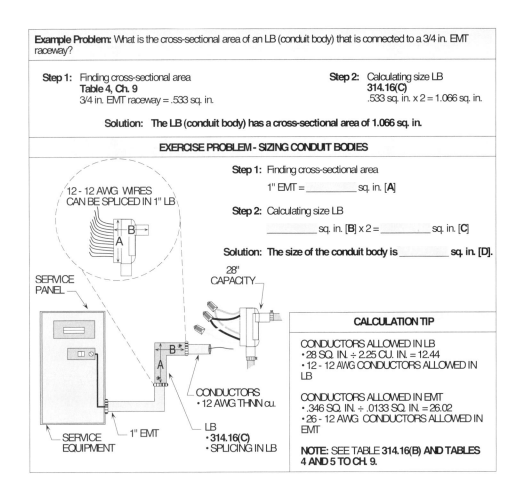

Figure 5-18. Conduit bodies containing 6 AWG or smaller conductors shall be at least twice the cross-sectional area of the largest conduit that it is connected to. This rule is designed to provide proper room in the LB.

Note that the sq. in. area of the 1 in. EMT and 12 AWG conductors are found in **Tables 4** and **5** to **Ch. 9** of the NEC.

Note that the 1 in. LB is listed and approved to contain splices, but only 12 can be spliced and not the 26 that the 1 in. EMT is capable of containing.

JUNCTION BOXES
314.28

Junction boxes housing conductors 4 AWG and larger that are pulled through raceways are sized based upon a straight pull or an angle pull.

Master Test Tip 29: In junction boxes having dimension over 6 ft., conductors must be cabled or racked up in an acceptable and approved manner per 314.28(B).

STRAIGHT PULLS
314.28(A)(1)

A straight pull (straight through) is where the raceway is connected to one side of the box and another raceway is connected to the opposite side, with the junction box used as a junction between the raceways. Junction boxes utilized in a straight pull shall be sized by multiplying the largest raceway in the run by the multiplier 8 per **314.28(A)(1)**.

Sometimes junction boxes will have more than two raceways connected to their sides (walls). Junction boxes with more than two raceways connected to their sides shall be sized by multiplying the largest raceway by 8.

See Figures 5-19(A) and (b) for a step-by-step procedure for sizing junction boxes for straight pulls.

Note that **110.70 thru 110.79** pertain to manholes. Questions and problems about manholes may be on the test; review these sections very carefully before taking the test.

Figure 5-19(a). Sizing a junction box based upon a straight pull.

For questions and problems pertaining to manholes see **110.70** thru **110.79**.

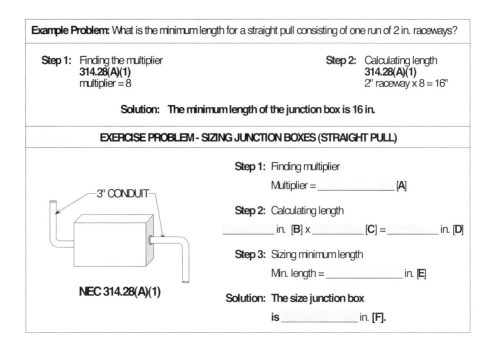

Figure 5-19(b). Sizing a junction box based upon a straight pull.

Master Test Tip 30: If permanent barriers are installed in junction boxes, each section is considered as a separate box per **314.28(D)**.

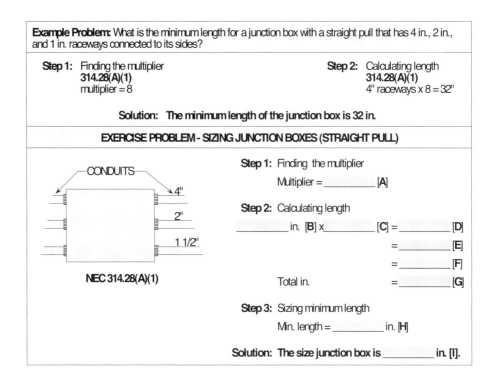

ANGLE PULLS
314.28(A)(2)

Note that junction boxes must be provided with compatible features and be suitable for the conditions of use in which they are installed. (See **314.28(C)**)

An angle pull is where the raceways enter and leave the junction box from the top wall to one of the side walls or from one of the side walls to the bottom wall. Junction boxes in angle pulls shall be sized by multiplying the largest raceway in the run by the multiplier 6 per **314.28(A)(2)**. A U pull is where the raceway leaves the same wall and forms a U pull configuration.

5-18

Raceways, Gutters, Wireways and Boxes

Junction boxes having more than one run of raceways connecting to their walls to form an angle pull shall be sized by multiplying the largest raceway by 6 and adding the remaining raceways to this value.

See Figures 5-20(A) and (b) for a step-by-step procedure for sizing junction boxes for angle pulls.

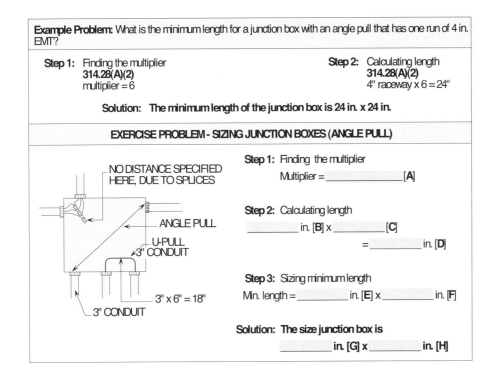

Figure 5-20(a). The size junction box for an angle pull with one conduit per wall can be found by multiplying the largest conduit by 6. Calculations will determine the size junction box that is needed.

Note that boxes containing splices do not appear to have a specific distance for spacing conduits. (See Exercise Problem in Figures 5-20(A) and (B))

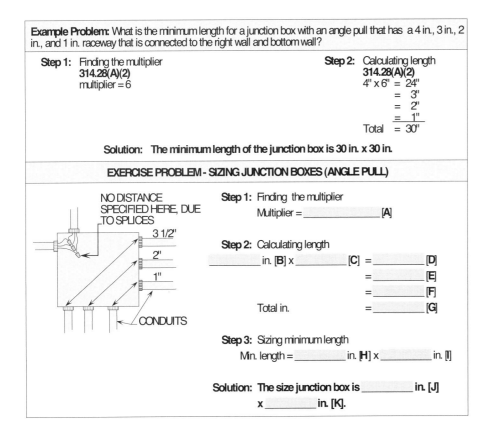

Figure 5-20(b). The size junction box for an angle pull can be found by multiplying the largest conduit by 6 and adding this number to the remaining conduits. Computations will determine the size junction box that is needed.

Master Test Tip 31: In calculating the size junction box, each row must be computed separately and the row producing the maximum distance must be used.

5-19

Master Test Tip 32: Conduits or tubing enclosing two or more conductors are allowed a 40 percent fill area per **Table 1** to **Chapter 9**.

SIZING CONDUITS OR TUBING
APPENDIX C

Conduits or tubing is used to enclose conductors to supply power to various types of electrical equipment. The same size conductors with the same type of insulation may be pulled through raceways. They may also be routed through conduits with different sizes and types of insulation. Tables in **Chapter 9** can be used to size conduits based on the types of insulation and size of conductors that are pulled through the conduit system.

ENCLOSING THE SAME SIZE CONDUCTORS
TABLES C1 THRU C12(A), APPENDIX C

Note that raceways can be classified as a raceway, nipple, or sleeve depending on their length as follows:
 (1). Raceways are over 24 in. in length.
 (2). Nipples are 24 in. or less in length.
 (3). Sleeves are 18 in. to 10 ft. in length.
See **Note 4** to **Table 1** in **Chapter 9** and **312.5(C), Ex.**

Raceways with conductors that have the same type of insulation and are the same size may be selected by using **Tables C1 through C12(A), to Appendix C**.

The procedure for selecting the size of the raceway is to take the size of the conductors and align them in the appropriate column of a particular Table and select the proper raceway based on the type insulation and number of conductors. This is the easiest and fastest method to use in determining the size of raceways housing the same size conductors.

See **Figure 5-21** for a step-by-step procedure for sizing raceways enclosing the same size conductors.

Figure 5-21. Tables C1 thru C12(A) in **Appendix C** may be used to size the conduit or tubing for enclosing the same size conductors.

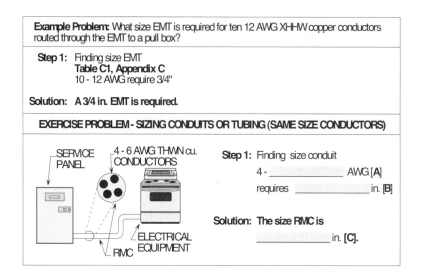

Master Test Tip 33: Raceways enclosing 2 conductors shall be permitted to have a 31 percent fill area per **Table 1** to **Chapter 9**.

ENCLOSING DIFFERENT SIZE CONDUCTORS
TABLE 5, TO CH. 9

Note that multiconductor cables having two or more conductors are calculated as a single conductor per **Note (9)** to **Table 1** in **Chapter 9**.

Conduits or tubing enclosing different size conductors are sized by selecting the square inch area of each conductor per **Table 5** and multiplying by the number. This total is used to select the size raceway per **Table 4** in **Chapter 9**. (See **Note 6** to **Table 1** of **Chapter 9**)

See **Figure 5-22** for an illustrated procedure for sizing raceways with different size conductors.

Raceways, Gutters, Wireways and Boxes

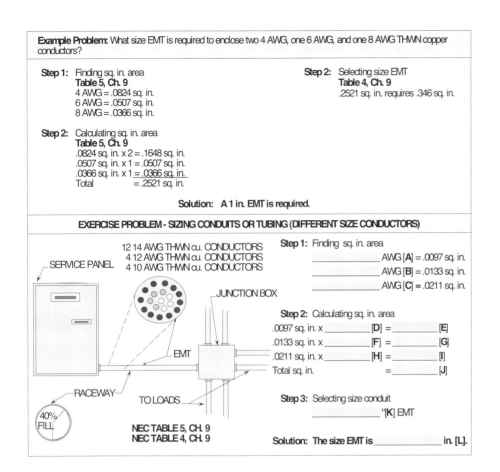

Figure 5-22. Table 5, Ch. 9 and Table 4, Ch. 9 can be used to size the conduit or tubing for enclosing different size conductors.

Master Test Tip 34: If three conductors are pulled through a conduit, jamming during the pull should be considered as outlined in Note (4) to Table 1 in Chapter 9.

SIZING NIPPLES
NOTE (4) TO TABLES, CH. 9

The difference between a nipple and a conduit is that a nipple is 24 in. or less in length; a conduit system is any length over 24 in.

A conduit system with more than two conductors is allowed 40 percent fill per **Table (4), Chapter 9**, while a nipple is permitted to have 60 percent fill per **Note (4) to Table 1 in Chapter 9**.

Nipples are sized by finding the square inch area of each conductor and multiplying this value by the number. The total 100 percent fill in square inches in **Table 5, Chapter 9** is selected based upon the size conduit, and this value is multiplied by 60 percent. If the total square inch area of all the conductors is less than or equal to the total of the square inch area produced by applying the 60 percent factor, the size of the nipple is selected based upon that conduit used in the calculation.

See Figure 5-23 for an illustrated procedure for sizing nipples when applying the 60 percent fill rule.

Master Test Tip 35: When determining ampacities of conductors in a nipple, the derating factors of Table 310.15(B)(2)(a) do not have to be applied for the total number of conductors exceeding three per Note (4) to Table 1 in Chapter 9.

Note that a conduit sleeve is 18 in. to 10 ft. in length, with one end connected to a box etc. and the other end having a protective bushing per 312.5(C), Ex.

GUTTER SPACE
312.7; 366.22; 376.22

Gutter space is used to enter or leave enclosures that enclose conductors that are connected to equipment terminals or tapped or spliced to serve other loads. Certain clearances and space requirements to protect the conductors and prevent overcrowding is essential per **312.7**, **366.22** and **376.22**.

Figure 5-23. Table 5, Ch. 9 and Table 4, Ch. 9 and Note (4) to Table 1, Ch. 9 are Tables that may be used to calculate the size of nipples.

Master Test Tip 36: For a nipple to be considered a nipple, it must be 24 in. or less in length per **Note (4)** to **Table 1, Chapter 9**.

Master Test Tip 37: Derating for four or more current-carrying conductors is not required for nipples per **Note (4)** to **Table 1, Chapter 9**.

Master Test Tip 38: Note that if a raceway was used instead of a nipple, the size EMT would be 1 1/4 in. per **Table C1** to **Appendix C**. A raceway is defined as over 24 in. in length.

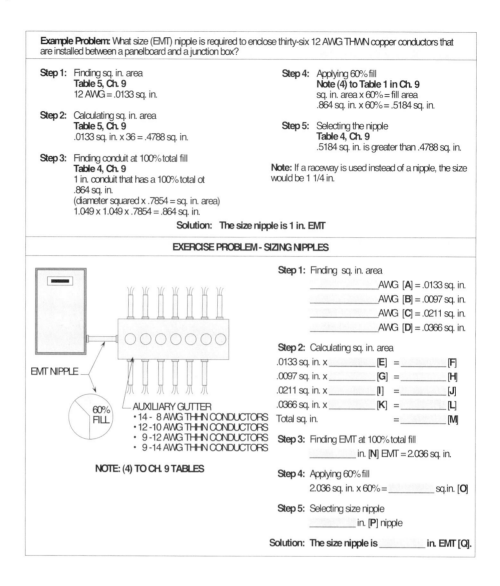

Master Test Tip 39: Auxiliary gutters and wireways are allowed to have a 20 percent fill area per **366.22(A)**.

Master Test Tip 40: Auxiliary gutters and wireways can contain 30 or less current-carrying conductors without applying derating factors per **366.22(A)**.

AUXILIARY GUTTERS
366.22(A)

The number of current-carrying conductors permitted in an auxiliary gutter without derating per **Table 310.15(B)(2)(a)** to **Tables 310.16 thru 310.19** is 30 or less per **312.5(A)**.

The size of an auxiliary gutter is determined by dividing the total square inch area of the conductors by 20 percent fill area. The total square inch area of the conductors is found by multiplying the square inch area per **Table 5, Ch. 9** of each conductor, which is based on the size and insulation of each conductor placed in the auxiliary gutter.

See Figure 5-24 for a detailed procedure for designing, sizing, and selecting auxiliary gutters.

PANELBOARDS
408.3(F)

Panelboards must comply with the provisions of **408.3(F)**, which requires the minimum gutter space in the panel to meet the clearances in **Table 312.6(A)** for L-bends (minimum width) and Table **312.6(B)** for S- or Z-bends (maximum width).

Raceways, Gutters, Wireways and Boxes

Electricians must enter panelboards using an L, S, or Z-bend so that there is adequate room to terminate the conductors without damaging the insulation, which later could create a ground-fault condition and cause a power failure.

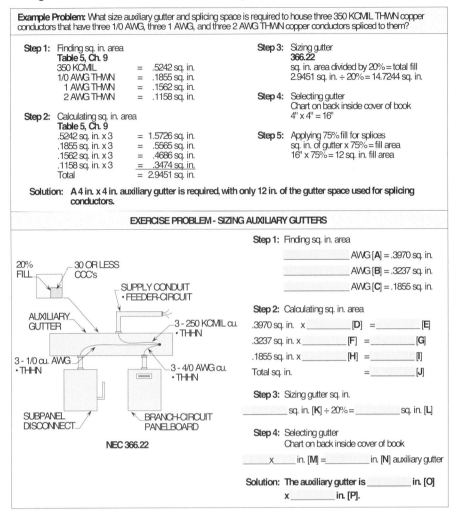

Figure 5-24. Table 5, Ch. 9 and Table 4, Ch. 9 and 366.22 are used to calculate the size auxiliary gutters.

Master Test Tip 41: Splices and taps may take up to 75 percent of the gutter or wireway fill area.

Note that enclosures housing OCPD's under certain conditions can have additional conductors feeding through or tapping off to other switches per **312.8**.

Panelboards utilizing S or Z-bends require greater clearances between the lugs and enclosure (walls), if such conductors are terminated using such bends. S or Z-bends are harder to handle, bend, shape, and make ready for termination. L-bends, due to their shape, are a lot easier to handle and connect than S or Z-bends.

If the lugs in the panelboard are removable, the clearance can be reduced from the dimension in **Table 312.6(B)** based on the number and size of conductors connected to each lug.

See Figures 5-25(A) and (b) for an illustrated procedure for determining the minimum clearance of conductors between lugs, enclosures and sides (walls) in a panelboard.

Master Test Tip 42: Single conductor cables rated 1/0 AWG thru 4/0 AWG, if installed in cable trays, must have a rung spacing of 9 in. or less.

SIZING CABLE TRAYS
ARTICLE 392

Cable trays can be used for the installation of a number of cables that are installed in a single run from one location to another. Steel or aluminum cable tray sections are used with the ends bolted together to form a single run of the required length. Cables routed in cable trays are considered supported where they are laid on racks, troughs, or hangers.

Master Test Tip 43: To select a proper size cable tray, match **392.10(A)(2)** to column 1 of **Table 392.10(A)** and pick the correct sq. in. area and cable tray width.

See **Figure 5-26** for an exercise problem on sizing cable tray.

Figure 5-25(a). Panelboards must comply with the provisions of **408.3(F)**, which requires the minimum gutter space in the board to meet the clearance rules in **Tables 312.6(A)** and **(B)** for L, S, or Z-bends.

Master Test Tip 44: Table 312.6(A) applies where the conductors do not enter or leave the enclosure through the wall opposite its terminal. Note that this space is the narrow width per **408.55**. (See Master Test Tip 45 below)

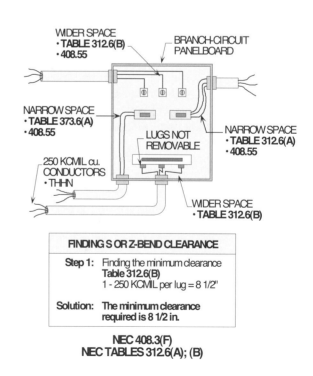

Figure 5-25(b). To prevent damaging conductors because of over bending, panelboards must comply with the provisions of **408.3(F)**, which requires the minimum gutter space in the board to meet the clearance rules in **Tables 312.6(A)** and **(B)** for L, S, or Z-bends.

Master Test Tip 45: Table 312.6(B) is used where the conductor enters or leaves the enclosure opposite its terminal. Note that this space is the wider width per **408.55**. (See Master Test Tip 44 above)

Note that test takers must review the **Ex.'s** to **408.55** before taking the Master examination.

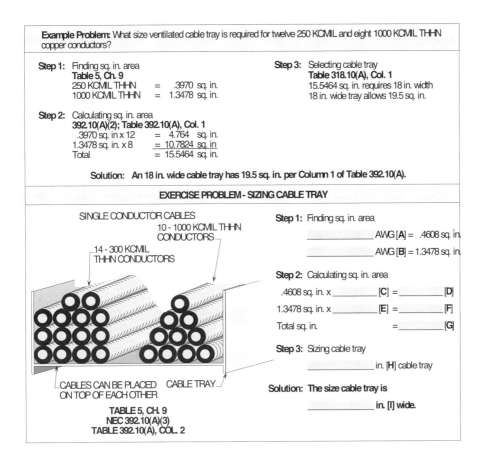

Figure 5-26. When installing cables in ladder or ventilated-trough trays, the total diameter of all cables 1000 KCMIL or larger, and less than 1000 KCMIL, must not exceed the width of the cable tray.

Master Test Tip 46: To select the size cable tray in the exercise problem, **392.10(A)(3)** must be aligned to **392.10(A)(3)** in Column 2 to Table **392.10(A)** of the NEC.

Note that for derating conductors installed in cable trays, see **392.11(A) thru (B)**.

Note that for sizing cable trays containing control and/or signal cables only, see **392.9(B) thru (D)**.

Note that a 1.00 sq. in. area metal cable tray will clear a 400 OCPD per **Table 392.7(B)** and **Table 250.122** of the NEC.

5-25

Name _____   Date _____

Chapter 5
Raceways, Gutters, Wireways, and Boxes

Section   Answer

1. Conductors passing through the box unbroken and not pulled into a loop or spliced together with scotchlocks shall be counted as two conductors each when determining the fill of the box.

    (a) True   **(b)** False

2. One or more equipment grounding conductors passing through or spliced together in the box shall be counted as one in determining the fill of the box.

    (a) True   **(b)** False

3. Usually, a conduit system is any run that is over 24 in. in length.

    (a) True   **(b)** False

4. A nipple shall be permitted to have a 40 percent fill area.

    (a) True   **(b)** False

5. A nipple, per the NEC, is 36 in. or less in length.

    (a) True   **(b)** False

6. Nails or insulated covered screws shall be permited to mount boxes where they pass through the inside of the box within 1/4 in. of the back or ends of the box.

    (a) True   **(b)** False

7. Boxes with threaded enclosures or hubs shall be considered supported by two or more conduits that are threaded wrench tight into the enclosure or hubs.

    (a) True   **(b)** False

8. If supported within 24 in. on two or more sides, enclosures with devices do not require additional supports.

    (a) True   **(b)** False

9. AC cables installed in attics above 7 ft. shall not be required to be protected.

    (a) True   **(b)** False

10. NM cables run within 5 ft. of a scuttle hole shall have guard strips if they are not routed on the sides of the ceiling joists.

    (a) True   **(b)** False

11. Generally, ceiling fans that are supported to listed octagon boxes shall not weigh more than _____ lbs.

    (a) 25   **(b)** 35   **(c)** 40   **(d)** 50

Section   Answer

12. Lighting fixtures that are supported by a octagon box shall not weigh more than:

    (a) 25 lbs.   (b) 35 lbs.   (c) 40 lbs.   (d) 50 lbs.

13. When cutting and threading rigid metal conduit in the field, a standard cutting die with a _____ in. taper per foot shall be used.

    (a) 1/4   (b) 1/2   (c) 3/4   (d) 1

14. EMT shall be supported within _____ ft. of each outlet box.

    (a) 3   (b) 4 1/2   (c) 5   (d) 10

15. Rigid metal conduit shall be supported every _____ ft. (General Rule).

    (a) 3   (b) 5   (c) 6   (d) 10

16. Flexible metal conduit shall be supported within _____ in. of each outlet box.

    (a) 6   (b) 12   (c) 18   (d) 24

17. Cable systems that are not run through the center of the framing member can be placed in a cut notch and protected by a steel plate (no Ex. applied) of _____ in.

    (a) 1/16   (b) 1/8   (c) 3/16   (d) 1/4

18. RNC shall be supported at each box and enclosure within _____ ft.

    (a) 3 1/2   (b) 3   (c) 5   (d) 6

19. AC cables installed in attics shall be protected by their conditions of use at or below _____ ft.

    (a) 6   (b) 8   (c) 7   (d) 10

20. NM cables secured to the face or placed and supported on running boards shall not be smaller than _____ AWG copper.

    (a) 10-3   (b) 8-3   (c) 6-3   (d) 4-3

21. What size octagon box with cable clamps is required for 2 - 12 AWG and 2 - 10 AWG w/ground NM cables that are spliced in the box between the panelboard and the loads?

    (a) 4 in. x 1 1/8 in. octagon box   (b) 4 in. x 1 1/4 in. octagon box
    (c) 4 in. x 1 1/2 in. octagon box   (d) 4 in. x 2 1/8 in. octagon box

22. What size junction box is required to contain four raceways with 10 - 12 AWG in each and two raceways with 8 - 12 AWG in each that are spliced together?

    (a) 6 in. x 4 in. x 4 in. junction box   (b) 6 in. x 6 in. x 4 in. junction box
    (c) 6 in. x 6 in. x 6 in. junction box   (d) 8 in. x 6 in. x 4 in. junction box

	Section	Answer

23. What is the minimum length of a junction box for an angle pull that has a 3 1/2 in., 2 in., and 1 in. raceway that is connected to the right wall and bottom wall?

    (a) 20 in. x 20 in.   **(b)** 24 in. x 24 in.
    (c) 26 in. x 26 in.   **(d)** 30 in. x 30 in.

24. What size EMT nipple is required to enclose 10 - 14 AWG, 6 - 12 AWG, and 6 - 10 AWG, and twelve 8 AWG THWN copper conductors?

    (a) 1 1/4 in. EMT nipple   **(b)** 1 1/2 in. EMT nipple
    (c) 2 in. EMT nipple   **(d)** 2 1/2 in. EMT nipple

25. Where a conductor loops through a box and is measured from the fitting in the box the loop shall be _____ inches.

    (a) 6   **(b)** 8   **(c)** 10   **(d)** 12

Feeder-Circuits and Branch-Circuits

6

Feeder-circuits are the conductors between the service equipment or the source of a separately derived system, and the final overcurrent protection devices protecting the branch-circuit conductors and equipment. Feeder-circuits are installed to supply subpanels in dwelling units, apartments, and commercial and industrial occupancies. Feeder-circuits are also installed in commercial and industrial occupancies and routed through the plant, where taps can be made to the feeder conductors to supply panelboards, disconnect switches, control centers, and other types of electrical equipment.

Branch-circuit conductors extend between the final overcurrent protection device protecting the circuit conductors and supplying power to equipment and outlets. The equipment supplied is either hard-wired or cord-and-plug connected to properly designed receptacle outlets.

Test takers must be very familiar with the requirements of feeder-circuits and branch-circuits because the Master electrician examination is loaded with questions and problems pertaining to the application of such rules.

Quick Reference

LOADS	6-1
CALCULATING AMPS	6-2
POWER FACTOR	6-2
NONCONTINUOUS OPERATED LOADS	6-4
CONTINUOUS OPERATED LOADS	6-5
DEMAND FACTORS	6-5
NEUTRAL	6-6
VOLTAGE DROP	6-8
RESIDENTIAL	6-10
RATINGS	6-15
PERMISSIBLE LOADS	6-15
CONDUCTORS	6-20
LIGHTING LOADS	6-28
ELECTRIC DISCHARGE LOADS	6-30
RECEPTACLE LOADS	6-36
INDIVIDUAL	6-37
MULTIWIRE BRANCH-CIRCUITS	6-38
COMMERCIAL COOKING EQUIPMENT	6-38
WATER HEATER LOADS	6-40
HEATING LOADS	6-41
AIR-CONDITIONING LOADS	6-42
MOTOR LOADS	6-44
TEST QUESTIONS	6-47

LOADS
ARTICLE 220, PARTS I; II; III

The requirements of **Article 220, Parts I, II,** and **III** shall be used to calculate the loads of electrical systems. These calculated loads are utilized to size the feeder-circuit conductors, OCPD's, and other pertinent elements. Demand factors can be applied to loads that are grouped, and percentages may be applied that will reduce the load. These percentages are based on how the equipment is used in the electrical system.

CALCULATING AMPS

Master Test Tip 1: For the purpose of calculating the loads for branch-circuits and feeder loads, system voltages of 120, 120/240, 208/120, 240, 347, 480/277, 480, 600/347 and 600 volts must be used per **220.5(A)**. **Note:** If the voltage in the test question or problem is listed 230/115 volts, then this system voltage must be used to calculate the loads.

If the VA is known, amps can be found by dividing the VA by the configuration of voltage for the feeder-circuit.

For example: What is the ampacity for a feeder-circuit load of 18,400 VA that is supplied by a 120/240 volt, single-phase system?

    Step 1: Finding amperage
        I = VA ÷ A
        I = 18,400 VA ÷ 240 V
        I = 76.7 A

    Solution: The feeder-circuit ampacity is 77 amps when rounded-up per 220.5(B).

For example: What is the ampacity for a feeder-circuit load of 18,400 VA that is supplied by a 120/208 volt, three-phase, four-wire system?

    Step 1: Finding amperage
        I = VA ÷ V x 1.732
        I = 18,400 VA ÷ 360
        I = 51.1 A

    Solution: The feeder-circuit ampacity is 51 amps. Note: The .1 amp is dropped per 220.5(B).

The feeder-circuit components shall be selected from the 51 amps per phase. The 51 amps per phase is connected to Phase A, Phase B, and Phase C. This calculation is verified by dividing the VA by the voltage per phase and dividing the total by 3.

For example: What is the ampacity for a feeder-circuit load of 18,400 VA that is supplied by a 120/208 volt, three-phase system?

    Step 1: Finding amperage
        I = VA ÷ V
        I = 18,400 VA ÷ 120 V
        I = 153.3
        I = 153.3 A ÷ 3
        I = 51.1 A

    Solution: The feeder-circuit ampacity is 51.1 amps. (See the Note in the solution above)

See Figure 6-1 for calculating amps based on volt-amps and voltage.

POWER FACTOR

Master Test Tip 2: When poor power factor is listed in questions and problems on the test, larger conductors must be provided to compensate or corrective capacitors used.

Power factor (PF) is the ratio of actual power used in a circuit to the apparent power drawn from the line. Actual power is the true power used to produce heat or work. Actual power is also known in the industry as real, true, or useful power.

    Terms
- Power factor   = PF
- Watts   = W
- Volts   = V
- Amps   = I

Feeder-Circuits and Branch-Circuits

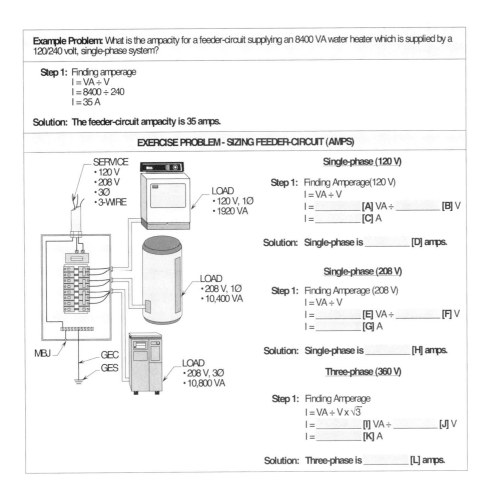

Figure 6-1. Finding the amps of a circuit based upon volt-amps and volts.

Master Test Tip 3: Before calculating the load of a circuit, determine if such load(s) are classified as noncontinuous, continuous or any combination of such. Also verify if the load(s) is qualified to have demand factors applied.

For example, ten pieces of cooking equipment having a total rating of 80,000 VA can have a demand factor of 65 percent applied, and the load after this application is 52,000 VA (80,000 VA x 65% = 52,000 VA).

ACTUAL POWER
SINGLE-PHASE

The actual power in watts in a pure resistance circuit is found by multiplying volts times amps.

For example: A 240 volt, single-phase feeder-circuit with a 51 amp load has a load of 12,240 watts.

    Step 1: W = V x I
            W = 240 V x 51 A
            W = 12,240

Solution: Actual power = 12,240 watts

If the circuit above has poor power factor of 75 percent due to inductive loads, the load in watts (actual power) will now be calculated as follows:

For example: A 240 volt, single-phase feeder-circuit of 51 amps with a power factor of 75 percent has a load of 9180 watts.

    Step 1: W = V x I x PF
            W = 240 V x 51 A x 75%
            W = 9180

Solution: Actual power = 9180 watts

Master Test Tip 4: Note that the computed watts in the "For example problem" will cause heating effects in components under certain conditions of use.

ACTUAL POWER THREE-PHASE

Master Test Tip 5: The voltage must be multiplied by the square root of 3 (1.732) when calculating loads for three-phase electrical systems. Note that 360 volts is used instead of the full square root of 3 (1.732 x 208 V =360.256 V).

The actual power in watts in a three-phase circuit with inductive loads is found by multiplying V x 1.732 x I x PF.

For example: The actual power in watts for a three-phase circuit with a load of 51 amps that is supplied by a 208 volt circuit with a power factor of 75 percent is 13,779 watts. (Using full square root of 3)

Step 1: W = V x 1.732 x I x PF
W = 208 V x 1.732 x 51 A x 75%
W = 13,780

Solution: Actual power = 13,780 watts

Master Note: The actual power is 18,373 watts (208 V x 1.732 x 51 A x 100% = 18,373 W) if the power factor is 100 percent instead of 75 percent.

See Figure 6-2 for calculating amps on a feeder-circuit having poor power factor.

Figure 6-2. Calculating the actual power in watts for a feeder-circuit based upon poor power factor.

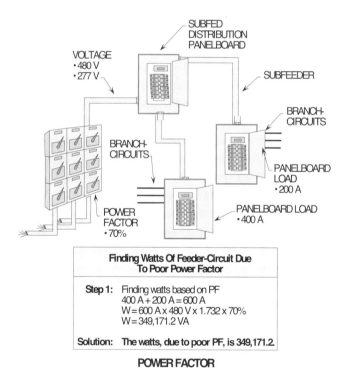

NONCONTINUOUS OPERATED LOADS
215.2(A)(1); 215.3; 408.30

Mater Test Tip 6: Demand factor is the ratio of the maximum demand of a system, or part of a system, to the total connected load of a system or the part of the system under consideration.

Noncontinuous operated loads shall be calculated at 100 percent, or demand factors can be applied to specific loads. Based upon their condition of use, the total VA rating for such load can be derived. Noncontinuous operated loads are classified as loads that do not operate for a period of time less than three hours per **Article 100**. Portable or fixed cord-and-plug connected or permanently connected loads can be classified as noncontinuous operated loads.

For example: What is the VA rating for a connected load of 34,500 VA to size the feeder-circuit at noncontinuous operation?

> **Step 1:** Finding VA
> **215.2(A)(1); 215.3; 408.30**
> VA = VA x 100%
> VA = 34,500 VA x 100%
> VA = 34,500

> **Solution:** The total VA for the feeder-circuit is 34,500 VA.

CONTINUOUS OPERATED LOADS
215.2(A)(1); 215.3; 408.30

Continuous operated loads are classified as loads that operate for a period of three hours or more. Continuous operated loads do not operate at varying or intermittent operation. A continuous operated load operates for a period of three hours or more without the circuit being interrupted. An industrial processing machine used in a facility to perform a work task for a work day of 8 hours falls under such use and classification.

The total VA rating for sizing conductors to such a machine is obtained by multiplying the continuous load by 125 percent. Feeder-circuit conductors may have to be increased in size due to voltage drop, ambient temperature, or too many current-carrying conductors in a raceway or cable.

Master Test Tip 7: OCPD's that supply loads that operate for three hours or more must have their rating in amps increased by 125 percent.

For example: What is the VA rating for a continuous operated processing machine with a connected load of 34,500 VA?

> **Step 1:** Finding VA
> **220.10; 215.2(A)(1)**
> VA = VA x 125%
> VA = 34,500 VA x 125%
> VA = 43,125 VA

> **Solution:** The total VA for the feeder-circuit is 43,125 VA.

Master Test Tip 8: The 80 percent derating of the OCPD for continuous loads is accomplished by multiplying the load in VA or amps by 125 percent per **215.2(A)(1)** and **215.3** for feeder-circuits, and **210.19(A)(1)** and **220.20(A)** for branch-circuits.

See Figure 6-3 for calculating the load in amps based upon continuous or noncontinuous operation.

DEMAND FACTORS
ARTICLE 220, PART II

Demand factors can be applied to the VA rating for specific loads, depending on their specific conditions of use. The percentages for which demand factors can be applied are listed in the NEC. Feeder-circuits supplying loads under specific condition of use can be reduced by applying demand factors to obtain the total rating. Loads that can be reduced by applying a demand factor are as follows:

Lighting	Table 220.42
Receptacles	Table 220.44
Dryers	Table 220.54
Ranges	Table 220.55
Kitchen equipment	Table 220.56
Dwelling units	Table 220.82
Existing dwelling units	Table 220.83
Multifamily dwelling units	Table 220.84
School loads	Table 220.86
Restaurants	Table 220.88
Farm loads	Table 220.102; Table 220.103

Master Test Tip 9: If six dryers on a test problem are supplied by a service or feeder, their total VA can be reduced by 75 percent per **Table 220.54**.

For example, a 50 percent demand factor can be applied to all VA of a receptacle load exceeding 10,000 calculated VA per **Table 220.44**.

Figure 6-3. Computing the load in amps for a feeder-circuit based upon continuous and noncontinuous operation.

Master Test Tip 10: In the Exercise problem, the neutral is not considered a current-carrying conductor per 310.15(B)(4)(c) to Table 310.16.

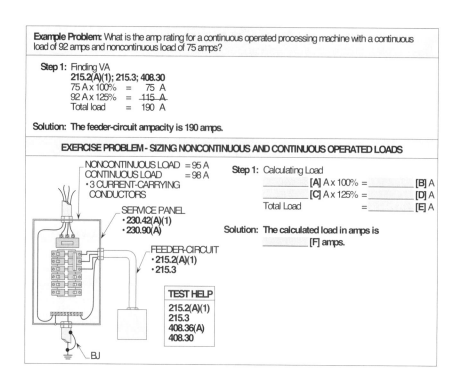

APPLYING DEMAND FACTORS FOR RECEPTACLE LOADS

Note that multioutlet assemblies can now be added to **220.42** and demand factors applied as listed in **Table 220.42**.

The VA rating for receptacle and multioutlet receptacle loads shall be calculated at 100 percent for the first 10,000 VA and the remaining VA calculated at 50 percent per **Table 220.44**. **Table 220.44** shall only be applied to receptacles that are used at noncontinuous operation. The demand factors of **Table 220.42** can be used to determine the load for receptacle outlets that are figured at 180 VA and then added to the lighting load.

NEUTRAL
ARTICLE 220, PART II; ARTICLE 310

Master Test Tip 11: The maximum unbalanced load between Phase A at 100 amps, Phase B at 95 amps and Phase C at 90 amps is 100 amps. Therefore, the neutral must be sized at 100 amps.

The grounded neutral conductor is intentionally grounded at the service equipment. The neutral shall be sized to carry the maximum unbalanced current. The largest between the neutral and any one ungrounded phase conductor is the maximum unbalanced current that must be carried. A demand factor can be applied to the neutral, under certain conditions of use. When sizing the neutral, also see **310.15(B)(4)**.

SIZING
220.61

The feeder neutral load shall be the maximum unbalanced load connected between the neutral and any one ungrounded conductor. The first 200 amps of neutral current shall be calculated at 100 percent. All resistive loads on the neutral exceeding 200 amps may have a demand factor of 70 percent applied, and this value added to the first 200 amps taken at 100 percent. All inductive neutral current shall be calculated at 100 percent with no demand factors applied. The feeder neutral load shall also be 70 percent of the demand load for cooking equipment or a dryer load that, for example, are installed in dwelling units. **(See Figure 6-4)**

For example: What is the neutral load for a feeder-circuit with a inductive load of 400 amps?	For example: What is the neutral load for a feeder-circuit with a resistive load of 400 amps?
Step 1: Calculating neutral load 220.61(C); 310.15(B)(4)(C) First 200 A x 100% = 200 A Remaining 200 A x 100% = 200 A Total load = 400 A	**Step 1:** Calculating neutral load 220.61(B); 310.15(B)(4)(C) First 200 A x 100% = 200 A Remaining 200 A x 70% = 140 A Total load = 340 A
Solution: The total inductive neutral load is 400 amps.	**Solution:** The total resistive neutral load is 340 amps.

For example: What is the neutral load for a 9 kW range?

Step 1: Finding demand
Table 220.55, Col. C
9 kW = 8 kW

Step 2: Calculating neutral load
220.61(A)(1)
8 kW x 70% = 5.6 kW

Solution: The total neutral range load is 5.6 kW.

NEC 220.61(A); (B); (C)

Figure 6-4. Calculating the feeder neutral load for an inductive, resistive, and range load.

The neutral current in amps for a three-wire, two-phase or five-wire, two-phase system shall be multiplied by 140 percent, which is derived by taking the square root of 2. By multiplying the neutral amps by 140 percent ($\sqrt{2}$ = 141%), the unbalanced current in the neutral will be collected from all ungrounded phase conductors. The neutral conductor is multiplied by 140 percent instead of 141 percent (rounded down) based on the phase conductor with the highest ampacity per **220.61(A)**. Basically, if calculated using this procedure, the neutral conductors are not overloaded by 120 volts being switched in and out of circuits at different intervals of time.

Master Test Tip 12: Note that a five-wire system consists of four 120 volt phase conductors and one oversized neutral (140% x larger phase in amps) per **220.61(C)**.

For example: What is the neutral load for a five-wire, two-phase feeder-circuit with each phase rated at 150 amps?

Step 1: Finding amperage
220.61(A), Ex.
150 A x 140% = 210 A

Solution: The neutral load is 210 amps.

DEMAND FACTORS
220.61(C)

A demand factor of 70 percent can be applied to all neutral loads exceeding 200 amps except for nonlinear loads. Nonlinear related loads shall be computed at 100 percent.

For example: What is the load for the neutral conductor exceeding 200 amps with more than 50 percent of the load harmonically related? The ungrounded phase conductors are carrying a total neutral load of 275 amps.

Step 1: Finding amperage
310.15(B)(4)(c)
Phases = 275 A

Step 2: Calculating amperage
220.61(C)
First 200 A x 100% = 200 A
Next 75 A x 100% = 75 A
Total load = 275 A

Solution: The neutral conductor load is 275 amps.

Master Test Tip 13: In a test problem having a total neutral load of 100 amps, if 150 amps of this load is harmonically related, the neutral is considered current-carrying per **310.15(B)(4)(c)**. Note that over 50% of the neutral load in amps is harmonically related.

6-7

Note that a neutral is considered a current-carrying conductor if the major portion of the neutral load contains harmonics. (See Master Tip 13)

The grounded neutral conductor is considered a current-carrying conductor due to the harmonic currents generated by these loads per **310.15(B)(2)(a)**, and **Table 310.15(B)(2)(a)** must be applied for four or more current-carrying conductors in a conduit, cable, etc.

For example: What is the neutral load for 120 volt loads with harmonic currents of 400 amps per phase?

Step 1: Finding amperage
310.15(B)(4)(c)
Ungrounded conductors = 400 A

Step 2: Calculating amperage
220.61(C)
400 A x 100% = 400 A

Solution: The neutral load is 400 amps. Note: No reduction of ampacity is permitted due to harmonic loads.

VOLTAGE DROP
ARTICLE 215

Conductors are sometimes increased in size to prevent excessive voltage drop because of long runs between the OCPD's and the load served. Due to the long runs, feeder-circuit conductors must be increased in size to compensate for poor voltage drop. The voltage drop on the feeder conductors should not exceed 3 percent at the farthest outlet supplying power to the loads. The voltage drop on the feeder and branch-circuit conductors should not exceed 5 percent overall.

SINGLE-PHASE CIRCUITS
215.2(A), FPN 2

Master Test Tip 14: If the VD is 3 percent on the feeder-circuit conductors and 2 percent on the branch-circuit conductors, the overall VD does not exceed 5 percent and is therefore permitted per the NEC.

The following formula shall be used to calculate the voltage drop in a two-wire or three-wire, single-phase or three-phase feeder-circuit or branch-circuit.

These values may be used when calculating the voltage drop:

VD = voltage drop
R = resistivity for conductor material
See **Chapter 9, Table 8 Column 6 or 8**
(uncoated ohm/MFT) use 12 AWG for cu. and 18 AWG for alu.
L = one-way length of circuit conductor in feet
I = current in conductor in amperage
CM = conductor area in circular mils
See **Chapter 9, Table 8**
1000 = length of conductors based on Table 8, Chapter 9

Resistive method
• VD = 2 x L x R x I ÷ 1000

Circular-mil method
• VD = 2 x R x I ÷ CM

Master Test Tip 15: If a VD problem is on the test, verify with the proctor in charge of the examination which method of computing VD must be used.

Conductors routed at greater lengths will have the resistance of each conductor increased, which will oppose the flow of current. By running larger conductors, the diameter of each conductor is increased in size, creating a greater path for the flow of current and less opposition to the movement of electrons, which keeps the voltage high at the conductor end.

As listed above, there are two methods that are used on the master test to calculate voltage drop in a feeder-circuit. One is the resistivity concept, which is considered the most accurate. The second is the circular-mil (CM) method which has been used for many years. The resistivity method is utilized by selecting the resistivity of the conductors. The CM method consists of selecting the CM rating of the conductors from **Table 8, Ch. 9** in the NEC. These values, whichever chosen, are inserted into the formula with other data, and the voltage drop computed. **(See Figures 6-5(a) and (b))**

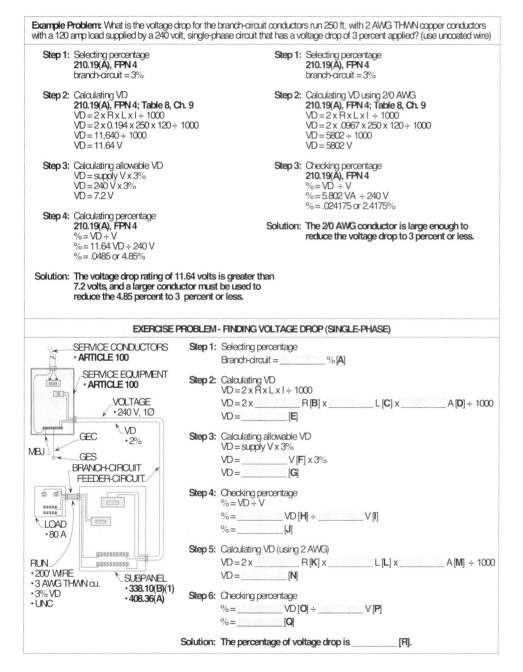

Figure 6-5(a). Test problem for computing voltage drop for single-phase branch-circuit conductors.

Master Test Tip 16: Note that the procedure for calculating VD in the exercise problem can be used for determining VD in branch-circuits or feeder-circuits.

Master Test Tip 17: Note that grounded conductors and EGC's in exercise problem 6-5(a) must be increased in size because of VD per **240.23, 215.2(A)(1)** and **250.122(B)**.

Figure 6-5(b). Test problem for finding the size conductors in CM for a three-phase feeder-circuit.

Master Test Tip 18: Note that the procedure used in the Exercise Problem to calculate the size conductors in CM to compensate for VD can be used to determine the VD in branch-circuits and feeder-circuits.

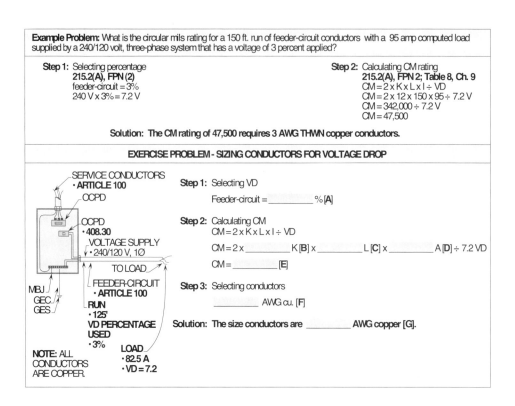

THREE-PHASE CIRCUITS
215.2(A), FPN 2

The voltage drop for three-phase circuits shall be calculated by multiplying the voltage drop by .866. The same formula for finding voltage drop in single-phase circuits shall be used to determine the voltage drop in three-phase circuits. Voltage drop multiplied by .866 is found by dividing $\sqrt{3}$ by 2 (1.732 ÷ 2 = .866). The .866 is basically produced from the additional conductor (third conductor), which is derived from a three-phase system instead of a two-phase system. **(See Figure 6-6)**

RESIDENTIAL

Master Test Tip 19: Note that the VA per sq. ft. per **Table 220.12** includes all general-purpose lighting and receptacle outlets for the dwelling unit. Review the rules of the single (star-shaped) asterisk below the Table.

Branch-circuits utilized in residential occupancies are computed differently from those for commercial and industrial locations. The general-purpose circuits shall be computed at 3 VA per square foot of the dwelling unit.

For example, a dwelling unit of 2000 sq. ft. has a volt-amp rating of 6000 VA (2000 sq. ft. x 3 VA = 6000 VA). This total VA rating is used to supply all the general-purpose lighting and receptacle outlets in the dwelling unit.

Another example would be the two 20 amp small appliance circuits used to supply countertop receptacles and other wall outlets in the kitchen, pantry, dining room, and breakfast room. Individual circuits such as a 12 kW range circuit can be computed at 8 kW per **Table 220.55**. Other loads in the dwelling can be reduced in VA due to operation and use.

Feeder-Circuits and Branch-Circuits

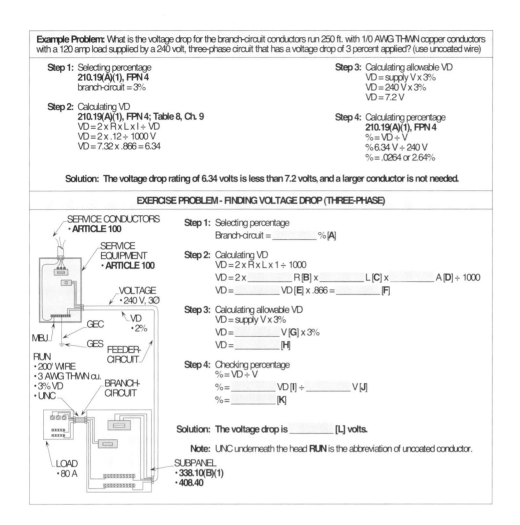

Figure 6-6. Test problem when determining voltage drop for a three-phase branch-circuit.

Master Test Tip 20: Note that 2.64 percent is found by multiplying .0264 by 100 (.0264 x 100 = 2.64%).

GENERAL-PURPOSE CIRCUITS
220.12; TABLE 220.12

The calculation for loads in various occupancies are based on VA (volt-amperes) per square foot. In calculating the VA per square foot, the outside dimensions of the building shall be used. They do not include the area of open porches and attached garages with dwelling unit occupancies. However, if there is an unused basement, it should be assumed that it will be finished later, so it should be included in the calculation so that the capacity of the wiring system will be adequate to serve such loads at a later date.

The load values used in **Table 220.12** are considered at 100 percent power factor. If less than 100 percent power factor is present because of equipment operation Such power factor of the equipment must be figured in to take care of the additional current values.

Master Test Tip 21: The number of 15 or 20 amp branch-circuits for a dwelling unit is determined by multiplying the square footage by 3 VA per sq. ft. per **220.12** and **Table 220.12** and dividing by the size overcurrent protection device times the voltage of the circuit. **(See Figure 6-7)**

SMALL APPLIANCE CIRCUITS
210.52(A); 220.52(B)

A minimum of two 20 amp, 1500 VA small appliance circuits are required to supply receptacle outlets that are located in the kitchen, pantry, breakfast room, and dining room. The two small appliance circuits shall be routed to the kitchen countertop(s) and the outlets proportioned among the two circuits as evenly as possible to prevent unbalanced loading of the circuits. Unbalanced loading can trip open the overcurrent protection device due to too many portable appliances being plugged into the same small appliance circuit. **(See Figure 6-8)**

Figure 6-7. Determining the number of 15 amp, two-wire circuits to supply power to the general-purpose lighting and receptacle outlets throughout the dwelling unit.

Figure 6-8. Determining the number of outlets that are permitted on a 20 amp small appliance circuit.

Master Test Tip 22: The two small appliance circuits must be supplied by 12-2 AWG with ground nonmetallic sheathed cable.

Master Test Tip 23: The receptacle for the laundry circuit must be of the single receptacle type unless a duplex is used to cord-and-plug connect a washing machine and gas dryer per **210.8(A)(2), Ex. 2** and **(A)(5), Ex. 2**.

LAUNDRY CIRCUIT
210.50(C); 220.52(B); 210.52(F)

At least one 20 amp, 1500 VA laundry circuit is required to supply receptacle outlets in the laundry room. All laundry equipment shall be located within 6 ft. of the receptacle outlet per **210.50(C)**. Sometimes a 20 amp duplex receptacle can be used to cord-and-plug connect a washing machine and gas dryer. No other outlets shall be supplied by this 20 amp, two-wire small appliance circuit. **(See Figure 6-9)**

INDIVIDUAL CIRCUITS
210.19(A)(1); 210.20(A)

Individual branch-circuits in dwelling units shall be calculated at 100 percent or 125 percent of the full-load amps of the appliance served or demand factors can be applied based upon the type of appliances utilized.

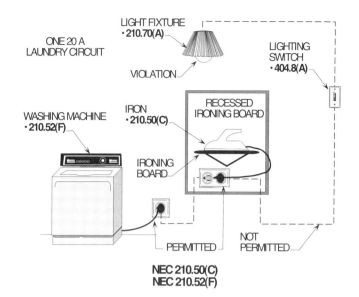

Figure 6-9. At least one 20 amp, 1500 VA laundry circuit is required to supply receptacle outlets in the laundry room.

RESIDENTIAL COOKING EQUIPMENT
220.55

The procedure for calculating the branch-circuit loads for ranges, cooktops, and ovens are determined by applying the demand factors in **Table 220.55**. The demand factors are listed in Columns A, B, and C and are based on the size of the range, cooktop, or oven. When the kW rating exceeds 12 or there's more than one unit, one of the Notes is used in conjunction with **Table 220.55**.

Master Test Tip 24: Note that the demand factors listed in **Table 220.55** can be used to calculate the kW ratings of cooking equipment, which can be utilized to size the components of services, feeders, and branch-circuits.

COLUMN C
TABLE 220.55

Column C of **Table 220.55** is used for cooking equipment rated from over 8.75 kW to 12 kW. These units are usually rated 9 to 12 kW.

COLUMN A
TABLE 220.55

Column A to **Table 220.55** is used for cooking equipment rated less 3.5 kW. These units vary in rating and have limited use.

COLUMN B
TABLE 220.55

Column B to **Table 220.55** is used for cooking equipment rated from 3.5 kW to 8.75 kW. These units are many of the types in use today.

Master Test Tip 25: There are very few pieces of cooking equipment rated at less than 3 1/2 kW that are used in modern day dwelling units.

NOTE 1
TABLE 220.55

Note 1 to **Table 220.55** is applied where the kW rating of the cooking equipment is over 12 kW but not over 27 kW. All kW ratings that exceed 12 kW are multiplied by 5 percent. The kW rating of one range is listed in **Column C** of **Table 220.55** and is multiplied by this total. Units of sizes larger than 12 kW are usually combination units found in kitchens with limited space.

Master Test Tip 26: It is permissible per Note 3 to Table 220.55 to add all pieces of cooking equipment with ratings over 1 3/4 kW through 8 3/4 kW together and multiply by the percentage of **Column A or B**, whichever produces the smaller kW rating can be used. This calculation may be allowed by the testing authority if it provides the smaller kW rating of all the methods available in **Table 220.55** and Notes.

NOTE 2
TABLE 220.55

For cooking equipment of unequal values rated over 12 kW to 27 kW in **Column C, Note 2** is applied and computed by adding the kW ratings of all units and dividing by the number of units; all ranges below 12 kW are computed at 12 kW. When an average rating is found, the number of units is increased 5 percent for each kW exceeding 12 kW to derive the allowable kW.

NOTE 4
TABLE 220.55

Note 4 to Table 220.55 can be applied to a counter-installed cooktop with one or two wall-mounted ovens. The total kW of each cooking unit is totaled, and all kW exceeding 12 kW is multiplied by 5 percent. The kW rating of one range, not two or three, listed in **Column C** of **Table 220.55** is multiplied by this total. The advantage of this rule is it allows one branch-circuit to be run, instead of two or three individual circuits, one to each unit.

SIZING TAPS
210.19(A)(4), Ex. 1

Master Test Tip 27: A plug and receptacle may be used as the disconnecting means for appliances installed in dwelling units. (See **422.33(A)** for specific rules)

A tap to connect a cooktop and oven(s) can be made from a 50 amp branch-circuit if the tap conductors are sized from the kW rating of each piece of cooking equipment per Note 4 to **Table 220.55**. Anytime taps are made from larger conductors with smaller conductors, the tap must comply with **240.21(A)**.

For example, can a 14 AWG tap be made from a 12 AWG branch-circuit and be down sized and used as a switch leg? Naturally, the answer is no, because the tap cannot meet the provisions of **240.21(A)** thru **(G)** for making an approved tap. However, **240.21(A)** and **210.19(A)(4), Ex. 1** applied together permit a smaller tap to be made for cooking equipment circuits tapped from conductors of a larger sized branch-circuit.

See Figure 6-10 for calculating cooking equipment loads for taps per **Ex. 1 to 210.19(A)(4)**.

DRYER EQUIPMENT LOADS
220.54

When sizing the branch-circuit load on the master examination, dryer equipment loads shall be 5000 VA or the nameplate rating, whichever is greater. If four or fewer dryers are installed, the load shall be calculated at 100 percent. When installing five or more dryers, the load shall be calculated by the percentages listed in **Table 220.54** based on the number of dryers being installed. The load for 4500 VA dryer shall be calculated at 5000 VA, and this rating is used to size the branch-circuit to supply the receptacle to cord-and-plug connect the unit. Dryers installed in dwelling units shall comply with the provisions of **220.14(B)** and **220.54**. Due to more than one family using dryers, dryer loads shall be calculated at 100 percent for noncontinuous operation and at 125 percent for continuous operation if they are installed as commercial dryers in laundry facilities.

Feeder-Circuits and Branch-Circuits

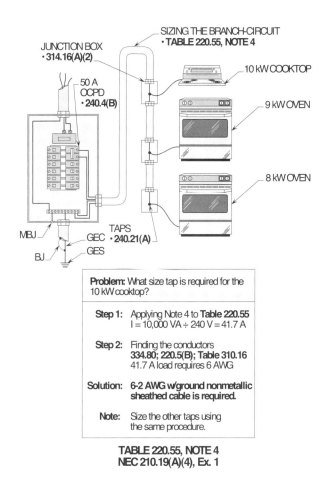

Figure 6-10. Determining the load in kW and amps to size the elements for a branch-circuit from which taps to cooking equipment are made.

Master Test Tip 28: Note that nonmetallic sheathed cable must be terminated as 60°C terminals per **334.80** and Column 2 of **Table 310.16**.

RATINGS
210.3

Branch-circuits shall be classified by the rating or setting of the overcurrent protection device protecting the circuit. Two or more outlets can be protected by the following overcurrent protection devices based upon the calculated load per **210.19(A)(1)** and **210.20(A)**:

(1) 15 amp
(2) 20 amp
(3) 30 amp
(4) 40 amp
(5) 50 amp

Master Test Tip 29: In industrial facilities, multiwire circuits can be used on nonlighting outlet loads if proper supervision, maintenance, and qualified personnel serve such circuits per **210.3, Ex.**

Note that A/C room units can be connected by cord-and-plug if their ampere rating does not exceed 50 percent of the general-purpose branch-circuit per **440.62(C)**.

PERMISSIBLE LOADS
210.23(A)(1); (A)(2)

The rating or setting of the overcurrent protection device shall not be exceeded by the load on an individual branch-circuit. If the load is continuous, the rating of the OCPD must be derated by 80 percent and not loaded past this reduced ampacity value. A fastened-in-place appliance can be connected to a general-purpose circuit if its amp rating does not exceed 50 percent of the branch-circuit.

6-15

15 AND 20 AMP BRANCH-CIRCUITS
210.23(A)

Master Test Tip 30: When classifying the rating of branch-circuits, it is the size of the OCPD that determines such ratings per **210.3**.

A 15 or 20 amp branch-circuit shall be permitted to supply lighting units and/or utilization equipment in residential, commercial, or industrial locations.

For example: What is the total VA rating for a 15 or 20 amp, 120 volt, two-wire branch-circuit supplying a noncontinuous load?

> **Step 1:** Finding VA
> **210.23(A)**
> VA = 15 A x 120 V
> VA = 1800 VA

Solution: The total VA is 1800.

> **Step 1:** Finding VA
> **210.23(A)**
> VA = 20 A x 120 V
> VA = 2400 VA

Solution: The total VA is 2400.

For example: What is the total VA rating for a 15 or 20 amp, 240 volt, three-wire branch-circuit supplying a noncontinuous load?

Master Test Tip 31: Branch-circuit conductors must have an allowable ampacity large enough to supply the load per **210.19(A)(1)**.

> **Step 1:** Finding VA
> **210.23(A)**
> VA = 15 A x 240 V
> VA = 3600

Solution: The total VA is 3,600.

> **Step 1:** Finding VA
> **210.23(A)**
> VA = 20 A x 240 V
> VA = 4800

Solution: The total VA is 4800.

Master Test Tip 32: OCPD's supplying electrical circuits over 600 volts are not required under certain conditions of use to be derated by 80 percent per **230.208(B)**.

The rating of any one cord-and-plug connected utilization equipment shall not exceed 80 percent of the branch-circuit rating if connected to a general-purpose circuit supplying two or more outlets.

For example: What size overcurrent protection device is required to be installed for a 7 amp, 120 volt, single-phase compactor that is cord-and-plug connected per **422.16(B)(2)**?

> **Step 1:** Finding amperage of 15 or 20 amp branch-circuit
> **210.23(A)(1)** (If continuous)
> A = 15 A x 80%
> A = 12 A
> A = 20 A x 80%
> A = 16 A

210.23(A)(1) (If noncontinuous)
A = 15 A x 100%
A = 15 A
A = 20 A x 100%
A = 20 A

Solution: The OCPD of 15 amp can be used.

The 7 amp full-load current rating of the compactor does not exceed the (15 amp x 80% = 12 amps) loading range of the 15 amp overcurrent protection device. The 7 amp full-load current rating of the compactor does not exceed the (20 amp x 80% = 16 amps) loading range of the 20 amp overcurrent protection device. Therefore, the 20 amp overcurrent protection device can be used to supply the 7 amp compactor (compactor must have OLP).

Master Test Tip 33: Note that the overload protector in the compactor protects the windings of the motor and conductors of the circuit per **430.32(A)(1)**.

Fixed appliances (fastened-in-place) are permitted to draw up to 50 percent of the rating of a branch-circuit supplying two or more general-purpose outlets that serve lighting and receptacle loads.

For example: Can an 8 amp, 120 volt, single-phase A/C window unit be connected to an existing branch-circuit?

Note that the conductors and OCPD for the A/C window unit must be calculated per **440.32** and **440.22(A)** of the NEC.

Step 1: Finding A of BC
210.23(A)
A = 1/2 of 20 A OCPD
A = 10 A

Step 2: Calculating A for A/C unit
210.23(A); 440.62(B); (C); 440.32
A = 8 A x 125%
A = 10 A

Step 3: Verifying permissive A
210.23(A); 440.62(B); (C); 210.3
A = 20 A OCPD x 80%
A = 16 A

Solution: Yes, the A/C window unit rated at 8 amps can be connected to the 20 amp branch-circuit.

A 20 amp overcurrent protection device is allowed to protect a fastened-in-place appliance with a rating of 10 amps or less after applying the 50 percent rule. The remaining 50 percent of the overcurrent protection device is allowed to protect lighting and/or cord-and-plug connected appliances. **(See Figure 6-11)**

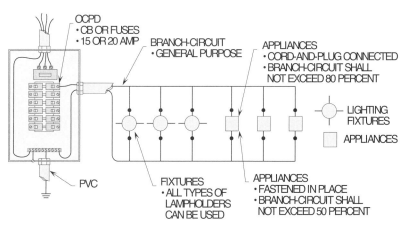

Figure 6-11. A 20 amp overcurrent protection device is permitted to protect a fastened-in-place appliance with a rating of 10 amps or less after applying the 50 percent rule. The remaining 50 percent of the overcurrent protection device can be used to supply lighting and/or cord-and-plug connected appliances.

30 AMP BRANCH-CIRCUIT
210.23(B)

A 30 amp branch-circuit shall be permitted to be installed to supply fixed lighting units with heavy-duty lampholders in other than a dwelling unit(s) or utilization equipment in any occupancy. The rating of any individual cord-and-plug connected appliance shall not exceed 80 percent of the branch-circuit rating. The rating of any individual cord-and-plug connected appliance shall draw no more than 24 amps (30 amp x 80% = 24 amps) if it's connected to a 30 amp overcurrent protection device. **(See Figure 6-12)**

Figure 6-12. A 30 amp branch-circuit shall be permitted to be installed to supply fixed lighting units with heavy-duty lampholders in other than a dwelling unit(s) or utilization equipment in any occupancy.

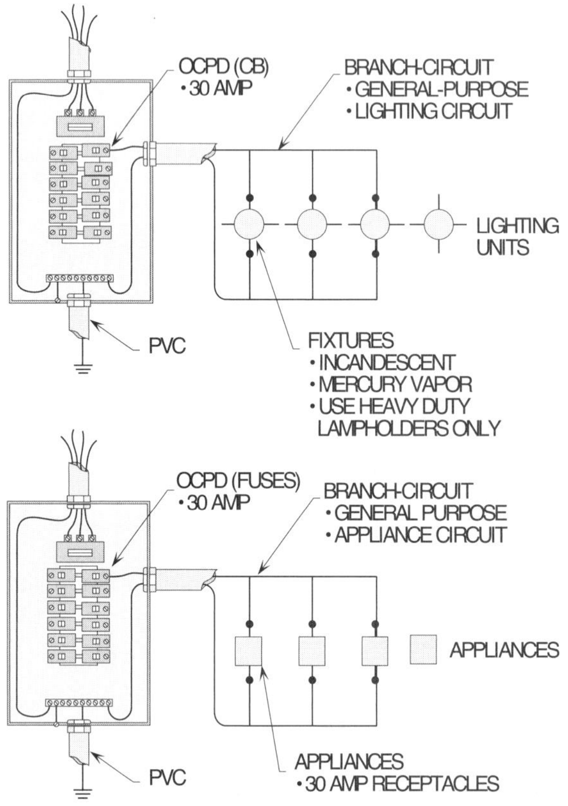

Master Test Tip 34: Heavy duty lampholders of the admedium type shall be rated not less than 660 W and for other types at least 750 W per 210.21(A).

Master Test Tip 35: If the dishwasher in the "For example problem" is cord-and-plugged connected, the receptacle must be of the single outlet type per 210.21(B)(1).

For example: What is the load for a 23 amp dishwasher used at continuous operation? (The branch-circuit is a separate circuit)

    Step 1: Finding amperage
        210.19(A)(1); 210.20(A)
        A = 23 A x 125%
        A = 28.75

    Step 2: Finding branch-circuit
        210.23(B)
        28.75 A requires 30 A

    Solution: The branch-circuit load is 28.75 amps.

A 30 amp individual branch-circuit can supply a single appliance that is used for continuous operation in any type occupancy.

40 AND 50 AMP BRANCH-CIRCUITS
210.23(C)

A 40 or 50 amp branch-circuit shall be permitted to supply cooking appliances that are fastened-in-place in any occupancy. Fixed lighting units with heavy-duty lampholders or infrared heating units shall be permitted for such circuits except for other than dwelling units. Equipment such as a water heater, dryer, or heating unit shall be permitted to be supplied by a 40 or 50 amp branch-circuit. **(See Figure 6-13)**

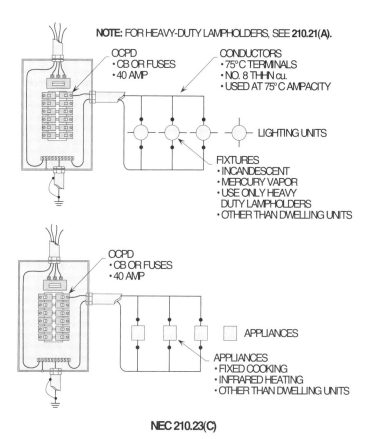

Figure 6-13. A 40 or 50 amp branch-circuit shall be permitted to supply cooking appliances that are fastened-in-place in any occupancy.

Master Test Tip 36: Note that branch-circuits supplying general-purpose lighting and receptacle outlets in dwelling units are not required to be derated is required per **408.30**.

For example: What is the load for a 37 amp water heater used at continuous operation in a commercial building?

Step 1: Finding amperage
422.13; 422.10(A)
A = 37 A x 125%
A = 46.25

Solution: The branch-circuit load is 46.25 amps.

BRANCH-CIRCUITS LARGER THAN 50 AMPS
210.23(D)

Branch-circuits larger than 50 amps shall supply only nonlighting outlet loads. A maximum load of 50 amps shall be used for multioutlet branch-circuits installed for lighting units. A combination of loads exceeding 50 amps can be used for multioutlet branch-circuits that are not connected to lighting units.

Master Test Tip 37: All conductors with the "W" (60°C or 75°C) rated insulation are allowed to be installed in dry, damp, or wet locations per **310.8(A) thru (D)**.

CONDUCTORS
ARTICLE 310

Master Test Tip 38: Note that the ampacities of 90°C insulated conductors are used for adjustment or correction factors (derating) purposes only per **Table 310.16** and **Table 310.15(B)(2)(a)**.

Branch-circuit conductors used for general wiring shall be rated for the following insulations per **Table 310.16** and **Table 310.13**:

(1) 60°C
(2) 75°C
(3) 90°C

The temperature rating of conductors and their conditions of use are listed in **Table 310.13**. Not more than three current-carrying conductors in a raceway or cable run in an ambient temperature of not more than 30°C or 86°F shall be used for branch-circuit conductors based on **Table 310.16**. The types of insulation available are listed in **Table 310.16** for copper and aluminum conductors. Conductor ampacities are determined by condition of use and by the terminal ratings of OCPD's and equipment per **110.14(C)(1)(a)** and **(C)(1)(b)**.

60°C CONDUCTORS
TABLE 310.16, COLUMN 2; 110.14(C)(1)(a); (C)(1)(b)

The following type of conductors are permitted to be pulled in conduit or installed in cables. The capacities of such conductors are selected from **Table 310.16**, Column 2 based on 60°C ampacities:

(1) TW
(2) UF

Master Test Tip 39: If a 100 amp circuit is required to supply power to a 100 amp piece of equipment, 1 AWG THHN copper conductors (110 A) are needed at 60°C ampacities. **Note:** That is why 14 AWG through 1 AWG is listed in 110.14(C)(1)(a).

The temperature rating of the conductors are rated at 60°C per **Table 310.13** and **Table 310.16** and must be terminated to 60°C terminals and 60°C ampacities must be used.

Overcurrent protection devices rated at 100 amps or less are permitted to be terminated with conductors 14 AWG through 1 AWG with 60°C ampacities per **110.14(C)(1)(a)**. The maximum size of 1 AWG must be used so that a 60°C conductor ampacity (**Table 310.16**) will mate to 100 amp terminals of OCPD's and equipment.

> **For example:** What size THHN copper conductor is required to supply a piece of equipment operating at 90 amps? (Load already calculated at continuous or noncontinuous operation)
>
> **Step 1:** Finding load
> Table 310.13; Table 310.16
> Load = 90 A
>
> **Step 2:** Finding conductor A to 60°C
> Table 310.16
> 95 A = 2 AWG
>
> **Solution: The conductor is required to be 2 AWG THHN copper.**

If the question or problem calls for nonmetallic sheathed cable (romex or rope) or UF cable, the ampacity for each conductor is rated at 60°C per **334.80** and **340.80**. However, for conductors in these cables with insulation rated at 90°C, use such higher ampacities for derating purposes only. **(See Figure 6-14)**

Feeder-Circuits and Branch-Circuits

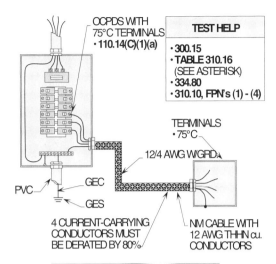

Figure 6-14. The ampacity for each conductor is rated at 60°C when installing nonmetallic sheathed cable (romex or rope) or UF cable per **334.80** and **340.80**. However, for the purpose of derating, conductors in these cables with insulation rated at 90°C can use the higher ampacities for derating to usable 60°C ampacities.

Note that NM cable is not blanketed over with insulation per **334.80**.

75°C CONDUCTORS
TABLE 310.16, COLUMN 3; 110.14(C)(1)(b)

The following types of conductors are permitted to be used if Column 3 for 75°C ampacities in **Table 310.16** is applied:

(1) FEPW
(2) RH
(3) RHW
(4) THHW
(5) THW
(6) THWN
(7) XHHW
(8) USE
(9) ZW

Master Test Tip 40: A 3 AWG THWN/THHN copper conductor rated at 110 amps per Column 4 of **Table 310.16** must be terminated at 100 amps for 75°C terminals and 85 amps for 60°C terminals per Columns 2 and 3 of **Table 310.16**. (See **110.14(C)(1)(a)**)

Overcurrent protection devices rated over 100 amps are permitted to be terminated with conductors larger than 1 AWG with 75°C ampacities per **110.14(C)(1)(a)**. Conductors shall be permitted to be installed with higher temperature ratings if the ampacities are matched to the terminals of the overcurrent protection device and equipment. To do so, the terminals shall be marked with the following temperature ratings:

(1) 60°C
(2) 60°C / 75°C

Note that 60°C ampacities shall only be applied to terminals marked 60°C, and 75°C ampacities shall only be applied to terminals marked 75°C. Notice that 60°C ampacities shall be applied for conductors 14 AWG through 1 AWG. Terminals of 75°C can be connected to 75°C or 90°C conductors and use their ampacities as listed in the 75°C column.

For example: What size amperage rating is allowed for a 4 AWG copper conductor using the ampacities of the 60°C column and 75°C column?

    Step 1: Finding amperage
        Table 310.16, Col.'s 2 and 3
        60°C = 70 A
        75°C = 85 A

    Solution: The amperage rating is 70 amps for 60°C and 85 amps for 75°C. Note: The 60°C terminal rating reduces the 4 AWG THHN copper conductor to only 70 amps and not 85 or 95 amps.

90°C CONDUCTORS
TABLE 310.16, COLUMN 4; TABLE 310.13

Master Test Tip 41: Overcurrent protection devices rated at 100 amps or less shall be terminated with conductors 14 AWG through 1 AWG and 60°C ampacities used per **110.14(C)(1)(a)** if not otherwise marked. Overcurrent protection devices rated over 100 amps shall be terminated with conductors larger than 1 AWG and 75°C ampacities used per **110.14(C)(1)(b)**.

The following types of conductors are allowed to be used to answer questions and problems appearing on the master examination. See **Table 310.16, Col. 4** and **Table 310.13** for ampacities and conditions of use for 90°C rated conductors:

  (1) TBS   **(18)** See list in **Table 310.13**
  (2) SA
  (3) SIS
  (4) FEP
  (5) FEPB
  (6) MI
  (7) RHH
  (8) RHW-2
  (9) THHN
  (10) THHW
  (11) THW-2
  (12) THWN-2
  (13) USE-2
  (14) XHH
  (15) XHHW
  (16) XHHW-2
  (17) ZW-2

Master Test Tip 42: Note that conductors with insulations rated over 600 volts can be terminated to 90°C terminals and equipment per **110.40**.

Conductors with the 90°C rated insulation can be connected to terminals rated 60°C, 75°C, and 90°C. Devices and equipment with 90°C terminals shall be used for connecting higher ampacity conductors that are rated 90°C. However, such OCPD's and equipment that mate (matched) are not available today. Therefore, most testing agencies do not have problems on the examination using 90°C ampacities based 90°C rated equipment.

For example: What is the ampacity for a 6 AWG THHN copper conductor connected to a 60°C circuit breaker installed in the service panelboard?

    Step 1: Finding amperage and condition of use
        Table 310.16, Col. 4; Table 310.13
        A = 75

    Step 2: Finding amperage
        Table 310.16, Col. 2
        A = 55

    Solution: The ampacity is limited to 55 amps because of the 60°C terminals.

Feeder-Circuits and Branch-Circuits

The following ampacity values shall be used for the temperature ratings of terminals using a 1/0 AWG THHN copper conductor:

(1) 60°C = 125 amps
(2) 75°C = 150 amps
(3) 90°C = 170 amps **(Note: No equipment with 90°C terminals is available)**

A load of 125 amps or less connected to 60°C terminals can be supplied by 1/0 AWG THHN copper conductors. A load of 150 amps or less on 75°C terminals can be served by a 1/0 AWG THHN copper conductor. A load of 170 amps or less on 90°C terminals can be served by 1/0 AWG THHN copper conductors. The ampacities are matched to the terminals of the overcurrent protection device and equipment. **(See Figure 6-15)**

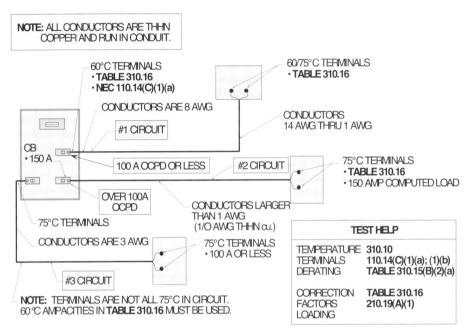

Figure 6-15. Determining the ampacity rating of terminals and sizing conductors based on 75°C ampacities from Columns 2 and 3 in Table 310.16. This rule prevents terminals from being overloaded.

Note 1: In **Figure 6-15**, the CB rating ahead of circuit 1 is 40 amps and the CB rating ahead of circuit 3 is 100 amps.

Note 2: In **Figure 6-15**, the calculated load of circuit 1 is 40 amp and the computed load of circuit 3 is 100 amps.

DERATING BY TABLE 310.15(B)(2)(a) AMBIENT TEMPERATURES

When there are not more than three current-carrying conductors in a raceway or cable, the allowable ampacities listed in **Table 310.16** may be used. If four or more current-carrying conductors or surrounding temperature exceeding 86°F are present, derating factors must be applied, based upon their conditions of use. See the top of **Table 310.16** for these conditions that must be applied before selecting the allowable ampacities of such conductors.

Master Test Tip 43: After the adjustment or correction factors have been applied, the calculated load at 100 percent or 125 percent or both must not exceed the derated ampacity of the conductors. **(Note: See Master Test Tip 44.)**

The ampacity of conductors may be derated at least two times if any one or all of the following rules are applied when determining the allowable ampacities:
(1) More than three current-carrying conductors
  • Adjustment factors
(2) Ambient temperature exceeds 86°F
  • Correction factors

Note that the calculation must be supplied after application of derating factor(s).

More than three current-carrying conductors in a raceway shall be derated (adjustment factors) by the percentages listed in **310.15(B)(2)(a)**, and **Table 310.15(B)(2)(a)**. Conductors routed through ambient temperatures exceeding 86°F shall be derated by the percentages according to the Ampacity Correction Factors of **Table 310.16**. Continuous duty loads (three hours or more) shall be multiplied by 125 percent per **210.19(A)(1)**, or the overcurrent protection device must be derated by 80 percent per **408.30**. OCPD's of the branch-circuit must be selected by the rules listed in **210.20(A)**.

Master Test Tip 44: A 90 amp overcurrent protection device is permitted to be used, because **240.4(B)** allows the next size OCPD above 85 amps (4 AWG cu.) to be used. After derating the overcurrent protection device per **408.30** by 80 percent, the remaining ampacity is larger than 70.5 amps (56.4 A x 125% = 70.5 amps). This 72 amp rating (90 A x 80% = 72 A) allows the calculated load of 70.5 amps to operate and complies with the rules of the NEC. **(See Figure 6-16 and Master Test Tip 43)**

For example: What size overcurrent protection device is required to serve a computed continuous load of 70.5 amps using a 4 AWG THHN cu. conductors with an ampacity of 85 amps?

Step 1: Finding ampacity
Table 310.16
A = 90(THHN)
4 AWG = 85 A = 90 A OCPD

Step 2: Calculating ampacity
384-16(d) ('99 NEC); **408.30**
90 A x 80% = 72 A
(72 A supplies 70.5 A)

Solution: A 90 amp overcurrent protection device is required.

Figure 6-16. More than three current-carrying conductors in a raceway shall be derated by the percentages listed in **Table 310.16** and **Table 310.15(B)(2)(a)**. Conductors routed through ambient temperatures exceeding 86°F shall be derated by the percentages according to the Ampacity Correction Factors of **Table 310.16**. Continuous operation (three hours or more) shall be multiplied by 125 percent per **210.19(A)(1)** or the overcurrent protection device must be derated by 80 percent per **384.16(D)** of the 1999 NEC. Note that the 80% rule is provided by multiplying the continuous load by 125% (1 ÷ 80% = 1.25). Section **408.30** refers to **Article 220** for rules to calculate loads. To satisfy this rule, **Article 220** refers to **210.19(A)(1)** for branch-circuits, **215.2(A)(1)** for feeders, and **230.42(A)(1)** for services.

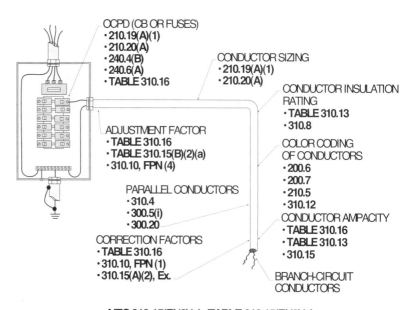

VOLTAGE LIMITATIONS
ARTICLE 210, PART A

The following voltage limitations are divided into three categories, and each category of voltage ratings is designed to supply certain loads:

(1) 120 volts between conductors
(2) 277 volts-to-ground
(3) 600 volts between conductors

120 VOLTS BETWEEN CONDUCTORS
210.6(B)

120 volts between conductors is not a restricted voltage and can be used to supply the following loads in any type of occupancy:

(1) Terminals of lampholders applied within their voltage ratings.

(2) Ballasts for fluorescent or high intensity discharge (HID) lighting units.

(3) Cord-and-plug connected or permanently connected appliances.

Cord-and-plug connected or permanently (hard-wired) connected appliances rated over 1440 VA are usually supplied by individual circuits. These appliances are generally connected to general-purpose circuits that supply more than one outlet, etc. Such appliances can be any one of the following types of equipment: **(See Figure 6-17)**

(1) Heating units
(2) A/C units
(3) Ranges
(4) Water heaters
(5) Processing machine
(6) Etc.

Master Test Tip 45: In dwelling units and guest rooms of hotels and motels the voltage of circuits supplying lighting units must not exceed 120 volts between conductors per **210.6(A)(1)**.

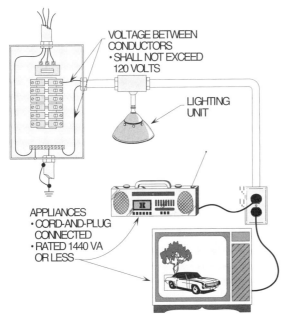

Figure 6-17. Cord-and-plug connected or permanently (hard-wired) connected appliances rated over 1440 VA are usually supplied by individual circuits.

Master Test Tip 46: Individual circuits of 120 volts supplying loads that are cord-and-plug connected must be rated less than 1/4 HP or 1440 VA.

Note that a lighting unit(s) is the same thing as a luminaire(s) (lighting fixtures).

277 VOLTS-TO-GROUND
210.6(C); 225.7(C)

Master Test Tip 47: Listed electric-discharge or incandescent luminaires (lighting fixtures) are no longer required to be installed at a minimum height of 8 ft. above finished grade when supplied by 277 volt, single-phase, branch-circuits. (See **210.6(C)**.)

Circuits exceeding 120 volts between conductors and not exceeding 277 volts-to-ground shall be permitted to supply any one of the following types of electrical apparatus: **(See Figures 6-18(a) through (d))**

  (1) Listed electric-discharge luminaires (lighting fixtures).
  (2) Listed incandescent luminaires (lighting fixtures).
  (3) Mogul-base screw-shell lampholders.
  (4) Other than screw-shell type lampholders applied within their voltage ratings.
  (5) Ballasts for fluorescent or high intensity discharge (HID) lighting units.
  (6) Cord-and-plug connected or permanently connected appliances or other utilization equipment.

Figure 6-18(a). This illustration is a diagram of a 120/208 volt, three-phase, four-wire circuit.

Figure 6-18(b). This illustration is a diagram of 120/240 volt, single-phase, three-wire circuit.

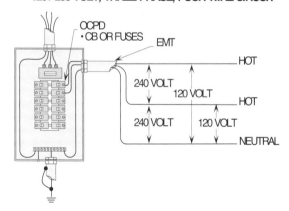

Figure 6-18(c). This illustration is a diagram of a 277 volt, single-phase, two-wire circuit.

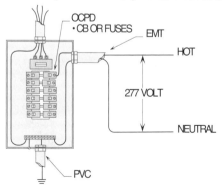

6-26

Feeder-Circuits and Branch-Circuits

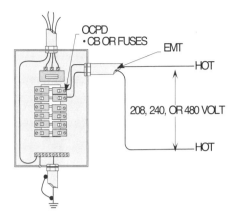

Figure 6-18(d). This illustration is a diagram of a 208, 240, or 480 volt, single-phase, two-wire circuit.

NEC 210.6(C)
NEC 225.7(C)

Luminaires (lighting fixtures) for illumination are permitted in outdoor areas of industrial establishments, office buildings, schools, stores, and other commercial or public buildings where the lighting fixtures are supplied by 480/277 volt circuits per **225.7(C)**. However, fixtures shall not be located within 3 ft. from windows, platforms, fire escapes, etc.

600 VOLTS BETWEEN CONDUCTORS
210.6(D); 225.7(D)

Circuits exceeding 277 volts-to-ground and not exceeding 600 volts between conductors are permitted to supply the following types of electrical apparatus:

(1) Ballasts for electric-discharge luminaires (fixtures) where such units are mounted by one of the following methods:

 (a) At a height not less than 22 ft. on poles or similar structures for the illumination of outdoor areas, such as highways, roads, bridges, athletic fields, or parking lots.

 (b) At a height not less than 18 ft. on other structures, such as tunnels.

(2) Utilization equipment that is cord-and-plug connected or permanently connected.

These circuits are usually derived by the following type of electrical systems:

(1) Ungrounded 480 volt, three-wire systems
(2) Corner grounded 480 volt, three-wire systems

Ungrounded 480 volt, three-wire systems will measure 480 volts-to-ground if one of their phase legs is accidentally grounded. In this case, it becomes a corner grounded system and the power is not lost. Corner grounded 480 volt, three-wire systems will have 480 volts-to-ground on one phase because one leg is intentionally grounded. Circuits exceeding 277 volts and not exceeding 600 volts between conductors shall be permitted to supply the auxiliary equipment of electric-discharge lamps as permitted in **225.7(D)**.

Master Test Tip 48: Replacement ballast for all fluorescent fixtures installed indoors must be of the thermal protected type per **410.73(E)(1)**.

Master Test Tip 49: HID discharge lighting ballasts that are mounted in lighting units or installed remotely must be thermally protected per **410.73(F)(4)**.

Master Test Tip 50: Open-circuit voltage of electric-discharge lighting units exceeding 1000 volts must not be installed in or on dwelling units.

DETERMINING AMPERAGE

The amperage for single-phase branch-circuits is determined by dividing the VA by the supply voltage (I = VA ÷ V). The amperage for three-phase branch-circuits is determined by dividing the VA by the supply voltage times the square root of 3 (I = VA ÷ V x 1.732). The 1.732 is determined by the $\sqrt{3}$. These amperage ratings are used to select the conductors and overcurrent protection devices and other elements of branch-circuits and feeders.

FINDING AMPERAGE
SINGLE-PHASE CIRCUITS

Master Test Tip 51: Unless otherwise specified on the test, all calculating must be done by dividing VA ratings of equipment by 120 V, 120/240 V, 120/208 V, 277/480 V, 347/600 V, 240 V, 480 V, 600 V per **220.5(A)**.

The amperage for single-phase branch-circuits is determined by dividing the VA rating of electrical equipment by the supply voltage of the circuit.

For example: What is the amperage rating of a 2400 VA calculated load connected to a 120 volt, single-phase branch-circuit?

Step 1: Finding amperage
I = VA ÷ V
I = 2400 VA ÷ 120 V
I = 20 A

Solution: The branch-circuit amperage is 20 amps.

FINDING AMPERAGE
THREE-PHASE CIRCUITS

The amperage for three-phase feeder-circuits is determined by dividing the VA rating for the electrical equipment by the supply voltage times 1.732. The amperage is evenly distributed on Phases 1, 2, and 3 if the VA is divided by the voltage times 1.732.

For example: What is the amperage rating of an 8960 VA computed load connected to a 208 volt, three-phase feeder-circuit? (See **220.5(A)** and **(B)**)

Step 1: Finding amperage
I = VA ÷ V x 1.732
I = 8960 VA ÷ 208 V x 1.732
I = 24.9 A (Rounded up to 25A)

Solution: The feeder-circuit amperage is 25 amps.

COMMERCIAL AND INDUSTRIAL

Master Test Tip 52: If the VA rating of a continuous load is increased by 125 percent, the loading of the OCPD is derated by 80 percent per **210.19(A)(1), 210.20(A)(1), 408.30,** and **220.10.**

Branch-circuits and feeders in commercial and industrial locations are calculated differently than those utilized in residential dwelling units. Most loads in commercial and industrial locations are used continuously for three hours or more without being interrupted. Therefore, such circuits supplying these loads must be calculated at 125 percent of their rating. However, there are certain types of loads that operate at noncontinuous operation and others will be thermostatically controlled.

LIGHTING LOADS
ARTICLE 220, PARTS I; II

Incandescent or electric-discharge lighting units can be installed in or on a premise. Such a load must be calculated at noncontinuous operation (100 percent) or continuous operation. Noncontinuous operated loads are calculated at 100 percent or with demand factors if they are used for less than three hours at any given time. Continuous operated loads are calculated at 125 percent when they are used for more than three hours without employing an OFF or cycle period to interrupt the continuous operation.

Feeder-Circuits and Branch-Circuits

NONCONTINUOUS DUTY
210.19(A)(1); 210.20(A)

Noncontinuous lighting shall be calculated at 100 percent of the total VA or amperage rating of the branch-circuit. Lighting units and cord-and-plug connected table lamps or floor lamps used at various intervals of time are allowed to be installed or connected to these circuits. The branch-circuit or feeder will never be overloaded during its time of use if proper calculations are made and the correct size elements selected.

For example: How many outlets (1.5 per outlet) shall be permitted to be connected to a 20 amp branch-circuit used at noncontinuous operation?

Step 1: Finding amperage of outlets
220.14(I); 210.19(A)(1); 210.20(A)
A = 180 VA x 100% ÷ 120 V
A = 1.5

Step 2: Finding number of outlets
210.11(A)
= 20 A OCPD ÷ 1.5 A
= 13 (Round down 13.3 to 13)

Solution: The number of outlets permitted on a 20 amp OCPD is 13.

CONTINUOUS DUTY
210.19(A)(1); 210.20(A)

Continuous lighting loads shall be calculated at 125 percent of the total VA rating of the branch-circuit or feeder. Incandescent or electric-discharge lighting units shall be permitted to be installed in or on commercial or industrial buildings for continuous operation of lighting loads.

For example: How many lighting outlets are permitted to be installed on a 20 amp branch-circuit used at continuous operation?

Step 1: Finding amperage of outlets
220.14(I); 210.19(A)(1); 210.20(A)
A = 180 VA x 125% ÷ 120 V
A = 1.875

Step 2: Finding number of outlets
210.11(A)
= 20 A OCPD ÷ 1.875 A
= 10 (Round down 10.7 to 10)

Solution: The number of continuous operated lighting outlets permitted on a 20 amp OCPD is 10.

OTHER LOADS
210.19(A)(1); 210.20(A)

The rating of the branch-circuit overcurrent device serving continuous loads, such as store lighting and similar loads, shall not be less than the noncontinuous load plus 125 percent of the continuous load. (In some cases, demand factors can be applied.)

Master Test Tip 53: Applying the same procedure (15 A OCPD ÷ 1.5 A = 10), 10 outlets are permitted on a 15 amp branch-circuit. The limitation of outlets on branch-circuits alleviates the nuisance tripping of the OCPD.

Note that a total lighting and receptacle load (General-purpose) of 100,000 VA, which is used in patient care rooms of a hospital, can have a demand applied to such load as follows:

• Applying demand
220.42; 220.44; Table 220.42
First 50,000 VA x 40% = 20,000 VA
Next 50,000 VA x 20% = 10,000 VA
Total = 30,000 VA

Solution: The demand factor load is 30,000 VA.

Note: See Master Test Tip 55.

Master Test Tip 54: Applying the same procedure as in the "For Example" problem (15 A OCPD ÷ 1.875 A = 8), eight outlets are permitted on a 15 amp OCPD. The limitation of outlets on a branch-circuit alleviates the nuisance tripping of the OCPD.

Master Test Tip 55: Lighting loads in dwelling units, hospitals, hotels, motels, and warehouses can have demand factors applied as outlined in **Table 220.42**.

Master Test Tip 56: The minimum branch-circuit and feeder conductor size, without the application of any adjustment or correction factors, shall have an allowable ampacity equal to or greater than the noncontinuous load plus 125 percent of the continuous load. **(See Figure 6-19)**

Master Test Tip 57: The neutral conductor is not considered a current-carrying conductor when designed to carry only the unbalanced load from resistive loads connected to the ungrounded phase conductors.

Figure 6-19. Both the OCPD and conductors shall be calculated at 100 percent for noncontinuous loads and 125 percent for continuous loads. These values are added together to obtain the total load. Note that, when appropriate, demand factors can be applied to specific loads.

Master Test Tip 58: The connected load is the total VA or amps on the nameplate of the equipment. To determine the calculated load, the VA or amps on the nameplate of the equipment has to be multiplied by 125 percent for continuous loads and 100 percent for noncontinuous loads **(See Master Test Tips 43 and 44)**

For example: What is the maximum continuous operated load that can be connected to a 20 amp branch-circuit?

Step 1: Finding amperage (using 80% rule)
408.30; 220.10; 210.19(A)(1); 215.2(A)(1)
A = 20 A x 80%
A = 16 A

Solution: The branch-circuit load is 16 amps.

For example: What size overcurrent protection device is allowed to serve a continuous operated load of 14 amps and noncontinuous load of 2.5 amps?

Step 1: Finding amperage (using 100% and 125% rule)
210.19(A)(1); 210.20(A)
14 A x 125% = 17.5 A
2.5 A x 100% = 2.5 A
Total load   = 20 A

Note: To obtain total load, the 125% rule is applied and not the 80% rule.

Solution: A 20 amp overcurrent protection device is required.

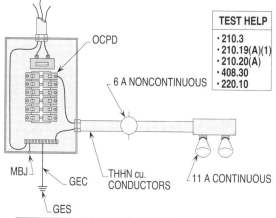

TEST HELP
• 210.3
• 210.19(A)(1)
• 210.20(A)
• 408.30
• 220.10

What size OCPD and THHN cu. conductors are required based upon an ambient temperature of 135° F?

Step 1: Finding load
210.19(A)(1); 210.20(A)
Conductors   OCPD
11 A x 125%   = 13.75 A
 6 A x 100%   = 6.00 A
17 A - connected load = 19.75 A - calculated load

Step 2: Applying correction factors per **Table 310.16** using 12 AWG THHN cu.
30 A x 71% = 21.3 A

Step 3: Finding conductor and OCPD
Table 310.16; 240.4(B); 240.6(A)
21.3 A supplies 17 A load (connected load)
19.75 A (calculated load) requires 20 A OCPD
20 A is the next size above 17 A and 19.75 A

Solution: 12 AWG THHN cu. conductors and 20 amp OCPD is required.

NEC 210.19(A)(1); 210.20(A)

ELECTRIC DISCHARGE LOADS
220.4(B)

Branch-circuit overcurrent protection devices shall be sized at 125 percent of the VA or amperage rating of each ballast where they are installed to supply the ballasts of electric discharge lighting units. Electric discharge lighting units loads are classified as follows:

(1) Fluorescent
(2) Mercury vapor
(3) High pressure sodium
(4) Low pressure sodium
(5) Metal halide

For example: What is the lighting load for a branch-circuit or feeder-circuit supplying twelve 1.5 amp ballasts that serve twenty-four F25, CW lamps used at continuous operation?

Step 1: Finding amperage
220.4(B)
I = ballast A x No. of outlets
I = 1.5 A x 12
I = 18 A

Step 2: Applying percentage
210.19(A)(1); 210.20(A)
I = 18 A x 125%
I = 22.5 A

Solution: The branch-circuit or feeder-circuit load is 22.5 amps.

Master Test Help: The total wattage of each lamp of an inductive lighting load shall not be calculated at 125 percent to determine the total lighting load. This type of calculation does not comply with the NEC and usually requires a larger lighting load than the VA rating of each ballast times the number times 125 percent.

Master Test Tip 59: The neutral conductor shall be considered a current-carrying conductor if it is installed for electric discharge lighting loads unless loading does not comply with **310.15(B)(4)(c)** to ampacity Tables 0-2000 volts.

OUTLETS
220.14(H); (I); (L)

Branch-circuits or feeder circuits of 120 volts shall be calculated at 180 VA (1.5 A) per outlet. This rule applies on the examination, where the VA rating of each ballast or the wattage of each incandescent bulb is unknown. Heavy-duty lampholders of 120 volts shall be calculated at 600 VA or 5 A each. Lampholders connected to a branch-circuit having a rating in excess of 20 amps shall have a rating of not less than 660 watts or 5.5 amps if of the admedium type, and not less than 750 watts or 6.25 amps if of any other type, per **210.21(A)**. Outlets are calculated at noncontinuous operation and continuous operation times their VA rating to determine the load. OCPD's and conductors for continuous loads shall be calculated at 125 percent of the load served. **(See Figure 6-20)**

Master Test Tip 60: If the lighting or receptacle outlet rating is not given, the minimum rating used for each outlet is 180 VA per **220.14(L)**.

For example: What size overcurrent protection device and THHN copper conductors are required for a branch-circuit supplying 11 outlets used at noncontinuous operation to serve luminaires (lighting fixtures)?

Step 1: Finding VA
210.19(A)(1); 210.20(A)
VA = No. of outlets x VA x 100%
VA = 11 x 180 VA x 100%
VA = 1980

Step 2: Finding amperage
210.11(A)
I = VA ÷ V
I = 1980 VA ÷ 120 V
I = 16.5

Step 3: Finding conductor and OCPD
Table 310.16; 240.4(D); 240.4(B); 240.6(A)
12 AWG THHN cu. = 30 A on 20 A OCPD

Solution: The branch-circuit conductors are sized 12 AWG, and the overcurrent protection is 20 amps.

Master Test Tip 61: Branch-circuits supplied by 14 AWG conductors must be protected by a 15 amp OCPD, 12 AWG by a 20 amp OCPD, and a 10 AWG by a 30 amp OCPD per **240.4(D)**. Also review **240.21(A)** thru **(G)** for exceptions to this rule. (See Figure 6-27 where the OCPD can be sized larger than the ampacity of the conductors.)

For example: What size overcurrent protection device and THHN copper conductors are required for a branch-circuit supplying 9 outlets used at continuous operation to serve lighting fixtures?

Step 1: Finding VA
210.19(A)(1); 210.20(A)
VA = No. of outlets x VA x 125%
VA = 9 x 180 VA x 125%
VA = 2025

Step 2: Finding amperage
210.11(A); 220.5(A)
I = VA ÷ V
I = 2025 VA ÷ 120 V
I = 16.9

Step 3: Finding conductor and OCPD
Table 310.16; 240.4(D); 240.4(B); 240.6(A)
12 AWG cu. = 30 A
OCPD = 20 A

Solution: The branch-circuit conductors are sized 12 AWG, and the overcurrent protection device is 20 amps, per asterisks to Table 310.16.

Figure 6-20. Outlets are calculated at noncontinuous operation (100 percent) and continuous operation (125 percent) times their VA rating to determine the branch-circuit load. Overcurrent protection devices and conductors shall be calculated in the same manner as the outlets.

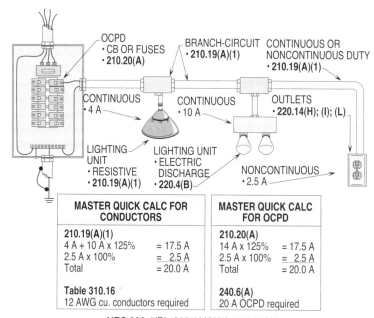

SHOW WINDOW LOADS
220.14(G); 220.43(A); 210.62

Master Test Tip 62: The greater between the number of outlets times 125 percent or the 200 VA per linear foot times 125 percent must be used for the calculation of a show window.

Show window lighting loads shall be calculated at a minimum of 180 VA per outlet for branch-circuits or feeder-circuits, if the VA is unknown. The total VA rating for show window lighting loads shall be multiplied by 125 percent for each classified circuit. Show window lighting loads are multiplied by 125 percent because they can operate for three hours or more to provide illumination for advertisement.

For example: What is the number of outlets permitted to be installed on a 20 amp branch-circuit supplying lighting loads in a show window?

Step 1: Finding No. of outlets
220.14(G)(1); (B)(11); 210.19(A)(1); 220.5(B)
No. of outlets = device x 120 V ÷ VA x 125%
No. of outlets = 20 A x 120 V ÷ 180 VA x 125%
No. of outlets = 10.7 (Round down to 10)

Solution: The number of outlets permitted to be installed on a 20 amp branch-circuit is 10. Note: Testing agencies usually allow 180 VA to be used if the load for such units is unknown.

See Figure 6-21 for calculating the load for show windows based upon either linear foot or individual outlets.

MASTER QUICK CALC	MASTER QUICK CALC
Number Of Receptacle Outlets 210.62.	Finding Load in VA 220.14(G)(2)
# = 80'/12' = 6.7	80 ft. x 200 = 16,000 VA
# = 7 RECEPTACLE OUTLETS	Finding Continuous Load 210.19(A)(1)
	16,000 VA x 125% = 20,000 VA

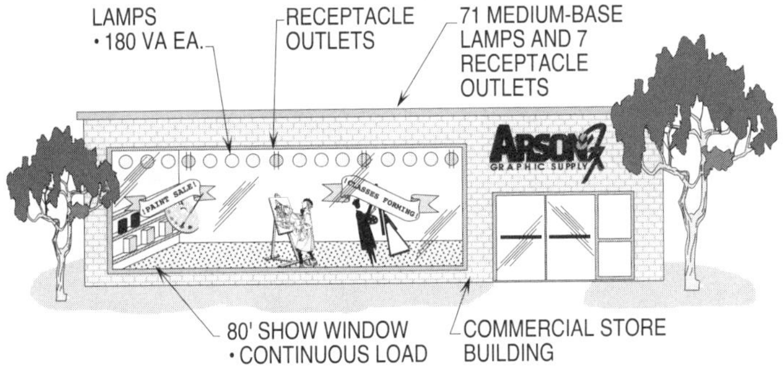

Figure 6-21. Calculating the load for a show window using the linear foot calculation per **220.14(G)(1), 220.19(A)(1),** and **225.2(A)(1)**.

LIGHTING TRACK LOADS
220.43(B)

Lighting track loads shall be calculated at 150 VA for each 2 ft. of track to determine the load of the branch-circuits. To properly balance the load, the load of the VA rating is divided between the number of circuits supplying the length of lighting track. (recommended)

By using one of the following methods (recommended), the load on multicircuit lighting track can be balanced as adequately as possible:

Calculating 2 circuits
2 branch-circuits
150 VA for each 2'
VA = 150 VA ÷ 2
VA = 75

Calculating 3 circuits
3 branch-circuits
150 VA for each 3'
VA = 150 VA ÷ 3
VA = 50

The 150 VA rule (multicircuit track lighting) shall not apply to dwelling units per **220.43(B)**. For dwelling units, the load for track lighting is included in the 3 VA per sq. ft. per **Table 220.12**. **(See Figure 6-22)**

Master Test Tip 63: Note that the 150 VA for each 2 ft. of lighting track does not apply to dwelling units or guest rooms in hotels or motels per **220.43(B)**.

Master Test Tip 64: The length of the lighting track is not limited by the 150 VA for each 2 ft. rule for calculating load values per **220.43(B)**.

Figure 6-22. The 150 VA for each 2 ft. of lighting track shall be used to compute the load for service or feeder-circuit loads. The load of the track should be divided as evenly as possible on each circuit supplying the track.

Note that lighting track must be supplied by a branch-circuit having a rating not more than that of the track per **410.101(B)**.

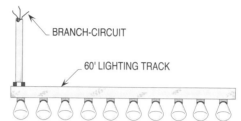

NOTE: DIVIDING THE LOAD IS A RECOMMENDED PROCEDURE ON THE TEST. TEST QUESTIONS OR PROBLEMS WILL STATE WHEN TO APPLY THIS METHOD.

What is the VA rating of 60 ft. of lighting track and for two or three circuits with the load divided?

Finding VA of track

Step 1: Calculating VA
220.43(B)
VA = Length ÷ 2' × 150 VA
VA = 60' ÷ 2' × 150 VA
VA = 4500

Solution: **VA = 4500**

Dividing loads

Step 1: Calculating loads
Two circuits   Three circuits
4500 VA ÷ 2 = 2,250 VA   4500 VA ÷ 3 = 1500 VA

Solution: **For two-circuits 2250 VA is used and for three-circuits 1500 VA is used.**

NEC 220.43(B)
NEC 410.10(B)

Master Test Tip 65: Where multicircuit lighting track systems are installed, the loads requirements of **220.43(B)** is considered to be divided equally between the supply circuits.

Master Test Tip 66: Each sign and outline lighting system must be installed with a disconnecting means to open all ungrounded (HOT) phase conductors per **600.6**.

Master Test Tip 67: The disconnecting means in Master Test Tip 66 must be within sight (50') from the sign or outline lighting system that it is designed to disconnect. (See **600.6(A)**)

SIGN LOADS
ARTICLE 600, PART I

Each commercial building and each commercial occupancy accessible to pedestrians from a sidewalk, street, etc. shall be provided with a sign circuit. Signs are considered to be continuous operation (three hours or more) when they are installed for commercial occupancies.

SIZE REQUIRED
600.5(A)

Each commercial building and each commercial occupancy shall have at least one 20 amp branch-circuit to supply an outlet for a sign or outline lighting that is located in an accessible location. This 20 amp branch-circuit shall have no other loads connected to the overcurrent protection device protecting such circuits.

RATING
600.5(B)(1); (2)

Branch-circuits that supply signs and outline lighting systems containing incandescent, fluorescent, and high-intensity discharge forms of illumination are limited to 20 amps or less. Branch-circuits that supply transformers for neon tubing installations shall be limited to 30 amps or less.

For example, transformers installed for channel letters and ballasts for electric discharge lamps shall be rated 20 amps or less for branch-circuits. Transformers installed only to connect branch-circuits for neon or channel letters shall be rated 30 amps or less.

Feeder-Circuits and Branch-Circuits

CALCULATED LOAD
210.19(A)(1)

The service or feeder calculation for a sign shall be calculated at a minimum of 1200 VA, and the load is multiplied by 125 percent for continuous operation if the sign burns for three hours or more. When sizing the load from the overcurrent protection device, the load shall be multiplied by a 125 percent to obtain the load based upon continuous operation.

For example: What is the total load allowed for a continuous operated wall sign supplied by a 20 amp, 120 volt branch-circuit?

Step 1: Finding amperage
600.5(A)
OCPD = 20 A

Step 2: Calculating amperage
600.5(A); 210.19(A)
20 A x 80% = 16 A
16 A x 125% = 20 A

Step 3: Selecting OCPD
600.5(A); 210.20(A)(1)
16 A load is allowed

Solution: The size overcurrent protection device required is 20 amps for a sign that burns for three hours or more at 16 amps or less.

See Figure 6-23 for calculating the load for a sign.

Master Test Tip 68: The bonding jumper used with nonmetallic conduit must be at least 14 AWG cu., per **600.7(D)**.

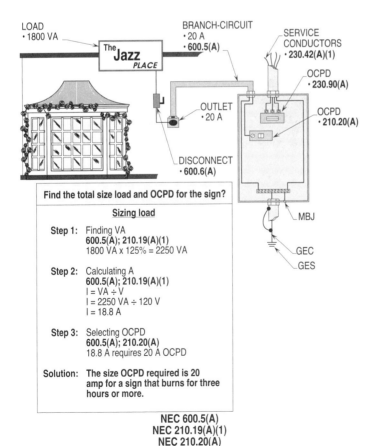

Figure 6-23. Sign loads that can operate for three hours or more shall be calculated at 125 percent, and elements selected from this value.

Master Test Tip 69: The load for a sign must be calculated at a minimum of 1200 VA per **600.5(A)** and **220.14(F)**.

RECEPTACLE LOADS
ARTICLE 220; PARTS I; II

Receptacle outlets shall be permitted to be installed for cord-and-plug connected appliances, utilization equipment, and table or floor lamps in dwelling units, apartments, condominiums, townhouses, and commercial or industrial locations. Cord-and-plug connected loads shall be connected in an arrangement so they won't load up the branch-circuit. Cord-and-plug connected loads are of the general-purpose type. Cord-and-plug connected items shall be installed and located in such a manner to prevent the use of extension cords except for temporary use.

Master Test Tip 70: Receptacles rated at 15 or 20 amp can be connected to a 20 amp general-purpose circuit per **210.21(B)(3)** and **Table 210.21(B)(3)**.

GENERAL PURPOSE
220.14(I)

A general purpose branch-circuit shall be calculated at 180 VA each if they supply more than one outlet for the connection of cord-and-plug connected items. The number of outlets shall be calculated at 180 VA times the noncontinuous operation at 100 percent. For continuous operation the 180 VA shall be increased by 125 percent. Note that overcurrent protection devices and conductors shall be sized at 125 percent of the load. **(See Figure 6-24)**

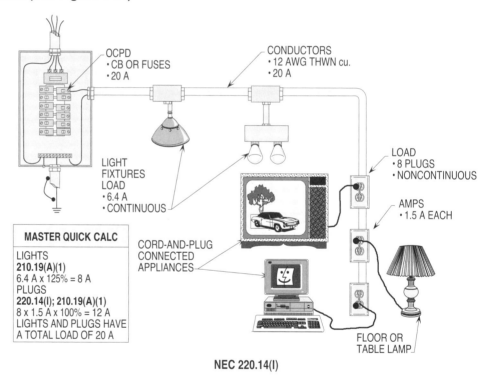

Figure 6-24. A general purpose branch-circuit shall be calculated at 180 VA for each outlet where supplying more than one outlet utilizing cord-and-plug connected items. The number of outlets shall be calculated at 180 VA times noncontinuous operation at 100 percent (or with demand factors) and continuous operation at 125 percent.

Master Test Tip 71: Note that 50 percent of a 15 or 20 amp circuit can be used to supply a small fixed appliance per **210.23(A)**. For application of this rule, see Figure 6-11 and example problem on page 6-17.

For example: What is the load for eight receptacle outlets that are supplying cord-and-plug connected loads used at noncontinuous operation?

Step 1: Finding VA
Load = No. of outlets x 180 VA x 100%
Load = 8 x 180 VA x 100%
Load = 1440 VA

Solution: The load is 1440 VA.

For example: How many receptacle outlets operating noncontinuously can be connected to a 20 amp general purpose branch-circuit?

Step 1: Finding amperage of outlets
220.14(I)
180 VA ÷ 120 V = 1.5 A

Step 2: Finding number of outlets
220.14(I)
20 A OCPD ÷ 1.5 A x 100% = 13 (Round 13.3 down to 13)

Solution: The number of outlets permitted on a 20 amp OCPD is 13.

For example: How many receptacle outlets operating continuously can be connected to a 20 amp general purpose branch-circuit?

Step 1: Finding amperage of outlets
220.14(I)
180 VA ÷ 120 V = 1.5 A

Step 2: Finding number of outlets
220.14(I)
20 A OCPD ÷ 1.5 A x 125% = 10 (Round 10.7 down to 10)

Solution: The number of outlets permitted on a 20 amp OCPD is 10.

INDIVIDUAL
210.19(A)(1); 210.20(A)

Individual cord-and-plug loads connected to outlets served by branch-circuits shall be determined by multiplying the noncontinuous operated loads by 100 percent and continuous operated loads by 125 percent. The amount of amperage for each cord-and-plug connected load shall be found by applying the power formula. The following abbreviations shall apply for the specific rules shown when using the power formula to calculate loads in VA or amps:

I	= amps	I	= VA ÷ V
P	= volt-amps	VA	= I x V
E	= voltage	V	= VA ÷ I

For example: What size THWN copper conductors and overcurrent protection device is required for an individual branch-circuit to a hot tub with a nameplate current rating of 42 amps? (Note that terminals are 75°C rated.)

Step 1: Finding amperage
680.9
42 A is the circuit current rating

Step 2: Calculating load
680.9
42 A x 125% = 52.5 A

Step 3: Selecting conductors
Table 310.16
52.5 A requires 6 AWG THWN cu.

Step 4: Selecting OCPD
110.14(C)(1)(a); 240.4(B); 680.9
52.5 A load requires 60 A OCPD
65 A conductor allows 70 A OCPD

Solution: A 60 amp OCPD and 6 AWG THWN copper conductors are permitted.

See **Figure 6-25** for calculating the load to an individual cord-and-plug connected load.

Master Test Tip 72: Cord-and-plug equipment shall not exceed 80 percent of the branch-circuit rating per **210.23(A)**.

Master Test Tip 73: When making calculations in this book, the variation of the power formula is used since the NEC recognizes volt-amps (VA) for load calculations per Examples in Ch. 9.

Master Test Tip 74: Note that a hot tub must have a disconnecting means located within sight (50') from the tub's location per **680.12** and **Article 100**.

Note that **680.41** (except for single family dwelling units) requires a clearly labeled emergency shutoff or control switch within 5 ft. of a hot tub or spa.

Figure 6-25. Individual loads shall be calculated at 100 percent for noncontinuous and 125 percent for continuous, times the nameplate FLA of the appliance.

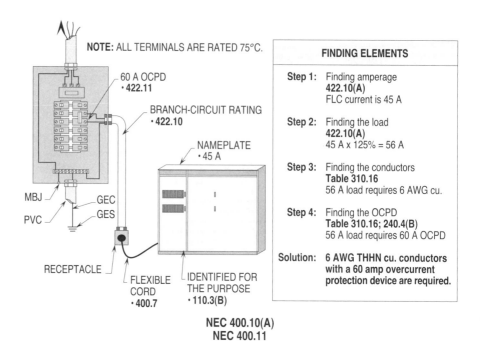

Master Test Tip 75: Section **250.148** also requires EGC's to be pigtailed and bonded properly to metal boxes such as fixture boxes, device boxes, junction boxes, etc.

MULTIWIRE BRANCH-CIRCUITS
210.4(A) THRU (D)

Multiwire branch-circuits are used to supply power to line-to-neutral loads only when more than one hot is sharing the neutral. All circuits shall originate from the same panelboard and shall be required to be switched by individual circuit breakers. Multiwire branch-circuits shall be supplied by double pole circuit breakers or individual circuit breakers with approved handle ties to provide safety from electrical shock while serving the circuits or equipment.

Section **300.13(B)** makes it mandatory that the neutrals be pigtailed to prevent different voltage levels between the hot ungrounded conductors should the neutral become loose due to bad connections, etc.

Multiwire branch-circuits can serve power to one outlet for an individual piece of equipment with the other circuit(s) supplying power to a number of outlets that are used for receptacles, lights, and other cord-and-plug connected equipment. **(See Figure 6-26)**

COMMERCIAL COOKING EQUIPMENT
210.19(A)(1); 215.2(A)(1); TABLE 220.56

Master Test Tip 76: The overcurrent protection device shall be calculated at 125 percent and the conductors shall be calculated at 125 percent. The OCPD may have to be increased or decreased in size per **240.4(A)** thru **(G)** or **240.4(B)**.

Commercial cooking equipment shall be calculated at noncontinuous operation (100 percent) or continuous operation (125 percent) to determine the branch-circuit load. When installing more than one piece of cooking equipment, the demand factor shall be selected from **Table 220.56**.

For example: What is the load for two 10 kW cooking units that are supplied by a 240 volt, single-phase branch-circuit when one is used at noncontinuous operation and the other one used at continuous operation?

Step 1: Finding kW for OCPD
210.19(A)(1); 215.2(A)(1)
10 kW x 100% = 10 kW

Step 2: Finding kW for conductor
210.19(A)(1); 215.2(A)(1)
10 kW x 125% = 12.5 kW

Solution: The noncontinuous operated load is 10 kW and the continuous operated load is 12.5 kW.

See Figure 6-27 for calculating the load for a commercial cooking unit.

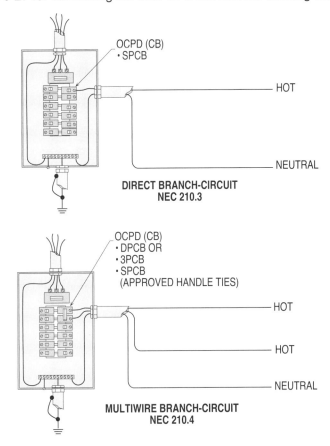

Figure 6-26. Multiwire branch-circuits can be used to supply power to line-to-neutral loads only when more than one hot is sharing the neutral.

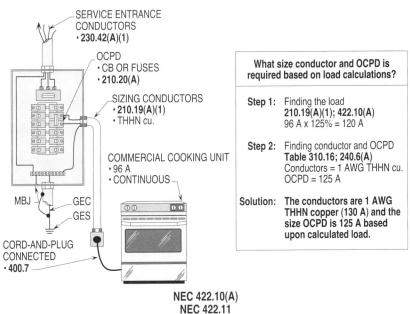

Figure 6-27. Commercial cooking equipment shall be calculated at non-continuous operation (100 percent) or continuous operation (125 percent) to determine the branch-circuit load.

WATER HEATER LOADS
ARTICLE 422, PART III

Master Test Tip 77: Section **422.31(B)** requires a water heater rated over 300 VA to be provided with a disconnecting means to open all ungrounded conductors.

Water heaters are designed with elements to heat water at different stages of use. For larger amounts of hot water, more of the elements must be connected into the circuit to heat the water to replace that which was used. When smaller amounts of hot water are needed, fewer elements are used in the circuit to heat the water.

CONDUCTORS
422.13; 422.10(A)

Storage-type water heaters having a capacity of 120 gallons or less shall have a rating not less than 125 percent of the nameplate rating to size the branch-circuit conductors.

For example: What size conductors are required to supply power to a 240 volt, single-phase water heater pulling 5000 VA?

    Step 1: Finding amperage per **220.5(B)**
    5000 VA ÷ 240 V = 21 A (Round 20.8 A up to 21 A)

    Step 2: Finding the loads
    422.13; 422.10(A); 220.5(B)
    21 A x 125% = 26 A (Round 26.25 A down to 26 A)

    Step 3: Finding the conductors
    Table 310.16; 240.4(D)
    26 A load requires 10 AWG cu.

Solution: The size conductors required are 10 AWG copper.

OVERCURRENT PROTECTION
422.13; 422.10(A)

Master Test Tip 78: Water heaters must be provided with a temperature limiting device, which is in addition to its control thermostat.

Overcurrent protection devices shall be sized not less than 125 percent of the heating load to prevent tripping open the OCPD and disconnecting all the elements connected into the circuit. Approximately 100 percent of the water heater's connected load is pulled from resistance heating elements.

For example: What size overcurrent protection device is required to supply power to a 240 volt, single-phase water heater pulling 5000 VA at continuous operation?

    Step 1: Finding load for OCPD
    422.13; 422.10(A)
    21 A x 125% = 26 A

    Step 2: Finding the OCPD
    422.13; 240.4(B)
    26 A load requires a 30 A OCPD

Solution: The overcurrent protection device is 30 amps.

DISCONNECTING MEANS
422.31(B)

A disconnecting means shall not be required to be installed at the water heater when the overcurrent protection device is readily accessible. The overcurrent protection device shall be permitted to serve as the disconnect where installed in a service panel that is readily accessible and is located outside or inside the building. An accessible cord-and-plug can serve as the disconnecting means for cord-and-plug connected water heaters where readily accessible per **422.33(A)**. However, such units must be listed for cord-and-plug connection by a qualified testing laboratory per **110.3(B)**. **(See Figure 6-28)**

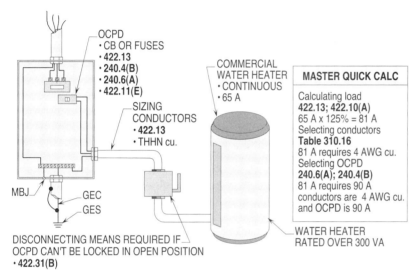

Figure 6-28. A disconnecting means is not required to be installed at the water heater when the overcurrent protection device is readily accessible.

HEATING LOADS
ARTICLE 424, PART I

Heating elements in heating units are rated at 5 kW each. The elements are stacked in the heating unit to provide the rated kW for a particular size occupancy. Two stacked elements provide a 10 kW heating unit. Three stacked elements provide a 15 kW heating unit and so forth. Branch-circuit conductors and overcurrent protection devices shall be sized and selected based on the kW rating of each heating unit plus the blower motor per **424.3(B)**. (See **Table 220.3**)

Master Test Tip 79: Field-wired conductors between the heating unit and the supplementary OCPD's shall not be less than 125 percent of the load served per **424.22(E)**.

Heating units with a number of elements must be subdivided so that they can more easily be supplied from a local panelboard or from the service equipment. Individual branch-circuits or a feeder-circuit is utilized in routing power to the heating unit.

CONDUCTORS
Table 220.3; 424.3(B)

Heating units shall be calculated at 125 percent of the heating element load plus 125 percent of the blower motor load if both are present. To determine the branch-circuit conductors, the heating elements and blower motor load shall be calculated at 125 percent. This calculated value in amps is used to select the ampacity of the conductors.

Master Test Tip 80: Heating equipment rated more than 48 amps shall have their heating elements subdivided, and such load shall not exceed 48 amps per **424.22(B)**.

For example: What size THWN copper conductors are required for a 25 kW, 240 volt, single-phase heating unit with a 4 amp blower motor?

- **Step 1:** Finding amperage
  25 kVA x 1000 ÷ 240 V = 104 A

- **Step 2:** Finding the load
  424.3(B); Table 220.3
  104 A + 4 A x 125% = 135 A

- **Step 3:** Finding conductors based on load
  Table 310.16
  135 A load requires 1/0 AWG THWN cu. rated at 150 amps

Solution: 1/0 AWG THWN copper conductors are required.

OVERCURRENT PROTECTION
424.3(B); Table 240.4(G)

To determine the size of the OCPD, a heating unit load shall be calculated at 125 percent of the heating element load plus 125 percent of the blower motor load. The next higher size rated overcurrent protection device shall be permitted to be installed where the load in amps does not correspond to a standard OCPD per **240.4(B)**. The next higher prevents the nuisance tripping of the OCPD if the heating unit requires all the 5 kW elements to be activated to satisfy the heating load called for by the thermostat.

Master Test Tip 81: Resistance heating elements rated at 48 amps shall be protected by an OCPD rated no greater than 60 amps (48 A x 125% = 60 A) per **424.22(B)**.

For example: What size overcurrent protection device is required for a 25 kW, 240 volt, single-phase heating unit with a 4 amp blower motor?

- **Step 1:** Finding load for OCPD
  424.3(B); Table 240.4(G)
  104 A + 4 A x 125% = 135 A

- **Step 2:** Finding OCPD based on load
  240.6(A); 240.4(B)
  150 A is the next higher standard size

Solution: The size OCPD is 150 amps based on the load.

DISCONNECTING MEANS
424.19

A disconnecting means shall be installed for a self-contained heating unit with a controller that energizes the circuits to the heating elements and blower motor. A fused or nonfused disconnect, an automatic breaker, or a nonautomatic circuit breaker used as the disconnecting means for a heating unit shall be located within sight and within 50 ft. of the heating unit per **Article 100, 424.19,** and **430.102. (See Figure 6-29)**

AIR-CONDITIONING LOADS
ARTICLE 440, PART III

Master Test Tip 82: OCPD's for A/C units shall be increased by 175 percent to 225 percent to allow the compressor to start and run per **440.22(A)**.

Air-conditioning units are rated in Btu or tons. One ton contains 12,000 Btu. A five ton compressor in a central air-conditioning unit has 60,000 Btu. (5 x 12,000 Btu. = 60,000 Btu.). The air-conditioner load is calculated by the square footage and construction of the premises. The size air-conditioner rated in Btu. is selected from this calculation. The conductors shall be sized large enough to carry the load of the compressor, and the overcurrent protection device shall be selected to allow the compressor to start without tripping open, due to inrush current from high head pressures.

Feeder-Circuits and Branch-Circuits

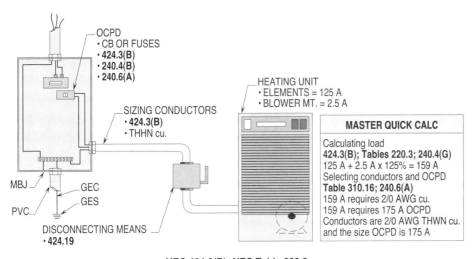

Figure 6-29. A disconnecting means shall be installed for a self-contained heating unit with a controller that energizes the circuits to the heating elements and blower motor.

CONDUCTORS
440.32; Table 220.3

Branch-circuit conductors supplying a single motor-compressor shall have an ampacity not less than 125 percent of either the motor-compressor rated load current or the branch-circuit selection current, whichever is greater.

The branch-circuit conductors shall be sized large enough to prevent damage to the insulation of the conductor caused by an overload. Overload relays are installed and are usually adjusted to trip open on overloads exceeding 140 percent of full-load current rating of the compressor. This type of condition is compensated for sizing and selecting the conductors at 125 percent of the compressor's full-load current rating. The condenser-motor load is added at 100 percent if present on the same branch-circuit.

Master Test Tip 83: Branch-circuit selection current is the value in amps to be used instead of the rated FLA in determining the ratings of certain components of the supply circuit per **440.2**.

OVERCURRENT PROTECTION
440.22(A); Table 240.4(G)

To allow the compressor to start and run, the overcurrent protection device must be sized properly. To determine the rating of the overcurrent protection device, a rating or setting not exceeding 175 percent of the motor-compressor's rated full-load current shall be used or the branch-circuit selection current, whichever is greater. The overcurrent protection device shall be permitted to be increased to 225 percent of the full-load current if an air-conditioner unit will not start using 175 percent. Condenser motors shall be calculated at 100 percent and added to this 125 percent total for central A/C units if they are not figured with the nameplate rating.

Master Test Tip 84: To size the OCPD, the A/C unit's compressor shall be calculated at 175 percent of its FLA and this value added to 100 percent of the condenser's FLA and the rating selected per **440.22(A)**.

Air-conditioning units installed on the roof or on the side of premises where the sun goes down may have trouble starting and operating during hot summer months. Overcurrent protection devices can be sized by the listing on the nameplate of the unit. Fuses and circuit breakers (HACR) shall be sized and installed where such sizes are specifically listed on the nameplate of the unit.

For example: What is the minimum and maximum size overcurrent protection device required to supply power to a 240 volt, single-phase air-conditioning unit with a compressor rating of 55 A plus a 2.5A condenser motor? **(See Figure 6-30)**

6-43

DISCONNECTING MEANS
440.14

The disconnecting means shall be located within 50 ft. and within sight from the air-conditioning unit per **Article 100**. The disconnecting means shall be permitted to be installed on or within the air-conditioning unit. The disconnecting means shall be permitted to be installed where capable of being locked in the open position where the conditions of maintenance and supervision ensure that only qualified personnel will service the equipment. **Note:** This Ex. to **440.14** only applies to industrial compressors and not A/C units. A disconnecting means is not required within sight of such equipment. **(See Figure 6-30)**

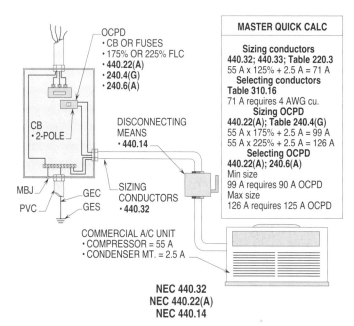

Figure 6-30. The disconnecting means shall be located within 50 ft. and within sight from the air-conditioning unit per **440.14** and **Article 100**.

Master Test Tip 85: Note that the conductors supplying an A/C unit do not have to be increased to 175 percent or 225 percent per **440.32**.

MOTOR LOADS
ARTICLE 430, PARTS II AND IV

Master Test Tip 86: The size of the conductors supplying power to a motor is determined by multiplying the motor's FLC in amps, which is selected from **Table 430.248** for single-phase and **Table 430.250** for three-phase per **430.6(A)(1)**.

Motors are rated in horsepower. The amount of work that a motor can perform depends upon its horsepower rating class, code letter, or design letter. A motor with a high horsepower rating can do more work than a motor with a low horsepower rating. The overcurrent protection devices and conductors used to supply power to motors are sized and based on the horsepower, voltage, and number of phases. The starting current of a motor is based on its code letter per **Table 430.7(B)** or Design letter per **Tables 430.251(A) and (B)**.

CONDUCTORS
430-22(A); Table 220.3

Branch-circuit conductors supplying a single motor shall have an ampacity not less than 125 percent of the motor's full-load current rating. Single-phase motors shall have their full-load current ratings selected from **Table 430.248** per **430.6(A)(1)**. Three-phase motors shall have their full-load current ratings selected from **Table 430.250** per **430.6(A)(1)**.

OVERCURRENT PROTECTION
430.52; TABLE 430.52

When power is applied to the windings of a motor and the motor starts, the motor has an inrush current. The amount of current required to drive a load is called running current. Inrush currents of the motor shall be held by an overcurrent protection device sized large to allow the motor to accelerate the driven load. The inrush current of most motors is four to six times the running current of the motor, based on its code letter per **Table 430.7(B)** or Design letter per **Tables 430.251(A) and (B)**. Note that Design E motors can have 8 1/2 to 15 times the FLA of the motor per 2002 NEC.

Master Test Tip 87: If the problem on the master test calls for a minimum size OCPD, then the rating of the OCPD must not exceed **Table 430.52** per **430.52(C)(1)**.

Overcurrent protection devices shall not exceed the percentages listed in **Table 430.52**. Section **430.52** requires the overcurrent protection device to be based on the percentages for the type of device used in the columns of **Table 430.52**. If the percentage does not correspond to a standard device listed in **240.6(A)**, the higher standard size rating shall be permitted to be used per **430.52(C)(1), Ex. 1**. If the motor will not start and run, the percentage may be increased per **430.52(C)(1), Ex. 2**. The percentages used to size these overcurrent protection device are as follows:

Master Test Tip 88: If a problem on the master test calls for the maximum size OCPD to start the motor then the percentages list for each type OCPD in **430.52(C)(1), Ex. 2** must not be exceeded.

(1) Nontime-delay fuse - shall not exceed 400 percent of the full-load current for fuses rated 600 amps or less per **430.52(C)(1), Ex. 2(a)**.

(2) Time-delay fuse - shall not exceed 225 percent of the full-load current per **430.52(C)(1), Ex. 2(B)**.

(3) Circuit breaker - shall not exceed 400 percent of the full-load current for ratings of 100 amps or less or shall not exceed 300 percent of the full-load current for ratings over 100 amps per **430.52(C)(1), Ex. 2(c)**.

(4) Instantaneous trip circuit breaker - shall not exceed 1300 or 1700 percent of the full-load current per **430.52(C)(3), Ex. 1**.

DISCONNECTING MEANS
430.102; 430.107

A disconnecting means shall be located within sight from the controller and shall disconnect the controller to allow personnel to service the motor without the danger of the branch-circuit accidentally being energized. To deenergize the circuit to the motor, the disconnecting means shall be installed by the following procedures:

Master Test Tip 89: If the problem on the master test doesn't call for a minimum or maximum size OCPD, the next size above the percentages listed in **Table 430.52** must be selected per **430.52 (C)(1), Ex. 1**.

(1) The disconnecting means shall be located adjacent to the controller within sight and located within 50 ft. Same rule can be applied for the motor.

(2) The disconnecting means shall be capable of being locked in the open position if located adjacent to the controller. Under this rule, the motor does not have to be located within 50 ft. and within sight of the controller. (**See Ex. to 430.102(B)**)

(3) An additional disconnecting means shall be provided within 50 ft. of the motor and within sight if the disconnecting means by the controller cannot be locked in the open position and the motor is located out of sight or further than 50 ft. (**See Figure 6-31**)

Figure 6-31. A disconnecting means shall be located within sight from the controller and motor and shall disconnect the power to allow personnel to service the motor or controller without the danger of the branch-circuit being accidentally energized.

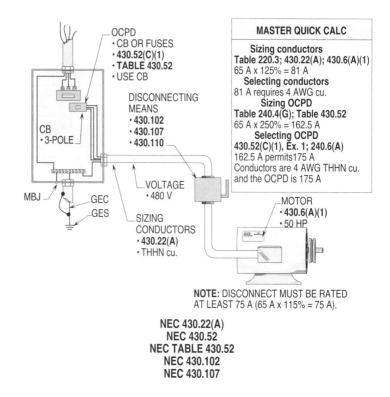

APPLYING EXCEPTION TO 430.102(B)

The exception to **430.102(B)** recognizes situations where installing the disconnecting means within sight of the motor would be impractical or increase the risk of hazards. In an industrial installation where there are written procedures and only qualified employees provide maintenance, the disconnecting means is not required to be within sight of the motor per **430.102(B), Ex. (See Figure 6-32)**

Figure 6-32. The disconnecting means is not required within sight of the motor per **430.102(B), Ex.**

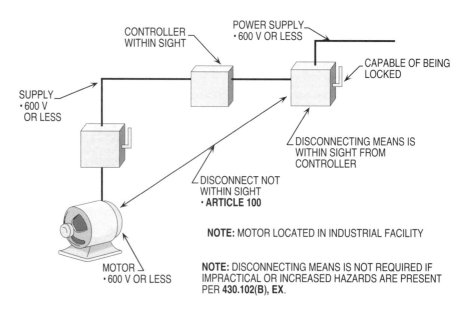

Name _____ Date _____

Chapter 6
Feeder-Circuits and Branch-Circuits

Section   Answer

1. Continuous loads shall be classified as loads that operate for a period of two hours or more.

     (a) True   **(b)** False

2. All inductive neutral current (over 200 amps) for feeder-circuits shall be permitted to have a demand factor applied.

     (a) True   **(b)** False

3. The neutral conductor for a feeder-circuit shall not be considered a current-carrying conductor when carrying only the unbalanced current from other ungrounded phase conductors.

     (a) True   **(b)** False

4. A minimum of three 20 amp, 1500 VA small appliance circuits shall be required to supply receptacle outlets that are located in the kitchen, pantry, breakfast room, and dining room.

     (a) True   **(b)** False

5. At least one 20 amp, 1500 VA laundry circuit shall be required to supply receptacle outlets in the laundry room.

     (a) True   **(b)** False

6. A tap to connect a cooktop and one or more ovens can be made from a 40 amp branch-circuit when the tap conductors are sized from the kW rating of each piece of cooking equipment per Note 4 to **Table 220.55**.

     (a) True   **(b)** False

7. Dryer equipment loads shall be calculated with at least 5000 VA or the nameplate rating, whichever is larger, when sizing the branch-circuit or feeder load.

     (a) True   **(b)** False

8. Conductors with 90°C rated insulation can be connected to terminals rated at 60°C or 75°C.

     (a) True   **(b)** False

9. Continuous lighting loads (branch-circuits) in commercial buildings shall be calculated at 125 percent of the total VA or amperage rating of the unit.

     (a) True   **(b)** False

10. Show window lighting loads shall be calculated at a minimum of 200 VA per outlet for branch-circuits if the VA is unknown.

     (a) True   **(b)** False

Section	Answer

11. When installing four or fewer dryers the load shall be calculated at _____ percent.

    (a) 75   **(b)** 85   **(c)** 90   **(d)** 100

12. A fastened-in-place appliance shall be permitted to be connected to a general-purpose circuit if its amp rating does not exceed _____ percent of the branch-circuit.

    (a) 50   **(b)** 65   **(c)** 75   **(d)** 80

13. A _____ or _____ amp branch-circuit shall be permitted to supply cooking appliances that are fastened-in-place in any occupancy.

    (a) 20 or 30   **(b)** 30 or 40   **(c)** 40 or 50   **(d)** 50 or 60

14. Overcurrent protection devices rated over _____ amps shall be permitted to be terminated with conductors larger than 1 AWG with 75°C ampacities.

    (a) 100   **(b)** 125   **(c)** 150   **(d)** 200

15. The general-purpose circuits shall be calculated at _____ VA per sq. ft. of the dwelling unit.

    (a) 2   **(b)** 3   **(c)** 3.5   **(d)** 4

16. The rating of any one cord-and-plug connected utilization equipment shall not exceed _____ percent of the branch-circuit rating if connected to a general-purpose circuit supplying two or more outlets.

    (a) 50   **(b)** 75   **(c)** 80   **(d)** 100

17. Branch-circuits larger than _____ amps shall supply only nonlighting outlet loads.

    (a) 20   **(b)** 30   **(c)** 40   **(d)** 50

18. 60°C ampacities shall only be applied to 60°C ampacities per **Table 310.16** where conductors _____ through _____ are installed.

    (a) 14 AWG through 2 AWG   **(b)** 14 AWG through 1 AWG
    (c) 12 AWG through 2 AWG   **(d)** 12 AWG through 1 AWG

19. Cord-and-plug connected or permanently (hard-wired) connected receptacles rated over _____ VA are usually supplied by individual circuits.

    (a) 1440   **(b)** 1660   **(c)** 2250   **(d)** 2440

20. Lighting track loads shall be calculated at _____ VA for each 2 ft. of track to determine the load for branch-circuits.

    (a) 150   **(b)** 180   **(c)** 200   **(d)** 220

21. How many 15 amp circuits are permitted to supply power to the general-purpose lighting and receptacle outlets in a 3000 sq. ft. dwelling unit?

    (a) 3   **(b)** 4   **(c)** 5   **(d)** 8

	Section	Answer

22. How many outlets are permitted to be connected to a 20 amp branch-circuit used at continuous operation?

      (a) 5   **(b)** 7   **(c)** 9   **(d)** 10

23. What is the neutral load for a feeder-circuit with a resistive load of 300 amps?

      (a) 230 A   **(b)** 250 A   **(c)** 270 A   **(d)** 300 A

24. What is the voltage drop for a single-phase system using the resistivity method with the following characteristics?

- VA is held to 3%
- Voltage is 240
- Length is 300 ft.
- Load is 180 amps
- OCPD is 200 amps
- Conductors are 3/0 AWG THWN copper

      (a) 6.2835 volts   **(b)** 6.5664 volts
      (c) 6.7382 volts   **(d)** 6.8224 volts

Generators and Transformers

CHAPTER 7

GENERATORS
- LOCATION OF GENERATORS 7-2
- NAMEPLATE MARKINGS 7-2
- OVERCURRENT PROTECTION 7-2
- AMPACITY OF CONDUCTORS FROM GENERATORS 7-6
- PROTECTION OF LIVE PARTS 7-8
- GUARDS FOR ATTENDANTS 7-8
- BUSHINGS 7-9

TRANSFORMERS
- CALCULATING PRI. AND SEC. CURRENTS 7-10
- FINDING AMPERAGE 7-11
- INSTALLING TRANSFORMERS 7-11
- LOCATION 7-12
- OVERCURRENT PROTECTION 7-14
- GUARDING 7-21
- VENTILATION 7-22
- GROUNDING 7-22

LOCATION OF TRANSFORMER VAULTS 7-23
- WALLS, ROOF, AND FLOOR 7-23
- DOORWAYS 7-23
- VENTILATION OPENINGS 7-24
- DRAINAGE 7-24
- WATER PIPES AND ACCESSORIES 7-25
- STORAGE IN VAULTS 7-25

TEST QUESTIONS 7-27

In addition to the requirements of **Article 445**, generators must comply with the requirements of other sections of the NEC, most notably, **Articles 215, 230, 250, 700, 701, 702,** and **705**.

Article 230 deals with generators when they are used to supply service related equipment, while **Article 215** is used for generators supplying feeder-circuit equipment. **Article 250** addresses the special grounding techniques based upon where the generator is installed and used. **Article 700** contains the rules for generators that are utilized to supply power to emergency systems. **Articles 701** and **702** pertain to generators that serve legally required and optional standby systems. **Article 705** is applied when generators are connected in parallel with the utility power sources and used to serve as an interconnected electric power production source.

Transformers per **Article 450** must be sized with enough capacity to supply power to loads and allow loads with high inrush currents to start and run. In addition, they must be protected by properly sized OCPD's and be equipped with conductors having allowable ampacity ratings large enough to supply the loads. OCPD's must be designed and installed in such a manner to safely protect the conductors and windings of such power sources from dangerous short-circuits, ground-faults, and overloads.

Note that questions and problems pertaining to generators are nearly always on the master electrician examination. A thorough study of these rules will serve as an aid to help answer such questions and problems correctly.

LOCATION OF GENERATORS
445.10

One of the first requirements is that the generator be suitable for the location where it is installed. Basically, standard-type generators are designed to operate indoors in dry places. The requirements of **430.14** must be utilized to help protect the elements of generators. If generators are installed in hazardous locations, the requirements of **Articles 500** thru **503, 510** thru **517, 520, 530, and 695** must also be complied with.

NAMEPLATE MARKINGS
445.11

To aid test takers, every generator problem on the master examination having a nameplate shall contain the following information so proper answers can be derived:

(1) The manufacturer's name,
(2) The operating frequency,
(3) Number of phases,
(4) Power factor,
(5) Rating in kVA or kW, with the corresponding volts and amperes,
(6) Rated Rpm's, and
(7) Insulation type, ambient temperature, and time rating.

On the examination, such information is used to solve generator questions and problems in residential, commercial, and industrial applications.

OVERCURRENT PROTECTION FOR GENERATORS
445.12

Constant-potential generators, except for AC generators and their exciters, shall be protected from excessive current by circuit breakers or fuses. AC generators are exempt from the need of overcurrent protection. This is due to their impedance, which limits short-circuit currents to a value that is not damaging to their windings. All generator exciters are usually separately excited. DC as well as AC units are normally operated without overcurrent protection. Apply this information when taking the examination. **(See Figure 7-1)**

CONSTANT-VOLTAGE GENERATORS
445.12(A)

Master Test Tip 1: Generators, on the test, usually produce a constant voltage and must be protected from overloads. This may be accomplished by designing and using overcurrent protective devices such as fuses, circuit breakers, etc. or by inherent protection.

The basic rule requires DC generators to have overcurrent protection. However, AC generators may be so designed that on a high overload the voltage of the generator falls off, thereby reducing the overload current to a safe value. For this reason, the examination does not require overload protection for AC generators unless specifically stated.

There are some installations where OLP can be omitted. In some cases, it is considered better to risk damage to the exciter rather than have the generator shut down through the operation of an exciter OCPD. (Watch for this on the test). **(See Figure 7-2)**

Generators and Transformers

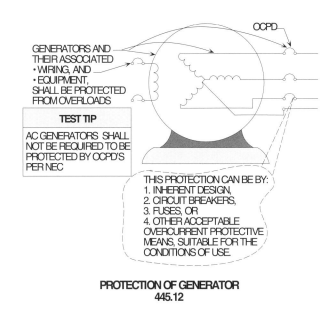

Figure 7-1. Generators and their elements must be protected from overloads, short-circuits, and ground-faults.

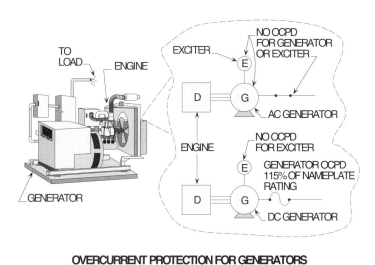

Figure 7-2. This figure illustrates when overcurrent protection is required for generators.

TWO-WIRE GENERATORS
445.12(B)

Two-wire DC generators are only required to have overcurrent protection in one wire, so the overcurrent protection device will be set off by the entire current and not the current in the shunt coil. However, the overcurrent device may not, under any circumstances, open the shunt coil. (Watch for this type of problem on the master's test.)

Master Test Tip 2: Test takers must be aware that the NEC does not allow an overcurrent protection device in the generator positive lead only, because an overcurrent device in the positive lead would not always be actuated by the entire current generated. Note that an overcurrent protection device is not permitted for the shunt field. If the shunt field circuit were to open and the field was at full strength, a dangerously high voltage would be induced, which might damage the generator. **(See Figure 7-3)**

Figure 7-3. This figure illustrates overcurrent protection for two-wire generators.

Note that the OCPD shall open the shunt circuit of DC generators.

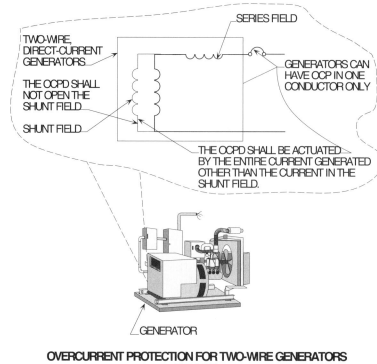

GENERATORS RATED AT 65 VOLTS
445.12(C)

A generator operating at 65 volts or less, and driven by an individual motor, is considered adequately protected by the motor overcurrent protection device, where such unit will open the circuit, if the unit is delivering not more than 150 percent of full-load current. **(See Figure 7-4)**

Figure 7-4. This figure illustrates the protection of generators operating at 65 volts or less.

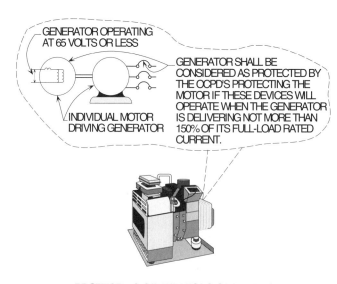

Master Test Tip 3: If the fuse or circuit breaker protecting the motor is set to operate when the generator is 50 percent or less overloaded, no protection is required in the generator leads. However, if the generator voltage is above 65 volts, it is the intent of the NEC to require separate overcurrent protection for the generator.

Generators and Transformers

BALANCER SETS
445.12(D)

Balancer sets consist of two smaller DC generators used with a larger two-wire generator. The two balancer generators are connected in series across the two-wire main generator lines. A neutral tap is brought from the midpoint connection between the two balancer generators. Note that each of the two balancer generators carries about one-half of any unbalanced load condition.

With such an arrangement, where there is heavy unbalance conditions in the load, the balancer generators may become overloaded, while there is no overload on the main generator. The balancer generators shall be equipped with an overload device that will actuate the main generator disconnect if the balancer generator should become overloaded. **(See Figure 7-5)**

Master Test Tip 4: Balancer sets shall be required to be equipped with overload devices that disconnect the three-wire system in case of an excessive unbalanced condition. Three-wire direct-current generators shall be provided with overcurrent protection devices, one in either armature lead arranged to disconnect the three-wire circuit in case of heavy overloads or extreme unbalance conditions.

Figure 7-5. This figure illustrates balancer sets used with a generator to disconnect the system if an excessive unbalanced current condition should occur.

Note that two DC balancing sets are used with the generator to detect excessive unbalanced current flow. Test takers must be able to recognize such a condition on a test question or problem.

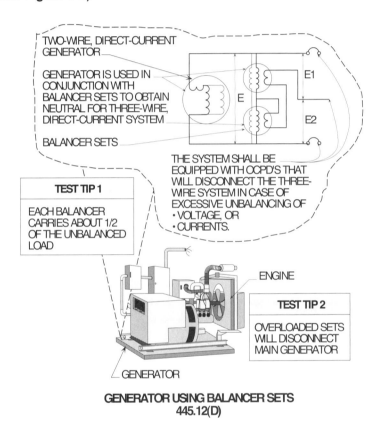

GENERATOR USING BALANCER SETS
445.12(D)

THREE-WIRE DC GENERATORS
445.12(E)

As in two-wire generators, the overcurrent protection device protecting a three-wire generator must be capable of taking the full generator current. When equalizer leads are provided, the overcurrent protection device, if not properly installed in the circuit, might take only a part of the generator current. To help solve this problem, three-wire DC generators operating in parallel are equipped with two equalizer leads. The overcurrent protection devices shall be so placed in the circuit that they will take the full generator current without tripping open the circuit.

A two-pole breaker placed ahead of the junction of the main and equalizer leads will provide such protection. However, a four-pole circuit breaker with two poles for the main leads and two poles for the equalizer leads can also be used, provided such CB is actuated by the full current flow of the generator. **(See Figure 7-6)**

Master Test Tip 5: Test takers must note that these generators, which are either shunt wound or compound wound, must be provided with overcurrent protection devices in each armature lead. Such devices must sense the entire armature current and be multipole devices that open all the poles in the event of an overcurrent condition.

Figure 7-6. This figure illustrates the overcurrent protection requirements for three-wire DC generators.

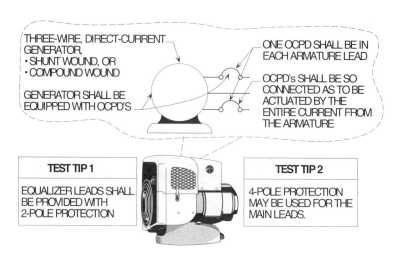

PROTECTING A THREE-WIRE DC GENERATOR
445.12(E)

EXCEPTIONS (A) THRU (E)
445.12(A) THRU (E), Ex.

There are some cases in which, if a generator fails, it would be less of a hazard than disconnecting it if an overcurrent condition occurs. In such instances, the AHJ may allow the generator to be connected to a supervision panel and annunciator or alarm, instead of requiring overcurrent protection. (Watch for this question or problem on the master's test.)

Note that the Ex. to **445.12(A)** thru **(E)** will permit the windings of such generators to actually burn out and fail.

AMPACITY OF CONDUCTORS FROM GENERATORS
445.13

Ungrounded phase conductors from a generator shall be rated at not less than 115 percent of the nameplate current rating. Neutrals can be computed and sized according to **220.61** and **Table 310.16**, or the over 2000 volt Tables. Phase conductors must be capable of carrying ground-fault currents as well as normal currents. For the neutral to accomplish this task, it is also required to be sized in accordance with **250.24(C)(1)** and **(C)(2)**.

Master Test Tip 6: In the For Example problem, a 125 amp OCPD can be selected per **Table 310.16** and **240.6(A)**.

For example, a generator with a 100 amp output shall have conductors with an ampacity of at least 115 amps (100 A x 115% = 115 A). **Note,** this calculation requires that 2 AWG THWN copper conductors be used. **(See Figure 7-7)**

PREVENTING OVERLOAD CONDITIONS
445.13, Ex.

In some cases, the conductors can be protected at 100 percent of the rated generated current. However, to do so, the design or operation of the generator shall be such as to prevent overloading. Therefore, an ampacity of 100 percent loading is all that is permitted.

For example, a generator with a 100 amp output shall have conductors with an ampacity of at least 100 amps (100 A x 100% = 100 A), and the load limited to this value. **(See Figure 7-8)**

Generators and Transformers

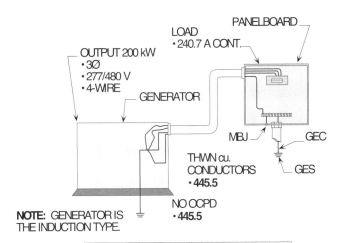

SIZING CONDUCTORS FOR A GENERATOR AT 115%
445.13

Figure 7-7. This figure illustrates the procedure for sizing conductors from a generator to a load when applying 115 percent multiplier.

Master Test Tip 7: If an OCPD was specified on a test problem, the size of such device in Figure 7-7 must be rated at 300 amp per **240.4(B)** and **240.6(A)**.

Note that an OCPD at the generator's output is required for these conductors if the conductors are not sized at 115 percent (or greater) of the generator's output.

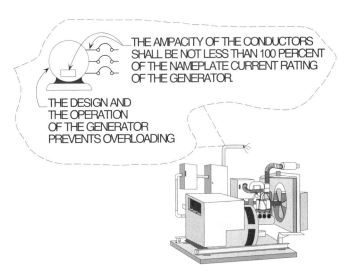

SIZING CONDUCTORS FOR A GENERATOR AT 100%
445.13, Ex.

Figure 7-8. This figure illustrates the procedure for sizing the conductors from a generator to a load when applying 100 percent multiplier.

Note that questions and problems on the test pertaining to generator protection must be carefully reviewed before an answer is given. Check the type of generator used, for the type determines whether OCPD is needed at the generator's output.

OCPD'S PROVIDED
445.13, Ex.

Where an integral overcurrent protection device (OCPD) is provided by the manufacturer with the generator and such leads are connected to the device, the generator is considered protected from overload conditions.

For example, if stated in a question on the master test that an OCPD rated at 100 amps is provided by the manufacturer and comes with the generator, then the conductors rated at 100 amps are considered protected from overload conditions and no additional protection is required. See **110.14(C)(1)(a)** and **(C)(1)(b)** for terminating conductors. **(See Figure 7-9)**

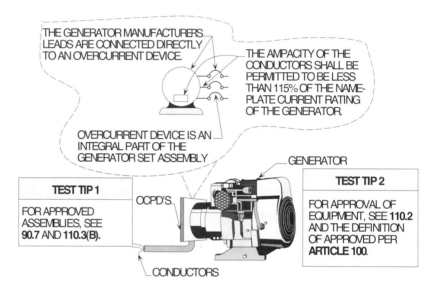

Figure 7-9. This figure illustrates the requirements of overcurrent protection for generator and conductors where provided by the manufacturer.

PROTECTION OF LIVE PARTS
445.14

Live parts of generators operating at more than 50 volts-to-ground shall not be exposed to accidental contact when accessible to unqualified persons. The basic rule is that live parts of generators shall not be exposed to accidental contact. Such live parts are as follows:

(1) Brushes,
(2) Collector rings, and
(3) Other live parts.

Master Test Tip 8: Note that any exposed live part(s) must be guarded per **110.27** of the NEC.

See **Figure 7-10** for a detailed illustration pertaining to this rule.

GUARDS FOR ATTENDANTS
445.15

If generators operate at more than 150 volts-to-ground, no live parts are permitted to be exposed to contact by unqualified personnel. Section **430.233** requires insulating mats or platforms to be provided around motors. Note that these protective items are also required for generators if the generator voltage is greater than 150 volts-to-ground. **(See Figure 7-11)**

Generators and Transformers

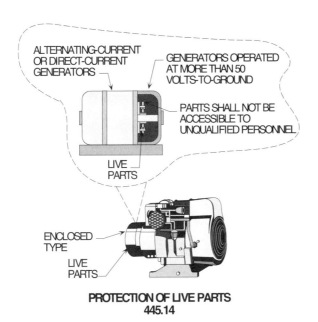

Figure 7-10. The above illustrates the requirements for generators with exposed parts.

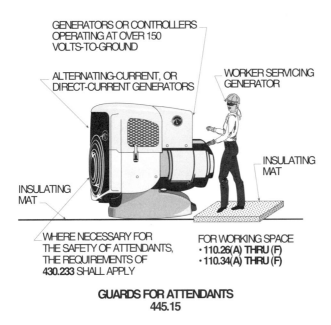

Figure 7-11. Insulating mats or platforms shall be provided for attendants servicing a generator if the voltage is greater than 150 volts-to-ground.

Master Test Tip 9: Generators and controllers shall be guarded against accidental contact only by location as specified in **430.232**, and because adjustments or other attendance may be necessary during the operation of the apparatus, suitable insulating mats or platforms shall be provided so that the attendant cannot readily touch live parts unless standing on the mats or platforms.

BUSHINGS
445.16

Where wires pass through an opening in an enclosure, conduit box, or barrier, a bushing must be used to protect the conductors from the edges of an opening having sharp edges.

The bushings are required to have smooth, well-rounded surfaces where they may be in contact with the conductors. If used where oils, grease, or other contaminants may be present, the bushing must be made of a material that will not be deleteriously affected. **(See Figure 7-12)**

Figure 7-12. This figure illustrates the rules for installing bushings to protect conductors passing through the openings of enclosures.

Master Test Tip 10: Unused openings in enclosures shall be effectively closed per **110.12(A), 312.5(A),** and **314.17(A).**

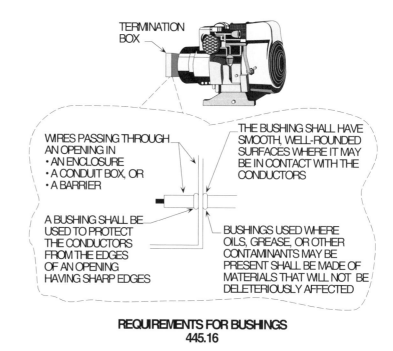

CALCULATING PRI. AND SEC. CURRENTS OF XFMR'S

The transformers primary amp rating shall be equivalent to the amps of the connected load when such is used as feeder-circuits to supply the primary of a transformer to step up or step down the voltage. To determine the FLA of a transformer, the kVA of the transformer must be divided by the voltage x 1.732 if the supply is three-phase. **(See Figures 7-13(a) and (b))**

Figure 7-13(a). Determining the amps of a single-phase transformer by dividing the VA of the XFMR by the primary (Pri.) and secondary (Sec.) voltage.

Master Test Tip 11: Note that connections from the secondary side of XFMR's can be 10 ft. up to 25 ft. in length per **240.21(C)(2),(C)(3),** and **(C)(6).** If a facility is classified as a supervised industrial installation, connections from the secondary of XFMR's can be 100 ft. and up to any length per **240.92(B)(1)(1)** thru **(D)**

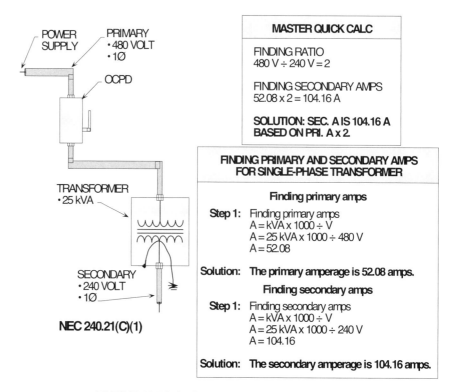

7-10

Generators and Transformers

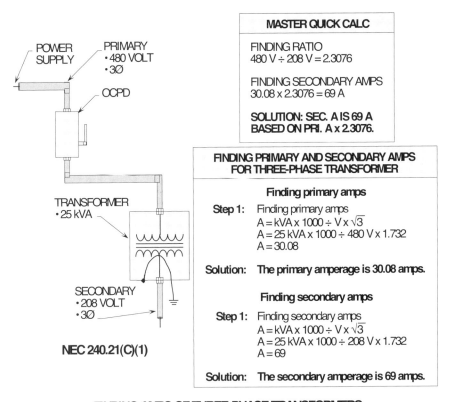

Figure 7-13(b). Determining the amps of a three-phase transformer by dividing the kVA of the XFMR by the primary and secondary voltage.

Master Test Tip 12: When using the 10 ft. connection rule to supply power to a power panel, such connected conductors do not have to terminate to a main OCPD.

FINDING AMPERAGE

The kVA or amp rating for the primary or secondary of a transformer can be determined for a single-phase system by applying the following formula:

kVA = volts x amps ÷ 1000
Amps = kVA x 1000 ÷ volts

The following formula shall be applied to determine the ratio of a transformer having a 480 volt primary and 240 volt secondary:

primary ÷ secondary
480 V ÷ 240 V
2:1 ratio

The amp rating for the primary or secondary can be determined for a three-phase system by applying the following formula:

kVA = volts x 1.732 x amps ÷ 1000
Amps = kVA x 1000 ÷ volts x 1.732

Master Test Tip 13: Test questions and problems on the master's examination having values of VA or watts must be converted to amps using the formulas shown in the text.

INSTALLING TRANSFORMERS
ARTICLE 450

Transformers shall be installed and protected per **Article 450**. Transformers shall be permitted to be installed inside or outside of buildings based upon their design and type.

Note that the OCPD's for XFMR's are now contained in Tables. Review **Tables 450.3(A)** and **(B)** very carefully before taking the master's examination.

Master Test Tip 14: Dry type XFMR's rated over 35,000 volts must be installed in vaults per **110.26(A)** and **(F)**. (Also, see **450.21(C)**)

LOCATION
450.13

Transformers shall be located where readily accessible to qualified personnel for inspection and maintenance. Where it is necessary to use a ladder, lift, or bucket truck to get to a transformer, it is not considered readily accessible. See definition of *readily accessible* and *accessible*, in **Article 100** of the NEC. **(See Figure 7-14)**

Figure 7-14. The general rule of **450.13** requires transformers to be readily accessible for maintenance, repair, and service.

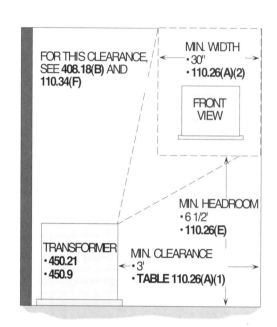

Master Test Tip 15: Note that it would be a violation if a problem on the test showed a panelboard mounted directly above a XFMR located on the floor per **110.26(A)** and **110.26(E)**.

TRANSFORMERS MUST BE READILY ACCESSIBLE
NEC 450.13

There are two exceptions to the accessibility rules, and they are explained in the next two headings. (Watch for these exceptions in questions on the master test.)

HUNG FROM WALL OR CEILING
450.13(A)

(1) For dry-type transformers not over 600 volts, located on open walls or steel columns, such units do not have to be readily accessible. It is permissible to gain access to this type of installation using a portable ladder or bucket lift. **(See Figure 7-15)**

MOUNTED IN CEILING
450.13(B)

Master Test Tip 16: The two alternate methods of installations are for dry-type transformers. They do not apply to oil or askarel-filled transformers. This rule is due to the damage of oil or threat of fire if a rupture in the case should occur.

(2) Dry-type transformers not over 600 volts and 50 kVA or less may be installed in suspended ceiling spaces of buildings. The transformers cannot be permanently closed in and there must be some access to the transformers, but they do not have to be readily accessible per **Article 100** of the NEC. It was not clear in the 1993 or previous editions of the NEC if dry-type transformers not exceeding 600 volts, nominal, and rated 50 kVA or less were permitted to be installed in the space above suspended ceilings with removable panels,

even if the transformer was accessible and provided with proper working clearances. Note that the space where the transformer is installed shall comply with the ventilation requirements of **450.9** and also meet the design rules of **450.21(A)**. If such ceiling space is used as a return air space for air-conditioning, **300.22(C)** must be reviewed and the provisions of this section shall be complied with. **(See Figure 7-16)**

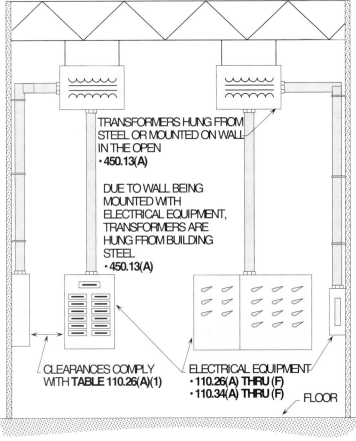

Figure 7-15. Transformers hung from wall or ceiling do not have to be readily accessible.

Master Test Tip 17: Before taking the master's examination, review **450.13(A)** and **(B)** pertaining to clearances, **450.21(A)** and **(B)** for clearances from combustible material, and **450.9** for ventilation rules. For XFMR's installed in suspended ceilings, see **300.22(C)** of the NEC.

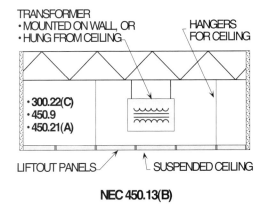

Figure 7-16. Transformers mounted in ceilings do not have to be readily accessible.

Master Test Tip 18: Note that the XFMR in the suspended ceiling must be listed per **450.13(B)** and **110.3(B)**. Note that a general-purpose XFMR would not be acceptable for such use. **(See 300.22(C))**

OVERCURRENT PROTECTION
450.3(A); (B)

There are two sets of rules for providing overcurrent protection of transformers that are rated 600 volts or less or rated over 600 volts. The OCPD may be placed in the primary only or in the primary and secondary.

PRIMARY ONLY - OVER 600 VOLTS
450.3(A)

Master Test Tip 19: When individual OCPD's are installed in the primary side of the XFMR, there must be proper supervision and maintenance availability in the facility. Review **450.3(A)** and **Table 450.3(A)** very carefully before attempting to answer questions or problems on the master's examination. **(See Example problem in Figure 7-17)**

The term *primary* is often inferred in the field as being the high side, and the term *secondary* as the low side of the transformer. This is really not the proper terminology. For testing purposes, the primary is the input side of the transformer and the secondary is the output side. Thus, voltage has nothing to do with the terms, whether high or low.

Each transformer has to be protected by an overcurrent device in the primary side. If the overcurrent protection is fuses, they shall be rated at not greater than 250 percent (2.5 times) of the rated primary current of the transformer. When circuit breakers are used, they shall be set at not greater than 300 percent (3 times) of the rated primary current. **(See Figure 7-17)**

This overcurrent protection device may be mounted in the vault or at the transformer, if approved for such purpose. It may also be mounted in the panelboard and designed to protect the windings and circuit conductors supplying the transformer.

In the case of a vault, the OCPD can be installed outdoors on a pole, with a disconnecting means installed in the vault to disconnect supply conductors.

TABLE 450.3(A)

Master Test Tip 20: Note that **450.3(A)** and **450.3(B)** protect the windings of the XFMR, and **240.4, 240.21, 240.100** and **240.101** protect the conductors.

When 250 percent (2.5 times) of the rated primary current of the transformer does not correspond to a standard rating of a fuse, the next higher standard rating shall be permitted per **240.6(A)**.

TABLE 450.3(A)

As provided in the provisions listed in **450.3(A)(2)** below. The provisions of **450.3(A)** and **Table 450.3(A)** cover transformers installed in supervised locations.

PRIMARY AND SECONDARY - OVER 600 VOLTS
450.3(A) AND TABLE 450.3(A)

A transformer over 600 volts, nominal, having an overcurrent device on the secondary side rated to open at not greater than the values listed in **Tables 450.3(A)**, or a transformer equipped with a coordinated thermal overload protection provided by the manufacturers, shall not be required to have an individual protection in the primary. However,

Generators and Transformers

if an individual OCP is not provided, a feeder or service overcurrent device rated or set to open at not greater than the values listed in **Tables 450.3(A)** shall be provided.

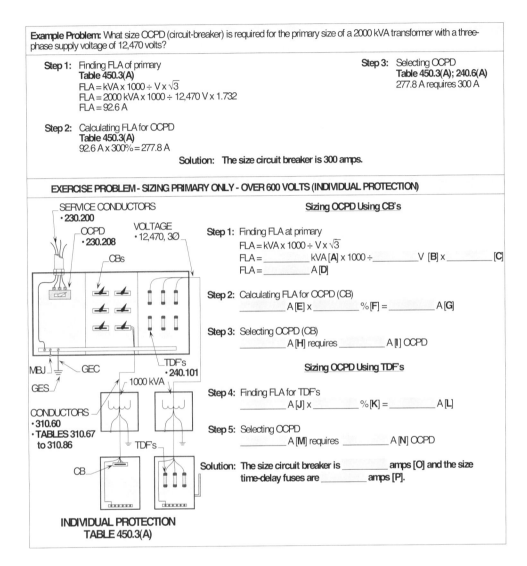

Figure 7-17. If the overcurrent protection is fuses, they shall be rated not greater than 250 percent (2.5 times) of the rated primary current of the transformer. When circuit breakers are used, they must be set not greater than 300 percent (3 times) of the rated primary current.

Master Test Tip 21: On the Master's test, conductors that are connected from the secondary side of the XFMR as shown in Figure 7-17 are sized either on the output of the XFMR or the load served.

NONSUPERVISED LOCATIONS (ANY LOCATION)
450.3(A); TABLE 450.3(A)

Overcurrent protection for a nonsupervised location may be placed in the primary and secondary side of high-voltage transformers, if the OCPD's are designed and installed according to the provisions listed in **Table 450.3(A)**.

If the secondary voltage is 600 volts or less, the OCPD and conductors on the secondary side must be sized at 125 percent of the FLC rating. OCPD's sized at 125 percent of the FLC in amps protects the conductors and windings of the transformer from dangerous overload conditions. With higher voltage on the secondary side of the transformer, the percentages for sizing the OCPD's can be selected from **Table 450.3(A)** based upon the particular voltage level. **(See Figure 7-18)**

Master Test Tip 22: In nonsupervised locations (any location in Table), the OCPD per **Table 450.3(A)** can be rounded up to the next size above the XFMR's FLC rating in amps times such percentages. **(See example problem in Figure 7-18)**

Figure 7-18. Determining the amps in the primary and secondary side of a transformer in a nonsupervised location.

Master Test Tip 23: In the example problem, a 1300 amp OCPD was selected based on the **FPN** to **240.6(A)** and **(B)**. The FPN does not require the next standard size listed in **240.6(A)**, which is 1600 amps.

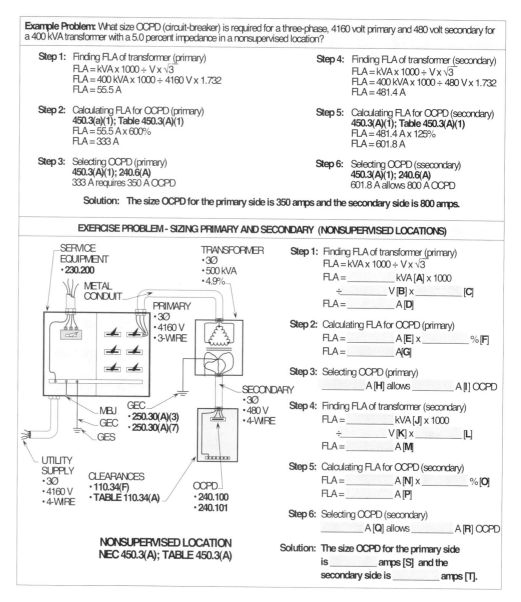

SUPERVISED LOCATIONS
450.3(A); TABLE 450.3(A)

Master Test Tip 24: In supervised locations, the OCPD per **Table 450.3(A)** must be rounded down to the next size below the XFMR's FLC rating in amps times such percentages. **(See Example problem in Figure 7-19)**

Overcurrent protection may be placed in the primary and secondary side of high-voltage transformers if the OCPD's are designed and installed according to the provisions listed in **Table 450.3(A)**.

Where the facility has trained engineers and maintenance personnel, the OCPD for the secondary can be sized at not more than 250 percent of the FLC in amps for voltages of 600 volts or less. With higher voltage on the secondary side of the transformer, the percentages for sizing the OCPD's are selected from **Table 450.3(A)** based upon the particular voltage level. **(See Figures 7-19)**

Generators and Transformers

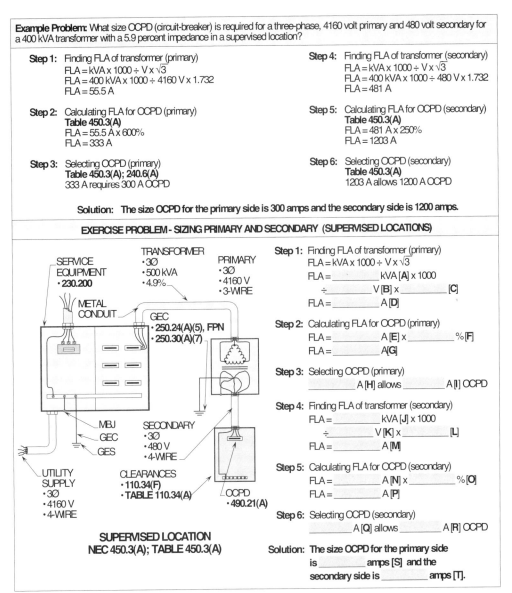

Figure 7-19. Determining the amps in the primary and secondary side of a transformer in a supervised location.

Master Test Tip 25: The OCPD in the three-phase, 480 volt panelboard that is in the exercise problem must not be rounded up to the next higher size. Note that it can be rounded up for nonsupervised locations. **(See Master Test Tip 24)**

Note: See Figures 7-20(a) and (b) for certain design conditions that permit the primary OCPD to be used to protect the primary and secondary sides of two-wire to two-wire connected transformers and three-wire to three-wire delta-connected transformers per **240.4(F)** and **240.21(C)(1)**.

PRIMARY ONLY - 600 VOLTS OR LESS
450.3(B); TABLE 450.3(B)

A transformer 600 volts or less, nominal, having an individual overcurrent device on the primary side shall be sized at no more than 125 percent of the transformer's full-load current rating. Note that with the OCPD and conductors sized at 125 percent or less of the transformer's FLC, the supply conductors and transformer windings, in most cases, are considered protected from overload conditions. **(See Figure 7-21)**

Master Test Tip 26: Note that the CB providing individual OCP in the primary must be rounded down below the 125 percent or selected at 125 percent of the primary's FLC in amps.

Figure 7-20(a). Determining the OCPD for a single-phase, two-wire to two-wire system.

Note: In the example problems in Figure 7-20(a), the secondary panelboard rated at 150 amps is considered protected from overloads by the 70 amp OCPD in the primary side. (70 A x 2 = 140 A is below 150 A)

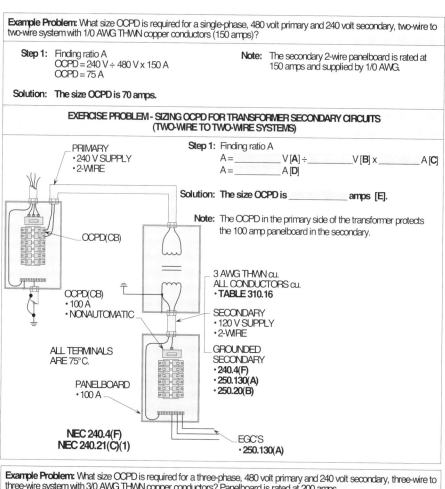

Figure 7-20(b). Determining the OCPD for a three-phase, three-wire to three-wire system.

Master Test Tip 27: Note that there must not be any line to neutral loads supplied from the secondary 3-wire panelboard.

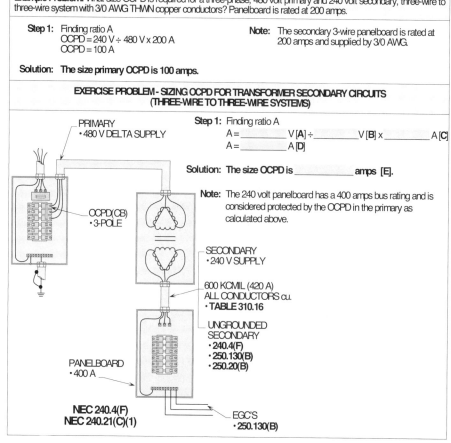

Generators and Transformers

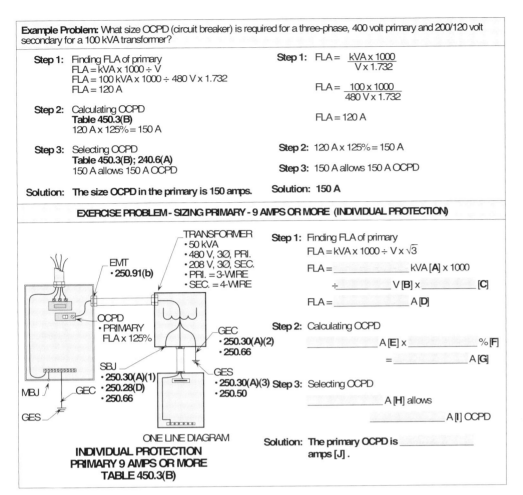

Figure 7-21. A transformer 600 volts or less, nominal, having an individual overcurrent protection device on the primary side shall be sized at no more than 125 percent of the transformer's full-load current rating.

Master Test Tip 28: There are three levels of current that must be found when sizing and selecting the OCPD to be placed in the primary side of the XFMR. Level 1 consist of primary current of 9 amps or more. Level 2 is when the primary current is less than 9 amps but 2 amps or more. In level 3, the primary current is less than 2 amps. (See Master Test Tip 30 and Figures 7-22(a) thru (c)).

9 AMPS OR MORE
TABLE 450.3(B)

Where the rated primary current of a transformer is 9 amps or more and 125 percent of this current does not correspond to a standard rating of a fuse or circuit breaker, the next size may be used per **240.6(A)**. Where the rated primary current of a transformer is less than 9 amps but more than 2 amps, an overcurrent device rated or set at no more than 167 percent of primary current may be used. When the rated primary current of a transformer is less than 2 amps, an overcurrent device rated or set at not more than 300 percent shall be used. **(See Figures 7-22(A), (B) and (c))**

PRIMARY AND SECONDARY - 600 VOLTS OR LESS
TABLE 450.3(B)

Combination protection can be provided for both the primary and secondary sides of a transformer. A current value of 250 percent of the rated primary current of the transformer may be used if 125 percent of the rated primary current of the transformer is not sufficient to allow loads with high inrush currents to start and operate. However, the secondary overcurrent protection device must be sized at 125 percent of the rated secondary full-load current of the transformer. Where the rated secondary current of a transformer is less than 9 amps, an overcurrent device rated or set at no more than 167 percent of secondary current may be used. **(See Figure 7-23)**

Master Test Tip 29: Signs indicating the presence of live exposed parts of transformers, or other suitable markings, shall be used in areas where transformers are located.

Figure 7-22(a). Where the rated primary current of a transformer is 9 amps or more and 125 percent of this current does not correspond to a standard rating of a fuse or circuit breaker, the next size may be used per **240.6(A)**.

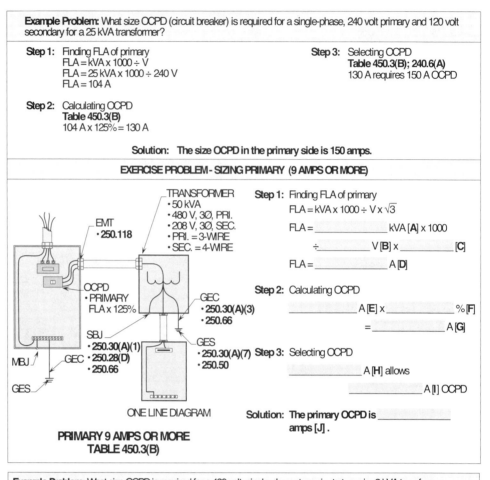

Figure 7-22(b). Where the rated primary current of a transformer is less than 9 amps but more than 2 amps, an overcurrent device rated or set at not more than 167 percent of primary current may be used.

Master Test Tip 30: There are two levels of current to be found. The first level is 9 amps or more, and the second level is less than 9 amps. **(See Master Test Tip 29 and Figure 7-22(a))**

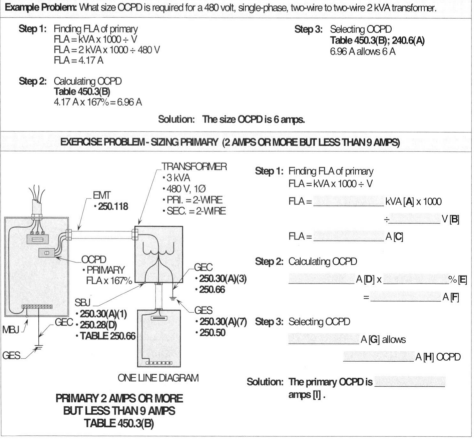

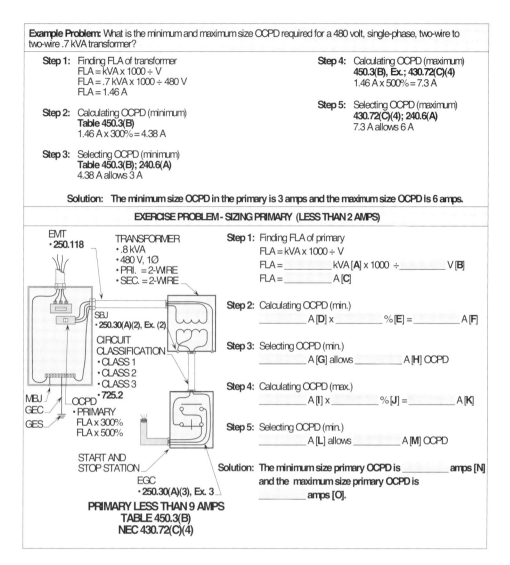

Figure 7-22(c). When the rated primary current of a transformer is less than 2 amps, an overcurrent protection device rated or set at not more than 300 percent shall be used.

Master Test Tip 31: If the XFMR has a primary of less than 2 amps and it is classified as a control XFMR, the OCPD in the primary side can be sized at 500 percent of the primary's FLC in amps per **430.72(C)(4)**.

9 AMPS OR MORE
TABLE 450.3(B)

Where the rated secondary current of a transformer is 9 amps or more and 125 percent of this current does not correspond to a standard rating of a fuse or circuit breaker, the next available size may be used per **240.6(A)**.

GUARDING
450.8

Transformers should be isolated in a room or accessible only to qualified personnel to prevent unauthorized persons from making accidental contact with live parts. To safeguard live parts from possible damage, the transformer may be elevated. The following are acceptable means of safeguarding live parts as required in **110.27(A)(4)** and **110.34(E)**.

(1) Transformers should be isolated in a room or accessible only to qualified personnel.

(2) Permanent partitions or screens may be installed to prevent unauthorized access.

(3) Transformers shall be elevated at least 8 ft. above the floor to prevent unauthorized personnel from contact.

Master Test Tip 32: XFMR's that are totally enclosed except for ventilating openings shall be considered guarded.

Figure 7-23. Sizing OCPD for the primary and secondary side of a transformer rated 600 volts or less.

Master Test Tip 33: Note that a minimum size OCPD is permitted in the primary (125 percent of the primary amps or less than 125 percent of the Primary amps). A next size above 125 percent of the primary amps is permitted, and a maximum size OCPD of 250 percent or less of primary amps is permitted per **Table 450.3(B)**.

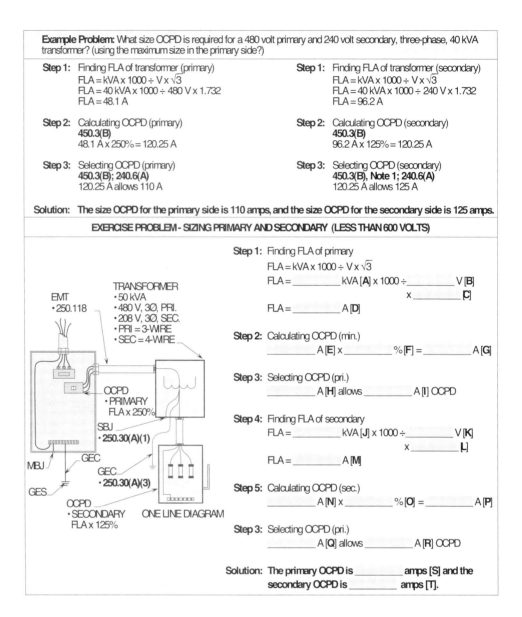

VENTILATION
450.9

Transformers shall be located and installed in rooms or areas that are not subjected to exceedingly high temperatures to prevent overheating and possible damage to windings. Transformers with ventilating openings shall be installed so that the ventilating openings are not blocked by walls or other obstructions that could block air flow.

GROUNDING
450.10

Master Test Tip 34: Note that XFMR's with exposed live parts rated over 600 volt must have a sign on the fence stating **DANGER - HIGH-VOLTAGE - KEEP OUT**, per **110.34(C)**.

Transformer cases shall be grounded per **250.110**. Transformers enclosed by fences or guards shall be grounded. Live parts require a guard such as a fence to guard against unauthorized entry and to protect the general public and unqualified persons from dangerous energized electrical parts. Such fences shall be grounded to prevent metal elements from being accidentally energized with voltage. **(See Figure 7-24)**

Generators and Transformers

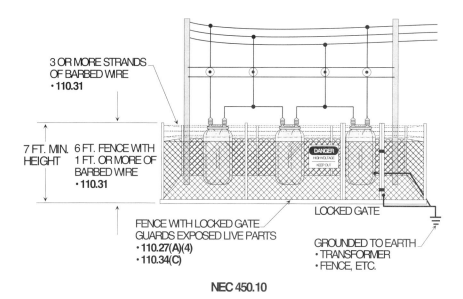

Figure 7-24. Transformers enclosed by fences or guards shall be grounded. Live parts require a guard such as a fence to guard against unauthorized entry and to protect the general public and unqualified persons from dangerous energized electrical parts.

LOCATION OF TRANSFORMER VAULTS
450.41

Vaults are used to house dry-type transformers that are rated over 35 kV or transformers filled with combustible material used as an aid in cooling their windings.

To prevent unauthorized entry and contain fire hazards, vaults are required to be designed and built with specific rules and regulations.

Wherever possible, transformer vaults must be located at an outside wall of the building. This rule is intended to allow ventilation directed to the outside without using ducts, flues, etc. per **450.41**.

Master Test Tip 35: The clearances for open type XFMR's (not totally enclosed) are found in **450.21(A)** and **(B)**. Note that subpart (A) deals with XFMR's rated at 112 1/2 kVA, and **Subpart (B)** with XFMR's rated over 112 1/2 kVA.

WALLS, ROOF, AND FLOOR
450.42

The rules for construction of vaults are set forth in this section. Floor, walls, and roof must be of fire-resistant material, such as concrete, and capable of withstanding heat from a fire within for at least three hours. A 6 in. thickness is specified for the walls and roof. The floor, when laid and in contact with the earth, shall be at least 4 in. thick. Walls, roof, and floor shall have at least a 3 hour fire rating. **(See Figure 7-25)**

DOORWAYS
450.43

The door to a transformer vault shall be built according to the standards of the National Fire Protection Association, which requires a 3 hour fire rating. The door sill shall be at least 4 in. high. This is to prevent any oil that may accumulate on the floor from running out of the transformer room and moving to other areas. Doors shall be kept locked at all times to prevent access of unqualified persons to the vault.

Master Test Tip 36: Personnel doors shall be swing out, and doors shall be equipped with panic bars, pressure plates or other devices that open under simple pressure.

Figure 7-25. Floors, walls, and roof shall be of fire-resistant material, such as concrete, and capable of withstanding heat from a fire within for at least three hours. A 6 in. thickness is specified for the walls and roof. The floor, when laid and in contact with the earth, shall be at least 4 in. thick.

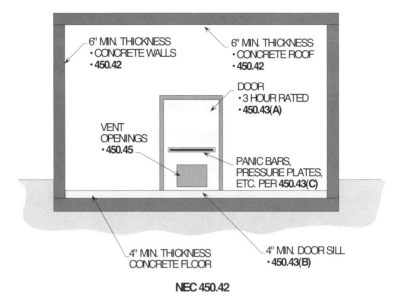

VENTILATION OPENINGS
450.45

Where ventilation is directed to the outside, without the use of ducts or flues, the vent opening shall have an area of at least 3 square inches for each kVA of transformer capacity, but never less than 1 square foot in area. The vent opening shall be fitted with a screen or grating, and also with an automatic closing damper. If ducts are used in the vent system, the ducts shall have sufficient capacity to maintain a suitable vault temperature. **(See Figure 7-26)**

Figure 7-26. When ventilation is directed to the outside, without the use of ducts and flues, the vent opening shall have an area of at least 3 sq. in. for each kVA of transformer capacity, but never less than 1 square foot in area.

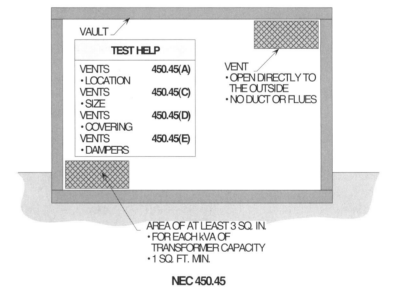

DRAINAGE
450.46

Drains should be provided to drain off oil that might accumulate on the floor due to a leak in a transformer caused by an accident. This rule is designed to prevent a fire hazard from occurring.

WATER PIPES AND ACCESSORIES
450.47

Piping for fire protection within the vault or piping to water-cooled transformers may be present in a vault. No other piping or duct system shall enter or pass through a vault. Valves or other fittings of a foreign piping or duct system are never permitted in a vault containing transformers. **(See Figure 7-27)**

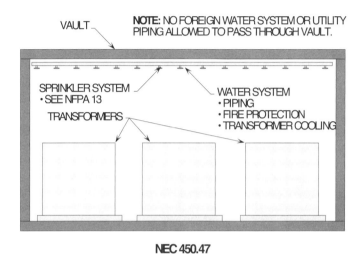

Figure 7-27. Piping for fire protection within the vault or piping to water-cooled transformers may be present in a vault. No other piping or duct system shall enter or pass through a vault.

STORAGE IN VAULTS
450.48

No storage of any kind may be kept in a vault other than the transformers and equipment necessary for their operation. This typically means the transformer vaults shall not be used for warehouses or storage areas but to contain transformers and accessories only. The reason the vault must be kept clear is due to the high-voltage and safety needed for personnel serving such equipment. Also, consideration must be given to certain material being a threat of fire under certain conditions. **(See Figure 7-28)**

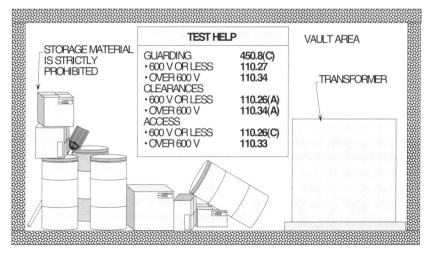

Figure 7-28. No storage of any kind shall be put in a vault other than the transformers and equipment necessary for their operation.

Name _____ Date _____

Chapter 7
Generators and Transformers

Section   Answer

1. Standard-type generators are usually designed to operate in dry places.

    (a) True   **(b)** False

2. DC generators shall not be required to be provided with overcurrent protection.

    (a) True   **(b)** False

3. Balancer sets consist of one smaller DC generator used with a larger two-wire generator.

    (a) True   **(b)** False

4. Where an integral overcurrent protection device is provided by the manufacturer with the generator and such leads are connected to the device, the generator shall be considered protected from overload conditions.

    (a) True   **(b)** False

5. Live parts to generators shall not be exposed to accidental contact.

    (a) True   **(b)** False

6. Transformers shall be connected in parallel and switched as a unit provided each transformer has overcurrent protection.

    (a) True   **(b)** False

7. Transformers enclosed by fences or guards shall not be required to be grounded.

    (a) True   **(b)** False

8. Walls, the roof, and floor for a transformer vault shall, at the least, have a 3 hour fire rating.

    (a) True   **(b)** False

9. Doors to transformer vaults shall be kept locked at all times to prevent access of unqualified persons to the vault area.

    (a) True   **(b)** False

10. Foreign piping or duct systems shall be permitted to be installed in a transformer vault.

    (a) True   **(b)** False

11. Ungrounded (phase) conductors from a generator shall be rated at no less than _____ percent of the nameplate current rating.

    (a) 100   **(b)** 110   **(c)** 115   **(d)** 125

Section	Answer

12. Conductors shall be permitted to be protected at _____ percent of the rated generated current if overload conditions for generators are to be prevented.

    (a) 100   **(b)** 110   **(c)** 115   **(d)** 125

13. Live parts of generators operated at more than _____ volts-to-ground shall not be exposed to accidental contact where accessible to unqualified personnel.

    (a) 50   **(b)** 120   **(c)** 250   **(d)** 600

14. Transformers shall be elevated at least _____ ft. above the floor to prevent unauthorized personnel from contact.

    (a) 6   **(b)** 8   **(c)** 10   **(d)** 15

15. All indoor dry-type transformers of over _____ volts shall be required to be installed in a vault.

    (a) 15,000   **(b)** 25,000   **(c)** 35,000   **(d)** 50,000

16. The walls and roof for a transformer vault shall have, at least, a _____ in. thickness.

    (a) 2   **(b)** 3   **(c)** 4   **(d)** 6

17. The door sills for a transformer vault shall be at least _____ in. high.

    (a) 2   **(b)** 3   **(c)** 4   **(d)** 6

18. Dry-type transformers rated 112 1/2 kVA or less shall be separated at least _____ in. from the combustible material where the voltage is over 600 volts.

    (a) 6   **(b)** 12   **(c)** 18   **(d)** 24

19. Askarel-insulated transformers of over _____ kVA shall be furnished with a relief vent such as a chimney.

    (a) 25   **(b)** 35   **(c)** 50   **(d)** 75

20. The floor for a transformer vault shall be at least _____ in. thick.

    (a) 2   **(b)** 4   **(c)** 6   **(d)** 12

21. What is the primary and secondary amperage for a 20 kVA, 480/240 volt, single-phase transformer?

    (a) Pri. = 38.4 A and Sec. = 78.8 A   **(b)** Pri. = 41.7 A and Sec. = 83.3 A
    (c) Pri. = 44.3 A and Sec. = 85.4 A   **(d)** Pri. = 48.2 A and Sec. = 88.5 A

	Section	Answer

22. What is the overcurrent protection device rating for the primary and secondary side of transformer with the following?

• 400 kVA transformer
• 4160/480 volts
• Three-wire to four-wire
• Three-phase
• Supervised location

        (a) Pri. = 250 A OCPD and Sec. = 900 A OCPD
        (b) Pri. = 250 A OCPD and Sec. = 1000 A OCPD
        (c) Pri. = 300 A OCPD and Sec. = 1000 A OCPD
        (d) Pri. = 300 A OCPD and Sec. = 1200 A OCPD

23. What is the individual overcurrent protection device rating for the primary side of a transformer with the following?

• 2 kVA transformer
• 480 volts
• Two-wire to two-wire
• Single-phase

        (a) 3 A OCPD   (b) 6 A OCPD   (c) 10 A OCPD   (d) 15 A OCPD

24. What is the minimum and maximum individual overcurrent protection device rating for the primary side of a transformer with the following?

• .7 kVA transformer
• 480 volts
• Two-wire to two-wire
• Single-phase

        (a) Min. = 3 A and Max. = 6 A
        (b) Min. = 3 A and Max. = 10 A
        (c) Min. = 6 A and Max. = 10 A
        (d) Min. = 10 A and Max. = 15 A

Motors and Compressors

8

On the master electrician examination, there are three currents that have to be found before calculating the size elements to makeup circuits supplying power to motors. The first current to be determined is the nameplate amps on the motor. This current rating in amps is used to size the overloads (OL's) to protect the motor windings and conductors. The second current to be found is the full-load amps (FLA) from **Table 430.248** for single-phase and **Table 430.250** for three-phase motors. This particular current rating in amps is used to size all the elements of the circuit except the overload protection. The third current that shall be found is the locked-rotor current (LRC) in amps from **Tables 430.251(A)** and **(B)**. The OCPD must be sized large enough to hold this current rating (LRC) in amps and permit the motor to start and run.

Article 440 deals with individual circuits or group installations having hermetically sealed motor compressors. The techniques for calculating the proper size conductors, disconnecting means, and controllers are discussed.

The conductors supplying power to HACR equipment shall be sized from the full-load amp (FLA) ratings of the compressor and condenser motor. These FLA ratings shall be increased by 125 percent per **Table 240.4(G)** and **440.32** to compensate for the starting periods and overload conditions.

The overcurrent protection devices (OCPD's) protecting the branch-circuits from short-circuit currents shall be sized from the provisions listed in **Table 220.3** and **440.22(A)**, which requires the FLA ratings to be increased from 175 percent (min.) up to 225 percent (max.) to allow the HACR equipment to start and run without tripping the OCPD ahead of the circuit.

Quick Reference

SIZING CONDUCTORS	8-2
SIZING THE BRANCH-CIRCUIT PROTECTIVE DEVICE	8-9
CODE LETTERS	8-16
LOCKED-ROTOR CURRENT UTILIZING HP	8-17
SIZING AND SELECTING OCPD'S	8-18
SINGLE-BRANCH CIRCUIT TO SUPPLY TWO OR MORE MOTORS	8-28
RUNNING OVERLOAD PROTECTION FOR THE MOTOR	8-32
SIZING THE CONTROLLER TO START AND STOP THE MOTOR	8-34
SIZING THE DISCONNECTING MEANS TO DISCONNECT BOTH THE CONTROLLER AND MOTOR	8-38
LOCATION OF THE DISCONNECTING MEANS FOR THE CONTROLLER AND MOTOR	8-42
SIZING CONDUCTORS FOR CONTROL CIRCUIT	8-45
SIZING OCPD FOR CONTROL CIRCUIT	8-47
MOTOR CONTROL AND MOTOR POWER CIRCUIT CONDUCTORS	8-50
CAPACITOR	8-51
RACEWAYS	8-52
MAGNETIC STARTER CONTACTOR AND ENCLOSURE	8-53
CLASS 1 CIRCUITS	8-54
CLASS 2 AND 3 CIRCUITS	8-54
DESIGNING MOTOR CURRENTS NOT LISTED IN TABLES	8-55
NAMEPLATE LISTING	8-55
AMPACITY RATING	8-56
DISCONNECTING MEANS	8-57
RATING AND INTERRUPTING CAPACITY	8-57
LOCATION	8-59
APPLICATION AND SELECTION	8-59
BRANCH-CIRCUIT CONDUCTORS	8-63
SINGLE MOTOR-COMPRESSORS	8-64
COMBINATION LOAD	8-64
CONTROLLERS FOR MOTOR COMPRESSORS	8-65
APPLICATION AND SELECTION	8-68
MOTOR COMPRESSORS AND EQUIPMENT ON 15 OR 20 AMP BRANCH-CIRCUIT - NOT CORD-AND-PLUG CONNECTED	8-68
CORD AND ATTACHMENT PLUG CONNECTED MOTOR COMPRESSORS AND EQUIPMENT ON 15 OR 20 AMP BRANCH-CIRCUITS	8-69
ROOM AIR-CONDITIONERS	8-70
BRANCH-CIRCUIT REQUIREMENTS	8-70
DISCONNECTING MEANS	8-70
TEST QUESTIONS	8-73

BRANCH-CIRCUIT AND FEEDER-CIRCUIT CONDUCTORS
430.1

On the Master's examination, the following branch-circuit and feeder-circuit elements of a motor system must be computed and sized based upon the characteristics of the motor:

(1) Motor branch-circuit and feeder conductors,
(2) Motor branch-circuit and feeder OCPD's,
(3) Motor overload protection,
(4) Motor control circuits,
(5) Motor controllers, and
(6) Motor disconnecting means.

SIZING CONDUCTORS FOR SINGLE MOTORS
430.22(A); Table 220.3

Master Test Tip 1: Each individual motor shall be calculated at 125 percent of the motor's FLC in amps. However, there are three exceptions where the sizing of the conductors in amps can be reduced from 125 percent to different percentages as listed in **Table 430.22(E)** per **430.22(E)**.

Branch-circuit conductors supplying a single motor shall have an ampacity not less than 125 percent of the motor's full-load current rating per **Tables 430.247, 430.248, 430.249** and **430.250**. For example, a 20 HP, 208 volt, three-phase motor per **Table 430.250** has a full-load current of 59.4 amps. The full-load amps (FLA) for sizing the conductors is determined by multiplying 59.4 A x 125%, which equals 74.25 amps.

For quick calcs, a motor has a starting current of four to six times the full-load current of the motor's FLA for motors marked with code letters A through G and 8 1/2 to 15 times for Design E or NEMA Design B, high-efficiency motors. Design B, C, and D motors have a starting current of about 4 to 10 times the full-load amps when starting and driving a motor load.

There are heating effects that develop on the conductors when motors are starting and accelerating the driven load. To eliminate such effects, the conductor's current-carrying capacity is increased by taking 125 percent of the motor's full-load current rating in amps. For example, a motor with a FLC rating of 42 amps shall have conductors with a current-carrying capacity of at least 52.5 amps (42 A x 125% = 52.5 A) to safely carry the starting load and protect insulation from overload conditions.

SIZING CONDUCTORS FOR SINGLE-PHASE MOTORS
430.22(A); Table 220.3

Section **430.6(A)(1)** of the NEC requires the full-load current in amps for single-phase motors to be obtained from **Table 430.248**. This FLC rating in amps is then multiplied by 125 percent per **Table 220.3** and **430.22(A)** to derive the total amps to select the conductors from **Table 310.16** to supply power to the motor windings. **(See Figure 8-1)**

SIZING CONDUCTORS FOR THREE-PHASE MOTORS
430.22(A)

Section **430.6(A)(1)** of the NEC requires the full load current in amps for three-phase motors to be obtained from **Table 430.250**. This FLC rating in amps is multiplied by 125 percent per **Table 220.3** and **430.22(A)** to derive the total amps to select the conductors for the branch-circuit supplying power to the motor windings. **(See Figure 8-2)**

Motors and Compressors

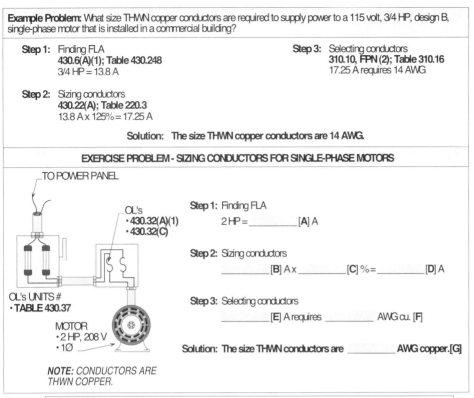

Figure 8-1. Determining the size branch-circuit conductors to supply single-phase motors.

Master Test Tip 2: Note that the 14 AWG THWN cu. conductors in the example problem are rated 20 amps and not 15 amps per **240.4(D)** and **240.4(G)**.

Figure 8-2. Determining the size branch-circuit conductors for supplying power to three-phase motors.

Master Test Tip 3: Note that the ampacity rating of the conductors do not have to be increased to match the size of the OCPD, if such OCPD is sized above 125 percent per **Table 220.3**, **430.22(A)** and **430.52(C)(1)** with **Exceptions**.

SIZING CONDUCTORS FOR MULTISPEED MOTORS
430.22(B); Table 220.3

The circuit conductors for multispeed motors shall be sized large enough between the source and the controller to supply the highest nameplate full-load current rating of the multispeed motor winding involved. A single OCPD is permitted to serve each speed for a multispeed motor per **Table 220.3** and **430.22(B)**. The speed with the greater amps is used to size the OCPD and conductors. Overload protection shall be provided for each speed to protect each winding from excessive current during an overload condition. **(See Figure 8-3)**

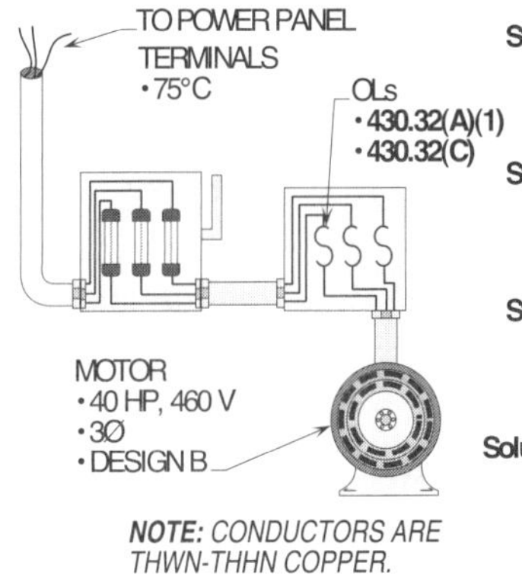

8-3

Figure 8-3. Determining the size branch-circuit conductors to supply multispeed motors.

Master Test Tip 4: Note that the conductors and OCPD for the branch-circuit supplying power to a multispeed motor shall be sized on the winding with the greater RPM.

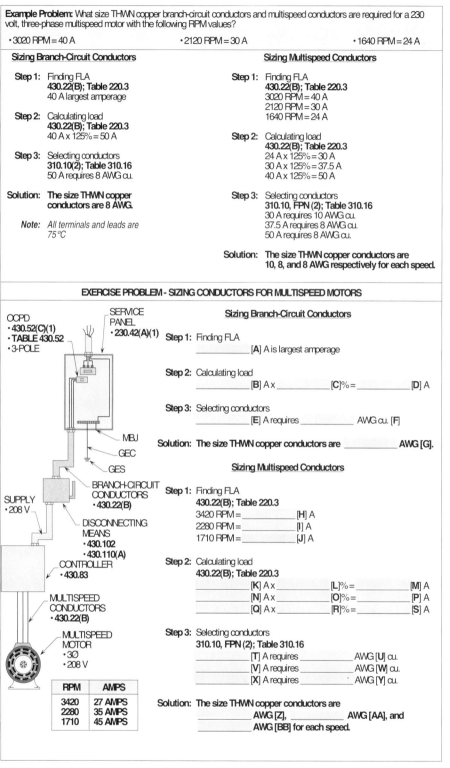

Example Problem: What size THWN copper branch-circuit conductors and multispeed conductors are required for a 230 volt, three-phase multispeed motor with the following RPM values?

- 3020 RPM = 40 A
- 2120 RPM = 30 A
- 1640 RPM = 24 A

Sizing Branch-Circuit Conductors

Step 1: Finding FLA
430.22(B); Table 220.3
40 A largest amperage

Step 2: Calculating load
430.22(B); Table 220.3
40 A x 125% = 50 A

Step 3: Selecting conductors
310.10(2); Table 310.16
50 A requires 8 AWG cu.

Solution: The size THWN copper conductors are 8 AWG.

Note: All terminals and leads are 75°C

Sizing Multispeed Conductors

Step 1: Finding FLA
430.22(B); Table 220.3
3020 RPM = 40 A
2120 RPM = 30 A
1640 RPM = 24 A

Step 2: Calculating load
430.22(B); Table 220.3
24 A x 125% = 30 A
30 A x 125% = 37.5 A
40 A x 125% = 50 A

Step 3: Selecting conductors
310.10, FPN (2); Table 310.16
30 A requires 10 AWG cu.
37.5 A requires 8 AWG cu.
50 A requires 8 AWG cu.

Solution: The size THWN copper conductors are 10, 8, and 8 AWG respectively for each speed.

EXERCISE PROBLEM - SIZING CONDUCTORS FOR MULTISPEED MOTORS

Sizing Branch-Circuit Conductors

Step 1: Finding FLA
_____ [A] A is largest amperage

Step 2: Calculating load
_____ [B] A x _____ [C] % = _____ [D] A

Step 3: Selecting conductors
_____ [E] A requires _____ AWG cu. [F]

Solution: The size THWN copper conductors are _____ AWG [G].

Sizing Multispeed Conductors

Step 1: Finding FLA
430.22(B); Table 220.3
3420 RPM = _____ [H] A
2280 RPM = _____ [I] A
1710 RPM = _____ [J] A

Step 2: Calculating load
430.22(B); Table 220.3
_____ [K] A x _____ [L] % = _____ [M] A
_____ [N] A x _____ [O] % = _____ [P] A
_____ [Q] A x _____ [R] % = _____ [S] A

Step 3: Selecting conductors
310.10, FPN (2); Table 310.16
_____ [T] A requires _____ AWG [U] cu.
_____ [V] A requires _____ AWG [W] cu.
_____ [X] A requires _____ AWG [Y] cu.

Solution: The size THWN copper conductors are _____ AWG [Z], _____ AWG [AA], and _____ AWG [BB] for each speed.

RPM	AMPS
3420	27 AMPS
2280	35 AMPS
1710	45 AMPS

SIZING CONDUCTORS FOR WYE-START AND DELTA RUN MOTORS
430.22(C); Table 220.3

The branch-circuit conductors for wye-start and delta run connected motors shall be selected based on the full-load current on the line side of the controller. The selection of conductors between the controller and the motor must be based on 58 percent (1÷1.732 = .58) of the motor's full-load current in amps times 125 percent. **(See Figure 8-4)**

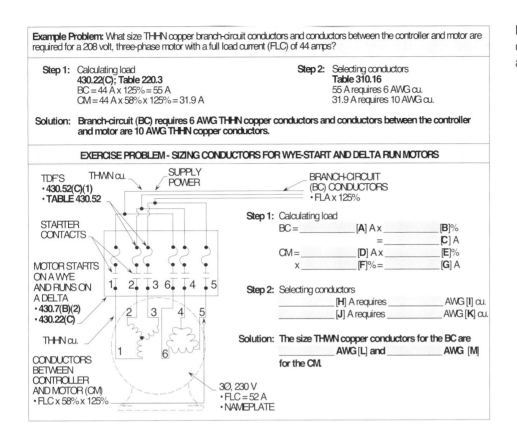

Figure 8-4. Determining the size conductors to supply motors starting on a wye and running on a delta.

SIZING CONDUCTORS FOR DUTY CYCLE MOTORS
430.22(E); Table 220.3

Conductors for a motor used for short-time, intermittent, periodic, or varying duty do not require conductors to be sized with a current-carrying capacity of 125 percent of the motor's full-load current in amps. **Table 220.3** and **Table 430.22(E)** permits the conductors to be sized with a percentage times the nameplate current rating based upon the duty cycle classification of the motor.

When sizing conductors to supply individual motors that are used for short-time, intermittent, periodic, or varying duty the requirements of **Table 430.22(E)** shall apply. Varying heat loads are produced on the conductors by the starting and stopping duration of operation cycles, which permits conductor sizing changes. In other words, such conductors are never subjected to continuous operation due to on and off periods and therefore conductors are never fully loaded for long intervals of time. For this reason, conductors can be down sized based on their condition of use. **(See Figure 8-5)**

Master Test Tip 5: For the definition of short-time, intermittent, periodic, or varying duty, see Duty in **Article 100** of the NEC.

SIZING CONDUCTORS FOR ADJUSTABLE SPEED DRIVE SYSTEMS
430.122(A); 430.2; Table 220.3

Power conversion equipment, when supplied from a branch-circuit includes all elements of the adjustable speed drive system. The rating in amps is used to size the conductors that are based upon the power required by the conversion equipment. When the power conversion equipment provides overcurrent protection for the motor, no additional overload protection is required.

Master Test Tip 6: Power conversion equipment contains solid state units that change the cycles or chop part of the wave forms to vary the speed of squirrel cage motors as needed for a particular application. **(See Figure 8-6)**

The disconnecting means can be installed in the line supplying the conversion equipment, and the rating of the disconnect shall not be less than 115 percent of the input current rating of the conversion unit. (See **430.110(A)** and **430.2**)

Power conversion equipment requires the conductors to be sized at 125 percent of the rated input of such equipment per **430.122(A)**.

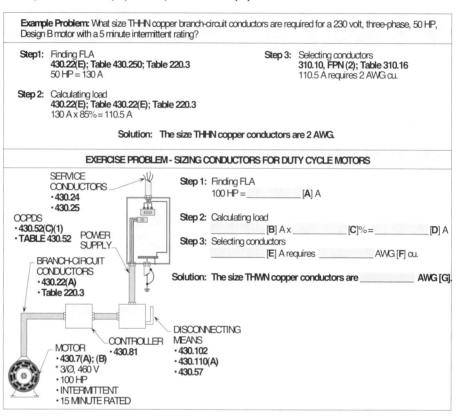

Figure 8-5. Determining the size conductors to supply duty cycle related motors.

Master Test Tip 7: When applying **Table 220.3** and **Table 430.22(B)**, the branch-circuit OCPD can serve as protection for the circuit and system per **430.33**.

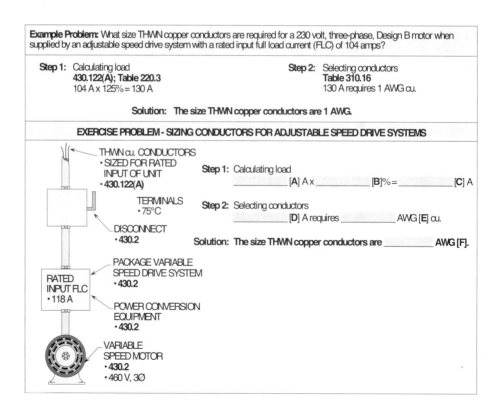

Figure 8-6. Determining the size conductors to supply power conversion equipment.

Master Test Tip 8: On the examination where a problem pertains to a drive system, calculate the load in amps to size circuit components by the drive input (FLA) and not the FLA of the motor.

Motors and Compressors

SIZING CONDUCTORS FOR PART-WINDING MOTORS
430.4; 430.22(D); Table 220.3

Induction or synchronous motors that have a part-winding start are designed so that at starting they energize the primary armature winding first. After starting, the remainder of the winding is energized in one or more steps. The purpose of this arrangement is to reduce the initial inrush current until the motor accelerates to its running speed.

The inrush current at start is locked-rotor current, and can be quite high. A standard part-winding-start induction motor is designed so that only half of its winding is energized to start, then, as it comes up to speed, the other half is energized, so both halves are energized and carry equal current.

Separate overload devices shall be used on a standard part-winding-start induction motor to protect the windings from excessive damaging currents. This means that each half of the motor winding has to be individually provided with overload protection. These requirements are covered in **430.32** and **430.37**. Each half of the windings has a trip current value that is one half of the specified running current. As required by **430.52**, each of the two motor windings shall have branch-circuit short-circuit and ground-fault protection that shall be calculated at not more than one half the percentages listed in **430.52** and **Table 430.52**. **(See Figure 8-7)**

Master Test Tip 9: Section **430.4, Ex.** permits a single device with one half rating, for both windings, provided that it will permit the motor to start and run. If a time-delay (dual element) fuse is used as a single device for both windings, its rating is permitted if it does not exceed 150 percent of the motor's full-load current in amps.

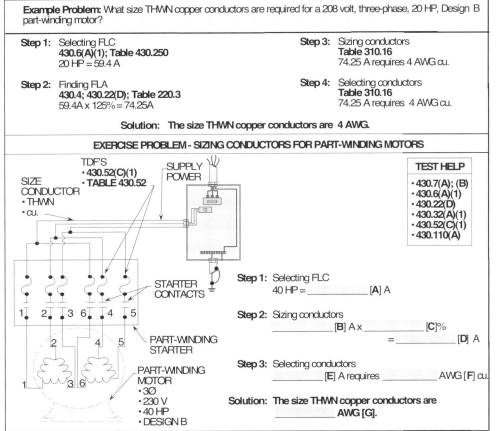

Figure 8-7. Determining the size conductors to supply part-winding motors.

Master Test Tip 10: Note that only half of the motor winding is used on starting, but all of the motor winding is used for running.

SIZING CONDUCTORS FOR SEVERAL MOTORS
430.24; Table 220.3

The full-load current rating of the largest motor shall be multiplied by 125 percent to select the size of conductors for a feeder supplying a group of two or more motors. The remaining motors of the group shall have their full-load current ratings added to this value, and this total amperage is then used to size the conductors. **(See Figure 8-8)**

Figure 8-8. Determining the size conductors for a feeder-circuit to supply several motors.

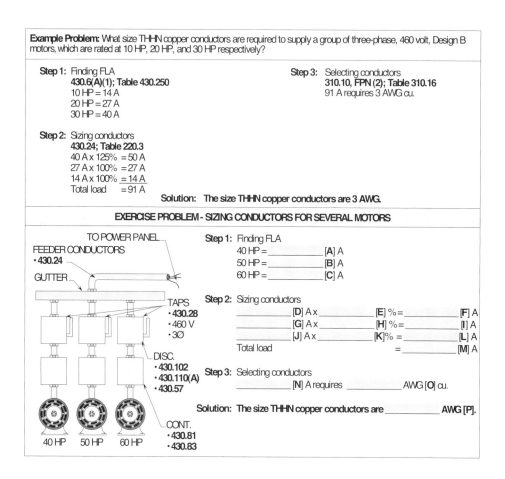

Master Test Tip 11: If the 20 HP motor rated at 27 amps was a varying duty, continuous rated motor, it would be the largest motor load at 54 amps (27 A x 200% = 54 A) per **Table 220.3** and **Table 430.22(E)** columns 1 and 5. Note that the 30 HP at 50 amps (40 A x 125% = 50A) would no longer be considered the largest motor load as before.

SIZING CONDUCTORS FOR DUTY CYCLE MOTORS
430.24, Ex. 1; Table 220.3

Table 430.22(E) shall be used for sizing the amperage rating for a motor that is classified as either short-time, intermittent, periodic, or varying duty. The amperage rating shall be based on 100 percent of the full-load current rating of the motor's nameplate if rated for continuous operation. The full-load current rating of the largest motor shall be multiplied by 125 percent to select the size of conductors for a feeder-circuit supplying a group of two or more motors. The remaining motors of the group shall have their total full-load current ratings added to the computed amps of the duty cycle motor or the amps of the 125 percent motor, whichever is greater. The feeder-circuit conductors shall be sized by this total full-load current. **(See Figure 8-9)**

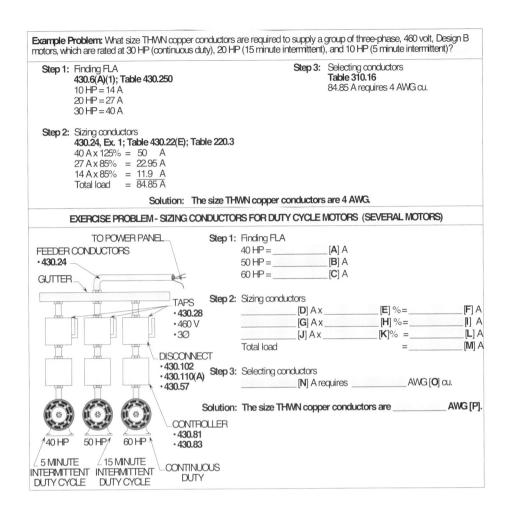

Figure 8-9. Determining the size conductors for a feeder-circuit to supply duty-cycle related motors.

Master Test Tip 12: When selecting the percentage that is permitted for a motor having a duty-cycle service, the following information must be known:

(1) Classification of service per Col. 1 of **Table 430.22(E)**.

(2) Minute rating of motor per Col.'s 2, 3 and 4 of **Table 430.22(E)**.

(3) See Col. 5 of **Table 430.22(E)**, for continuous rated motors based on duty-cycle service.

SIZING CONDUCTORS SUPPLYING MOTORS AND OTHER LOADS
430.25; Table 220.3

The motor load shall be calculated per **Table 220.3** and **430.22(A)** or **430.24**. If a test problem has combination loads that consist of one or more motor loads on the same circuit with lights, receptacles, appliances, or any combination of such loads, article 220 and other applicable articles shall be used to calculate other loads. The ampacity required for the feeder-circuit shall be equal to all the total loads involved. The OCPD's used to protect conductors and elements from short-circuits and ground-faults shall be sized per **430.62(A)** and **430.63**, and the larger protective device used. **(See Figures 8-10(a), (b), (c) and (d))**

SIZING THE BRANCH-CIRCUIT PROTECTIVE DEVICE
TABLE 430.52; 430.52(C)(1); (C)(3); Table 240.4(G)

The motor branch-circuit overcurrent device shall be capable of carrying the starting current of the motor. Short-circuit and ground-fault currents are considered as being properly taken care of when the overcurrent protection device does not exceed the values in **Table 430.52** as permitted by the provisions of **430.52(C)(1)** with Exceptions.

Different percentages are selected for particular devices based upon one of the four columns listed in **Table 430.52**. The percentages are used to size and select the proper size overcurrent protection device to allow a certain type of motor to start and run. The motor has a momentary starting current that is necessary for the motor to have power to start

Master Test Tip 13: In cases where the values for branch-circuit protective devices determined by **Table 430.52** do not correspond to standard sizes or ratings of fuses, nonadjustable circuit breakers, or thermal devices, or possible settings of adjustable circuit breakers adequate to carry the starting currents of the motor, the next higher size rating or setting may be used.

Stallcup's Master Electrician's Study Guide

Master Test Tip 14: If a test problem fails to call for the minimum size (rounded down), next size (rounded up), or maximum size (rounded down) OCPD, then round up and not down on the OCPD rating per **430.52(C)(1), Ex. 1** and **Table 240.4(G)** and **240.6(A)**.

and drive the connected load at the driven equipment. Note that the OCPD sized per **430.52(C)(1)** provides protection from short-circuits and ground-faults. Overload protection must be provided for conductors and motor windings per **430.32(A)** or **430.32(C)**. **(See Figure 8-11)**

Figure 8-10(a). Test problem for sizing and selecting OCPD based on loads.

Note: For selecting the OCPD based on the largest OCPD for any one motor on the feeder-circuit, see the exercise problem in Figure 8-10(b).

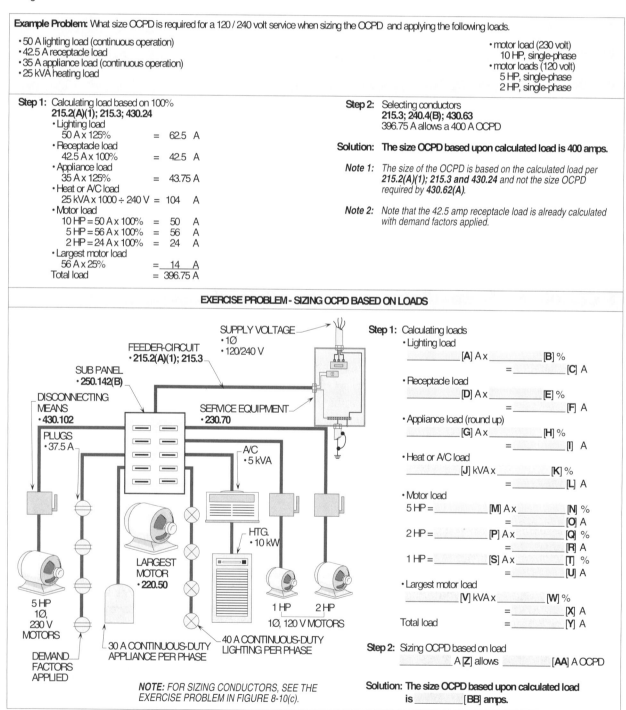

Note: For selecting the size OCPD based on the calculated load, see the exercise problem in **Figure 8-10(a)**.

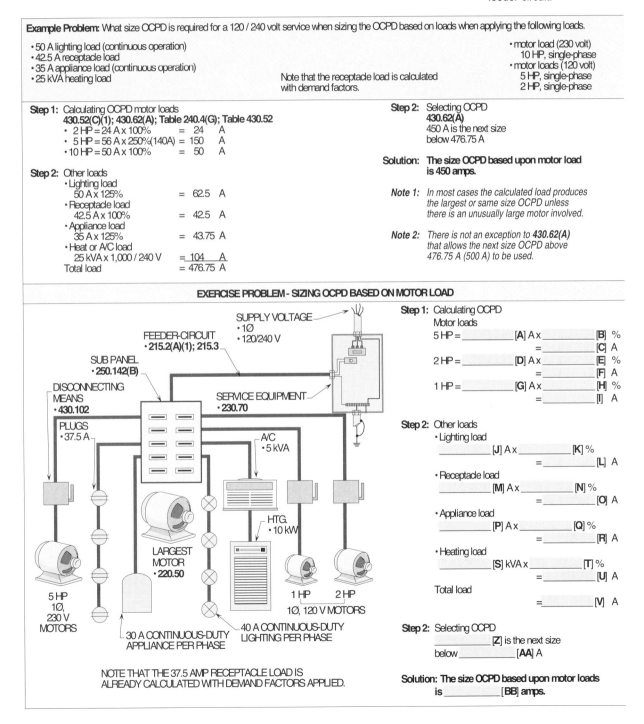

Figure 8-10(b). Test problem for sizing and selecting the OCPD based on the largest OCPD for any one motor on the feeder-circuit.

Figure 8-10(c). Test problem for sizing and selecting THWN copper conductors for a feeder supplying combination loads.

Note: For sizing the neutral conductor, see the exercise problem in **Figure 8-10(d)**.

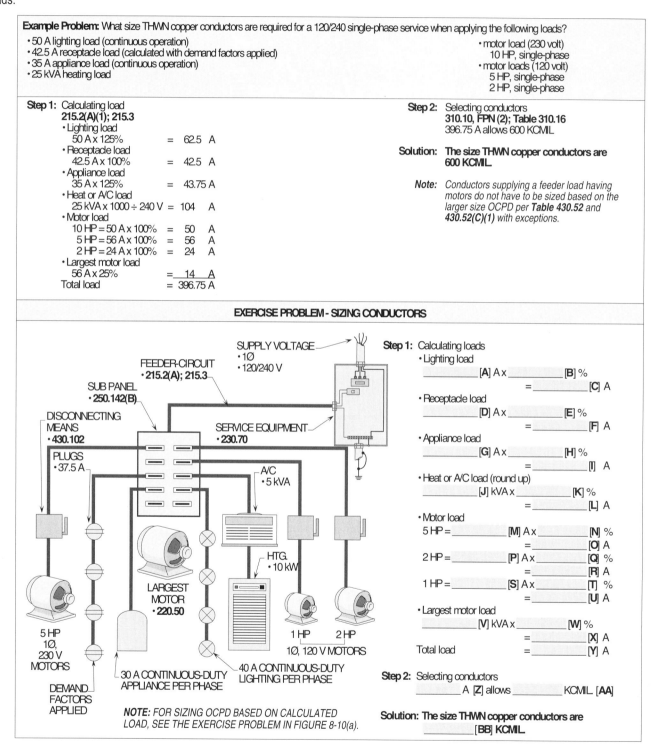

Note: For sizing the phase conductors, see the exercise problem in **Figure 8-10(c)**.

Figure 8-10(d). Test problem for sizing and selecting THWN copper neutral conductor.

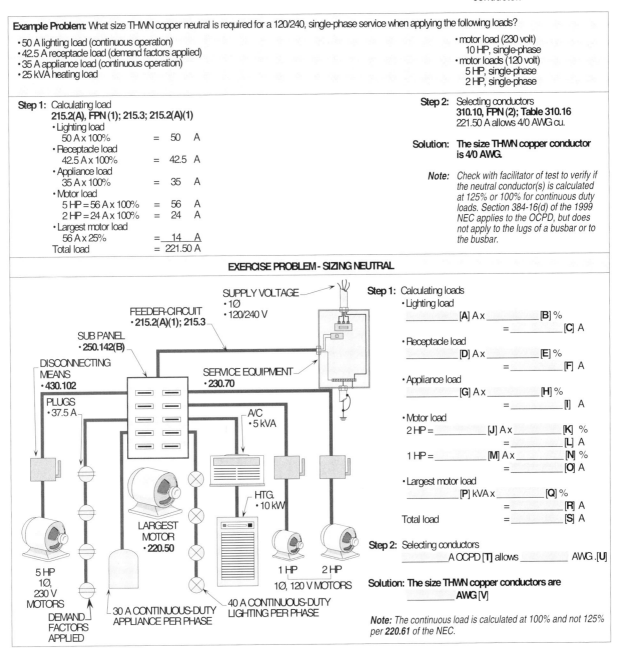

Note, when taking the master's examination, be sure and check with the testing authority and verify if they require the neutral to be calculated and sized at 100% or 125% for continuous duty loads.

Master Test Tip 15: Note that the amps of the 240 volt, single-phase heating load, the 230 volt, single-phase motor and the 230 volt, A/C unit in the exercise problem does not apply when calculating 120 volt loads for sizing the neutral.

8-13

Figure 8-11. Selecting the percentages to size the minimum, next size, and maximum size circuit breaker to allow a motor to start and run.

Master Test Tip 16: You can round down or up the size of the OCPD (min. size) when the test question or problem instructs you to do so. **(See Master Test Tip 17)**

APPLYING THE EXCEPTIONS 430.52(C)(1), Ex.'s 1 AND 2

There are Exceptions where the overcurrent protection device as specified in **Table 430.52** will not take care of the starting current to allow the motor to start and run. When the motor fails to start and run due to excessive inrush starting currents, one of the following exceptions can be applied:

APPLYING Ex. 1

If the values of the branch-circuit, short-circuit, and ground-fault protection devices determined from **Table 430.52** do not conform to standard sizes or ratings of fuses, non-adjustable circuit breakers, or possible settings on adjustable circuit breakers, it does not matter if they are capable or not capable of adequately carrying the load involved; the next higher setting or rating will be permitted. In other words, you can round up (next size) or round down (min. size) the size of the overcurrent protection device automatically by choice. **(See Figure 8-12 and Master Test Tip 17)**

Figure 8-12. When the percentages of **Table 430.52** times the full-load current of the motor in amps do not correspond to a standard size OCPD, the next higher size above this percentage may be used.

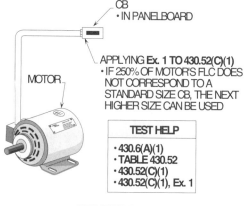

Master Test Tip 17: Round up to the next size OCPD when the test question or problem does not specify the minimum, next size, or maximum size OCPD. **(See Master Test Tip 16)**

APPLYING Ex. 2

If the ratings listed in **Table 430.52** and **Ex. 1 to 430.52(C)(1)** are not sufficient for the starting current of the motor. The following OCPD's with percentages shown can be used to start and run motors having high inrush starting currents: **(See Figure 8-13)**

Anytime the percentages of **Table 430.52** and **Ex. 1 to 430.52(C)(1)** won't allow the motor to start and run its driven load, the following percentages per **430.52(C)(1), Ex. 2** can be applied for different types of OCPD's:

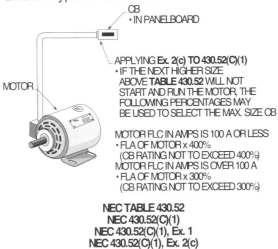

Figure 8-13. If the percentages of **Table 430.52** and **430.52(C)(1), Ex. 1** won't allow the motor to start and run the driven load, the maximum size CB of **430.52(C)(2), Ex. 2(c)** may be used.

Master Test Tip 18: The percentages to **430.52(C)(1), Ex.'s 2 (a), (b), (c) or (d)** must not be applied in a question or a problem on the test unless the maximum size OCPD is specified.

(1) When nontime delay fuses are used and they do not exceed 600 amperes in rating, it shall be permitted to increase the fuse size up to 400 percent of the full-load current, but over 400 percent must never be used.

(2) Time-element fuses (dual element) are not to exceed 225 percent of the full-load current, but they may be increased to up this percentage.

(3) Inverse time-element breakers shall be permitted to be increased in rating. However, they shall not exceed:

  (a) 400 percent of full-load current of the motor for 100 amperes or less, or

  (b) They may be increased to 300 percent where a full-load current is greater than 100 amperes.

See Figure 8-14 for a detailed illustration on selecting the percentage for sizing OCPD's.

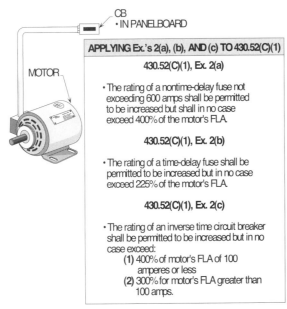

Figure 8-14. When the percentages of **Table 430.52** and **430.52(C)(1), Ex. 1** won't allow the motor to start and run, the maximum percentages of **430.52(C)(1), Ex.'s 2(a), (b), and (c)** may be applied.

Master Test Tip 19: Note that the size of the OCPD must never exceed the percentage listed in **430.52(C)(1), Ex.'s 2(a), (b), and (c)**.

USING INSTANTANEOUS TRIP CB's
430.52(C)(3), Ex. 1

Master Test Tip 20: For Design E or NEMA Class B high-efficiency motors, the setting on the instantaneous circuit breaker can be adjusted up to a maximum of 1700 percent of the motor's FLA to allow the motor to start and run.

An instantaneous trip circuit breaker shall be used only if it is adjustable, and it is part of a combination controller and overcurrent protection is provided in each controller. Such combination if used has to be an approved assembly. An instantaneous trip circuit breaker is allowed to have a damping device, to limit the inrush current when the motor is started.

If the specified setting in **Table 430.52** is found not to be sufficient for the starting current of the motor, the setting on an instantaneous trip circuit current may be increased, provided that in no instance it exceeds 1300 percent of the motor's full-load current rating for motors marked as Class B, C, or D.

See Figure 8-15 for adjusting the maximum trip settings on instantaneous trip circuit breakers to allow motors to start and accelerate their driven load.

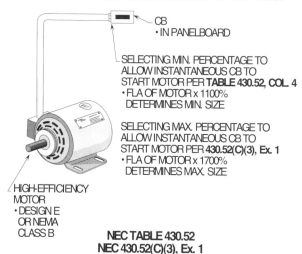

Figure 8-15. Determining the minimum and maximum size instantaneous trip circuit breaker to allow the motor to start and run the driven load.

CODE LETTERS
TABLES 430.7(B) AND 430.52

Code letters shall be installed on motors by manufacturers for calculating the locked-rotor current (LRC) in amps based upon the kVA per horsepower, which is selected from the motor's code letter. Overcurrent protection devices shall be set above the locked-rotor current of the motor to prevent the overcurrent protection device from opening when the rotor of the motor is starting. The following two methods shall be used to calculate the locked-rotor current of motors:

Master Test Tip 21: The LRC for a motor can be determined by selecting the kVA per HP based on the code letter from **Table 430.7(B)** and applying the following formula:

$$LRC = \frac{kVA \text{ per HP} \times HP \times 1000}{V \times 1.732}$$

Note: Use 1.732 x the voltage if the voltage of the motor is three-phase. **(See Master Test Tip 23)**

(1) Utilizing code letters to determine LRC per **Table 430.7(B)**
(2) Utilizing Design letters to determine LRC per **Tables 430.251(A)** and **(B)**

LOCKED-ROTOR CURRENT UTILIZING CODE LETTER
430.7(B)

Code letters shall be marked on the nameplate, and such letters are used for designing locked-rotor current. Locked-rotor current for code letters are identified in **Table 430.7(B)** in kVA (kilovolt-amps) per horsepower, which is based upon a particular code letter (A through V).

Motors and Compressors

LOCKED-ROTOR CURRENT UTILIZING HP TABLES 430.251(A) AND (B)

The locked-rotor current of a motor may be found in **Tables 430.251(A)** and **(B)**. The locked-rotor current for a single-phase and three-phase motor is determined from this Table based upon the phases, voltage, and horsepower rating of the motor. Note that on the master examination, code letters won't be found in **Tables 430.251(A)** and **(B)** because they won't be listed on the motor's nameplate anymore. Motors will be marked with a Design B, C, D, or E letter to indicate that locked-rotor currents are to be selected from **Tables 430.251(A)** and **(B)** based upon horsepower, phases, and voltages.

For example: What is the locked-rotor current rating for a three-phase, 460 volt, 30 horsepower, Design B motor?

Table method using Design letter

> **Step 1:** Finding LRC amps
> **Table 430.251(B)**
> 30 HP requires 218 A

Solution: The locked-rotor current is 218 amps.

See **Figure 8-16** for calculating and selecting the locked-rotor current of a motor based on Design letters.

Master Test Tip 22: Test takers must select the locked-rotor current rating from **Tables 430.251(A)** and **(B)** when using Design B, C, D, or E motors. The overcurrent protection device must be set above the locked-rotor current of the motor so the motor will start and run.

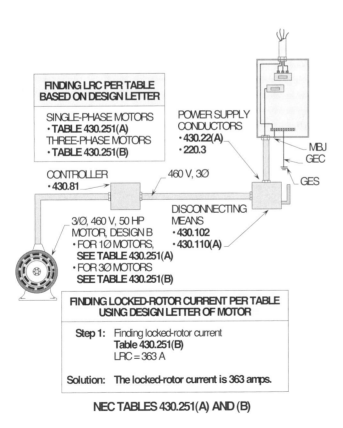

Figure 8-16. Tables 430.251(A) and (B) can be used to determine the LRC in amps for motors having Design letters.

Master Test Tip 23: If the motor in Figure 8-16 had a code letter B, the LRC would be 222 amps

$$LRC = \frac{3.54 \times 50 \times 1000}{460 \text{ V} \times 1.732} = 222 \text{ A}$$

Note: When finding LRC of a motor, test takers must read the questions carefully and determine if the motor is marked with a code letter or a design letter.

SIZING AND SELECTING OCPD'S
TABLE 430.52, COLUMNS 2, 3, 4, AND 5; Table 240.4(G)

Master Test Tip 24: The standard size OCPD's and those that are not standard can be found in **240.6(A)**.

The overcurrent protection device shall be sized for the starting current of the motor and selected to allow the motor to start and run. The OCPD per **Table 430.52** must protect the branch-circuit conductors from short-circuits and ground-faults. The following four overcurrent protection devices selected from **Table 430.52** will start most motors under normal starting conditions:

(1) Nontime-delay fuses per Column 2
(2) Time-delay fuses per Column 3
(3) Instantaneous trip circuit breakers per Column 4
(4) Inverse-time circuit breakers per Column 5

NONTIME-DELAY FUSES
TABLE 430.52, COLUMN 2

Nontime-delay fuses can be installed with instantaneous trip features to detect short-circuits and thermal characteristics to sense slow heat buildup in the circuit. Note that nontime-delay fuses are not equipped with time delay characteristics.

See Figure 8-17 for a detailed illustration of sizing nontime-delay fuses to allow motors to start and run.

Figure 8-17. If nontime-delay fuses are sized large enough, they will allow the motor to start and run.

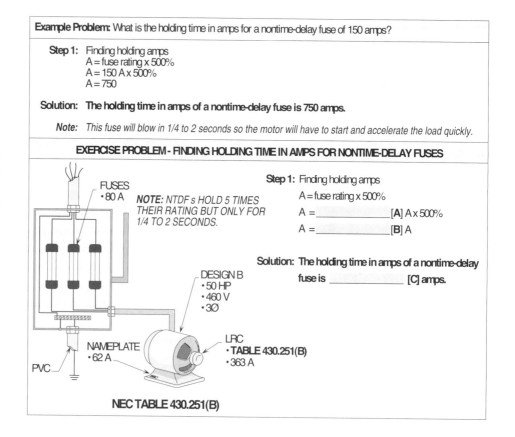

Master Test Tip 25: The benefit of a test taker knowing that a fuse holds five times its rating is, after sizing the fuse to start a motor, a fast check can be made to verify if it's large enough, for example, the 150 A fuse (150 A x 5 = 725 A) in the example problem will allow a 100 HP, three-phase, 460 volt motor with a LRC of 725 amps to start and run.

Motors and Compressors

TIME-DELAY FUSES
TABLE 430.52, COLUMN 3

Time-delay fuses are also equipped with instantaneous trip features to detect short-circuits and thermal characteristics to sense heat buildup in the circuit. Time-delay fuses are used because of their time-delay action to allow a motor to start. Note that time-delay fuses that are sized at 125 percent or less of the motor's FLC rating (nameplate) in amps can provide overload protection for the motor.

See **Figure 8-18** for sizing time-delay fuses to hold the motor's locked-rotor current.

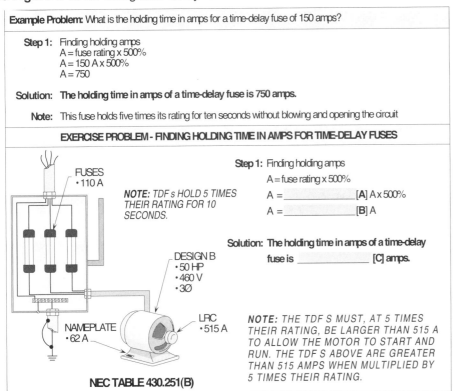

Figure 8-18. Time-delay fuses are equipped with a time lag that will allow the motor to start and run.

Master Test Tip 26: For the benefit of knowing that a fuse holds 5 times its rating, see **Master Test Tip 25** of this chapter.

INSTANTANEOUS TRIP CIRCUIT BREAKERS
TABLE 430.52, COLUMN 4

Instantaneous trip circuit breakers are installed with instantaneous values of current that will respond from a short-circuit only. Thermal protection is not provided for instantaneous trip circuit breakers. To allow motors with high inrush currents to start and run, certain types allow such settings to be adjusted from 0 to 1700 percent. **(See Figure 8-19)**

INVERSE-TIME CIRCUIT BREAKERS
TABLE 430.52, COLUMN 5

Inverse-time circuit breakers are designed with instantaneous trip features to detect short-circuits and thermal characteristics to sense slow heat in the circuit. If heat should occur in the windings of the motor, the instantaneous values or current will be detected by the thermal action of the circuit breaker and will trip open the circuit if it is sized properly. The magnetic action of the circuit breaker will clear the circuit if short-circuits or ground-faults should occur on the circuit elements or equipment served. **(See Figure 8-20)**

Master Test Tip 27: CB's with a slash marking (480/277 V) can only be used on grounded systems per **250.20(B)** and **240.85**.

Figure 8-19. An instantaneous circuit breaker with its rating set above the locked-rotor current of a motor will allow the motor to start and run.

Example Problem: What is the holding amps for an instantaneous trip circuit-breaker with a 200 amp frame size? (use low setting)

Step 1: Finding holding amps
A = frame size x 3
A = 200 A x 3
A = 600

Solution: **The setting in amps of an instantaneous trip circuit-breaker is 600 amps.**

EXERCISE PROBLEM - FINDING HOLDING TIME IN AMPS FOR AN INSTANTANEOUS CIRCUIT-BREAKER

Step 1: Finding holding amps (medium setting)
A = frame size x 5
A = ______________ **[A]** A x 5
A = ______________ **[B]** A

Solution: **The setting of an instantaneous trip circuit-breaker is ______________ [C] amps.**

NOTE: *THE 3-POLE INST. CB S WILL HOLD THEIR PRESET OR ADJUSTED RATING UNTIL IT IS EXCEEDED IN AMPS.*

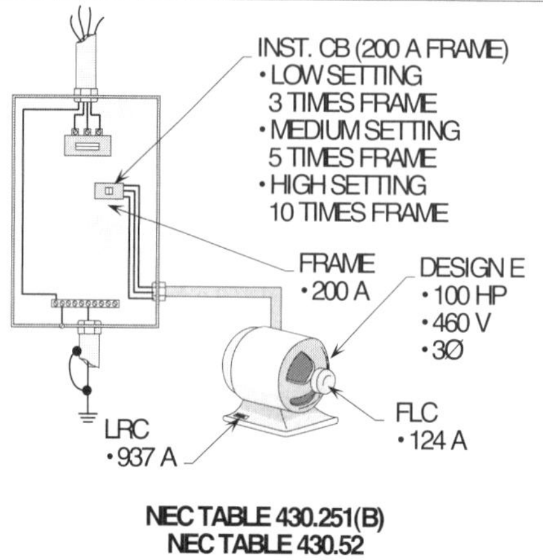

Figure 8-20. Circuit breakers sized at least three times their rating if they have an amp rating above the locked-rotor current of the motor will hold such LRC.

Master Test Tip 28: The benefit of knowing that a CB holds three times its rating is, after sizing the CB to start a motor, a fast check can be made to verify if it's large enough. For example, the 200 amp CB (200 A x 3 = 600 A) in the example problem will allow a 75 HP, three-phase, 460 volt motor with an LRC of 543 amps to start and run.

Example Problem: What is the holding amps for an inverse-time circuit-breaker with a 200 amp frame size?

Step 1: Finding holding amps
A = frame size x 3
A = 200 A x 3
A = 600

Solution: **The rating in amps of an inverse-time circuit breaker is 600 amps.**

EXERCISE PROBLEM - FINDING HOLDING TIME IN AMPS FOR INVERSE-TIME CIRCUIT BREAKERS

Step 1: Finding holding amps
A = frame size x 3
A = ______________ **[A]** A x 3
A = ______________ **[B]** A

Solution: **The holding amps of an inverse-time circuit breaker is ______________ [C] amps.**

Note: The CB in the panelboard is 3-pole.

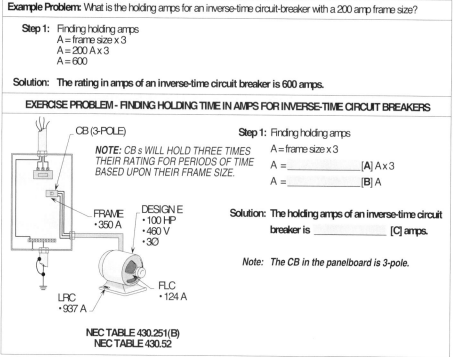

SIZING OCPD'S TO ALLOW MOTORS TO START AND RUN
430.52(C)(1); TABLE 430.52; Table 240.4(G)

The branch-circuit protection for a motor may be a fuse or circuit breaker located in the line at the point where the branch-circuit originates. The fuse or circuit breaker is located either at a service cabinet or distribution panel or in the motor control center. When there is only one motor on a branch-circuit, the fuse or circuit breaker is sized according to **Table 430.52** and **430.52(C)(1)** or by applying one of the exceptions.

Proper use of the Table will require explanation. There is the matter of "Design letters." A Design letter provides certain electrical characteristics of a particular motor that are needed to size the OCPD to permit the motor to start and accelerate its load. To apply **Table 430.52**, it is necessary to take the following four steps:

- **(1)** Select the phase of the motor
  - Single-phase
  - Three-phase

- **(2)** Select type of motor
  - Squirrel-cage induction
  - Wound-rotor
  - DC
  - Synchronous

Note that motors can be single-phase or three-phase types.

- **(3)** Select the Design letter of the motor
  - Design B
  - Design C
  - Design D
  - Design E

- **(4)** Select the type OCPD
  - Column 2 is for NTDF's
  - Column 3 is for TDF's
  - Column 4 is for CB's with instantaneous trip setting or adjustments
  - Column 5 is for CB's with both instantaneous fixed trip settings and thermal trip characteristics

See Figures 8-21(a) through (d) for sizing and selecting the OCPD's per **Table 430.52** to allow motors to start and run their driven loads. Note that the minimum and next size OCPD will be sized for a particular size motor.

Master Test Tip 29: If the type of motor is not specified on the test question or problem, the motor will be of the squirrel-cage type.

SIZING MAXIMUM OCPD
430.52(C)(1), Ex.'s 2(a) thru (c)

Where the rating specified in **Table 430.52** and **430.52(C)(1), Ex. 1** is not sufficient for the starting current of the motor, the following ratings (percentages) shall be applied:

- **(1)** Nontime-delay fuses (400 percent)
- **(2)** Time-delay fuses (225 percent)
- **(3)** Inverse-time circuit breaker (400 and 300 percent)
- **(4)** Instantaneous trip circuit breakers (0-1700 percent based on Design letter)

Master Test Tip 30: The **Ex. 2** to **430.52(C)(1)** can only be applied after the minimum size and next size OCPD per **430.52(C)(1) and Ex. 1** have failed to start the motor.

NONTIME-DELAY FUSES
430.52(C)(1), Ex. 2(a)

If the minimum or next size (rounded-up size to 400%) OCPD will not allow the motor to start and run, the maximum size rating of a nontime delay fuse not exceeding 600 amps shall be permitted to be increased but shall in no case exceed 400 percent of the full-load current of the motor. **(See Figure 8-22)**

TIME-DELAY FUSES
430.52(C)(1), Ex. 2(b)

To allow a motor to start and run, the rating of a time-delay fuse shall be permitted to be increased but shall in no case exceed 225 percent of the full-load current of the motor. **(See Figure 8-23)**

Figure 8-21(a). Determining the minimum and next size nontime-delay fuse to start and run a motor.

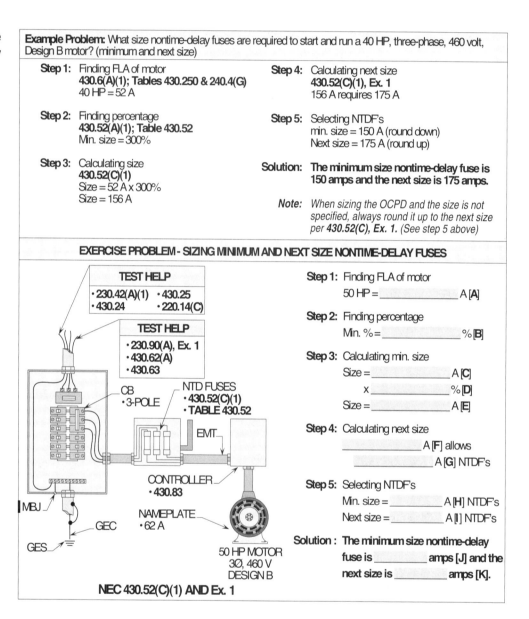

Note: In the exrecise problem, the minimum (min.) is the round down size (65 A x 300% = 195 A = 175 A) and next size is the round up size (65 A x 300 % = 195 A = 200 A).

Motors and Compressors

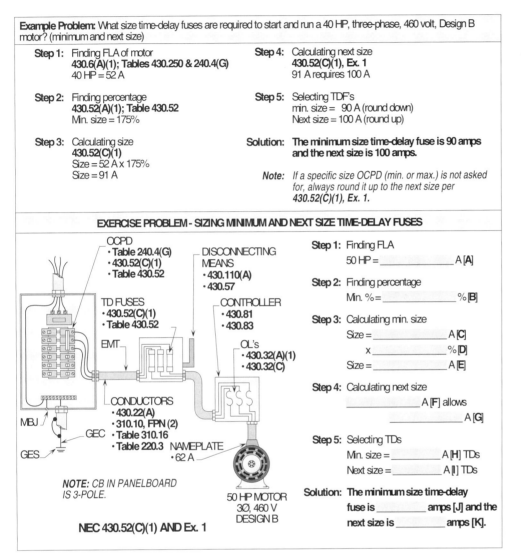

Figure 8-21(b). Determining the minimum and next size time-delay fuse to start and run a motor.

Review note below exercise problem 8-21(a) on page 8-22 of this book.

Figure 8-21(c). Determining the minimum and maximum setting for an instantaneous trip circuit breaker to start and run a motor.

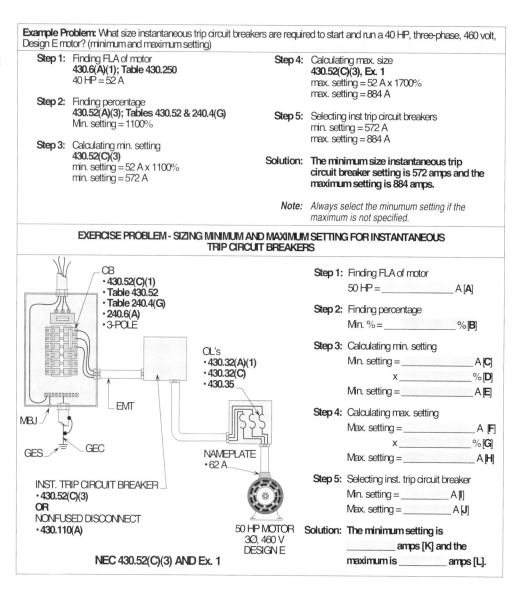

Review note below exercise problem 8-21(a) on page 8-22 of this book.

Motors and Compressors

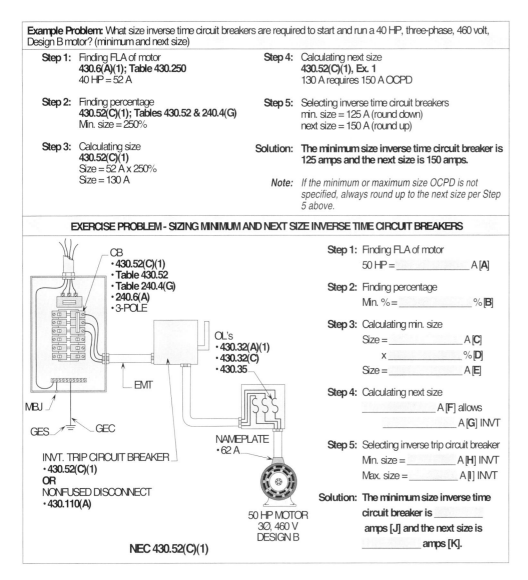

Figure 8-21(d). Determining the minimum and next size inverse-time circuit breaker to start and run a motor.

Review note below exercise problem 8-21(a) on page 8-22 of this book.

Figure 8-22. Nontime-delay fuses can be increased to a maximum of 400 percent of the motor's full-load current rating in amps.

Master Test Tip 31: When selecting the maximum size fuse in the example problem, the rating shall not exceed the computed amps of 208. Therefore, 200 amp NTDF's must be selected from **240.6(A)**.

Example Problem: What size nontime-delay fuse is required to start and run a 40 HP, three-phase, 460 volt, Design B motor? (maximum size)

Step 1: Finding FLA of motor
430.6(A)(1); Table 430.250
40 HP = 52 A

Step 2: Finding percentage
430.52(C)(1), Ex. 2(a); Tables 430.52 & 240.4(G)
Max. size = 400%

Step 3: Calculating max. size
430.52(C)(1), Ex. 2(a)
Max. size = 52 A x 400%
Max. size = 208 A

Step 4: Selecting NTDF's
Max. size = 200 A

Solution: The maximum size nontime-delay fuse is 200 amps.

EXERCISE PROBLEM - SIZING MAXIMUM SIZE NONTIME-DELAY FUSES

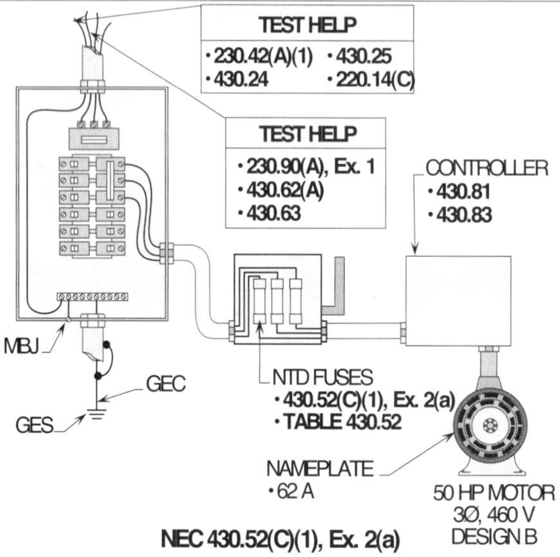

TEST HELP
• 230.42(A)(1) • 430.25
• 430.24 • 220.14(C)

TEST HELP
• 230.90(A), Ex. 1
• 430.62(A)
• 430.63

CONTROLLER
• 430.81
• 430.83

NTD FUSES
• 430.52(C)(1), Ex. 2(a)
• TABLE 430.52

NAMEPLATE
• 62 A

50 HP MOTOR
3Ø, 460 V
DESIGN B

NEC 430.52(C)(1), Ex. 2(a)

Step 1: Finding FLA of motor
50 HP = _____ A [A]

Step 2: Finding percentage
Max. size = _____ % [B]

Step 3: Calculating max. size
Max. size = _____ A [C]
x _____ % [D]
Max. size = _____ A [E]

Step 4: Selecting NTDFs
Max. size = _____ A [F] NTDF's

Solution: The maximum size nontime-delay fuse is _____ amps [G].

NOTE: CB in panelboard is 3-pole.

Figure 8-23. Determining the maximum size time-delay fuse to start and run a motor.

Master Test Tip 32: When selecting TDF's at the maximum rating per **430.52(C)(1), Ex. 2(b)**, the 225 percent multiplier shall not be exceeded. See steps 3 and 4 in the example problem for application of such rule.

Example Problem: What size time-delay fuse is required to start and run a 40 HP, three-phase, 460 volt, Design B motor? (maximum size)

Step 1: Finding FLA of motor
430.6(A)(1); Table 430.250
40 HP = 52 A

Step 2: Finding percentage
430.52(C)(1), Ex. 2(b); Tables 430.52 & 240.4(G)
Max. size = 225%

Step 3: Calculating max. size
430.52(C)(1), Ex. 2(b)
Max. size = 52 A x 225%
Max. size = 117 A

Step 4: Selecting NTDF's
Max. size = 110 A

Solution: The maximum size time-delay fuse is 110 amps.

EXERCISE PROBLEM - SIZING MAXIMUM SIZE TIME-DELAY FUSES

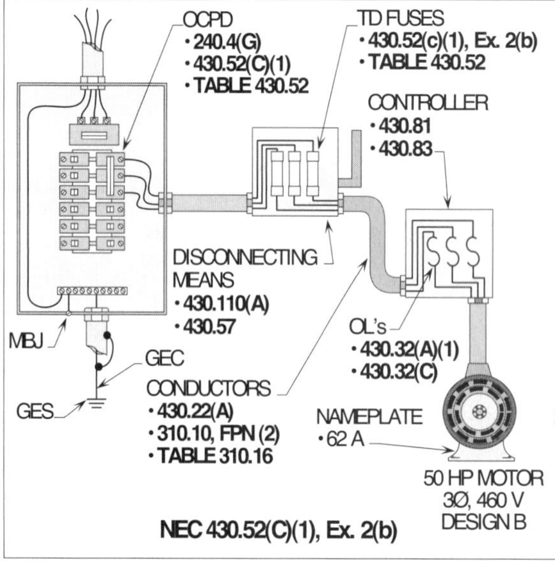

OCPD
• 240.4(G)
• 430.52(C)(1)
• TABLE 430.52

TD FUSES
• 430.52(c)(1), Ex. 2(b)
• TABLE 430.52

CONTROLLER
• 430.81
• 430.83

DISCONNECTING MEANS
• 430.110(A)
• 430.57

OL's
• 430.32(A)(1)
• 430.32(C)

CONDUCTORS
• 430.22(A)
• 310.10, FPN (2)
• TABLE 310.16

NAMEPLATE
• 62 A

50 HP MOTOR
3Ø, 460 V
DESIGN B

NEC 430.52(C)(1), Ex. 2(b)

Step 1: Finding FLA of motor
50 HP = _____ A [A]

Step 2: Finding percentage
Max. size = _____ % [B]

Step 3: Calculating max. size
Max. size = _____ A [C]
x _____ % [D]
Max. size = _____ A [E]

Step 4: Selecting TDF's
Max. size = _____ A [F]

Solution: The maximum size time-delay fuse is _____ amps [G].

Note: CB in panelboard is 3-pole.

Motors and Compressors

INVERSE TIME CIRCUIT BREAKERS
430.52(C)(1), Ex. 2(c)

The rating for inverse-time circuit breakers shall be permitted to be increased but shall in no case exceed 400 percent for a full-load current of 100 amps or less. Full-load current ratings greater than 100 amps shall be permitted to be increased 300 percent. **(See Figure 8-24)**

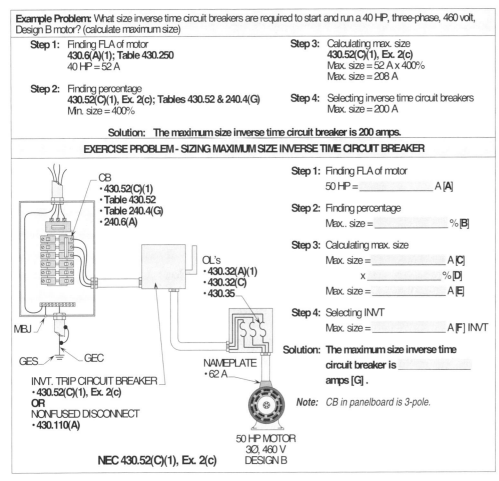

Figure 8-24. An inverse time circuit breaker can be increased to a maximum of 400 percent of the motor's full-load current rating in amps.

Master Test Tip 33: The only time that an OCPD can be rounded up to the next size above its multiplier per **Table 430.52** (based on the motor and the type of OCPD) is when the next size is selected per **430.52(C)(1), Ex. 1**.

SIZING OCPD FOR TWO OR MORE MOTORS
430.62(A); Table 240.4(G)

To determine the size overcurrent protection device to be installed for a feeder supplying two or more motors the following procedures must be applied:

(1) Apply **Table 430.52** to select the largest motor
(2) Size largest OCPD for any one motor of group
(3) Add FLA of remaining motors
(4) Do not exceed this value with OCPD rating

See Figure 8-25 for a detailed illustration for sizing OCPD for a feeder motor circuit.

The largest overcurrent protection device shall be selected based on the motor's full-load current rating in amps times the percentages selected in **Table 430.52**. The next higher standard size rating can be applied per **430.52(C)(1), Ex. 1**. However, the maximum size shall not be used if the motor will start. The full-load current ratings of the remaining motors are added to the rating of the largest overcurrent protection device. The next lower standard size overcurrent protection device shall be selected from this total per **240.6(A)**, if this value does not correspond to a standard OCPD.

Master Test Tip 34: When sizing the OCPD for a feeder circuit, the final step in selecting such protective device is to round it down. See example problem in Figure 8-26.

Figure 8-25. Determining OCPD for a feeder-circuit with several motors being protected from short-circuit and ground-fault conditions.

Master Test Tip 35: Note that two or more motors can be connected to a branch-circuit per **430.52** and **430.53** of the NEC. Don't get confused and think that, for two or more motors to be placed on a circuit, such circuit has to be classified as a feeder-circuit.

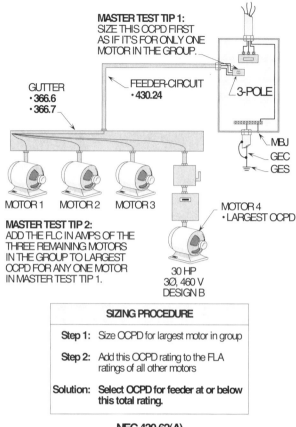

A larger overcurrent protection device is not allowed to be installed because there is no exception to **430.62(A)** to allow the next size above this rating to be selected. Note that **430.62(A)** only allows a smaller OCPD to be installed ahead of such feeder-circuit conductors. This smaller OCPD is the next size below the computation. **(Figure 8-26)**

SINGLE-BRANCH CIRCUIT TO SUPPLY TWO OR MORE MOTORS
430.52 AND 430.53; Table 240.4(G)

Section **430.53** permits two or more motors, or one or more motors, and other loads to be installed and connected to a branch-circuit that is protected by an individual overcurrent protection device.

MOTOR NOT OVER 1 HP
430.53(A)

Master Test Tip 36: Note that a 20 amp OCPD can only be used on 120 volt circuits. A 15 amp OCPD shall be used on circuits with motors rated over 120 volts.

Two or more motors may be installed without individual overcurrent protection devices if rated less than 1 horsepower each and the full-load current rating of each motor does not exceed 6 amps. Motors not rated over 1 horsepower shall be within sight of the motor, manually started, and portable. Section **430.32** shall be applied for running overload protection for each motor, if these conditions are not to be met. The overcurrent protection device rated at 20 or less can protect a 120 volt or less branch-circuit supplying these motors. Branch-circuits of 600 volts or less can be protected by a 15 amp or less OCPD. **(See Figure 8-27)**

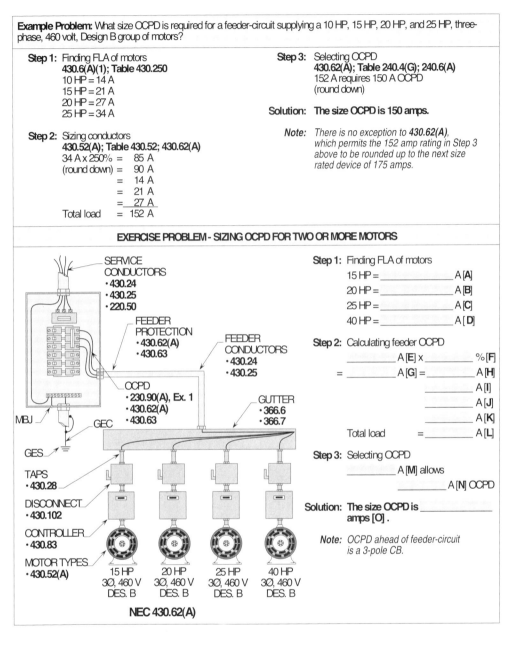

Figure 8-26. Sizing an OCPD for a feeder-circuit.

Figure 8-27. Determining the number of motors allowed on a 20 amp branch-circuit.

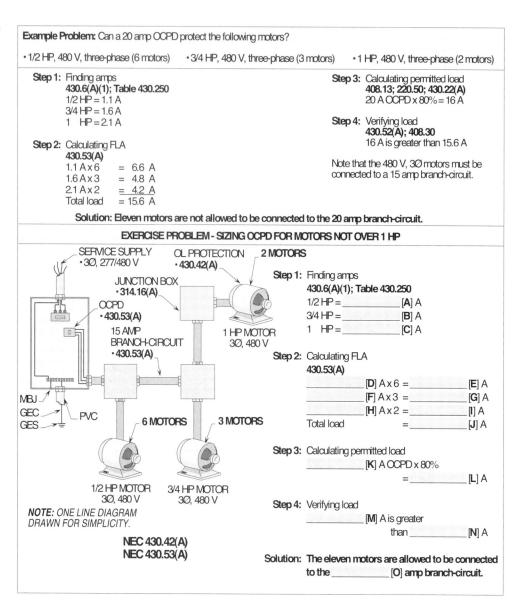

Master Test Tip 37: Note that the size of the OCPD in the example problem can be rated at 20 amps or less. However, the OCPD in the exercise problem is limited to 15 amps, due to the 480 volt circuit, and a 20 amp must not be used per **430.53(A)**.

SMALLEST RATED MOTOR PROTECTED
430.53(B)

The branch-circuit overcurrent protection device may be used to protect the smallest rated motor of the group for two or more motors of different ratings if the largest motor is allowed to start. The smallest rated motor of the group shall have the overcurrent protection device set no higher than allowed per **Table 430.52**. The smallest rated motor and other motors of the group shall be provided with overload protection if necessary per **430.32**. **(See Figure 8-28)**

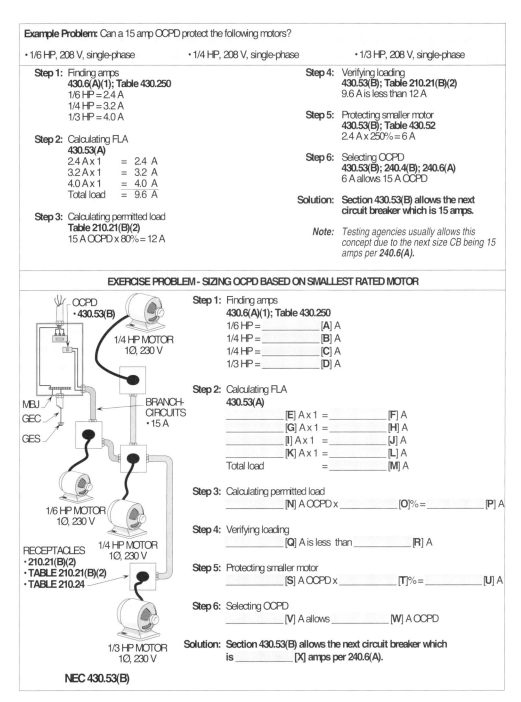

Figure 8-28. Determining the number of motors allowed on a 15 amp branch-circuit.

Master Test Tip 38: Note that the OCPD for the motor circuit in the example problem in Figure 8-28 is not limited to 15 amps as in the exercise problem in Figure 8-27. (See **430.53(B)** for further details of this rule.)

Figure 8-29. Taps can be made from a feeder-circuit when the proper size conductors and OCPD are sized and installed per **430.28**.

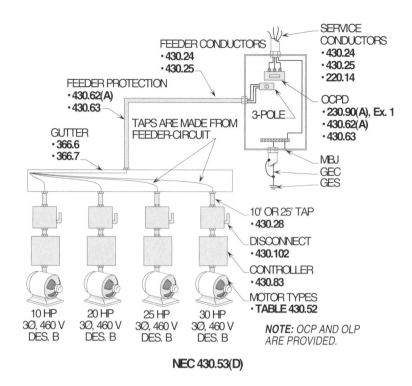

OTHER GROUP INSTALLATIONS
430.53(C)

Master Test Tip 39: Any number of motor taps may be installed where a fuse or circuit breaker is installed at the point where each motor is tapped to the line. This type of installation made from a feeder-circuit, which is considered a tap per **430.28** and **430.53(D)**. **(See Figure 8-29)**

Two or more motors of any size shall be permitted to be installed and connected to an individual branch-circuit. However, the largest of the group shall be protected by the percentages listed in **Table 430.52** for sizing and selecting fuses and circuit breakers to allow the motor to start and run. Each motor controller and component installed in the group shall be approved for such use. The following are elements that must be sized and selected properly:

(1) Overcurrent protection devices
(2) Controllers
(3) Running overload protection devices
(4) Tapped conductors

The elements may be installed as a listed factory assembly or field installed as separate assemblies listed for such conditions of use.

RUNNING OVERLOAD PROTECTION FOR THE MOTOR
430.32(A)(1); 430.32(C)

Master Test Tip 40: Note that there are methods of sizing the overloads for the protection of a motor and its circuit and they are as follows:
(1) Min. per **430.32(A)(1)**
(2) Max. per **430.32(C)**
(3) Shunt OL's per **430.35**

Devices such as thermal protectors, thermal relays, or fusetrons may be installed to provide running overload protection for motors rated more than 1 horsepower. The service factor or temperature rise of the motor shall be used when sizing and installing the running overload protection for motors. The running overload protection is set to open at 115 percent or 125 percent of the motor's full-load current. Under certain conditions of use, the running overload protection shall be set at 130 or 140 percent if the motor is not marked with a service factor or temperature rise. Time-delay fuses selected and sized at 115 or 125 percent provides overload or back up overload protection for supply conductors and motor windings.

Motors and Compressors

MINIMUM SIZE OVERLOAD PROTECTION
430.32(A)(1)

The amperage for full-load current ratings listed in **Tables 430.247** thru **430.250** shall not be used when sizing the running overload protection for motors. The full-load current in amps (nameplate) of the motor shall be used to size the setting of the running overload protection per **430.6(A)(2)**.

The running overload protection shall be selected and rated no larger than the following minimum percentage based upon the full-load current rating in amps listed on the motor's nameplate: **(See Figure 8-30)**

(1) Motors with a marked service factor not less than 1.15 use 125 percent x FLA
(2) Motors with a marked temperature rise not over 40°C use 125 percent x FLA
(3) All other motors use 115 percent x FLA

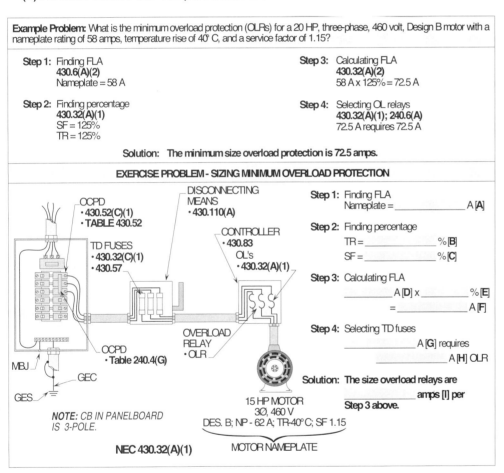

Figure 8-30. Determining the minimum size overloads based upon service factor and temperature rise of the motor.

Master Test Tip 41: Note, if the service factor on the motor's nameplate is 1.00 or 1.10, the OL's (for the protection of the motor's winding's and conductors of the supplying circuit) shall be calculated at 115 percent or less of the motor's FLC in amps per **430.32(A)(1)**.

MAXIMUM SIZE OVERLOAD PROTECTION
430.32(C)

The running overload protection (overload relay) shall be permitted to be selected at higher percentages if the percentages of **430.32(A)(1)** are not sufficient. The running overload protection shall be selected to trip or shall be rated no greater than the following percentages (nameplate per **430.6(A)(2)** full-load current rating: **(See Figure 8-31)**

(1) Motors with marked service factor not less than 1.15 use 140 percent x FLA
(2) Motors with a marked temperature rise not over 40°C use 140 percent x FLA
(3) All other motors use 130 percent x FLA

Figure 8-31. Determining the maximum size overloads based upon service factor and temperature rise of the motor.

Master Test Tip 42: Note that test takers must not apply the maximum percentages of **430.32(C)** unless the question or problem specifies such rating to be used.

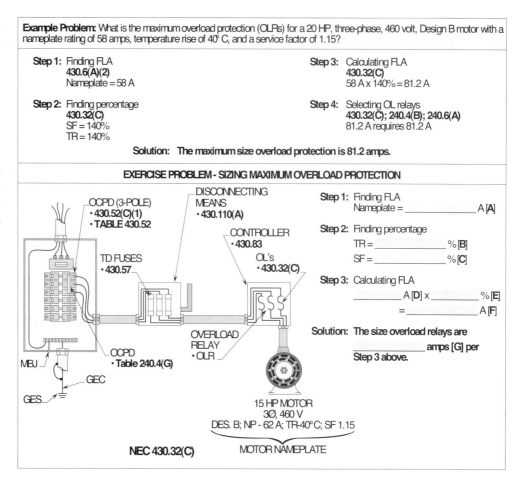

SIZING THE CONTROLLER TO START AND STOP THE MOTOR
430.81 AND 430.83

The sizes and types of motor controllers shall be required to be installed with a horsepower rating at least equal to the motor to be controlled. However, there is an Exception to this rule for motors rated at and below a certain horsepower. **(See Figure 8-34)**

STATIONARY MOTOR 1/8 HORSEPOWER OR LESS
430.81(A)

The branch-circuit protective device shall be permitted to serve as the controller where the motor is rated 1/8 horsepower or less. **(See Figure 8-32)**

> For example, motors less than 1/8 horsepower, where they are mounted stationary or permanent and the construction is such that if they fail during operation, the branch-circuit elements plus motor won't be damaged. In other words, the components of the circuit won't be burned out, etc.

PORTABLE MOTOR OF 1/3 HORSEPOWER OR LESS
430.81(B)

The controller shall be permitted to be an attachment plug and receptacle that are acceptable for use with portable motors rated 1/3 horsepower or less. **(See Figure 8-33)**

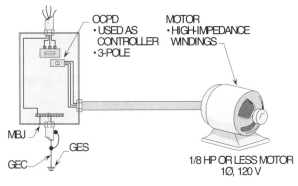

Figure 8-32. The branch-circuit OCPD can serve as a controller for motors rated 1/8 HP or less.

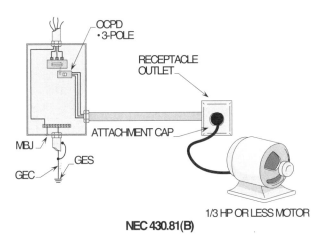

Figure 8-33. The controller for 1/3 HP motor or less can be an attachment cap and receptacle.

OTHER THAN HP RATED
430.83

The following five requirements allow other than horsepower rated controllers to be used for energizing and deenergizing circuits supplying motors:

- **(1)** Design E motors rated more than 2 horsepower (2002 NEC)
- **(2)** Stationary motors rated 2 horsepower or less (300 volts or less)
- **(3)** Inverse-time circuit breakers
- **(4)** Torque motors
- **(5)** Stationary motors rated 1/8 horsepower or less or portable motors rated 1/3 horsepower or less

Master Test Tip 43: If the Design E motor comes with a controller and disconnect rated for Design E motors and circuits, the 1.4 or 1.3 multiplier does not apply per 430.83(A)(1) of the 2002 NEC.

DESIGN E MOTORS
430.83(A)(1) PER 2002 NEC

The controller for Design E motors rated more than 2 horsepower shall be marked for use with Design E motors. For existing controllers, the horsepower rating of Design E controllers shall not be less than 1.4 times the motor horsepower for motors rated 3 through 100 horsepower. For motors rated over 100 horsepower, use 1.3 x HP. **(See 2002 NEC)**

Figure 8-34. Determining the size controller based on the HP of the motor.

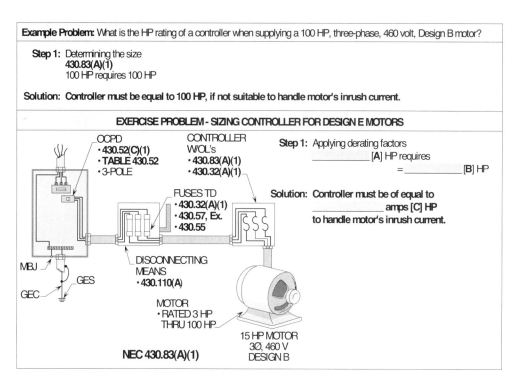

STATIONARY MOTORS
430.83(C)(1)

For a stationary motor rated 2 horsepower or less, the controller shall be permitted to be a general-use switch rated for at least twice the motor's full-load current. An AC general-use snap switch may be installed as the controller where the full-load current rating of the switch does not exceed 80 percent of the branch-circuit rating. **(See Figure 8-35)**

Figure 8-35. For stationary motors rated 2 HP or less, a general-use snap switch may be used if sized not less than twice the motor's full-load current in amps.

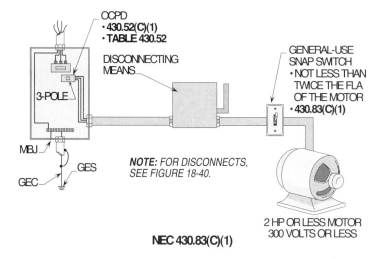

INVERSE-TIME CIRCUIT BREAKERS
430.83(A)(2); 430.83(C)(2)

Inverse-time circuit breakers shall only be permitted to be installed as a controller where rated in amps. If such circuit breaker is also used for motor overload protection, it shall be sized at 125 percent or less of the motor's nameplate current rating per **430.6(A)(1)** and **430.32(A)(1)**. **(See Figure 8-36)**

Motors and Compressors

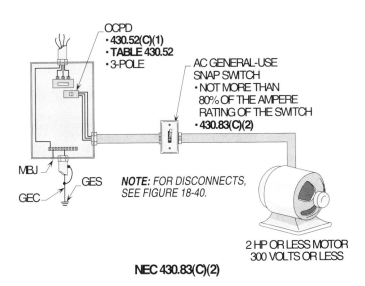

Figure 8-36. A circuit breaker rated at 125 percent or less of the motor's FLA can be used as a controller for the motor and also provide overload protection.

TORQUE MOTOR
430.83(D)

The motor controller for a torque motor shall have a continuous duty, full-load current rating not less than the nameplate current rating in amps of the motor. **(See Figure 8-37)**

Master Test Tip 44: If the motor controller is rated in horsepower and not marked or rated to determine the amperage or horsepower rating, use **Tables 430.247** thru **430.250**.

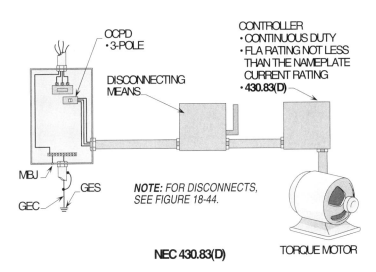

Figure 8-37. The controller for a torque motor shall be capable of holding the amps indefinitely.

STATIONARY AND PORTABLE MOTORS
430.81(A)

Stationary motors rated 1/8 horsepower or less and portable motors rated 1/3 horsepower of less shall be permitted to serve as controllers and shall not be required to be horsepower rated. These horsepower rated motors due to smaller locked-rotor currents can be disconnected by cord-and-plug connections.

SIZING THE DISCONNECTING MEANS TO DISCONNECT BOTH THE CONTROLLER AND MOTOR
430.109

Master Test Tip 45: Before sizing the disconnecting means for a question or problem on the examination, the disconnecting means must be classified as the general-purpose type or HP rated type.

The disconnecting means for motor circuits shall have an ampere rating of at least 115 percent of the full-load current rating of the motor per **430.110(A)**. The disconnecting means shall be horsepower rated and capable of deenergizing locked-rotor currents in amps per **Tables 430.251(A)** and **(B)** of the NEC.**(See Figure 8-38)**

OTHER THAN HP RATED
430.109

The following eight exceptions allow other than a horsepower rated disconnection means to be used to deenergize the power circuit to certain types of motors:

(1) System isolation equipment

(2) Stationary motors rated 1/8 horsepower or less

(3) Stationary motors rated 2 horsepower or less (300 volts or less)

(4) Stationary motors rated 2 horsepower to 100 horsepower

(5) Manual motor controller

(6) Cord-and-plug connected motors

(7) Torque motors

(8) Instantaneous trip circuit breakers

(9) Self-protected combination controller

DESIGN E MOTORS
430.109(A)(1) PER 2002 NEC

Master Test Tip 46: For further details on sizing the controller and disconnecting means, See Master Test Tip 43 on page 8-35 of this chapter.

The disconnecting means for Design E motors rated more than 2 horsepower shall be marked for use with Design E motors. The horsepower rating of Design E disconnecting means shall not be less than 1.4 times the motor's horsepower for motors rated 3 through 100 horsepower. For motor rated over 100 horsepower, use 1.3 x HP.

STATIONARY MOTORS
430.109(B)

For a stationary motor rated 1/8 horsepower or less, the branch-circuit overcurrent device shall be permitted to serve as the disconnecting means. This rule is allowed because the windings of such motors do not produce locked-rotor currents that are high enough to create damage to such motors, circuit conductors, or elements. **(See Figure 8-39)**

Motors and Compressors

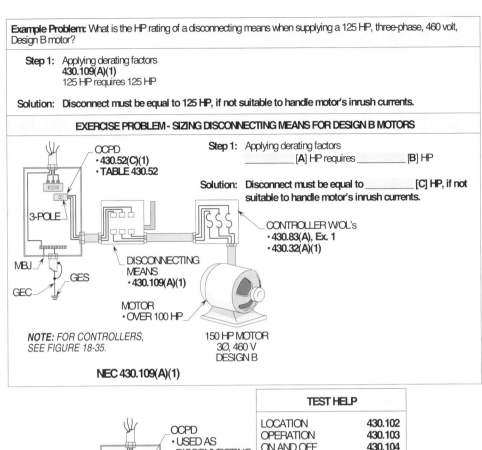

Figure 8-38. It may be necessary to increase the HP rating of a controller by 1.3, if it's not specifically marked for Design E use.

Figure 8-39. Motors rated 1/8 HP or less can be disconnected by the OCPD located in the panelboard that is used to supply the circuit.

STATIONARY MOTORS
430.109(C)

For a stationary motor rated 2 horsepower or less, the controller shall be permitted to be a general-use switch rated for at least twice the motor's full-load current in amps. An AC general-use snap switch may be installed as the controller where the full-load current rating of the switch does not exceed 80 percent of the branch-circuits rating. **(See Figure 8-40)**

MOTORS OVER 2 HP THROUGH 100 HP
430.109(D)

Motors rated over 2 horsepower through 100 horsepower shall be permitted to be installed with a separate disconnecting means (general-use switch) if the motor is equipped with an autotransformer-type controller and all the following conditions are complied with:

Master Test Tip 47: By derating the loading of the OCPD by 80 percent, the same results is accomplished as multiplying the load by 1.25:

$$\left(\frac{1}{80\%} = 1.25 \right)$$

(1) The motor drives a generator that is provided with overload protection.
(2) The controller is capable of interrupting the locked-rotor current of the motor.
(3) The controller is provided with a no-voltage release.
(4) The controller is provided with running overload protection not exceeding 125 percent of the motor's full-load current rating in amps.
(5) Separate fuses or an inverse-time circuit breaker is rated at 150 percent or more of the motor's full-load current in amps. **(See Figure 8-41)**

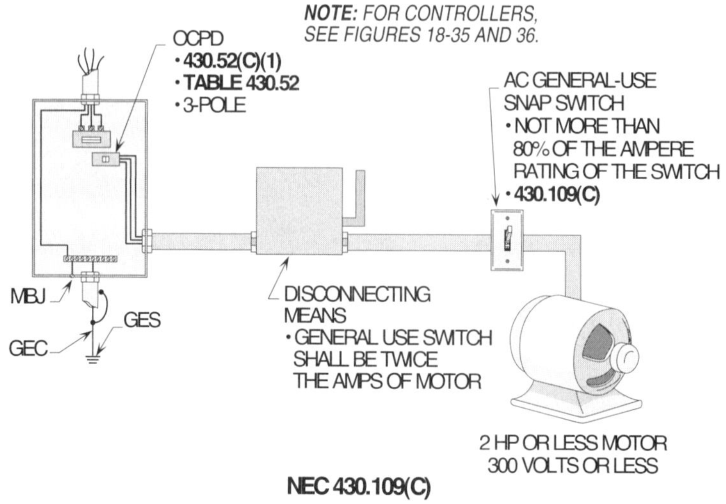

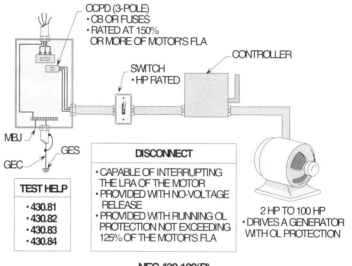

Figure 8-40. This illustration lists the rules pertaining to the disconnecting means for motors rated 2 HP or less.

Figure 8-41. This illustration lists the rules for a disconnecting means and controller used to disconnect and control motors rated 2 HP to 100 HP.

STATIONARY MOTORS
430.109(E)

The disconnecting means is permitted to be a general-use or isolating switch for DC stationary motors rated 40 horsepower or greater and AC motors rated 100 horsepower or greater. However, such disconnects must be plainly marked "Do not operate under load." **(See Figure 8-42)**

CORD-AND-PLUG CONNECTED MOTORS
430.109(F)

Cord-and-plug connected motors shall be permitted to serve as the disconnecting means for motors. Motors shall be permitted to be installed with a cord-and-plug used as a

Master Test Tip 48: Note that the 1.4 multiplier is derived by dividing 8.5 by 6 (8.5 ÷ 6 = 1.416, rounded down to 1.4). **(See Master Test Tip 49.)**

disconnecting means. A cord-and-plug-connected appliance shall not be required to have a horsepower-rated attachment plug and receptacle. The OCPD ahead of the supplying branch-circuit may be utilized for such purposes. For further information see **422.32** and **440.63**. **(See Figure 8-43)**

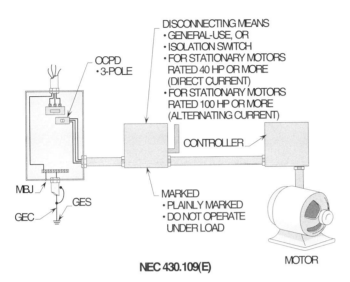

Figure 8-42. This illustration lists the rules for a disconnecting means used to disconnect motors rated at 40 HP or more.

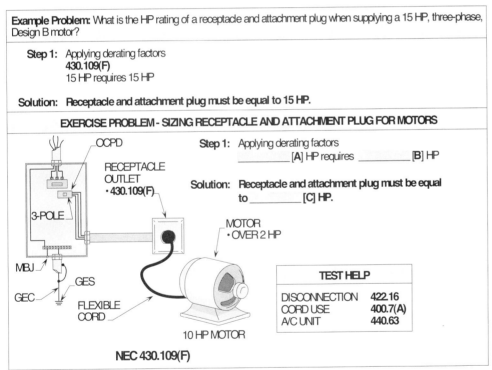

Figure 8-43. It may be necessary to increase the HP rating of a receptacle and attachment cap used as a disconnecting means for motors rated over 2 HP.

Master Test Tip 49: Note that a code letter motor per **Table 430.7(B)** has an LRC in amps of about 6 times its nameplate current in amps. However, Design E motors have a LRC in amps of about 8.5 times its nameplate current in amps. **(See Master Test Tip 48 for more details).**

TORQUE MOTOR
430.109(G)

The disconnecting means for a torque motor shall be permitted to be installed as a general-use switch. Such switch must be capable of handling the locked-rotor current in amps of the motor indefinitely. **(See Figure 8-44)**

Figure 8-44. The disconnecting means for a torque motor may be a general-use switch.

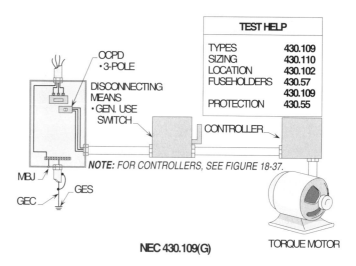

INSTANTANEOUS TRIP CIRCUIT BREAKER
430.111(B)(2); 430.52(C)(3)

An approved instantaneous trip circuit breaker, or INVT, that is part of a listed combination motor controller shall be permitted to serve as a disconnecting means. However, it shall be sized at least 125 percent of the motor's full-load current in amps. **(See Figure 8-45)**

Figure 8-45. The disconnecting means for a motor can be an approved instantaneous trip circuit breaker or an inverse time circuit breaker.

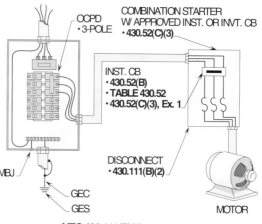

LOCATION OF THE DISCONNECTING MEANS FOR THE CONTROLLER AND MOTOR
430.102 AND 430.107

A motor and its driven machinery or load shall be installed within sight of the controller for the motor. This rule provides safety for electricians and maintenance personnel while servicing such machinery and circuit elements.

Master Test Tip 50: For the definition of within sight etc, see Article 100 of the NEC, which states that such motor shall be visible and within 50 ft. of the disconnecting means. (See **430.102(B), Ex.** and **Article 100** for an exception to such rule.)

WITHIN SIGHT
430.102; 430.106; AND 430.107

The disconnecting means shall be installed within sight of the motor controller. All of the ungrounded conductors shall be disconnected from both the motor and controller supplying the motor circuit. The disconnecting means shall be installed within sight of the motor and not more than 50 ft. from the motor as permitted in **Article 100**. If such disconnecting means is not installed within 50 ft. of the motor, other provisions for disconnecting the motor shall be made. The controller has a direct relationship to the disconnecting means and shall be installed within sight and within 50 ft. of the disconnecting means. The motor does not have a direct relationship with the controller. **(See Figure 8-46)**

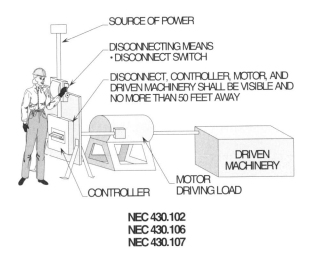

Figure 8-46. The disconnecting means shall be within sight and within 50 ft. of the motor and driven machinery.

Note that a disconnecting means is always required to be located in sight from the controller location. A single disconnecting means shall be permitted to be located adjacent to a group of coordinated controllers mounted adjacent one to another, such as on a multi-motor continuous process machine. For further information pertaining to sizing, selecting, and locating such controllers and disconnecting means, review **430.102**, **430.103**, **430.107**, and **430.109**.

LOCKED IN THE OPEN POSITION
430.102(B), Ex.

The **Ex. to 430.102(B)** allows the disconnect on the line side of the controller, if within sight and within 50 ft., and capable of being individually locked open, to serve as the disconnecting means for both the controller and motor. In this case, a disconnecting means shall not be required to be installed within sight and within 50 ft. of the motor. **Note:** A disconnect in a motor control center will satisfy this rule. **(See Figures 8-47(a) and (b))**

CANNOT BE LOCKED IN THE OPEN POSITION
430.102(B)

For motors rated 600 volts or less, an additional disconnecting means must be mounted by the motor and within sight where the disconnecting means installed by the controller cannot be locked in the open position. The controller disconnecting means for a motor branch-circuit over 600 volts is permitted to be located out of sight of the motor branch-circuit controller and motor. However, the controller shall have a warning label that marks and lists the location and identification of the disconnecting means. To completely satisfy this rule, such disconnecting means must be capable of being locked in the open position. **(See Figure 8-48)**

Master Test Tip 51: Note that the disconnecting means for a controller in a motor control center will satisfy the rule listed in **430.102(B), Ex.** Note that this rule only applies when qualified personnel are servicing the motor and written procedures are available. The rule applies when the installation of a disconnect for the motor presents a hazard.

Figure 8-47(a). Locating the disconnecting means to disconnect power conductors to motors rated 600 volts or less or for motors rated over 600 volts.

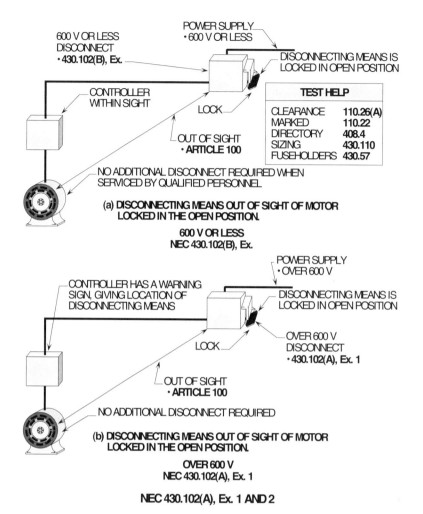

Figure 8-47(b). The locked door of a panelboard shall not serve as the required disconnecting means for a motor. However, an individual locked circuit breaker can serve as such.

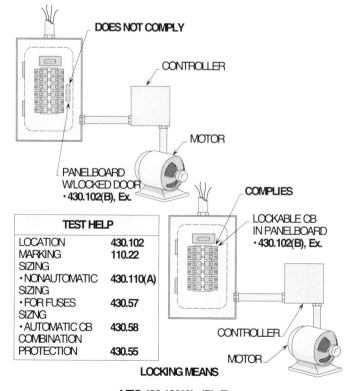

Motors and Compressors

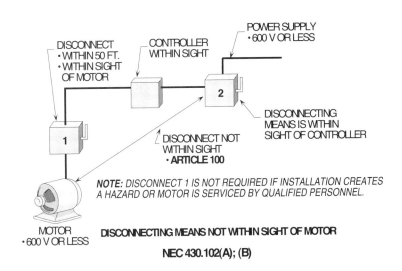

Figure 8-48. If the disconnecting means is within sight and 50 ft. of the controller and can't be locked in the open position, an additional disconnecting means shall be installed by the motor. Note that the motor is either out of sight or over 50 ft. from disconnecting means.

SIZING CONDUCTORS FOR CONTROL CIRCUIT
430.72(B)(1); (B)(2); TABLE 430.72(B)

A motor control circuit tapped on the load side of fuses and circuit breakers utilized for motor branch-circuits shall protect such conductors or supplementary protection devices must be provided.

SIZING CONTROL CIRCUIT CONDUCTORS
INSIDE ENCLOSURES
430.72(B)(2); TABLE 430.72(B), COL. B

Motor control circuit conductors that do not extend beyond the control equipment enclosure shall be permitted to be protected by the motor branch-circuit fuses or circuit breaker. **Table 430.72(B), Col. B** permits this type of installation where the devices do not exceed 400 percent of the ampacity rating of sizes 14 AWG and larger conductors. Overcurrent protection for conductors smaller than 14 AWG shall not exceed the values listed in **Table 430.72(B), Col. B**.

See Figure 8-49 for a test problem on how to size, select, and protect control circuit conductors with an OCPD ahead of the motor's branch-circuit conductors.

Master Test Tip 52: The rules and regulations for sizing control circuits that are tapped per **430.72** can only be applied if such circuits are tapped from the line side of the branch-circuit supplying the controller and motor.

SIZING REMOTE CONTROL CIRCUIT CONDUCTORS
430.72(B)(2); TABLE 430.72(B), COL. C

Motor control circuit conductors that extend beyond the control equipment enclosure can be protected by the motor branch-circuit fuses or circuit breaker. **Table 430.72(B), Col. C** permits this type of installation where the devices do not exceed 300 percent of the ampacity rating for conductors smaller than 14 AWG. In some cases, conductors can be smaller.

See Figure 8-50 for a test problem on how to size, select, and protect control circuit conductors with an OCPD ahead of the motor's branch-circuit conductors.

Figure 8-49. Test problem for calculating and sizing the control circuit conductors inside the enclosure of the controller cabinet.

Example Problem: What size THHN copper conductors are allowed in a controller where the upstream OCPD protecting the branch-circuit is rated 90 amps?

Step 1: Verifying OCPD upstream
430.72(B)(1); (B)(2); Table 430.72(B), Col. B
Upstream OCPD = 90 A

Step 2: Determining size conductors from Table
Table 430.72(B), Col. B
90 A allows 14 AWG cu.

Step 3: Calculating size conductors from Ex. 1
430.72(B)(1); (B)(2); Table 430.72(B), Col. B
- 14 AWG cu. = 25 A x 400% = 100 A OCPD
- 100 A OCPD allows 14 AWG cu.
- 90 A OCPD is allowed to protect 14 AWG cu.

Solution: **The size THHN copper control circuit conductors are 14 AWG copper based on 90 amp OCPD ahead of the motor's branch-circuit conductors.**

Note: *In Step 3, use* ***Table 310.17.***

EXERCISE PROBLEM - SIZING CONTROL CIRCUIT CONDUCTORS INSIDE ENCLOSURES

Step 1: Verifying OCPD upstream
OCPD = ________ [A] A

Step 2: Determining size conductors
________ [B] A OCPD allows ________ AWG [C] cu.

Step 3: Calculating size conductors from **430.71(B)(1); (B)(2)**
________ AWG [D] cu. = ________ [E] A x ________ [F] %
= ________ [G] A OCPD
________ [H] A OCPD allows ________ AWG [I] cu.
________ [J] A OCPD protects ________ AWG [K] cu.

Solution: **The size THHN copper control circuit conductors are ________ AWG [L] copper based on ________ [M] amp OCPD ahead of the motor's branch-circuit conductors.**

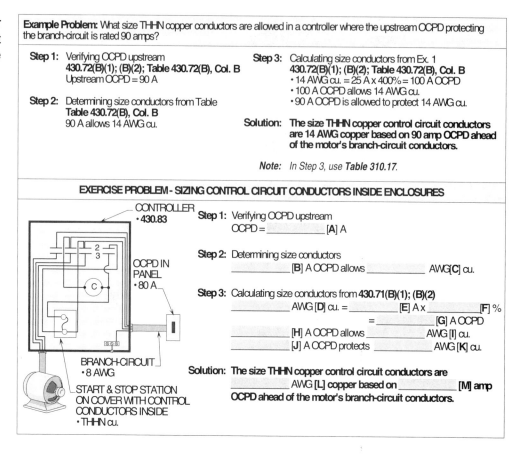

Figure 8-50. Test problem for calculating and sizing the control circuit conductors extending beyond the controller enclosure.

For more detailed information on selecting and using 60°C ampacities per **Tables 310.16** and **17**, see Master Test Tip 54 and 55.

Example Problem: What size THHN copper control circuit conductors are allowed to be routed remote from the magnetic starter where the upstream OCPD is rated 90 amps?

Step 1: Verifying OCPD upstream
430.72(B)(1); (B)(2); Table 430.72(B), Col. C
Upstream OCPD = 90 A

Step 2: Determining size conductors from Table
Table 430.72(B), Col. C
90 A allows 10 AWG cu.

Step 3: Calculating size conductors from **Ex. 2**
430.72(B)(1); (B)(2); Table 430.72(B), Note 3; Table 310.16
- 10 AWG cu. requires 30 A OCPD per **240.4(D)**
- 30 A OCPD x 300% = 90 A
- 90 A OCPD is allowed to protect 10 AWG cu.

Solution: **The size THHN copper control circuit conductors are 10 AWG copper based on 90 amp OCPD ahead of the motor's branch-circuit conductors.**

EXERCISE PROBLEM - SIZING REMOTE CONTROL CIRCUIT CONDUCTORS

Step 1: Verifying OCPD upstream
OCPD = ________ [A] A

Step 2: Determining size conductors
________ [B] A OCPD allows ________ AWG [C] cu.

Step 3: Calculating size conductors from **430.72(B)(1); (B)(2)**
________ AWG [D] cu. requires ________ [E] A OCPD
________ [F] A OCPD x ________ [G]%
= ________ [H] A OCPD
________ [I] A OCPD protects ________ AWG [J] cu.

Solution: **The size THHN control circuit conductors are ________ AWG [K] copper based upon a ________ [L] amp OCPD ahead of the motor's branch-circuit conductors.**

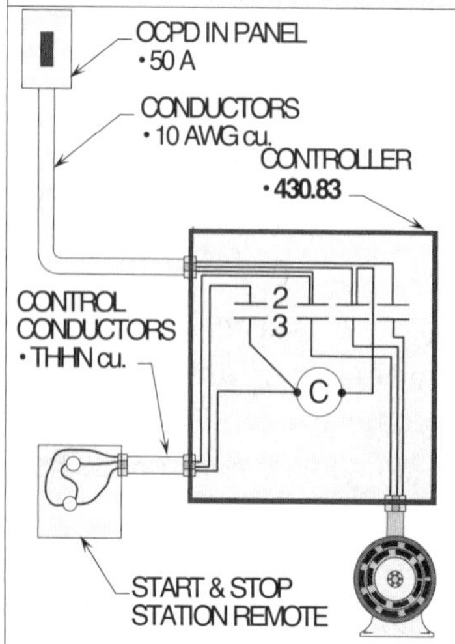

Motors and Compressors

CONTROL TRANSFORMER CIRCUIT CONDUCTORS
430.72(C)(4)

When a motor control circuit transformer is provided, the transformer and conductors shall be protected by the rules and regulations of **Article 430**. A fuse or circuit breaker shall be permitted to be installed in the secondary circuit of the transformer per **430.72(C)(4)**. The primary OCPD may be used to provide the protection for the conductors tapped from the secondary side of a control transformer, if its rating does not exceed the secondary-to-primary ratio of the transformer.

See Figure 8-51 for a test problem when sizing the control circuit conductors on the primary side of a control transformer.

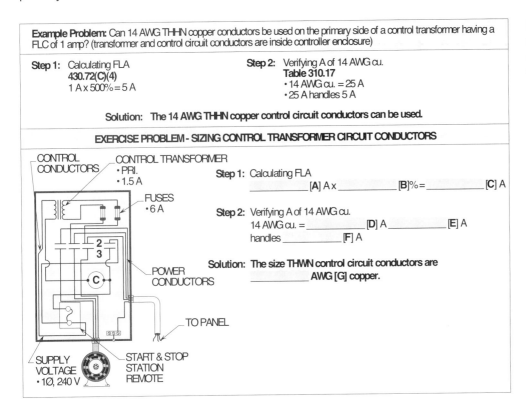

Figure 8-51. Test problem for calculating and sizing the control circuit conductors from the primary of a control transformer.

SIZING OCPD FOR CONTROL CIRCUIT
430.72(B)(1)

Conductors larger than 14 AWG are selected from **Tables 310.16** and **310.17** for motor control-circuit conductors that are tapped from a motor power circuit. Overcurrent protection for conductors smaller than 14 AWG shall not exceed the values listed in **Table 430.72(B), Col. A**. Conductors 18 AWG and 16 AWG shall be protected at the following amperage ratings:

   (1) 18 AWG shall be protected at 7 amps when used for remote-control circuits.
   (2) 16 AWG shall be protected at 10 amps when used for remote-control circuits.

Fuses selected at either 1 amp, 3 amp, 6 amp, or 10 amp are normally used to protect these conductors from short-circuits, ground-faults, and overloads. See **240.6(A)** for selection of such fuse sizes.

Master Test Tip 53: Note that 430.72(B) covers the overcurrent protection of conductors. Section 430.73 covers the mechanical protection of conductors. Section 430.74 pertains to the disconnecting of conductors.

For more detailed information on selecting and using 60°C ampacities per Tables 310.16 and 17, see Master Test Tip 54 and 55 on page 8-48.

PROTECTION OF CONDUCTORS
430.72(B)(2)

Master Test Tip 54: The free air ampacities of **Table 310.17** for 60°C wire are used to determine the ampacity ratings for the control circuit conductors. This method of installation provides more free space to dissipate the heat where control circuit conductors are installed in open air space of enclosures instead of enclosed raceways.

Motor control-circuit conductors that do not extend beyond the control equipment enclosure shall be permitted to be protected by the motor's branch-circuit fuses or circuit breakers. **Table 430.72(B), Col. B** permits this type of installation where the devices do not exceed 400 percent of the ampacity rating of size 14 AWG and larger conductors. Overcurrent protection for conductors smaller than 14 AWG shall not exceed the values listed in **Table 430.72(B), Col. B**. Conductors rated 18 AWG through 10 AWG can be protected with the following sized OCPD's:

(1) 18 AWG = 25 amps (7 A x 400% = 28 A and requires 25 A OCPD)
(2) 16 AWG = 40 amps (10 A x 400% = 40 A and requires 40 A OCPD)
(3) 14 AWG = 100 amps (25 A x 400% = 100 A and requires 100 A OCPD)
(4) 12 AWG = 120 amps (30 A x 400% = 120 A and requires 110 A OCPD)
(5) 10 AWG = 160 amps (40 A x 400% = 160 A and requires 150 A OCPD)
(6) 8 AWG and larger = 400 percent

See **Figure 8-52** for selecting such conductors based upon the OCPD rating.

Figure 8-52. Control circuit conductors located in controller and protected by the branch-circuit OCPD.

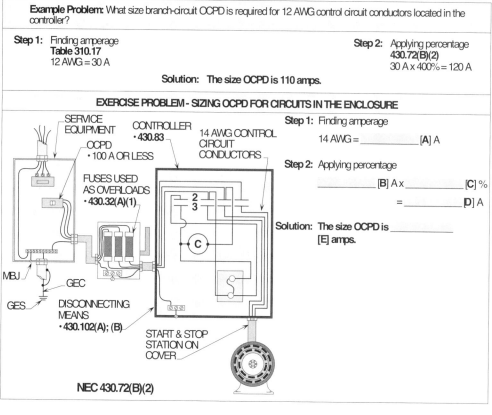

Master Test Tip 55: Note that, for protecting conductors that are located in the controller or motor control center, the 400 percent is multiplied times the ampacity of conductors from the 60° Column in **Table 310.17**.

PROTECTION OF CONDUCTORS
430.72(B)(1); (B)(2)

Motor control-circuit conductors that extend beyond the control equipment enclosure can be protected by the motor's branch-circuit fuse or circuit breaker. **Table 430.72(B), Column C** allows this type installation where the devices do not exceed 300 percent of the ampacity rating of size 14 AWG and larger conductors. Overcurrent protection for conductors smaller than 14 AWG shall not exceed the values listed in **Table 430.72(B), Column C**. Conductors rated 18 through 10 AWG can be protected with the following sized OCPD's:

(1) 18 AWG = 7 amps and requires 7 A OCPD
(2) 16 AWG = 10 amps and requires 10 A OCPD
(3) 14 AWG = 45 amps (15 A OCPD x 300% = 45 A and requires 45 A OCPD)
(4) 12 AWG = 60 amps (20 A OCPD x 300% = 60 A and requires 60 A OCPD)
(5) 10 AWG = 90 amps (30 A OCPD x 300% = 90 A and requires 90 A OCPD)
(6) 8 AWG and larger = 300 percent

Note that the above protection shall be required anytime the control circuit is used for remote-control of a coil in a motor controller enclosure.

See Figure 8-53 for selecting such conductors based upon the OCPD rating.

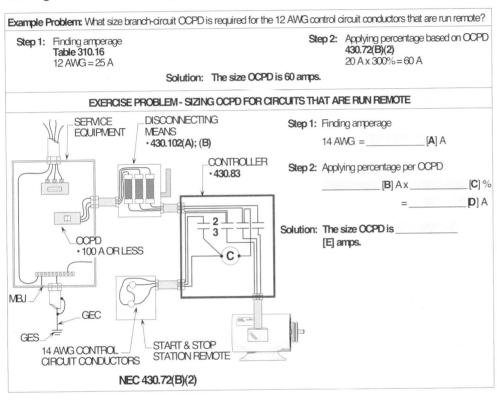

Figure 8-53. Control circuit conductors are run remote and protected by the branch-circuit OCPD in the panelboard.

PROTECTION OF CONDUCTORS
430.72(B), Ex. 2

The secondary conductors of the coil circuit may be protected by the primary side of the transformer. The transformer shall be protected per **450.3(B)** and **Table 450.3(B)**. A two-wire secondary for a transformer installed within the control starter enclosure can be protected per **240.4(F)** and **240.21(C)(1)**.

The secondary conductor ampacity shall be multiplied by the secondary-to-primary voltage ratio to provide protection in accordance with **Table 450.3(B)**. Where the rated primary current is 9 amps or greater and 125 percent of this current does not correspond to a standard rating of a fuse or circuit breaker, the next higher standard can be selected. Where the rated primary current is less than 9 amps, but is 2 amps or greater, an overcurrent protection device rated or set at not more than 167 percent of the primary current can be used. Where the rated primary current is less than 2 amps, an overcurrent protection device rated or set not greater than 300 percent to 500 percent can be used. **(See Figure 8-54)**

> For example, if the primary full-load current of a motor control transformer is less than 2 amps, the OCPD can be calculated and sized at 500 percent times such full-load current in amps per **430.72(C)(4)**.

Master Test Tip 56: Note that, for protecting the conductors that are located remotely from the controller or motor control center, the 300 percent is multiplied times the ampacity of ratings of the OCPD's per **240.4(D)**. Note that the 300 percent is multiplied times the OCPD of conductors rated at 15, 20, and 30 amps, respectively.

Figure 8-54. Control circuit conductors can be supplied by a control transformer and protected by fuses in the primary side.

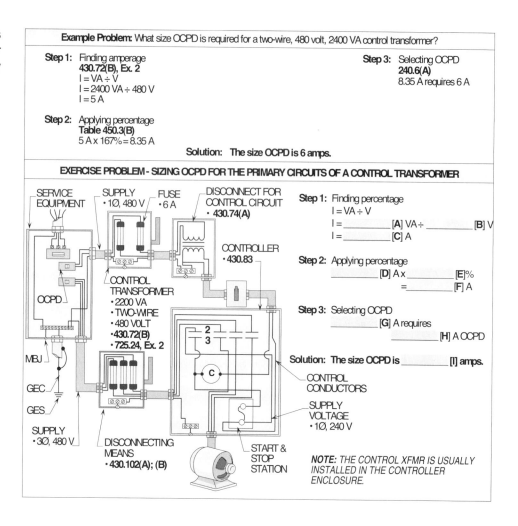

Master Test Tip 57: Section **430.72(C)** covers the protection of conductors. Section **430.74(A)** covers the disconnecting of power sources. Section **430.74(B)** deals with the location of the XFMR in the controller enclosure. **Table 450.3(B)** pertains to the minimum protection of the XFMR.

TRANSFORMER
430.72(C)(4)

If a motor control-circuit transformer is not provided, the transformer shall be protected by the rules and regulations of **Article 450**. However, a fuse or circuit breaker shall be permitted to be installed in the secondary circuit of the transformer per **430.72(C)**. The primary OCPD may be used to provide the protection for the conductors tapped from the secondary side of a control transformer, if its rating does not exceed the secondary-to-primary ratio of the transformer. **(See Figure 8-55)**

MOTOR CONTROL AND MOTOR POWER CIRCUIT CONDUCTORS
430.74

Motor control circuits shall be disconnected from all sources of supply when the disconnecting means is in the open position. The disconnecting means for the starter may be installed to serve as the disconnecting means for the motor circuit conductors if the control-circuit conductors are tapped from the line terminal of the magnetic starter. An auxiliary contact shall be installed in the disconnecting means of the controller or an additional disconnecting means shall be mounted adjacent to the controller to disconnect the motor control-circuit conductors if they are fed from another source and not tapped from the starter branch-circuit conductors. **(See Figure 8-56)**

Motors and Compressors

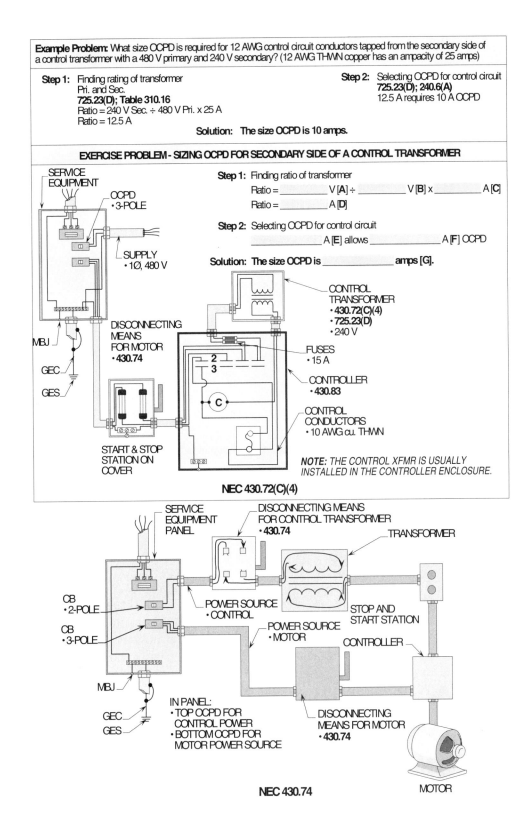

Figure 8-55. The primary OCPD may be used to provide the protection for the conductors tapped from the secondary side of a control transformer, it its rating does not exceed the secondary to primary voltage ratio of the transformer.

QUICK CALC 1
Exercise Problem in Figure 8-55
• $\dfrac{480\ V}{240\ V} = 2$
• 12 AWG = $\dfrac{25\ A}{2}$ = 12.5 A
• OCPD of 10 A is less than 12.5 A
• The 10 A OCPD is OK

Figure 8-56. One disconnecting means or a number of disconnects may be required to disconnect a power supply to a motor and control circuit.

Master Test Tip 58: In Figure 8-56, note that two separate disconnecting means shall be required to disconnect the power sources supplying the controller and motor per **430.74(A)**.

CAPACITOR
460.8

The ampacity of capacitor circuit conductors shall not be less than 135 percent of the rated current of the capacitor. The leads for a capacitor that supplies a motor shall not be less than one-third the ampacity of the motor circuit conductors. The larger of the above conductors must be used for the capacitor supply conductors. **(See Figure 8-57)**

Figure 8-57. There are two calculations that shall be performed, and one of them must be selected to size the capacitor circuit conductors.

Example Problem: What size THHN copper conductors are required to supply power to a 460 volt, three-phase, 10 kVA capacitor serving a 460 volt, three-phase, 50 HP motor? (motor supplied by 480 volts and 4 AWG THHN copper conductors)

Step 1: Calculating conductors at 1/3
480.8(A); Table 310.16
4 AWG = 85 A x 1/3 = 28 A

Step 2: Selecting conductors
310.10, FPN (2); Table 310.16
28 A requires 10 AWG cu.

Step 3: Calculating FLA of capacitor
480.8(A)
FLA = kVA x 1000 ÷ V x √3
FLA = 10 kVA x 1000 ÷ 480 V x 1.732
FLA = 12 A
FLA = 12 A x 135%
FLA = 16 A

Step 4: Selecting conductors
310.10, FPN (2); Table 310.16
16 A requires 14 AWG cu.

Solution: The size THHN conductors are 10 AWG copper.

EXERCISE PROBLEM - SIZING CAPACITOR CIRCUIT CONDUCTORS

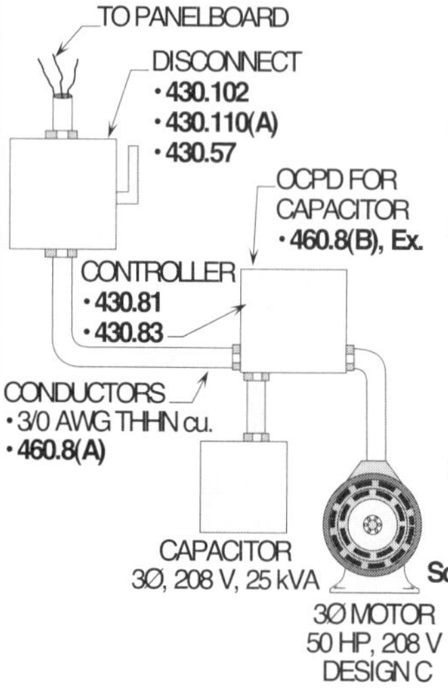

TO PANELBOARD
DISCONNECT
• 430.102
• 430.110(A)
• 430.57

CONTROLLER
• 430.81
• 430.83

OCPD FOR CAPACITOR
• 460.8(B), Ex.

CONDUCTORS
• 3/0 AWG THHN cu.
• 460.8(A)

CAPACITOR
3Ø, 208 V, 25 kVA

3Ø MOTOR
50 HP, 208 V
DESIGN C

Step 1: Calculating conductors at 1/3
3/0 AWG _____ [A] A ÷ _____ [B] = _____ [C] A

Step 2: Selecting conductors
_____ [D] A requires _____ AWG [E] cu.

Step 3: Calculating FLA of capacitor
FLA = kVA x 1000 ÷ V x √3
FLA = _____ [F] kVA x 1000 ÷ _____ [G] V x _____ [H]
FLA = _____ [I] A
FLA = _____ [J] A x _____ [K] %
FLA = _____ [L] A

Step 4: Selecting conductors
_____ [M] A requires _____ AWG [N] cu.

Solution: The size THHN conductors are _____ AWG [O] copper.

Master Test Tip 59: If the capacitor conductors are tapped from the load side of the OL's in the motor controller, such OL's can serve to protect the tapped conductors for the capacitors. However, a disconnecting means is required between controller and capacitor per **460.8(B), Ex.**

RACEWAYS
300.3(C)(1) AND 725.26(A); (B)

Conductors of different systems are allowed to occupy the same raceway without regard to AC or DC current per **300.3(C)(1)**. Conductors shall be insulated for the maximum voltage of any one conductor when occupying the same raceway. These conductors shall not exceed 600 volts. Class 1 conductors are permitted to occupy the same raceway when installed with the power conductors supplying the magnetic starter and motor per **725.26(A)** and **(B)**. For control circuit conductors to occupy the same raceway as the motor circuit conductors, they must be functionally associated with the motor system. **(See Figure 8-58)**

Master Test Tip 60: Note that a class 1, motor control circuit can occupy the same raceway as the power circuit conductors supplying a 480 volt motor. However, if the voltage on the Class 1 circuit is 30 volts, its insulation must be at least 480 volts. See **300.3(C)(1)**, **725.21(A)** and **725.26(B)**.

OCCUPYING THE SAME ENCLOSURE
300.3(C)(1) AND 300.3(C)(2)(c); (d)

Motor excitation, magnetic starter, control relay, or ammeter conductors may occupy the same enclosure if rated at 600 volts or less or over 600 volts. These conductors shall not be permitted to occupy the same raceway where installed as a combination of conductors of 600 volts or less with conductors of over 600 volts. The motor enclosure and the starter enclosure can contain motor excitation, control, relay, and ammeter conductors of 600 volts or less together with power conductors of over 600 volts. **(See Figure 8-59)**

Motors and Compressors

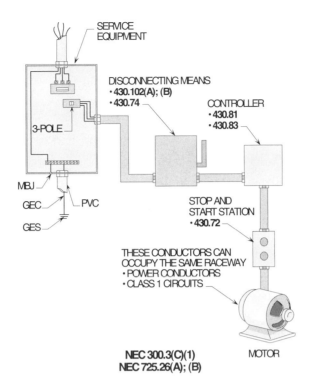

Figure 8-58. When functionally associated with motor operation, a Class 1 control circuit can occupy the same raceway as the motor's power circuit conductors.

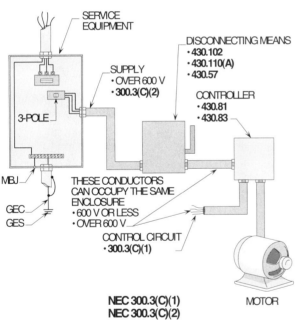

Figure 8-59. Over 600 volts and 600 volts or less conductors can occupy the same enclosure under certain conditions of use.

MAGNETIC STARTER CONTACTOR AND ENCLOSURE 725.28(A) AND (B)

When installing four or more current-carrying conductors that are continuous operated in a raceway for Class 1 remote-control, signal, and power-limited circuit conductors they may have to be derated by the derating factors of **Table 310.15(B)(2)(a)** to Ampacity Tables of 0 to 2000 volts per **725.28**. Conductors that are noncontinuous operated shall not be derated by the derating factors of **Table 310.15(B)(2)(a)**. When installing power and control conductors in the same raceway, **Table 310.15(B)(2)(a)** shall be applied to all the current-carrying conductors operating for three hours or more. **(See Figure 8-60)**

Master Test Tip 61: Control circuit conductors operating for three hours or more shall not be required to be derated per **Table 310.15(B)(2)(a)**, if their ampacity does not exceed 10 percent of the control conductors ampacity rating.

8-53

Figure 8-60. Control-circuit conductors that do not carry more than 10 percent of their ampacity are not considered current-carrying conductors.

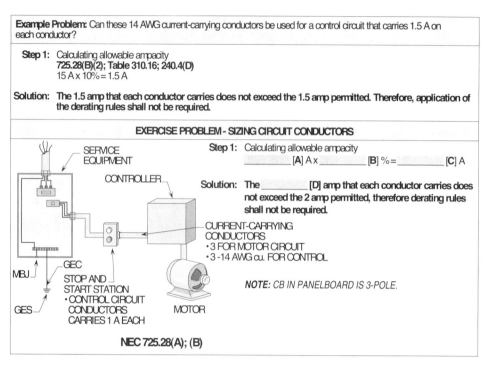

CLASS 1 CIRCUITS
725.21(A)

Class 1 circuits are divided into two types, power-limited and remote-control and signaling circuits. Power-limited Class 1 circuits are limited to 30 volts and 1000 volt-amperes. Class 1 remote-control and signaling circuits are limited to 600 volts, but there aren't any limitations on the power output of the source. Note that the rules pertaining to Class 1 circuits for motor control purposes are reviewed in this section.

Master Test Tip 62: Class 1 circuits can operate at 30 volts or less with a maximum output of 1000 VA. Class 1 circuits can also operate at over 30 volts up to 600 volts with no power limitations on its output source.

CLASS 2 AND 3 CIRCUITS
725.41(A); (B)

Class 2 and Class 3 circuits are defined by two Tables, one for AC current and one for DC current. In general, a Class 2 circuit operating at 24 volts with a power supply durably marked "Class 2" and not exceeding 100 volt-amperes is the type most commonly used.

For power systems that are inherently or not inherently protected, see **Tables 11(A)** and **11(B), Chapter 9.**

A Class 2 circuit is defined as that portion of the wiring system between the load side of a Class 2 power source and the connected equipment. Due to its power limitations, a Class 2 circuit is considered safe from a fire initiation standpoint and provides acceptable protection from electric shock.

A Class 3 circuit is defined as the portion of the wiring system between the load side of a Class 3 power source and the connected equipment. Due to its power limitations, a Class 3 circuit is usually considered safe from a fire initiation standpoint. Since higher levels of voltage and current for Class 3 circuits are permitted, additional safeguards are specified to provide protection from an electric shock hazard that might be encountered.

Power for Class 2 and Class 3 circuits is limited either inherently (in which no overcurrent protection is required) or by a combination of a power source and overcurrent protection scheme.

The maximum circuit voltage is 150 volts AC or DC for a Class 2 inherently limited power source and 100 volts AC or DC for a Class 3 inherently limited power source. The maximum circuit voltage is 30 volts AC and 60 volts DC for a Class 2 power source limited by overcurrent protection, and 150 volts AC or DC for a Class 3 power source limited by overcurrent protection. **(See Figure 8-61)**

For example, heating system thermostats are commonly Class 2 systems. However, the majority of small bell, buzzer, and annunciator systems are Class 2 circuits. Class 2 also includes small intercommunicating telephone systems in which the voice circuit is supplied by a battery and the ringing circuit by a transformer.

Class 2 and 3 systems do not require the same wiring methods as power, light, and Class 1 systems. However, for safety reasons, a 2 in. separation is required between these systems.

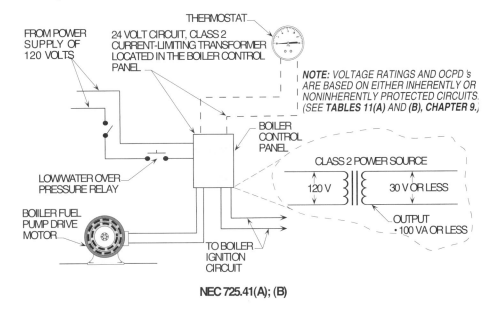

Figure 8-61. The maximum voltage is usually 30 volts AC or DC for a Class 2 inherently limited power source and 150 volts AC or DC for a Class 3 inherently limited power source.

DESIGNING MOTOR CURRENTS NOT LISTED IN TABLES 430.247 THRU 430.250

The following method can be used to determine the full-load current rating in amps for motors that are not listed in **Tables 430.247** thru **430.250**:

(1) The horsepower rating of a listed motor shall be selected which is below the unlisted motor.
(2) The motor's full-load current rating in amps shall be divided by its horsepower rating to obtain the multiplier.
(3) The multiplier times the HP of the unlisted motor derives FLC in amps for the unlisted motor.

Master Test Tip 63: The full-load current rating of the motor in amps is determined by multiplying these values by the horsepower rating of the unlisted motor. **(See Figure 8-62)**

NAMEPLATE LISTING
440.1

Those taking the master's examination must learn the differences between motors and hermetically sealed motor compressors and how to apply the rules for each. The overcurrent protection devices, running overload protection devices, conductors, disconnecting means, and controllers shall be sized and selected by the information provided on the nameplate listing for air-conditioning and refrigeration equipment. The information on the nameplate, if in a problem, is very important to test takers, therefore, nameplate values must be evaluated very carefully before sizing elements for such equipment.

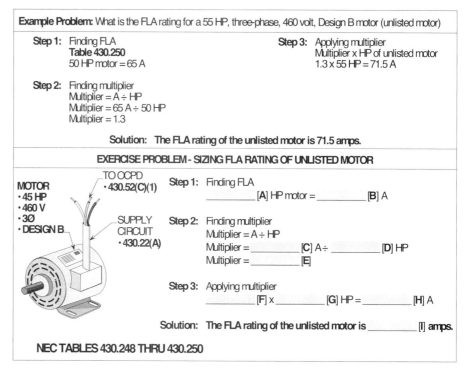

Figure 8-62. This illustrates the procedure for calculating the FLA of a motor not listed in **Tables 430.248** thru **430.250**. If such a problem is given on the test, solve it as shown.

AMPACITY RATING
440.6

The full-load current rating in amps listed on the nameplate of the motor-compressor shall be used to determine the branch-circuit conductor rating, short-circuit protection rating, motor overload protection rating, controller rating, or disconnecting means rating. The branch-circuit selection current (if greater) shall be applied instead of the full-load current rating. The full-load current rating shall be used to determine the motor's overload protection rating. The full-load current rating listed on the compressor nameplate shall be used when the nameplate for the equipment does not list a full-load current rating in amps based upon the BCSC. **(See Figure 8-63)**

Figure 8-63. If the branch-circuit selection current (BCSC) on the nameplate calls for a certain size circuit and OCPD, this rating shall be used instead of actually calculating such values.

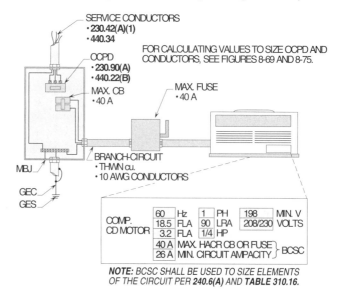

Master Test Tip 64: In Figure 8-63, the size OCPD shall be selected at 40 amp and the branch-circuit conductors shall be selected at 10 AWG THWN copper as specified by the nameplate rating of 40 amp (OCPD rating) and 26 amp (circuit rating), respectively.

8-56

Motors and Compressors

DISCONNECTING MEANS
440.11

The full-load current rating in amps, if listed on the nameplate or the nameplate branch-circuit selection current of the compressor, whichever is greater, shall be used to size the branch-circuit conductors and the disconnecting means to deenergize conductors and power to air-conditioners and refrigeration equipment.

RATING AND INTERRUPTING CAPACITY
440.12

To size the disconnecting means the full-load current rating (in amps) of the nameplate or the nameplate branch-circuit selection current of the motor compressor, whichever is greater, must be equivalent to 115 percent. A horsepower rated switch, circuit breaker, or other switches may be used as the disconnecting means per **430.109**. **(See Figure 8-64)**

Master Test Tip 65: Note that the disconnecting means shall be capable of deenergizing both running and LRC in amps of the motor involved per **440.12(A), (B), and (C)**.

Figure 8-64. The full-load current rating of the nameplate or the nameplate branch-circuit selection current of the compressor, whichever is greater, shall be sized at 115 percent to size the disconnecting means.

Master Test Tip 66: A minimum size disconnecting means is derived when multiplying the FLA by 115 percent. Therefore, on larger units, the 115 percent may not be of sufficient ampacity for opening the circuit under severe load conditions.

Master Test Tip 67: In the exercise problem, the nameplate on the A/C unit does not list BCSC values, therefore the size of circuit conductors and OCPD shall be calculated per **440.22(A)** and **440.32**. (See Figures 8-69 and 8-75.)

Example Problem: What size nonautomatic, nonfused, and horsepower rated disconnecting means is required for the following A/C unit?

A/C Unit
• 3Ø, 460 volt
• Compressor
  FLA 24 A
• Condenser
  FLA 2 A
  LRA 140 A

Note: For sizing the controller, see Figure 8-77.

Sizing Nonautomatic or Nonfused Disconnect

Step 1: Calculating disconnect
440.12(A)(1)
24 A + 2 A x 115% = 29.9 A

Step 2: Selecting disconnect
440.12(A)(1); 240.6(A); Chart
29.9 A requires 30 A CB
29.9 A requires 30 A disconnect

Solution: The minimum size nonautomatic CB is 30 amps and nonfused disconnect is 30 amps.

Sizing HP of Disconnect

Step 1: Calculating disconnect
440.12(A)(1); Table 430.250; Table 430.251(B)
26 FLA requires 20 HP
140 LRA requires 20 HP

Solution: The disconnecting means is required to be rated at least 20 HP.

Note: 24 A + 2 A = 6 A

EXERCISE PROBLEM - SIZING NONAUTOMATIC, NONFUSED, AND HORSEPOWER RATED DISCONNECTING MEANS

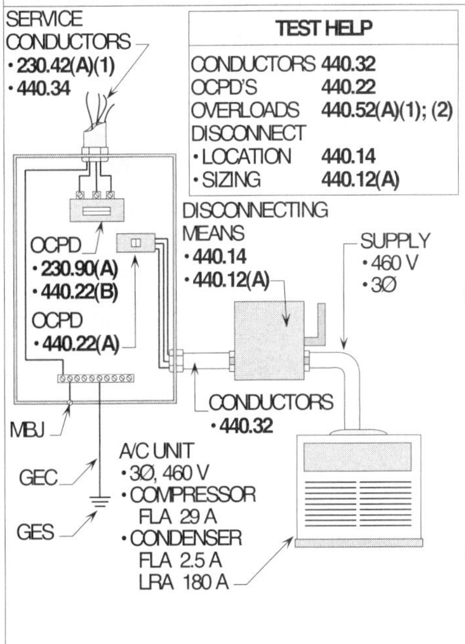

SERVICE CONDUCTORS
• 230.42(A)(1)
• 440.34

TEST HELP
CONDUCTORS 440.32
OCPD'S 440.22
OVERLOADS 440.52(A)(1); (2)
DISCONNECT
• LOCATION 440.14
• SIZING 440.12(A)

DISCONNECTING MEANS
• 440.14
• 440.12(A)

OCPD
• 230.90(A)
• 440.22(B)

OCPD
• 440.22(A)

SUPPLY
• 460 V
• 3Ø

CONDUCTORS
• 440.32

MBJ
GEC
GES

A/C UNIT
• 3Ø, 460 V
• COMPRESSOR
  FLA 29 A
• CONDENSER
  FLA 2.5 A
  LRA 180 A

Sizing Nonautomatic or Nonfused Disconnect

Step 1: Calculating disconnect
_____ [A] A + _____ [B] A
x _____ [C] % = _____ [D] A

Step 2: Selecting disconnect
_____ [E] A requires _____ [F] A CB
_____ [G] A requires _____ [H] A disconnect

Solution: The minimum size nonautomatic CB is _____ [I] amps and nonfused disconnect is _____ [J] amps.

Sizing HP of Disconnect

Step 1: Calculating disconnect
_____ [K] FLA requires _____ [L] HP
_____ [M] LRA requires _____ [N] HP

Solution: The disconnecting means is required to be rated at least _____ [O] HP.

Note: 29 A + 2.5 A = 31.5 A

8-57

Stallcup's Master Electrician's Study Guide

Master Test Tip 68: The circuit breaker shall be sized at 115 percent or more of the branch-circuit selection current if it were greater in rating so as to be capable of disconnecting the circuit safely.

The horsepower rating shall be selected from **Tables 430.247** thru **430.250** and mated to the values listed on the nameplate rating or branch-circuit selection current of the motor compressor or equipment, whichever is greater. The horsepower amperage rating for locked-rotor currents shall be selected from **Tables 430.251(A)** and **(B)**, if the nameplate lists the locked-rotor current. Note that the disconnecting means must be sized with enough capacity in horsepower that is capable of disconnecting the total locked-rotor amps. **(See Figure 8-65)**

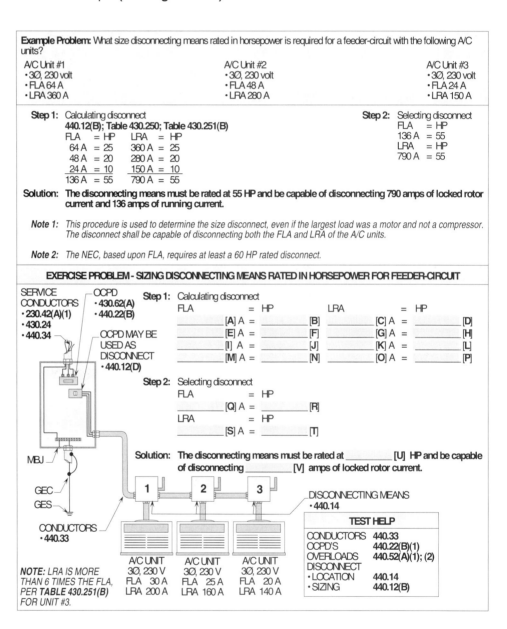

Figure 8-65. Selecting horsepower ratings to size disconnecting means based upon running and locked-rotor current in amps.

Master Test Tip 69: If only the FLA is listed on the A/C unit, then the disconnect shall be capable of deenergizing the total FLA rating. However, if both the FLA and the LRC are listed, then the disconnect has to open the circuit based on the greater of the two values.

The full-load current rating in amps of the nameplate may be used to size a circuit breaker at 115 percent or more to disconnect a hermetically sealed motor from the power circuit. **(See Figure 8-66)**

Two or more hermetic motors or combination loads such as hermetic motor loads, standard motor loads, and other loads must have their separate values totaled to determine the rating of a single disconnecting means. This total rating shall be sized at 115 percent to determine the size disconnecting means required to disconnect the circuits and elements in a safe and reliable manner. **(See Figure 8-67)**

Motors and Compressors

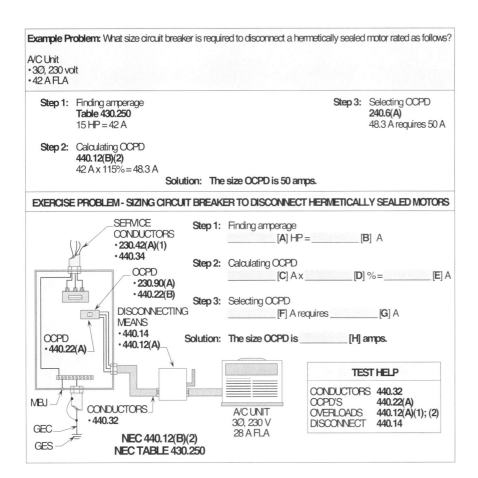

Figure 8-66. The full-load amp rating on the nameplate may be used to size a circuit breaker at 115 percent or more to disconnect a hermetically sealed motor from the power circuit.

Master Test Tip 70: Nonautomatic CB's and nonfused disconnects without overcurrent characteristics shall be sized and selected at 115 percent of the total FLA of the A/C unit. Note that such rating shall be sized at 115 percent or greater. For example, the rating of 48.3 amp in Figure 8-66 had to be rounded up to 50 amp instead of rounding down to 45 amp per **440.12(B)(2)** and **240.6(A)**.

LOCATION
440.14

The disconnecting means for air-conditioning or refrigeration equipment shall be located within sight and within 50 ft. and shall be readily accessible to the user. An additional circuit breaker or disconnecting switch shall be provided at the equipment if the air-conditioning or refrigeration equipment is not within sight or within 50 ft. The disconnecting means is permitted to be installed within or on the air-conditioning or refrigeration equipment. See **Ex. 1 to 440.14** for an exception to such rule. **(See Figure 8-68)**

APPLICATION AND SELECTION
440.22; Table 240.4(G)

The branch-circuit fuse or circuit breaker ratings for hermetically sealed motors must be sized with enough capacity to allow the motor to start and develop speed without tripping open the OCPD due to the momentary inrush current of the compressor and other elements. Maximum protection is always provided by the ratings and settings of the OCPD being sized with values as low as possible. Hermetic refrigerant motor-compressors are considered protected by properly sizing and selecting the ratings and settings of the OCPD's to protect the branch-circuit conductors and other elements in the circuit from short-circuit and ground-fault conditions. Overload protection is provided by OL's in the compressor.

8-59

Figure 8-67. Two or more hermetic motors or combination loads such as hermetic motor loads, standard motor loads, and other loads must have their separate values totaled to determine the rating of a single disconnecting means.

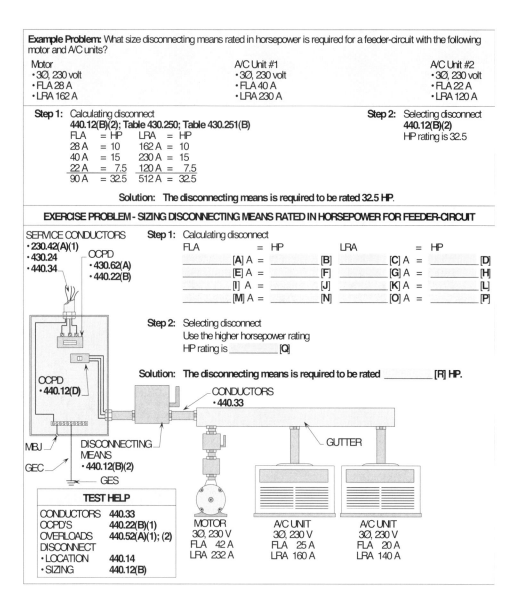

RATING AND SETTING FOR INDIVIDUAL MOTOR COMPRESSOR
440.22(A); Table 240.4(G)

The OCPD for hermetic sealed compressors shall be selected at 175 percent (for minimum) or 225 percent (for maximum) of the compressor FLA rating or the branch-circuit selection current, whichever is greater. **(See Figure 8-69)**

OCPD's for hermetic sealed compressors may be selected up to 225 percent to allow the motor to start if the compressor will not start and develop speed. The 225 percent can be applied only after the 175 percent has been used and the compressor fails to start and run.

RATING OR SETTING FOR EQUIPMENT
440.22(B); Table 240.4(G)

When sizing the overcurrent protection device the rating or setting must be selected and comply with the number of hermetic motors, or combination of hermetic motors, and standard motors installed on a circuit.

Master Test Tip 71: A normal circuit breaker shall not be installed if such equipment is marked for a particular fuse size or HACR circuit breaker rating. The branch-circuit conductors shall be protected only by that specified fuse size or HACR circuit breaker rating listed on the nameplate.

Motors and Compressors

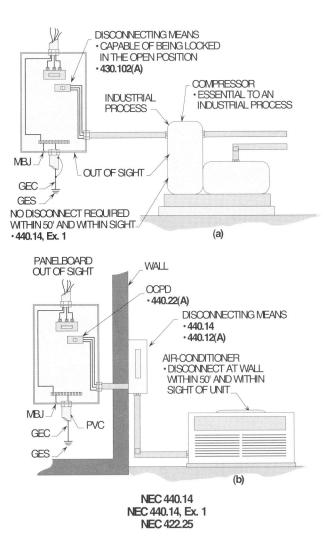

Figures 8-68(a) and (b). The disconnecting means for air-conditioning or refrigeration equipment shall be located within sight and within 50 ft. and shall be readily accessible to the user. An additional circuit breaker or disconnecting switch must be provided at the equipment if the air-conditioning or refrigeration equipment is not within sight or within 50 ft.

Master Test Tip 72: If the OCPD in the out of sight panelboard in Figure 8-68(b) can be locked open, it must not be used as a disconnecting means as it is in Figure 8-68(a). This is due to the equipment in Figure 8-68(b) being an A/C unit and not essential to an industrial process.

SIZING OCPD FOR TWO OR MORE HERMETIC MOTORS
440.22(B)(1); Table 240.4(G)

The OCPD for a feeder-circuit supplying two or more air-conditioning or refrigerating units shall be sized to allow the largest unit to start and allow the other units to start at different intervals of time. The full-load current rating (in amps) of the nameplate or the branch-circuit selection current rating (in amps) of the largest motor, whichever is greater, shall be sized at 175 percent if there are two or more hermetically sealed motors installed on the same feeder-circuit. **(See Figure 8-70)**

After applying 175 percent and the compressor motor fails to start and run, an OCPD may be selected up to 225 percent to allow the motor to start and develop speed. **(See Figure 8-71)**

SIZING OCPD FOR HERMETIC MOTOR AND OTHER LOADS WHEN A HERMETICALLY SEALED MOTOR IS THE LARGEST
440.22(B)(1); Table 240.4(G)

When installing hermetically sealed motors and other loads such as motors on the same circuit, and the largest motor of the group is hermetic, the same procedure used for two or more hermetic motors on a feeder-circuit shall be used to size the OCPD. The full-load current rating of the nameplate or the branch-circuit selection current rating of the largest hermetic motor, whichever is greater, shall be sized at 175 percent and the sum of the full-load current ratings of the other motors added to this largest hermetic motor load.

Figure 8-69. The OCPD for hermetic sealed compressors are selected at 175 percent (for minimum) or 225 percent (for maximum) of the compressor FLA rating or the branch-circuit selection current, whichever is greater.

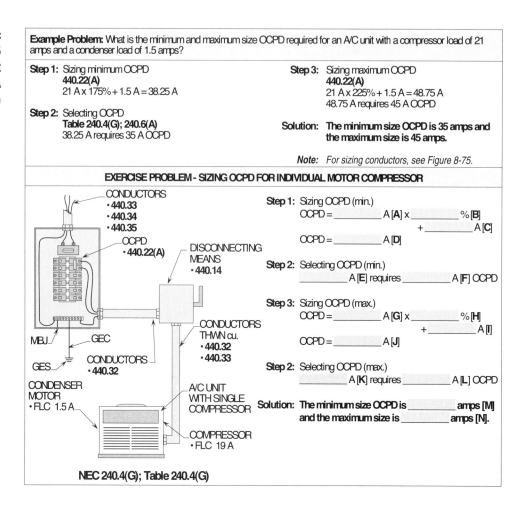

Master Test Tip 73: Note that the next larger standard size is not permitted to be selected, for there is not an exception to allow the next higher size per **440.22(B)(2)** or **430.62(A)**.

SIZING OCPD FOR HERMETIC MOTORS AND OTHER LOADS WHEN A MOTOR IS THE LARGEST
440.22(B)(2); Table 240.4(G)

When installing hermetically sealed motors and other loads such as motors on the same circuit, and the largest of the group is a motor, the overcurrent protection device shall be sized and selected based on the percentages from **Table 430.52**. The next size (round up) branch-circuit overcurrent protection device rating shall be used, and if the standard motor is the largest of the group, the sum of the full-load current ratings of the remaining hermetically sealed motors and other motors of the group added to this total to derive the proper size OCPD. The next lower standard size overcurrent protection device below this total sum shall be selected per **240.6(A)**. **(See Figure 8-72)**

USING 15 OR 20 AMP OCPD
440.22(B)(2), Ex. 1

Where the equipment will start, run, and operate on 15 or 20 amp, 120 volt, single-phase branch-circuit, or a 15 amp, 208 volt or 240 volt, single-phase branch-circuit, having a 15 or 20 amp overcurrent protection device. Such OCPD may be used to protect the branch-circuit. However, the values of the overcurrent protection device in the branch-circuit shall not exceed the values marked on the nameplate of the equipment. **(See Figure 8-73)**

Motors and Compressors

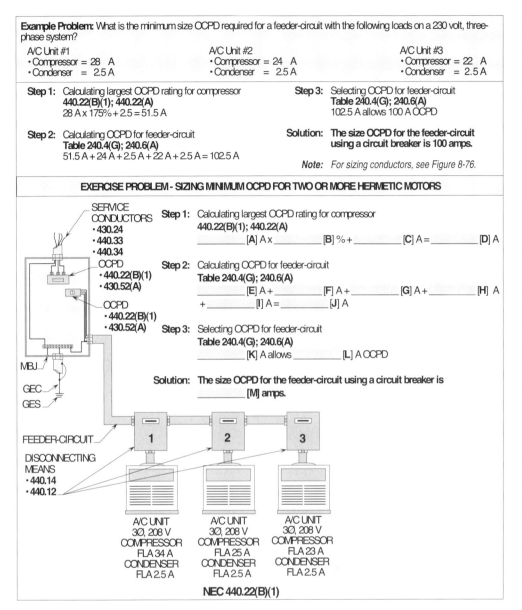

Figure 8-70. The full-load current rating of the nameplate or the branch-circuit selection current rating of the largest motor, whichever is greater, shall be sized at 175 percent if there are two or more hermetically sealed motors installed on the same feeder-circuit.

QUICK CALC 2

Sizing conductors in the Example Problem in Figure 8-70.

- Finding load in Amps
  440.33

28 A x 125%	=	35.0 A
	+	2.5 A
	+	24.0 A
	+	2.5 A
	+	22.0 A
	+	2.5 A
Total LD.	=	88.5 A

- Selecting Conductors
  Table 310.16
  88.5 A requires 3 AWG

- Conductors must be 3 AWG THWN cu.

Note: For a detailed illustration, see Figure 8-76.

USING A CORD-AND-PLUG CONNECTION NOT OVER 250 VOLTS
440.22(B)(2), Ex. 2

The rating of the overcurrent protection device shall be determined by using the rating of the nameplate of the cord-and-plug connected equipment having single-phase, 250 volt or less hermetically sealed motors. **(See Figure 8-74)**

Master Test Tip 74: Note that an A/C window unit supplied by a circuit over 250 volts shall not be cord and plug connected per **440.63**.

BRANCH-CIRCUIT CONDUCTORS
Table 220.3; 440.31

In general, to prevent conductors and motor elements of the branch-circuit from overheating, the conductors must be sized with enough capacity to allow a hermetic motor to start and run. To ensure adequate sizing, a derating factor of 80 percent shall be applied to the branch-circuit conductors or such conductors shall be sized at 125 percent of the load.

Figure 8-71. OCPD's for hermetically sealed motors may be selected up to 225 percent to permit the motor to start and develop speed.

Master Test Tip 75: Note that the OCPD for the feeder-circuit in the example problem in Figure 8-71 shall not be selected at 125 amps (rounded up), but selected at 110 amps (rounded down). There is no exception in **440.34**, which permits the next higher size above 116.5 amps.

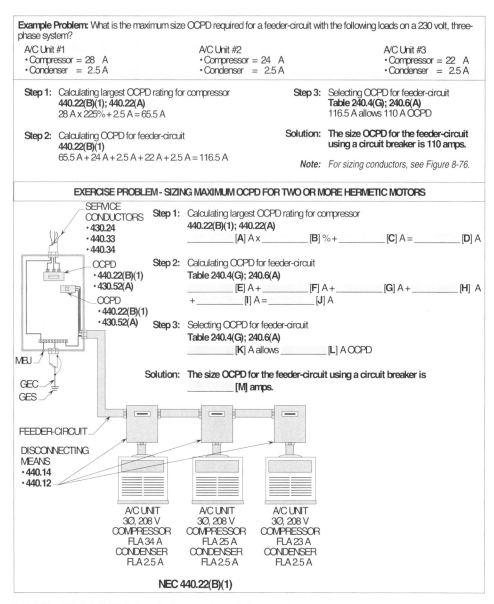

SINGLE MOTOR-COMPRESSORS
440.32

The conductors supplying power to an air-conditioning or refrigerating unit shall be sized to carry the load of the unit plus an overload for a period of time that won't damage the elements. The full-load current rating of the nameplate or branch-circuit selection current, whichever is greater, shall be sized at 125 percent to size and select the conductors supplying hermetically sealed motors. **(See Figure 8-75)**

COMBINATION LOAD
440.33; 440.34

Note that loads for service conductors must be computed per **230.42(A)(1)**.

Two or more motor compressors with motor loads plus other loads may be connected to a feeder-circuit or service conductors. The largest compressor or motor load is calculated at 125 percent plus 100 percent of the remaining compressors and motors. The other loads are computed at 125 percent for continuous and 100 percent for noncontinuous operation per **215.2(A)(1)** and these total values used to select conductors. **(See Figure 8-76)**

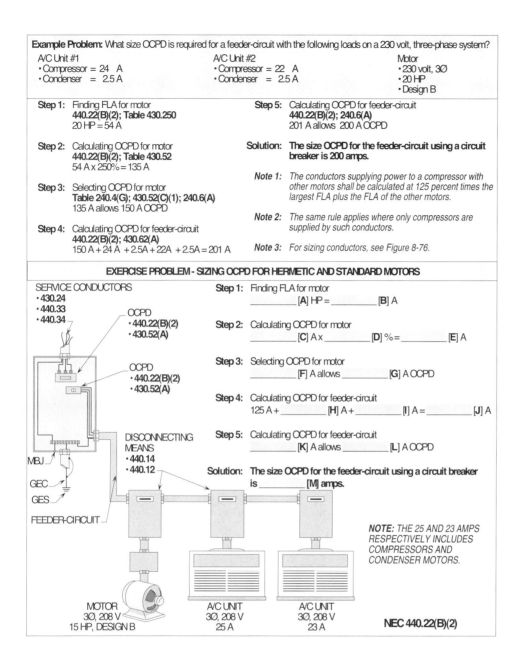

Figure 8-72. When installing hermetically sealed motors and other loads such as motors on the same circuit, and the largest in the group is a motor, the largest overcurrent protection device is sized and selected based on the percentages from **Table 430.52**.

QUICK CALC 3

Sizing conductors in the Example Problem in Figure 8-72.

- Finding load in amps
440.33

54 A x 125%	=	67.5 A
	+	24.0 A
	+	2.5 A
	+	22.0 A
	+	2.5 A
Total LD.	=	118..5 A

- Selecting Conductors
Table 310.16
118.5 A Requires 1 AWG cu.

- Conductors must be 1 AWG THWN cu.

CONTROLLERS FOR MOTOR COMPRESSORS
440.41

The circuit supply conductors are run from a motor controller and connected to the terminals of the compressor when the motor is controlled by a motor controller. The full-load current rating and locked-rotor current rating of the compressor motor must be sized at continuous operation to determine load values to select components.

MOTOR COMPRESSOR CONTROLLER RATING
440.41(A)

The full-load current rating of the nameplate or the branch-circuit selection current rating, whichever is greater, shall be used to size and select the motor controller. If necessary, the locked-rotor current rating of the motor may be used to size and select the motor controller. **(See Figure 8-77)**

Master Test Tip 76: Note that the motor controller must be sized and selected using the same procedure as used for the sizing of the disconnecting means.

Figure 8-73. Where the equipment will start, run, and operate on 15 or 20 amp, 120 volt, single-phase branch-circuit, or a 15 amp, 208 volt or 240 volt, single-phase branch-circuit, having a 15 or 20 amp overcurrent protection device, this OCPD may be used to protect the branch-circuit.

NEC 440.22(B)(2), Ex. 1

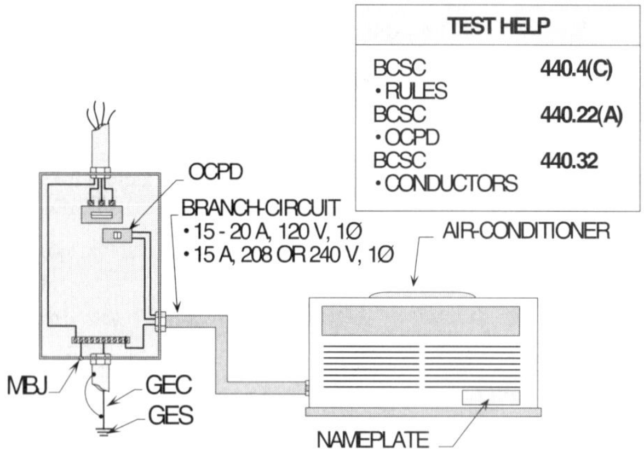

Figure 8-74. The rating of the overcurrent protection device shall be determined by using the rating of the nameplate of the cord-and-plug connected equipment if the hermetically sealed motor(s) is rated single-phase, 250 volts or less.

NEC 440.22(B)(2), Ex. 2

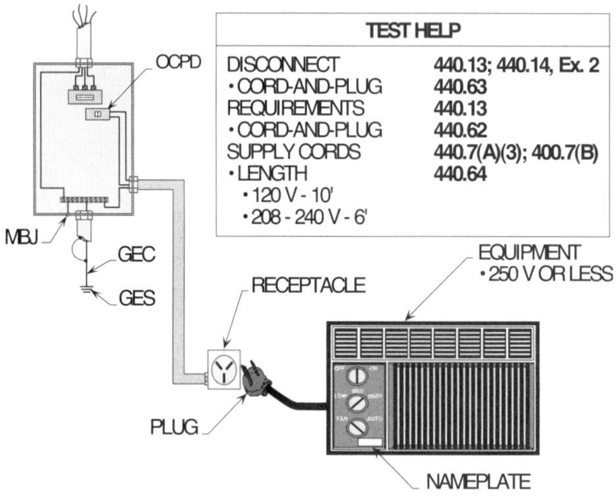

Figure 8-75. The full-load current rating of the nameplate or branch-circuit selection current, whichever is greater, shall be sized at 125 percent to size and select the conductors supplying hermetically sealed motors.

Example Problem: What size THHN copper conductors are required to supply power to a compressor with a full-load current of 33 amps and a condenser motor having a nameplate of 2 amps with terminals rated at 75° C?

Step 1: Calculating amps
Table 220.3; 440.32
33 A x 125% = 41.25 A
2 A x 100% = 2 A
Total load = 43.25 A

Step 2: Selecting conductors
310.10, FPN (2); Table 310.16
43.25 A requires 8 AWG cu.

Solution: **The size THHN copper conductors are 8 AWG.**

Note: *For sizing OCPD, see Figure 8-69.*

EXERCISE PROBLEM - SIZING CONDUCTORS FOR SINGLE MOTOR COMPRESSORS

Step 1: Calculating amps
_______ [A] A x _______ [B]% = _______ [C] A
_______ [D] A x _______ [E]% = _______ [F] A
Total load = _______ [G] A

Step 2: Selecting conductors
_______ [H] A requires _______ AWG [I] cu.

Solution: **The size THHN conductors are _______ AWG [J] copper.**

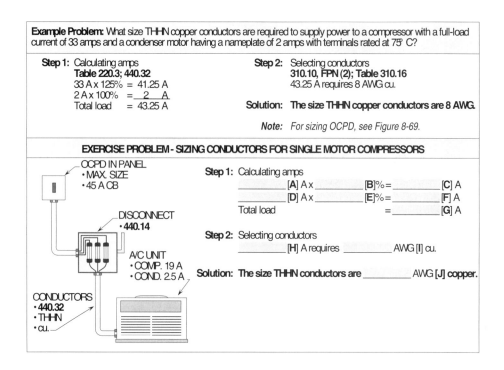

Motors and Compressors

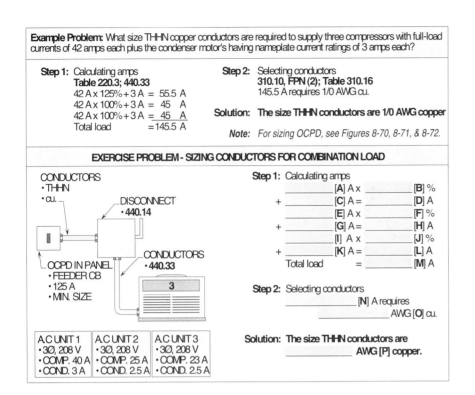

Figure 8-76. Two or more motor compressors with motor load(s) plus other loads may be connected to a feeder-circuit or service conductors. The largest compressor or motor load shall be calculated at 125 percent plus 100 percent of the remaining compressors and motors plus the other loads. Note that only compressor loads are illustrated in Figure 8-76.

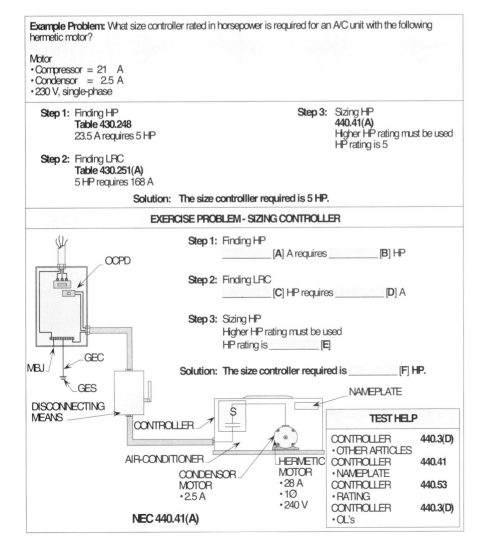

Figure 8-77. The full-load current rating of the nameplate or the branch-circuit selection current, whichever is greater, shall be used to size and select the motor controller.

Master Test Tip 77: For sizing the disconnecting means for the compressor load in the example problem in Figure 8-77, see Figure 8-64.

8-67

APPLICATION AND SELECTION
440.52

The overload relay for the motor compressor shall trip at not more than 140 percent of the full-load current rating. If a fuse or circuit breaker is used for the protection of the motor compressor, it shall trip at not more than 125 percent of the full-load current rating (in amps). **(See Figure 8-78)**

Figure 8-78. The overload relay for the motor compressor shall trip at not more than 140 percent of the full-load current rating. If a fuse or circuit breaker is used for the protection of the motor compressor, it shall trip at not more than 125 percent of the full-load current rating.

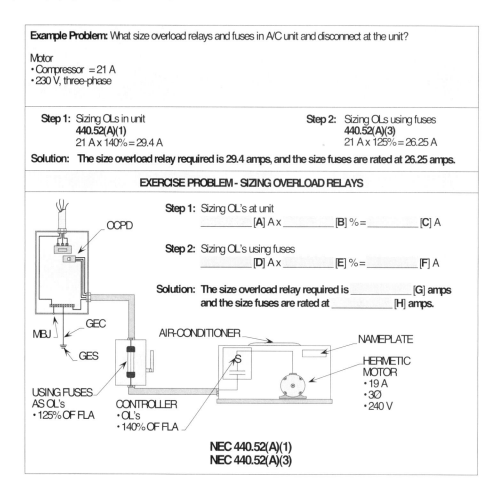

MOTOR COMPRESSORS AND EQUIPMENT ON 15 OR 20 AMP BRANCH-CIRCUIT - NOT CORD-AND-PLUG CONNECTED
440.54

Overload protection shall be provided for direct or fixed-wired motor compressors and equipment that is connected to 15 or 20 amp, 120 volt, single-phase branch-circuits. Note its 15 amp rating for 240 volt, single-phase branch-circuits.

When sizing separate overload relays, the full-load current rating of the hermetically sealed motor shall be selected at 140 percent. Hermetic motors shall be provided with fuses or circuit breakers that provide sufficient time delay to allow the motor to come up to running speed without tripping open the circuit due to the high inrush current. **(See Figure 8-79)**

Master Test Tip 78: Note that a three-phase A/C unit shall not be cord-and-plug connected per **440.55**.

Motors and Compressors

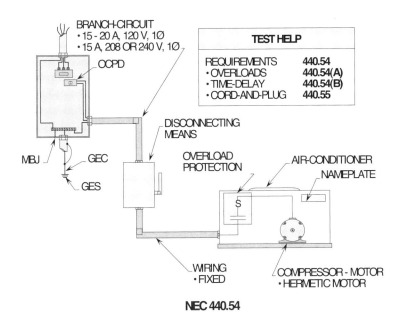

Figure 8-79. Overload protection shall be provided for direct or fixed-wired motor compressors and equipment that is connected to 15 or 20 amp, 120 volt, single-phase branch-circuits.

CORD AND ATTACHMENT PLUG CONNECTED MOTOR COMPRESSORS AND EQUIPMENT ON 15 OR 20 AMP BRANCH-CIRCUITS
440.55

When attachment plugs and receptacles are used for circuit connection, they shall be rated no higher than 15 or 20 amp, for 120 volt, single-phase circuits, or 15 amp, for 208 or 240 volt, single-phase branch-circuits. **(See Figure 8-80)**

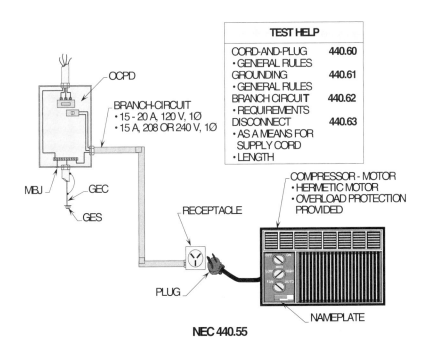

Figure 8-80. When attachment plugs and receptacles are used for circuit connections, they shall be rated no higher than 15 or 20 amp, for 120 volt, single-phase circuit, or 15 amp, for 208 or 240 volt, single-phase branch-circuits.

ROOM AIR-CONDITIONERS
440.60

Master Test Tip 79: A cord-and-plugged A/C window unit rated at 208 or 240 volts shall not be equipped with a cord over 6 ft. long per **440.64**.

Room air-conditioners are usually cord-and-plug connected when installed on 120/240 volt, single-phase systems. However, they may be hard-wired. Air-conditioners are always hard-wired when installed on three-phase, or electrical supply systems over 250 volts.

BRANCH-CIRCUIT REQUIREMENTS
440.62(A)(2)

The full-load current rating (in amps) of the room air-conditioner shall be marked on the nameplate and shall not operate at more than 40 amps on 250 volts. The branch-circuit overcurrent protection device shall be installed with a rating no greater than the circuit conductors' ampacity or the rating of the receptacle serving the unit, whichever is less. The ampacity of a cord-and-plug connected air-conditioning window unit shall not exceed 80 percent of the branch-circuit where no other loads are served. If other loads such as lighting fixtures are served by the branch-circuit, the cord-and-plug connected air-conditioner unit shall not exceed 50 percent of the branch-circuit. **(See Figure 8-81)**

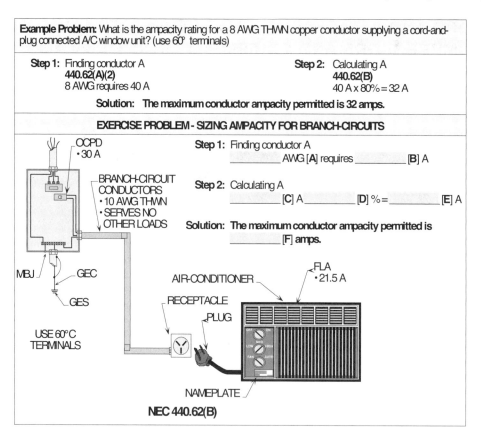

Figure 8-81. The ampacity of a cord-and-plug connected air-conditioning window unit shall not exceed 80 percent of the branch-circuit where no other loads are served.

DISCONNECTING MEANS
440.63

Master Test Tip 80: A cord-and-plugged A/C window unit rated at 120 volt, single-phase shall not be equipped with a cord over 10 ft. long per **440.64**.

A cord-and-plug may serve as the disconnecting means for the room air-conditioner if all the following conditions are complied with:

(1) Operates at 240 volts or less
(2) Controls are manually operated

(3) Controls are within 6 ft. of the floor
(4) Controls are readily accessible to user

Room air-conditioners may be hard-wired and located within sight of the service equipment, or they may be wired so that they are readily accessible to a disconnecting switch for the user. However, such switch must be located within sight and within the unit. **(See Figures 8-82(a) and (b))**

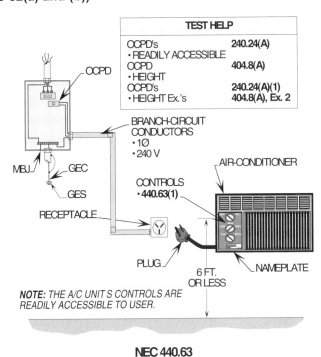

Figure 8-82(a). A cord-and-plug may serve as the disconnecting means for room air-conditioners if it operates at 250 volts or less, controls are manually operated, controls are within 6 ft. of the floor, and controls are readily accessible to the user.

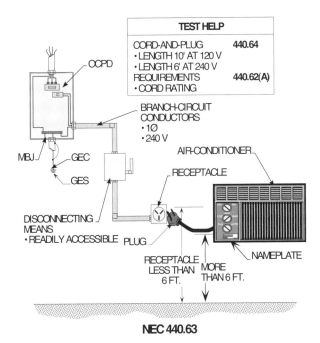

Figure 8-82(b). Under certain conditions, a cord-and-plug may serve as the disconnecting means even if the room air-conditioner's manual controls are above 6 ft. from the finished grade.

Name _____   Date _____

Chapter 8
Motors and Compressors

Section   Answer

1. For a stationary motor rated 1/8 horsepower or less, the branch-circuit overcurrent protection device shall be permitted to serve as the disconnecting means.

    (a) True   **(b)** False

2. An instantaneous trip circuit breaker that is part of a listed combination motor controller shall be permitted to serve as a disconnecting means.

    (a) True   **(b)** False

3. The general rule requires the disconnecting means for a motor to be installed within sight of the motor and not more than 50 ft. from the motor.

    (a) True   **(b)** False

4. Control circuit conductors and motor circuit conductors shall be permitted to occupy the same raceway if they are functionally associated with the motor system.

    (a) True   **(b)** False

5. A horsepower rated switch shall be permitted to be used as a disconnecting means for an air-conditioning unit. (motor 2 HP or less)

    (a) True   **(b)** False

6. Cord-and-plug connected room air-conditioners shall not be permitted to serve as the disconnecting means for hermetic motors.

    (a) True   **(b)** False

7. An additional circuit breaker or disconnecting means shall be provided at the equipment if an air-conditioning unit is not within sight or within 50 ft.

    (a) True   **(b)** False

8. Short-circuit and ground-fault protection can be provided by overload relays and thermal protectors for air-conditioning units.

    (a) True   **(b)** False

9. The ampacity of a cord-and-plug connected air-conditioning window unit shall not exceed 80 percent of the branch-circuit where no other loads are served.

    (a) True   **(b)** False

10. Room air-conditioners installed with flexible cords shall not be required to be a length that is limited to 12 ft. for 120 volt circuits.

    (a) True   **(b)** False

11. The branch-circuit protective device shall be permitted to serve as the controller for stationary motors where the motor is rated _____ horsepower or less.

    (a) 1/6   **(b)** 1/8   **(c)** 1/4   **(d)** 1/3

Section	Answer

12. An A/C general-use snap switch shall be permitted to be installed as the controller for a stationary motor where the full-load current of the switch does not exceed _____ percent of the branch-circuit rating.

(a) 50 percent (b) 75 percent (c) 80 percent (d) 90 percent

13. The disconnecting means for air-conditioning or refrigeration equipment shall be located within sight and within _____ ft. and shall be readily accessible to the user.

(a) 20 (b) 25 (c) 40 (d) 50

14. The full-load current rating of the room air-conditioner shall be marked on the nameplate and shall not operate at more than _____ amps on 250 volts.

(a) 20 (b) 30 (c) 40 (d) 50

15. A cord-and-plug shall not be permitted to serve as the disconnecting means if the room air-conditioner manual controller is above _____ ft.

(a) 5 (b) 6 (c) 10 (d) 12

16. Room air-conditioners installed with flexible cords shall be required to be a length that is limited to _____ ft. for 208 or 240 volt circuits.

(a) 5 (b) 6 (c) 10 (d) 12

17. The rating of the overcurrent protection device shall be determined by using the rating of the nameplate of the cord-and-plug connected equipment serving single-phase, _____ volt or less hermetic sealed rated motors.

(a) 120 (b) 250 (c) 480 (d) 600

18. Overload protection shall be provided for direct or fixed-wired motor compressors and equipment that are connected to _____ or _____ amp, 120 volt, single-phase branch-circuits.

(a) 15 or 20 (b) 20 or 30 (c) 30 or 40 (d) 40 or 50

19. Room air-conditioners shall be required to be grounded where operating over _____ volts-to-ground.

(a) 120 (b) 150 (c) 250 (d) 300

20. A cord-and-plug shall be permitted to serve as the disconnecting means for the room air-conditioner if it operates at _____ volts or less.

(a) 120 (b) 150 (c) 250 (d) 300

21. What size THWN copper conductors are required to supply an 30 HP, 40 HP, and 50 HP, 460 volt, three-phase, Design B motors?

(a) 1/0 AWG THWN copper conductors (b) 2/0 AWG THWN copper conductors
(c) 3/0 AWG THWN copper conductors (d) 4/0 AWG THWN copper conductors

	Section	Answer

22. What is the maximum size inverse-time circuit breaker for a 50 HP, 230 volt, three-phase, Design B motor?

    (a) 200 A OCPD   **(b)** 250 A OCPD
    (c) 300 A OCPD   **(d)** 350 A OCPD

23. What is the maximum size OCPD required for an A/C unit with a compressor rated at 20 amps and the condenser rated at 2.5 amps?

    (a) 30 A OCPD   **(b)** 45 A OCPD
    (c) 50 A OCPD   **(d)** 70 A OCPD

24. What size THHN copper conductors are required to supply an A/C unit with a compressor rated at 20 amps and the condenser rated at 2.5 amps?

    (a) 12 AWG THWN copper conductors   **(b)** 10 AWG THWN copper conductors
    (c) 8 AWG THWN copper conductors   **(d)** 6 AWG THWN copper conductors

Residential Calculations – Single-Family Dwellings

9

Quick Reference

APPLYING THE STANDARD CALCULATION	9-1
DEMAND FACTORS	9-2
MINIMUM SIZE SERVICE	9-9
APPLYING THE STANDARD CALCULATION	9-10
APPLYING THE OPTIONAL CALCULATION	9-28
APPLYING THE OPTIONAL CALCULATION FOR EXISTING UNITS	9-35
FEEDER TO MOBILE HOME - STANDARD CALCULATION	9-36
TEST QUESTIONS	9-43

Residential calculations are the most difficult to perform due to the rules and regulations of the NEC being more restrictive than those for commercial and industrial facilities. The standard or optional calculation shall be permitted to be used to calculate the loads to size and select the elements of the feeder or service. The optional calculation seems to be the favorite of most testing agencies. This is true because once loads are calculated, this calculation produce smaller VA or amps than the longer, complicated standard calculation. The procedures for laying out residential calculations will be different in some ways from those used for commercial and industrial.

APPLYING THE STANDARD CALCULATION PART II TO ARTICLE 220

Residential occupancies are known in the industry as dwelling units. This chapter mainly deals with calculation procedures for one and two family dwelling units.

When using the standard calculation for calculating loads for a residential occupancy, all loads are divided into three groups and four columns. The groups are as follows:

Group 1:
General lighting and receptacle loads
Group 2:
Small appliance loads
Group 3:
Special appliance loads

DEMAND FACTORS
ARTICLE 220, PART II

Master Test Tip 1: The total VA for general lighting and receptacle loads are derived by multiplying the square footage by 3 VA and applying demand factors where appropriate.

Master Test Tip 2: The small appliance circuits shall be permitted to be added to the VA per sq. foot calculation and demand factors applied.

Master Test Tip 3: Cooking equipment loads shall be permitted to have demand factors applied based upon the number of units.

Master Test Tip 4: When there are four or more fixed appliances, a demand factor of 75 percent shall be permitted to be applied to the total VA.

Master Test Tip 5: For four or fewer dryers, the total VA shall be calculated at 100 percent with no demand factors applied.

The following four loads are separated into two columns of loads and demand factors are applied accordingly:

Column 1:
    General lighting and receptacle loads and small appliance loads   **Table 220.42**
Column 2:
    Cooking equipment loads   **220.19; Table 220.55**
    Fixed appliance loads   **220.53**
    Dryer loads   **220.54; Table 220.54**

General lighting and receptacle loads, the small appliance circuits, plus the laundry circuit shall be permitted to have demand factors applied.

The total number of fixed appliance loads shall be permitted to have demand factors applied. The total number of ranges and dryers in a dwelling unit shall be permitted to be reduced by a percentage.

GENERAL LIGHTING AND RECEPTACLE LOADS AND SMALL APPLIANCE PLUS LAUNDRY LOADS
COLUMN 1 - TABLE 220.42; 220.52(A); (B)

The general lighting load for a dwelling unit shall be calculated by multiplying the square footage by 3 VA per sq. ft. per **Table 220.12**. The required square footage per unit load (volt-amps) is found in **Table 220.12**. All small appliance and laundry loads shall be computed at 1500 VA per **220.52(A) and (B)**. A demand factor shall be permitted per **Table 220.42**. The demand factors for the general lighting and receptacle loads per **Table 220.42** for dwelling units are as follows:

Volt-amps	
0 - 3000	at 100%
3001 - 120,000	at 35%
120,001 +	at 25%

See **Figure 9-1** for determining the general-purpose lighting and receptacle loads including the small appliance plus laundry loads.

COOKING EQUIPMENT LOADS
COLUMN 2 - 220.55; TABLE 220.55

The cooking equipment loads and Demand load 2 are separated from Group 3 and placed in Column 2 and demand factors applied accordingly.

The demand factors listed in **Table 220.55** apply to the demand loads for cooking equipment. The footnotes are based on the kW rating and number of units. Ranges, wall-mounted ovens and counter-mounted cooktops are units of cooking equipment per **Table 220.19**.

DEMAND LOAD 2
TABLE 220.55, COLUMN A

The nameplate (kW) wattage rating of a range shall be calculated by the number of cooking units times the percentage factor found in **Table 220.55**. This calculation is used to obtain the maximum demand load.

See **Figure 9-2** for an example and exercise problem for the demand loads to be applied per **Table 220.55, Col. A**, based upon the size and number of units.

Residential Calculations - Single-Family Dwellings

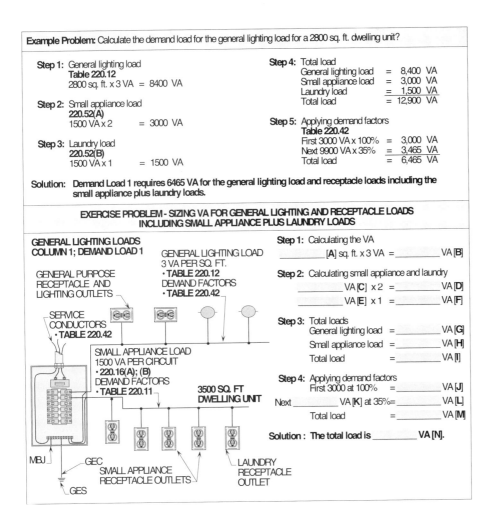

Figure 9-1. This figure illustrates an example and exercise problem for determining the general-purpose lighting and receptacle loads including small appliance plus laundry loads.

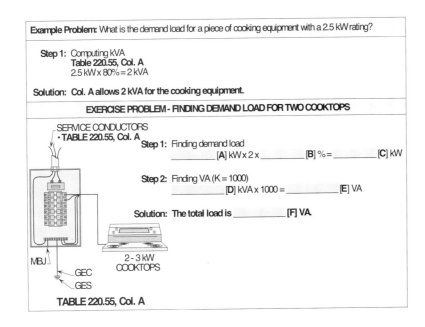

Figure 9-2. This figure illustrates an example and exercise problem for demand loads that shall be permitted to be applied per **Table 220.55, Col. A**.

DEMAND LOAD 2
TABLE 220.55, COLUMN B

The nameplate (kW) wattage rating of a range shall be calculated by the number of cooking units times the percentage factor applied in **Table 220.55**. This calculation derives the maximum demand load.

See **Figure 9-3** for an example and exercise problem of the demand loads to be applied per **Table 220.55, Col. B**, based upon the size and number of units.

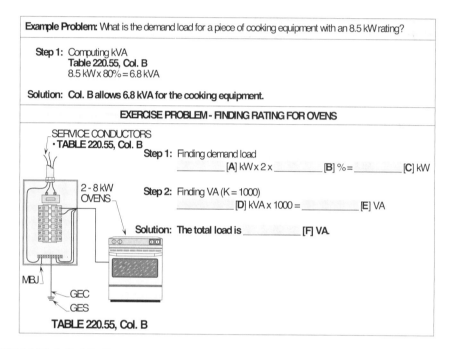

Figure 9-3. This figure illustrates an example and exercise problem for demand loads that shall be permitted to be applied per **Table 220.55, Col. B**.

DEMAND LOAD 2
TABLE 220.55, COLUMN C

The demand load of cooking equipment is calculated in kW per **Table 220.55**. The maximum demand for the size and number of ranges is already calculated for selecting the elements of the feeder or service. Therefore, cooking equipment does not need to be multiplied by the kW rating of the unit by a percentage until **Columns A** or **B** are utilized.

See **Figure 9-4** for an example and exercise problem on demand loads that shall be permitted to be applied per **Table 220.55, Col. C,** based upon the size and number of units.

DEMAND LOAD 2
TABLE 220.55, NOTE 1

For cooking equipment rated over 12 kW to 27 kW in **Column C, Note 1** shall be increased 5 percent for each kW over 12 kW to calculate the demand load.

See **Figure 9-5** for an example and exercise problem of the demand loads to be applied per **Table 220.55, Note 1**, based upon the size and number of units.

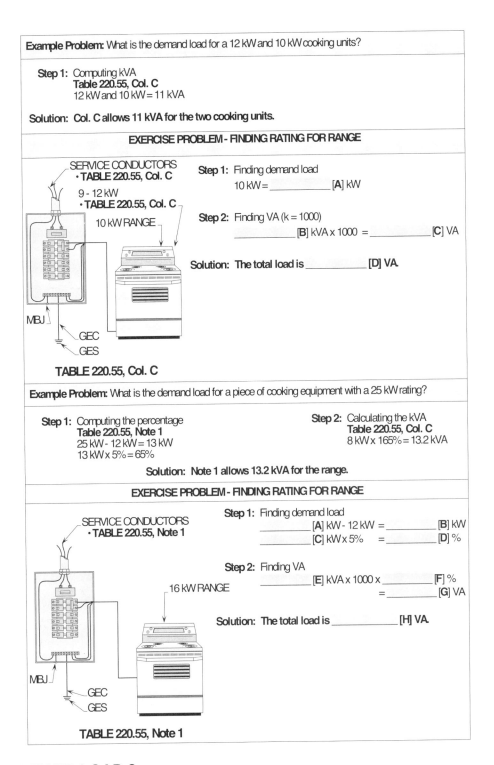

Figure 9-4. This figure illustrates an example and exercise problem for demand loads that shall be permitted to be applied per Table 220.55, Col. C.

Figure 9-5. This figure illustrates an example and exercise problem for demand loads that shall be permitted to be applied per Table 220.55, Note 1.

DEMAND LOAD 2
TABLE 220.55, NOTE 2

Cooking equipment of unequal values rated over 12 kW to 27 kW in **Column C, Note 2** shall be calculated by adding the kW ratings of all units and dividing by the number of units, all ranges below 12 kW shall be calculated at 12 kW. When an average rating is found, the number of units shall be increased by 5 percent for each kW exceeding 12 kW to derive the allowable kW.

See Figure 9-6 for an example and exercise problem of the demand loads to be applied per **Table 220.55, Note 2**, based upon the size and number of units.

9-5

Figure 9-6. This figure illustrates an example and exercise problem for demand loads that shall be permitted to be applied per **Table 220.55, Note 2**.

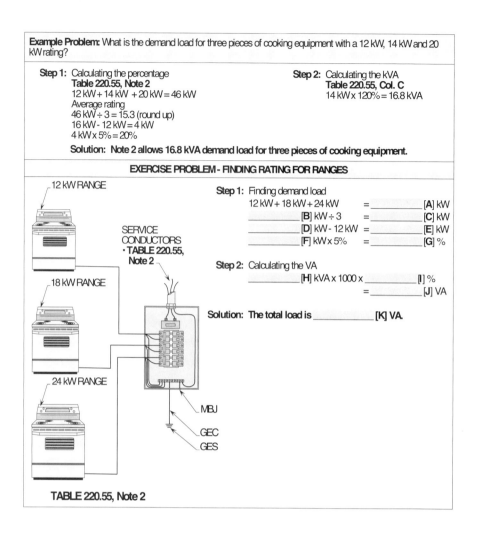

DEMAND LOAD 2
TABLE 220.55, NOTE 4

The demand load for a cooktop and two or less wall-mounted ovens shall be calculated by finding the average rating of each unit in kW. Each kilowatt that exceeds 12 kW shall be increased by 5 percent. The nameplate rating of the appliance shall be calculated per **Table 220.55, Column C. Sections 240.21(A)** and **210.19(A)(3), Ex. 1** list requirements for each piece of cooking equipment to be supplied by a tap. Each piece of cooking equipment shall be installed in the same room in order to comply with **Note 4 to Table 220.55**. The demand load for each kW exceeding 12 kW shall be found by multiplying the kW rating by 5 percent to determine the average rating per **Table 220.55, Column C**.

See **Figure 9-7** for an example and exercise problem of the demand loads to be applied per **Table 220.55, Note 4**, based upon the size and number of units.

Master Test Tip 6: Three or fewer fixed appliances shall be calculated at 100 percent of their total VA.

FIXED APPLIANCE LOAD
COLUMN 2 - 220.53

Cooking equipment loads, dryer equipment loads, air-conditioning loads and heating equipment loads shall not be applied to the fixed appliance load per **220.53**. The fixed appliance load for three or less fixed appliances shall be determined by adding wattage (volt-amps) values by the nameplate ratings for each appliance. However, these fixed appliance loads shall be permitted to have demand factors applied if there are four or more. All other fixed appliances of four or more grouped into a special appliance load shall be found by adding wattage ratings from appliance nameplates and multiplying the total wattage (volt-amps) by 75 percent to obtain demand load. Three or less fixed appliance loads shall be calculated at 100 percent. The 75 percent per **220.53** shall not be permitted to be applied due to the number.

Fixed appliance loads include dishwashers, compactors, disposals, water heaters, blower motors, water circulating pumps, etc. It shall be permitted to apply a demand factor to these loads to calculate the elements of the service or feeder equipment. However, the branch-circuit elements shall be calculated and sized according to other Sections in the NEC.

See Figure 9-8 for an example and exercise problem of demand factors to be applied per **220.53**.

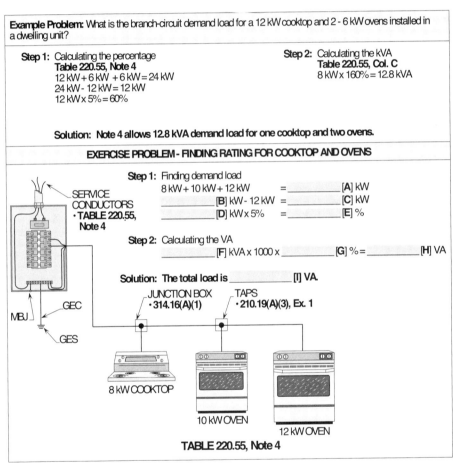

Figure 9-7. This figure illustrates an example and exercise problem for demand loads that shall be permitted to be applied per **Table 220.55, Note 4**.

DRYER LOAD
COLUMN 2 - 220.54; TABLE 220.54

The demand load for household dryers shall be calculated at 5 kVA or the nameplate rating, whichever is greater. Dryer equipment of four or fewer dryers shall be calculated at 100 percent of the nameplate rating. Dryer equipment of five or more dryers shall be permitted to have a percentage applied based on the number of units per **Table 220.54**.

See Figure 9-9 for a example and exercise problem of demand factors to be applied per **220.54** and **Table 220.54,** based upon the size and number of units.

Figure 9-8. This figure illustrates an example and exercise problem for demand factors that shall be permitted to be applied per **220.53**.

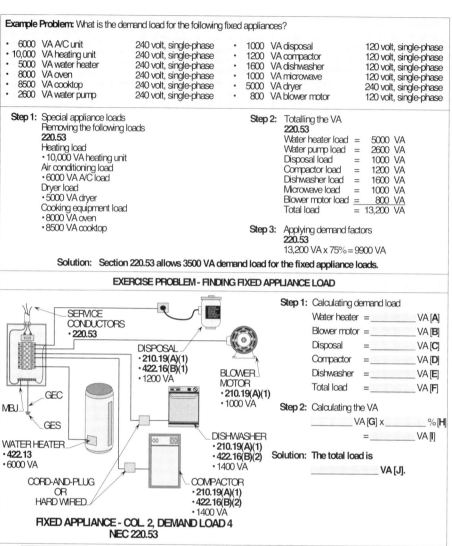

Master Test Tip 7: A demand factor of 75 percent shall be permitted to be applied to four or more fixed appliances installed in a dwelling unit.

Figure 9-9. This figure illustrates an example and exercise problem for demand factors that shall be permitted to be applied per **220.54** and **Table 220.54**.

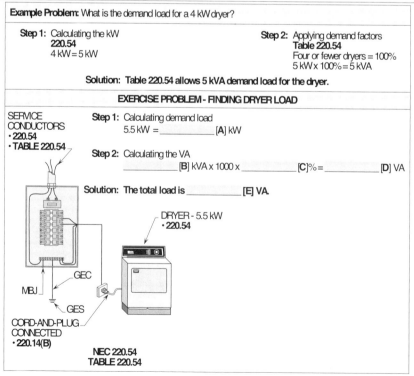

Residential Calculations - Single-Family Dwellings

LARGEST LOAD BETWEEN HEAT AND A/C
COLUMN 3 - 220.60

Section **220.60** shall be applied when determining the largest load between heat and A/C. The heating and A/C loads shall be computed at 100 percent, and the smaller of the two loads is dropped.

See **Figure 9-10** for an example and exercise problem of demand factors to be applied per **220.60**.

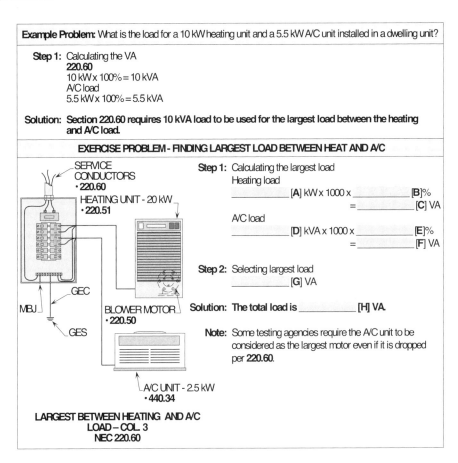

Figure 9-10. This figure illustrates an example and exercise problem for demand factors to be applied per **220.60**.

Note: Where a heat pump is involved, both the heat pump and the electric heating load shall be added together at 100 percent of its nameplate ratings.

LARGEST MOTOR LOAD
COLUMN 4 - 220.14(C); 220.50

The motor's total full-load current rating shall be calculated at 25 or 125 percent per **220.14(C)** and **220.50**, which is required per **430.24** and **430.25**, for calculating the load of one or more motors with other loads.

See **Figure 9-11** for an example and exercise problem of the demand factors to be applied in **220.14(C)** and **220.50** to determine the largest motor load.

Master Test Tip 8: The largest motor load shall be permitted to be selected from the A/C unit load, if not dropped, or from one of the fixed appliance motors.

MINIMUM SIZE SERVICE
230.42(B); 230.79(C)

Section **230.79(C)** requires the disconnecting means for a one family dwelling to have a minimum of 100 amps regardless of how few branch-circuits are installed in the service panelboard.

Figure 9-11. This figure illustrates an example and exercise problem for demand factors to be applied per **220.14(C)** and **220.50** when finding the VA for a single-phase, 230 volt motor.

Master Test Tip 9: The largest motor load shall be computed for both the ungrounded (phase) conductors and the grounded (neutral) conductor.

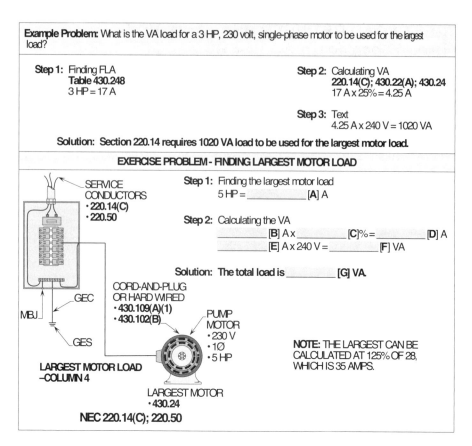

APPLYING THE STANDARD CALCULATION

The standard calculation for dwelling units are separated into four columns.

Col. 1	Col. 2	Col. 3	Col. 4
general lighting receptacle small appliance laundry	special appliance fixed appliance	largest load AC/heat	largest motor load

The elements of the service is determined by using the total load of these four columns.

APPLYING THE STANDARD CALCULATION SERVICE WITH GAS HEAT

When applying the standard calculation for dwelling units, all gas-related loads shall not be applied to the total load when calculating the amperage for the service. **(See Design Problem 9-1)**

Design Problems. Use Design Example Problem 9-1 to size the following components and elements of a dwelling unit with gas heat:

(1) What size nonmetallic-sheathed cable is required to supply the dishwasher?

   Step 1: Finding amperage
   210.23(A); 210.19(A)(1)
   I = VA ÷ V
   I = 1600 VA ÷ 120 V
   I = 13.33 A

Step 2: Calculating amperage
210.23(A); 210.19(A)(1)
13.33 A x 100% = 13.33 A

Step 3: Finding conductor size
Table 310.16, Footnotes; 240.4(D)
14 AWG cu. = 15 A
13.33 A is less than 15 A

Solution: The size nonmetallic-sheathed cable is 14 AWG.

(2) What size OCPD is required for the water heater?

Step 1: Finding amperage

I = VA ÷ V
I = 6000 VA ÷ 240 V
I = 25 A

Step 2: Calculating amperage
422.13
25 A x 125% = 31.25 A

Step 3: Finding size OCPD
240.4(B); 240.6(A)
31.25 A requires 35 A OCPD

Solution: The size OCPD is 35 amps.

(3) What size OCPD is required for the A/C unit using the minimum setting?

Step 1: Finding amperage

I = VA ÷ V
I = 6000 VA ÷ 240 V
I = 25 A

Step 2: Calculating amperage
440.22(A)
25 A x 175% = 43.75 A

Step 3: Selecting size OCPD
Table 240.4(G); 240.6(A)
43.75 A requires 40 A OCPD

Solution: The size OCPD is 40 amps.

(4) What size nonmetallic-sheathed cable is required to supply the A/C unit?

Step 1: Finding the amperage

I = VA ÷ V
I = 6000 VA ÷ 240 V
I = 25 A

Step 2: Calculating the amperage
440.32
25 A x 125% = 31.25 A

Step 3: Finding conductor size
  Table 310.16, 75°C Column
  10 AWG = 35 A
  31.25 A is less than 35 A

Solution: The size nonmetallic-sheathed cable is 10 AWG.

APPLYING THE STANDARD CALCULATION SERVICE WITH ELECTRIC HEATING

When applying the standard calculation for dwelling units, the electric heating shall be applied to the total load (if largest load between heat and A/C) when calculating the amperage for the service. **(See Design Problem 9-2)**

Design Problems. Use Design Example Problem 9-2 to size the following components and elements of a dwelling unit with electric heating:

(1) What size panelboard is required for the service equipment?

  Step 1: Finding panelboard size
    408.30; 408.36(A); Chart on back inside cover
    222 A requires 225 A panelboard

  Solution: The size panelboard required is 225 amps.

(2) What size overcurrent protection device is required for the service equipment?

  Step 1: Finding conductor size
    Table 310.16
    4/0 AWG cu. = 230 A

  Step 2: Finding OCPD size (based on load)
    240.4(B); 230.90(A); 240.6(A)
    230 A requires 225 A OCPD

  Solution: The size OCPD required is 225 amps.

(3) What size copper conductor is required for a driven rod to act as a supplemental ground?

  Step 1: Finding conductor size
    250.53(E); 250.104(A); 250.66(A); Table 250.66
    Driven rod requires 6 AWG cu.

  Solution: The size supplemental ground is 6 AWG copper.

(4) What size copper grounding electrode conductor is required to ground the service to the metal water pipe?

  Step 1: Finding conductor size
    250.52(A)(1); Table 250.66
    4/0 AWG cu. requires 2 AWG cu.

  Solution: The size grounding electrode conductor is 2 AWG copper.

APPLYING STANDARD CALCULATION
SERVICE WITH ELECTRIC HEATING AND HEAT PUMP

When applying the standard calculation for dwelling units, the electric heating and heat pump shall be applied to the total load (if largest load between heat and A/C) when calculating the amperage for the service. The heat pump shall also be used as the largest motor load for the ungrounded (phase) conductors. **(See Design Problem 9-3)**

Design Problems. Use Design Example Problem 9-3 to size the following components and elements of a dwelling unit with electric heating and heat pump:

(1) How many 15 amp two-wire circuits are required for the general lighting and receptacle loads?

> **Step 1:** Finding total VA
> **Table 220.12**
> 3000 sq. ft. x 3 VA = 9000 VA
>
> **Step 2:** Finding # of outlets
> **210.11(A)**
> # = VA ÷ CB x V
> # = 9000 VA ÷ 15 A CB x 120 V = 5

Solution: The number of outlets is 5.

(2) What size nonmetallic-sheathed cable is required to supply the branch-circuit for the cooktop and oven per **Note 2 to Table 220.55**?

> **Step 1:** Finding total kW
> Cooktop = 16,000 VA
> Oven   = 12,000 VA
> Total load = 28,000 VA
>
> **Step 2:** Finding demand factor
> **Table 220.55, Note 2**
> 28,000 VA ÷ 1000 = 28 kW
> 28 kW - 12 kW = 16 kW
> 16 kW x 5% = 80%
>
> **Step 3:** Applying demand factor
> **Table 220.55, Col. C**
> 8 kW x 180% = 14.4 kW
>
> **Step 4:** Finding amperage
>
> I = kVA x 1000 ÷ V
> I = 14.4 kVA x 1000 ÷ 240 V
> I = 60 A
>
> **Step 5:** Finding conductor size
> **Table 310.16**
> 6 AWG cu. = 65 A
> 60 A is less than 65 A

Solution: The size nonmetallic-sheathed cable is 6 AWG copper.

(3) What size overcurrent protection device is required for the branch-circuit to the cooktop and oven?

> **Step 1:** Finding conductor size
> **Table 310.16**
> 6 AWG cu. = 65 A
>
> **Step 2:** Finding OCPD size
> **240.4(B); 240.6(A); 210.19(A)(3), Ex. 1**
> 65 A requires 70 A OCPD
>
> **Solution: The size OCPD required is 70 amps.**

Note: Apply **210.19(A)(3), Ex. 1** where smaller taps are made to supply each piece of cooking equipment.

APPLYING STANDARD CALCULATION SERVICE WITH 120 VOLT A/C WINDOW UNIT

When applying the standard calculation for dwelling units, all 120 volt A/C window units shall be applied to the total load (if largest load between heat and A/C) when calculating the amperage for the service. The A/C window unit shall also be used as the largest motor load for the ungrounded (phase) conductors and grounded (neutral) conductor. **(See Design Problem 9-4)**

Design Problems. Use Design Example Problem 9-4 to size the following components and elements of a dwelling unit with 120 volt, A/C window unit:

(1) What size overcurrent protection device is required for the service-entrance conductors (maximum size)?

> **Step 1:** Finding amperage (based on conductors)
> **Table 310.16; 75°C Column; 110.14(C)(1)(b)**
> 300 KCMIL cu. = 285 A
>
> **Step 2:** Finding OCPD size
> **230.90(A); 240.4(B); 240.6(A)**
> 285 A requires 300 A OCPD
>
> **Solution: The size OCPD is 300 amps.**

(2) What size panelboard is required for the service equipment?

> **Step 1:** Finding panelboard size
> **408.30; 408.36(A)**; Chart on inside back cover
> 265 A requires 300 A panelboard
>
> **Solution: The size panelboard required is 300 amps.**

(3) What size nonmetallic-sheathed cable is required for the feeder-circuit to a panelboard supplying the 10 kW heating unit and 18 amp A/C central unit downstairs?

    Step 1: Calculating load in VA
        215.2(A)(1); 220.51; 220.60; 440.34
        Heating unit load
        10,000 VA x 100% = 10,000 VA
        A/C unit load
        18 A x 240 V x 100% = 4320 VA

    Step 2: Selecting largest load
        220.60
        Heating unit
        10,000 VA is largest load

    Step 3: Calculating total VA
        Table 220.3, 424.3(B)
        Heating unit load
        10,000 VA x 125% = 12,500 VA
        Blower motor load
        800 VA x 125%   =   1000 VA
        Total load   = 13,500 VA

    Step 4: Finding amperage

        I = VA ÷ V
        I = 13,500 VA ÷ 240 V
        I = 56.25 A

    Step 5: Selecting conductor size
        Table 310.16, 60°C Column
        4 AWG cu. = 70 A
        56.25 A is less than 70 A

Solution: The size nonmetallic-sheathed cable required is 4 AWG copper.

(4) What size overcurrent protection device is required for the feeder-circuit to the 10 kW heating unit and the feeder-circuit to the 18 amp A/C unit upstairs? Size the overcurrent protection device located in the panelboard by the heating unit to serve the A/C unit.

<u>Heating unit</u>

    Step 1: Selecting conductor size
        Table 310.16, 60°C Column
        4 AWG cu. = 70 A

    Step 2: Selecting OCPD size
        240.4(B); 240.6(A)
        70 A requires 70 A OCPD

Solution: The size OCPD is 70 amps.

A/C unit

Step 1: Calculating amperage
440.22(A)
18 A x 175% = 31.5 A

Step 2: Selecting OCPD size
Table 240.4(G); 240.6(A)
31.5 A requires 30 A OCPD

Solution: The size OCPD is 30 amps.

APPLYING STANDARD CALCULATION DIVIDING VA ON LINE AND NEUTRAL

When applying the standard calculation for dwelling units, the load in VA should be balanced when determining the ungrounded (phase) conductors and grounded (neutral) conductor. **(See Design Problem 9-5)**

Design Problems. Use Design Example Problem 9-5 to size the following components and elements of a dwelling unit when dividing VA on line and neutral:

(1) What size nonmetallic-sheathed cable is required to supply power to the branch-circuit for the dryer?

Step 1: Finding amperage
220.14(B); 220.54; Text
I = VA ÷ V
I = 5000 VA ÷ 240 V
I = 20.8 A

Step 2: Selecting conductor size
334.80; Table 310.16, 60°C column; 110.14(C)(1)(a)
10 AWG NMC cu. = 30 A

Solution: The size nonmetallic-sheathed cable required is 10-2 AWG w/ground.

(2) What size overcurrent protection device is required for the dryer?

Step 1: Selecting size conductor
Table 310.16, Footnotes
10-2 AWG NMC requires 30 A OCPD

Solution: The size OCPD required is 30 amps.

(3) What size nonmetallic-sheathed cable is required for the branch-circuit to the central A/C unit?

Step 1: Finding amperage
Table 220.3; 440.32
I = VA ÷ V
I = 6000 VA ÷ 240 V
I = 25 A
25 A x 125% = 31.25 A

Step 2: Selecting size conductor
Table 310.16, Footnotes; 60°C column
8 AWG NMC cu. = 40 A
31.25 is less than 40 A

Solution: The size nonmetallic-sheathed cable required is 8-2 AWG w/ground.

(4) What size overcurrent protection device is required for the central A/C unit? (Use the maximum setting)

Step 1: Calculating amperage
Table 240.4(G); 440.22(A)
25 A x 225% = 56.25 A

Step 2: Selecting OCPD size
240.4(B); 240.6(A)
56.25 A requires 50 A OCPD

Solution: The size OCPD required is 50 amps.

(5) What size nonmetallic-sheathed cable is required for the branch-circuit supplying the oven?

Step 1: Finding kW
Table 220.55, Col. C
12,000 VA = 8 kW

Step 2: Finding amperage
I = kW x 1000 ÷ V
I = 8 kW x 1000 ÷ 240 V
I = 33 A

Step 3: Selecting conductor size
334.80; Table 310.16; 60°C Column; 110.14(C)(1)(a)
8 AWG NMC cu. = 40 A
33 A is less than 40 A

Solution: The size nonmetallic-sheathed cable required is 8-2 AWG w/ground.

(6) What size overcurrent protection device is required for the oven?

Step 1: Selecting conductor size
Table 310.16; 60°C Column
8 AWG NMC cu. = 40 A

Step 2: Selecting OCPD size
240.4(B); 240.6(A)
40 A requires 40 A OCPD

Solution: The size OCPD required is 40 amps.

Design Example Problem 9-1. What is the load in VA and amps for a residential dwelling unit with the following loads?

Given Loads:	120 V, single-phase loads	240 V, single-phase loads
General lighting and receptacle load • 3000 sq. ft. dwelling unit • 2 small appliance circuits • 1 laundry circuit	• 2600 VA water pump • 1200 VA disposal • 1400 VA compactor • 1600 VA dishwasher • 1200 VA microwave	• 6000 VA A/C unit • 10,000 VA heating unit (gas) • 6000 VA water heater • 12,000 VA oven (gas) • 16,000 VA cooktop (gas) • 5000 VA dryer • 800 VA blower motor

Sizing ungrounded (phase) conductors = •
Sizing grounded (neutral) conductor = √

Column 1
Calculating general lighting and receptacle load

Step 1: General lighting and receptacle load
Table 220.12
3000 sq. ft. x 3 VA   =   9,000 VA

Step 2: Small appliance and laundry load
220.52(A); (B)
1500 VA x 2   =   3,000 VA
1500 VA x 1   =   1,500 VA
Total load   =   4,500 VA

Step 3: Applying demand factors
Table 220.42
General lighting load   =   9,000 VA
Small appliance load   =   3,000 VA
Laundry load   =   1,500 VA
Total load   =   13,500 VA

First 3000 VA x 100%   =   3,000 VA
Next 10,500 VA x 35%   =   3,675 VA
Total load   =   6,675 VA • √

Column 2
Calculating cooking equipment load

Step 1: Applying demand factors
Table 220.55, Note 1
Total kW rating   =   0

Calculating fixed appliance load

Step 1: Applying demand factors
220.53
Water heater   =   6,000 VA
Water pump   =   2,600 VA √
Disposal   =   1,200 VA √
Compactor   =   1,400 VA √
Dishwasher   =   1,600 VA √
Microwave   =   1,200 VA √
Blower motor   =   800 VA
Total load   =   14,800 VA
14,800 VA x 75% =   11,100 VA •

Calculating dryer load

Step 1: Applying demand factors
220.54; Table 220.54
5000 VA x 100%   =   5,000 VA •

Column 3
Largest load between heating or A/C

Step 1: Selecting the largest load
220.60
A/C load
6000 VA x 100%   =   6,000 VA •

Column 4
Calculating largest motor load

Step 1: Selecting largest motor load
220.14(C); 220.50
Water pump & A/C Unit
6000 VA x 25%   =   1500 VA
2600 VA x 25%   =   650 VA •

Calculating ungrounded (phase) conductors

General lighting load   =   6,675 VA •
Cooking equipment   =   0 VA •
Fixed appliance load   =   11,100 VA •
Dryer load   =   5,000 VA •
Heating load   =   6,000 VA •
Largest motor load   =   1500 VA •
Six total loads   =   30,275 VA

Finding amps for ungrounded (phase) conductors - Phases A and B

I = VA ÷ V
I = 30,275 VA ÷ 240 V
I = 126 A

Grounded (neutral) conductors

Column 1
General lighting load and demand load 1

220.61
6,675 VA √   =   6,675 VA √

Column 2
Cooking equipment load and demand load 2

(Use 70% of cooking load)
220.61
Total kW rating   =   0 VA √

Column 2
Fixed appliance load and demand load 3

(Use 75% of 120 volt load)
220.53
Water pump   =   2,600 VA
Disposal   =   1,200 VA
Compactor   =   1,400 VA
Dishwasher   =   1,600 VA
Microwave =   1,200   VA
Total load =   8,000   VA
8000 VA x 75%   =   6,000 VA √

Column 2
Dryer load and demand load 4

(Use 70% of dryer load)
220.61
5000 VA x 70%   =   3,500 VA √

Column 4
Largest motor load

(Use 25% of largest motor load)
6000 VA x 25%   =   1500 VA √

Calculating grounded (neutral) conductor

General lighting load   =   6,675 VA √
Cooking equipment load   =   0 VA √
Dryer load   =   3,500 VA √
Fixed appliance load   =   6,000 VA √
Largest motor load   =   650 VA √
Five total loads   =   16,825 VA

Finding amps for grounded (neutral) conductor

I = VA ÷ V
I = 16,825 VA ÷ 240 V
I = 70 A

Finding conductor size

Table 310.16 and 310.15(B)(6)
Ungrounded (phase) conductors - Phases A and B
1 AWG THWN copper
310.15(B)(6) allows 2 AWG THWN copper
Grounded (neutral) conductor
4 AWG THWN copper

Residential Calculations - Single-Family Dwellings

Design Exercise Problem 9-1. What is the load in VA and amps for a residential dwelling unit with the following loads?

Given loads:

General lighting and receptacle load
- 2800 sq. ft. dwelling unit
- 2 small appliance circuits
- 1 laundry circuit

120 V, single-phase loads
- 1400 VA microwave
- 1200 VA disposal
- 1400 VA compactor
- 1600 VA dishwasher

240 V, single-phase loads
- 5000 VA dryer
- 6000 VA A/C unit
- 20,000 VA heating unit (gas)
- 6000 VA water heater
- 10,000 VA oven (gas)
- 9000 VA cooktop (gas)
- 1200 VA blower motor
- 1000 VA water pump

Sizing ungrounded (phase) conductors = •
Sizing grounded (neutral) conductor = √

Column 1
Calculating general lighting and receptacle load

Step 1: General lighting and receptacle load
Table 220.12
2800 sq. ft. x 3 VA = _____ VA **[A]**

Step 2: Small appliance and laundry load
220.52(A); (B)
1500 VA x 2 = _____ VA **[B]**
1500 VA x 1 = _____ VA **[C]**
Total load = _____ VA **[D]**

Step 3: Applying demand factors
Demand load 1; Table 220.42
First 3000 VA x 100% = 3,000 VA
Next 9900 VA x 35% = _____ VA **[E]**
Total load = _____ VA **[F]** •√

Column 2
Calculating cooking equipment load

Step 1: Applying demand factors for phases
Demand load 2; Table 220.55, Note 1
= _____ VA **[G]** •

Step 2: Applying demand factors for neutral
220.61
= _____ VA **[H]** √

Column 2
Calculating dryer load

Step 1: Applying demand factors for phases
Demand load 3; Table 220.54
5000 VA x 100% = _____ VA **[I]** •

Step 2: Applying demand factors for neutral
220.61
5000 VA x 70% = _____ VA **[J]** √

Column 2
Calculating fixed appliance load

Step 1: Applying demand factors for phases
Demand load 4; 220.53
1400 VA x 75% = _____ VA **[K]**
1200 VA x 75% = _____ VA **[L]**
1400 VA x 75% = _____ VA **[M]**
1600 VA x 75% = _____ VA **[N]**
1200 VA x 75% = _____ VA **[O]**
1000 VA x 75% = _____ VA **[P]**
6000 VA x 75% = _____ VA **[Q]**
Total load = _____ VA **[R]** •

Step 2: Applying demand factors for neutral
Demand load 4; 220.61
1400 VA x 75% = _____ VA **[S]**
1200 VA x 75% = _____ VA **[T]**
1400 VA x 75% = _____ VA **[U]**
1600 VA x 75% = _____ VA **[V]**
Total load = _____ VA **[W]** √

Column 3
Largest load between heating and A/C load

Step 1: Selecting largest load
Demand load 5; 220.60
A/C unit
6000 VA x 100% = _____ VA **[X]** •

Column 4
Calculating largest motor load

Step 1: Selecting largest motor load for phases
220.14(C); 430.24
6000 VA x 25% = _____ VA **[Y]** •

Step 2: Selecting largest motor load for neutral
220.14(C); 430.24
1400 VA x 25% = _____ VA **[Z]** √

Calculating ungrounded (phase) conductors

- General lighting load = _____ VA **[AA]** •
- Cooking equipment load = _____ VA **[BB]** •
- Dryer load = _____ VA **[CC]** •
- Appliance load = _____ VA **[DD]** •
- A/C load = _____ VA **[EE]** •
- Largest motor load = _____ VA **[FF]** •
- Six total loads = _____ VA **[GG]**

Calculating grounded (neutral) conductor

- General lighting load = _____ VA **[HH]** √
- Cooking equipment load = _____ VA **[II]** √
- Dryer load = _____ VA **[JJ]** √
- Appliance load = _____ VA **[KK]** √
- Largest motor load = _____ VA **[LL]** √
- Five total loads = _____ VA **[MM]**

Finding amps for ungrounded (phase) conductors - Phases A and B

I = VA ÷ V
I = _____ VA **[NN]** ÷ 240 V
I = _____ A **[OO]**

Finding amps for grounded (neutral) conductor

I = VA ÷ V
I = _____ VA **[PP]** ÷ 240 V
I = _____ A **[QQ]**

Finding conductor size

Table 310.16 and 310.15(B)(6)

Ungrounded (phase) conductors - Phases A and B
_____ AWG THWN copper **[RR]**
310.15(B)(6) allows _____ AWG THWN copper **[SS]**
Grounded (neutral) conductor
_____ AWG THWN copper **[TT]**

Design Example Problem 9-2. What is the load in VA and amps for a residential dwelling unit with the following loads?

Given Loads:	120 V, single-phase loads	240 V, single-phase loads
General lighting and receptacle load • 3000 sq. ft. dwelling unit • 2 small appliance circuits • 1 laundry circuit	• 2600 VA water pump • 1200 VA disposal • 1400 VA compactor • 1600 VA dishwasher • 1200 VA microwave	• 6000 VA A/C unit • 10,000 VA heating unit • 6000 VA water heater • 12,000 VA oven • 16,000 VA cooktop • 5000 VA dryer • 800 VA blower motor

Sizing ungrounded (phase) conductors = •
Sizing grounded (neutral) conductor = √

Column 1
Calculating general lighting and receptacle load

Step 1: General lighting and receptacle load
Table 220.12
3000 sq. ft. x 3 VA = 9,000 VA

Step 2: Small appliance and laundry load
220.52(A); (B)
1500 VA x 2 = 3,000 VA
1500 VA x 1 = 1,500 VA
Total load = 4,500 VA

Step 3: Applying demand factors
Table 220.42
General lighting load = 9,000 VA
Small appliance load = 3,000 VA
Laundry load = 1,500 VA
Total load = 13,500 VA

First 3,000 VA x 100% = 3,000 VA
Next 10,500 VA x 35% = 3,675 VA
Total load = 6,675 VA • √

Column 2
Calculating cooking equipment load

Step 1: Applying demand factors
Table 220.55, Note 2
Total kW rating
12 kW + 16 kW = 28 kW
28 kW - 12 kW = 16 kW
16 kW x 5% = 80% (round up)
Table 220.55, Col. C
Total kVA rating
11 kW x 180% x 1,000 = 19,800 VA •

Calculating fixed appliance load

Step 1: Applying demand factors
220.53
Water heater = 6,000 VA
Water pump = 2,600 VA √
Disposal = 1,200 VA √
Compactor = 1,400 VA √
Dishwasher = 1,600 VA √
Microwave = 1,200 VA √
Blower motor = 800 VA
Total load = 14,800 VA
14,800 VA x 75% = 11,100 VA •

Calculating dryer load

Step 1: Applying demand factors
220.54; Table 220.54
5000 VA x 100% = 5,000 VA •

Column 3
Largest load between heating or A/C

Step 1: Selecting the largest load
220.60
Heating load
10,000 VA x 100% = 10,000 VA •

Column 4
Calculating largest motor load

Step 1: Selecting largest motor load
220.14(C); 430.24
Water pump
2600 VA x 25% = 650 VA •

Calculating ungrounded (phase) conductors

General lighting load = 6,675 VA •
Cooking equipment = 19,800 VA •
Fixed appliance load = 11,100 VA •
Dryer load = 5,000 VA •
Heating load = 10,000 VA •
Largest motor load = 650 VA •
Six total loads = 53,225 VA

Finding amps for ungrounded (phase) conductors - Phases A and B

I = VA ÷ V
I = 53,225 VA ÷ 240 V
I = 222 A

Grounded (neutral) conductor

Column 1
General lighting load and demand load 1

220.61
6675 VA √ = 6,675 VA √

Column 2
Cooking equipment load and demand load 2

(Use 70% of cooking load)
220.61
19,800 VA x 70% = 13,860 VA √

Column 2
Fixed appliance load and demand load 3

(Use 75% of 120 volt loads)
220.53
Water pump = 2,600 VA
Disposal = 1,200 VA
Compactor = 1,400 VA
Dishwasher = 1,600 VA
Microwave = 1,200 VA
Total load = 8,000 VA
8000 VA x 75% = 6,000 VA √

Column 2
Dryer load and demand load 4

(Use 70% of dryer load)
220.61
5000 VA x 70% = 3,500 VA √

Column 4
Largest motor load

(Use 25% of largest motor load)
2600 VA x 25% = 650 VA √

Calculating grounded (neutral) conductor

General lighting load = 6,675 VA √
Cooking equipment load = 13,860 VA √
Dryer load = 3,500 VA √
Fixed appliance load = 6,000 VA √
Largest motor load = 650 VA √
Five total loads = 30,685 VA

Finding amps for grounded (neutral) conductor

I = VA ÷ V
I = 30,685 VA ÷ 240 V
I = 128 A

Finding conductor size

Table 310.16 and 310.15(B)(6)
Ungrounded (phase) conductors - Phases A and B
4/0 AWG THWN copper
310.15(B)(6) allows 3/0 AWG THWN copper
Grounded (neutral) conductor
1 AWG THWN copper

Residential Calculations - Single-Family Dwellings

Design Exercise Problem 9-2. What is the load in VA and amps for a residential dwelling unit with the following loads?

Given loads:

General lighting and receptacle load

- 2800 sq. ft. dwelling unit
- 2 small appliance circuits
- 1 laundry circuit

120 V, single-phase loads

- 1400 VA microwave
- 1200 VA disposal
- 1400 VA compactor
- 1600 VA dishwasher

240 V, single-phase loads

- 5000 VA dryer
- 6000 VA A/C unit
- 20,000 VA heating unit
- 6000 VA water heater
- 10,000 VA oven
- 9000 VA cooktop
- 1200 VA blower motor
- 1000 VA water pump

| Sizing ungrounded (phase) conductors | = • |
| Sizing grounded (neutral) conductor | = √ |

Column 1
Calculating general lighting and receptacle load

- **Step 1:** General lighting and receptacle load
  Table 220.12
  2800 sq. ft. x 3 VA = _____ VA **[A]**

- **Step 2:** Small appliance and laundry load
  220.52(A); (B)
  1500 VA x 2 = _____ VA **[B]**
  1500 VA x 1 = _____ VA **[C]**
  Total load = _____ VA **[D]**

- **Step 3:** Applying demand factors
  Demand load 1; **Table 220.42**
  First 3000 VA x 100% = 3,000 VA
  Next 9900 VA x 35% = _____ VA **[E]**
  Total load = _____ VA **[F]** • √

Column 2
Calculating cooking equipment load

- **Step 1:** Applying demand factors for phases
  Demand load 2; **Table 220.55, Col. C**
  9 kW and 10 kW = _____ VA **[G]** •

- **Step 2:** Applying demand factors for neutral
  220.61
  11,000 VA x 70% = _____ VA **[H]** √

Column 2
Calculating dryer load

- **Step 1:** Applying demand factors for phases
  Demand load 3; **Table 220.54**
  5000 VA x 100% = _____ VA **[I]** •

- **Step 2:** Applying demand factors for neutral
  220.61
  5000 VA x 70% = _____ VA **[J]** √

Column 2
Calculating fixed appliance load

- **Step 1:** Applying demand factors for phases
  Demand load 4; **220.53**
  1400 VA x 75% = _____ VA **[K]**
  1200 VA x 75% = _____ VA **[L]**
  1400 VA x 75% = _____ VA **[M]**
  1600 VA x 75% = _____ VA **[N]**
  1200 VA x 75% = _____ VA **[O]**
  1000 VA x 75% = _____ VA **[P]**
  6000 VA x 75% = _____ VA **[Q]**
  Total load = _____ VA **[R]** •

- **Step 2:** Applying demand factors for neutral
  Demand load 4; **220.61**
  1400 VA x 75% = _____ VA **[S]**
  1200 VA x 75% = _____ VA **[T]**
  1400 VA x 75% = _____ VA **[U]**
  1600 VA x 75% = _____ VA **[V]**
  Total load = _____ VA **[W]** √

Column 3
Largest load between heating and A/C load

- **Step 1:** Selecting largest load
  Demand load 5; **220.60**
  Heating unit
  20,000 VA x 100% = _____ VA **[X]** •

Column 4
Calculating largest motor load

- **Step 1:** Selecting largest motor load for phases
  220.14(C); 430.24
  1400 VA x 25% = _____ VA **[Y]** •

- **Step 2:** Selecting largest motor load for neutral
  220.14(C); 430.24
  1400 VA x 25% = _____ VA **[Z]** √

Calculating ungrounded (phase) conductors

- General lighting load = _____ VA **[AA]** •
- Cooking equipment load = _____ VA **[BB]** •
- Dryer load = _____ VA **[CC]** •
- Appliance load = _____ VA **[DD]** •
- Heating load = _____ VA **[EE]** •
- Largest motor load = _____ VA **[FF]** •
- Six total loads = _____ VA **[GG]**

Calculating grounded (neutral) conductor

- General lighting load = _____ VA **[HH]** √
- Cooking equipment load = _____ VA **[II]** √
- Dryer load = _____ VA **[JJ]** √
- Appliance load = _____ VA **[KK]** √
- Largest motor load = _____ VA **[LL]** √
- Five total loads = _____ VA **[MM]**

Finding amps for ungrounded (phase) conducors - Phases A and B

I = VA ÷ V
I = _____ VA **[NN]** ÷ 240 V
I = _____ A **[OO]**

Finding amps for grounded (neutral) conductor

I = VA ÷ V
I = _____ VA **[PP]** ÷ 240 V
I = _____ A **[QQ]**

Finding conductor size

Table 310.16 and **310.15(B)(6)**

Ungrounded (phase) conductors - Phases A and B
_____ AWG THWN copper **[RR]**

310.15(B)(6) allows _____ AWG THWN copper **[SS]**

Grounded (neutral) conductor
_____ AWG THWN copper **[TT]**

Stallcup's Master Electrician's Study Guide

Design Example Problem 9-3. What is the load in VA and amps for a residential dwelling unit with the following loads?

Given Loads:	120 V, single-phase loads	240 V, single-phase loads
General lighting and receptacle load • 3000 sq. ft. dwelling unit • 2 small appliance circuits • 1 laundry circuit	• 2600 VA water pump • 1200 VA disposal • 1400 VA compactor • 1600 VA dishwasher • 1200 VA microwave	• 6000 VA A/C unit (heat pump) • 10,000 VA heating unit • 6000 VA water heater • 12,000 VA oven • 16,000 VA cooktop • 5000 VA dryer • 800 VA blower motor

Sizing ungrounded (phase) conductors = •
Sizing grounded (neutral) conductor = √

Column 1
Calculating general lighting and receptacle load

Step 1: General lighting and receptacle load
Table 220.12
3000 sq. ft. x 3 VA   =   9,000 VA

Step 2: Small appliance and laundry load
220.52(A); (B)
1500 VA x 2   =   3,000 VA
1500 VA x 1   =   1,500 VA
Total load   =   4,500 VA

Step 3: Applying demand factors
Table 220.42
General lighting load   =   9,000 VA
Small appliance load   =   3,000 VA
Laundry load   =   1,500 VA
Total load   =   13,500 VA

First 3000 VA x 100%   =   3,000 VA
Next 10,500 VA x 35%   =   3,675 VA
Total load   =   6,675 VA •√

Column 2
Calculating cooking equipment load

Step 1: Applying demand factors
Table 220.55, Note 2
Total kW rating
12 kW + 16 kW   =   28 kW
28 kW - 12 kW   =   16 kW
16 kW x 5%   =   80% (round up)
Table 220-55, Col. C
Total kVA rating
11 kW x 180% x 1,000   =   19,800 VA •

Calculating fixed appliance load

Step 1: Applying demand factors
220.53
Water heater   =   6,000 VA
Water pump   =   2,600 VA√
Disposal   =   1,200 VA√
Compactor   =   1,400 VA√
Dishwasher   =   1,600 VA√
Microwave   =   1,200 VA√
Blower motor   =   800 VA
Total load   =   14,800 VA
14,800 VA x 75%   =   11,100 VA •

Calculating dryer load

Step 1: Applying demand factors
220.54; Table 220.54
5000 VA x 100%   =   5,000 VA •

Column 3
Largest load between heating or A/C

Step 1: Selecting the largest load
220.60
Heating load
16,000 VA x 100%   =   16,000 VA •

Column 4
Calculating largest motor load

Step 1: Selecting largest motor load
220.14(C); 430.24
Heat pump
6000 VA x 25%   =   1,500 VA •

Calculating ungrounded (phase) conductors

General lighting load   =   6,675 VA •
Cooking equipment   =   19,800 VA •
Fixed appliance load   =   11,100 VA •
Dryer load   =   5,000 VA •
Heating load   =   16,000 VA •
Largest motor load   =   1,500 VA •
Six total loads   =   60,075 VA

Finding amps for ungrounded (phase) conductors - Phases A and B

I = VA ÷ V
I = 60,075 VA ÷ 240 V
I = 250 A

Grounded (neutral) conductor

Column 1
General lighting load and demand load 1

220.61
6675 VA√   =   6,675 VA√

Column 2
Cooking equipment load and demand load 2

(Use 70% of cooking load)
220.61
19,800 VA x 70%   =   13,860 VA√

Column 2
Fixed appliance load and demand load 3

(Use 75% of 120 volt loads)
220.53
Water pump   =   2,600 VA
Disposal   =   1,200 VA
Compactor   =   1,400 VA
Dishwasher   =   1,600 VA
Microwave   =   1,200 VA
Total load   =   8,000 VA
8000 VA x 75%   =   6,000 VA√

Column 2
Dryer load and demand load 4

(Use 70% of dryer load)
220.61
5000 VA x 70%   =   3,500 VA√

Column 4
Largest motor load

(Use 25% of largest motor load)
2600 VA x 25%   =   650 VA√

Calculating grounded (neutral) conductor

General lighting load   =   6,675 VA√
Cooking equipment load   =   13,860 VA√
Dryer load   =   3,500 VA√
Fixed appliance load   =   6,000 VA√
Largest motor load   =   650 VA√
Five total loads   =   30,685 VA

Finding amps for grounded (neutral) conductor

I = VA ÷ V
I = 30,685 VA ÷ 240 V
I = 128 A

Finding conductor size

Table 310.16 and 310.15(B)(6)
Ungrounded (phase) conductors - Phases A and B
250 KCMIL copper
310.15(B)(6) allows 4/0 AWG THWN copper
Grounded (neutral) conductor
1 AWG THWN copper

Residential Calculations - Single-Family Dwellings

Design Exercise Problem 9-3. What is the load in VA and amps for a residential dwelling unit with the following loads?

| Sizing ungrounded (phase) conductors = • |
| Sizing grounded (neutral) conductor = √ |

Given loads:

General lighting and receptacle load
- 2800 sq. ft. dwelling unit
- 2 small appliance circuits
- 1 laundry circuit

120 V, single-phase loads
- 1400 VA microwave
- 1200 VA disposal
- 1400 VA compactor
- 1600 VA dishwasher

240 V, single-phase loads
- 5000 VA dryer
- 6000 VA A/C unit (heat pump)
- 20,000 VA heating unit
- 6000 VA water heater
- 10,000 VA oven
- 9000 VA cooktop
- 1200 VA blower motor
- 1000 VA water pump

Column 1
Calculating general lighting and receptacle load

Step 1: General lighting and receptacle load
Table 220.12
2800 sq. ft. x 3 VA = _____ VA **[A]**

Step 2: Small appliance and laundry load
220.52(A); (B)
1500 VA x 2 = _____ VA **[B]**
1500 VA x 1 = _____ VA **[C]**
Total load = _____ VA **[D]**

Step 3: Applying demand factors
Demand load 1; **Table 220.42**
First 3000 VA x 100% = 3,000 VA
Next 9900 VA x 35% = _____ VA **[E]**
Total load = _____ VA **[F]** •√

Column 2
Calculating cooking equipment load

Step 1: Applying demand factors for phases
Demand load 2; **Table 220.55, Col. C**
9 kW and 10 kW = _____ VA **[G]** •

Step 2: Applying demand factors for neutral
220.61
11,000 VA x 70% = _____ VA **[H]** √

Column 2
Calculating dryer load

Step 1: Applying demand factors for phases
Demand load 3; **Table 220.54**
5000 VA x 100% = _____ VA **[I]** •

Step 2: Applying demand factors for neutral
220.61
5000 VA x 70% = _____ VA **[J]** √

Column 2
Calculating fixed appliance load

Step 1: Applying demand factors for phases
Demand load 4; **220.53**
1400 VA x 75% = _____ VA **[K]**
1200 VA x 75% = _____ VA **[L]**
1400 VA x 75% = _____ VA **[M]**
1600 VA x 75% = _____ VA **[N]**
1200 VA x 75% = _____ VA **[O]**
1000 VA x 75% = _____ VA **[P]**
6000 VA x 75% = _____ VA **[Q]**
Total load = _____ VA **[R]** •

Step 2: Applying demand factors for neutral
Demand load 4; **220.61**
1400 VA x 75% = _____ VA **[S]**
1200 VA x 75% = _____ VA **[T]**
1400 VA x 75% = _____ VA **[U]**
1600 VA x 75% = _____ VA **[V]**
Total load = _____ VA **[W]** √

Column 3
Largest load between heating and A/C load

Step 1: Selecting largest load
Demand load 5; **220.60**
Heating unit
26,000 V x 100% = _____ VA **[X]** •

Column 4
Calculating largest motor load

Step 1: Selecting largest motor load for phases
220.14(C); 430.24
6000 VA x 25% = _____ VA **[Y]** •

Step 2: Selecting largest motor load for neutral
220.14(C); 430.24
1400 VA x 25% = _____ VA **[Z]** √

Calculating ungrounded (phase) conductors

- General lighting load = _____ VA **[AA]** •
- Cooking equipment load = _____ VA **[BB]** •
- Dryer load = _____ VA **[CC]** •
- Appliance load = _____ VA **[DD]** •
- Heating load = _____ VA **[EE]** •
- Largest motor load = _____ VA **[FF]** •
- Six total loads = _____ VA **[GG]**

Calculating grounded (neutral) conductor

- General lighting load = _____ VA **[HH]** √
- Cooking equipment load = _____ VA **[II]** √
- Dryer load = _____ VA **[JJ]** √
- Appliance load = _____ VA **[KK]** √
- Largest motor load = _____ VA **[LL]** √
- Five total loads = _____ VA **[MM]**

Finding amps for ungrounded (phase) conductors - Phases A and B

I = VA ÷ V
I = _____ VA **[NN]** ÷ 240 V
I = _____ A **[OO]**

Finding amps for grounded (neutral) conductor

I = VA ÷ V
I = _____ VA **[PP]** ÷ 240 V
I = _____ A **[QQ]**

Finding conductor size

Table 310.16 and 310.15(B)(6)

Ungrounded (phase) conductors - Phases A and B
_____ KCMIL THWN copper **[RR]**

310.15(B)(6) allows _____ KCMIL THWN copper **[SS]**

Grounded (neutral) conductor
_____ AWG THWN copper **[TT]**

Design Example Problem 9-4. What is the load in VA and amps for a residential dwelling unit with the following loads?

Given Loads:	120 V, single-phase loads	240 V, single-phase loads
General lighting and receptacle load • 3000 sq. ft. dwelling unit • 2 small appliance circuits • 1 laundry circuit	• 1200 VA disposal • 1400 VA compactor • 1600 VA dishwasher • 1200 VA microwave • 16 A A/C window unit	• 10 kW heating unit (downstairs) • 10 kW heating unit (upstairs) • 18 A A/C central unit (downstairs) • 18 A A/C central unit (upstairs) • 6000 VA water heater • 12,000 VA oven • 16,000 VA cooktop • 5000 VA dryer • 800 VA blower motor

Sizing ungrounded (phase) conductors = •
Sizing grounded (neutral) conductor = √

Column 1
Calculating general lighting and receptacle load

Step 1: General lighting and receptacle load
Table 220.12
3000 sq. ft. x 3 VA = 9,000 VA

Step 2: Small appliance and laundry load
220.52(A); (B)
1500 VA x 2 = 3,000 VA
1500 VA x 1 = 1,500 VA
Total load = 4,500 VA

Step 3: Applying demand factors
Table 220.42
General lighting load = 9,000 VA
Small appliance load = 3,000 VA
Laundry load = 1,500 VA
Total load = 13,500 VA

First 3000 VA x 100% = 3,000 VA
Next 10,500 VA x 35% = 3,675 VA
Total load = 6,675 VA • √

Column 2
Calculating cooking equipment load

Step 1: Applying demand factors
Table 220.55, Note 2
Total kW rating
12 kW + 16 kW = 28 kW
28 kW - 12 kW = 16 kW
16 kW x 5% = 80% (round up)
Table 220.55, Col. C
Total kVA rating
11 kW x 180% x 1,000 = 19,800 VA •

Calculating fixed appliance load

Step 1: Applying demand factors
220.53
Water heater = 6,000 VA
Water pump = 2,600 VA √
Disposal = 1,200 VA √
Compactor = 1,400 VA √
Dishwasher = 1,600 VA √
Microwave = 1,200 VA √
Blower motor = 800 VA
Total load = 14,800 VA
14,800 VA x 75% = 11,100 VA •

Calculating dryer load

Step 1: Applying demand factors
220.54; Table 220.54
5000 VA x 100% = 5,000 VA •

Column 3
Largest load between heating or A/C

Step 1: Selecting the largest load
220.21
Heating load
20,000 VA x 100% = 20,000 VA •

Column 4
Calculating largest motor load

Step 1: Selecting largest motor load
220.14(C); 220.50
A/C unit
1920 VA x 25% = 480 VA •

Calculating ungrounded (phase) conductors

General lighting load = 6,675 VA •
Cooking equipment = 19,800 VA •
Fixed appliance load = 11,100 VA •
Dryer load = 5,000 VA •
Heating load = 20,000 VA •
Largest motor load = 480 VA •
Six total loads = 63,055 VA

Finding amps for ungrounded (phase) conductors - Phases A and B

I = VA ÷ V
I = 63,055 VA ÷ 240 V
I = 263 A

Grounded (neutral) conductor

Column 1
General lighting load and demand load 1

220.61
6675 VA √ = 6,675 VA √

Column 2
Cooking equipment load and demand load 2

(Use 70% of cooking load)
220.61
19,800 VA x 70% = 13,860 VA √

Column 2
Fixed appliance load and demand load 3

(Use 75% of 120 volt loads)
220.53
Water pump = 2,600 VA
Disposal = 1,200 VA
Compactor = 1,400 VA
Dishwasher = 1,600 VA
Microwave = 1,200 VA
Total load = 8,000 VA
8000 VA x 75% = 6,000 VA √

Column 2
Dryer load and demand load 4

(Use 70% of dryer load)
220.61
5000 VA x 70% = 3,500 VA √

Column 4
Largest motor load

(Use 25% of largest motor load)
1920 VA x 25% = 480 VA √

Calculating grounded (neutral) conductor

General lighting load = 6,675 VA √
Cooking equipment load = 13,860 VA √
Dryer load = 3,500 VA √
Fixed appliance load = 6,000 VA √
Largest motor load = 480 VA √
Five total loads = 30,515 VA

Finding amps for grounded (neutral) conductor

I = VA ÷ V
I = 30,515 VA ÷ 240 V
I = 127 A

Finding conductor size

Table 310.16 and **310.15(B)(6)**
Ungrounded (phase) conductors - Phases A and B
300 THWN copper
310.15(B)(6) allows 250 KCMIL THWN copper
Grounded (neutral) conductor
1 AWG THWN copper

Residential Calculations - Single-Family Dwellings

Design Exercise Problem 9-4. What is the load in VA and amps for a residential dwelling unit with the following loads?

Sizing ungrounded (phase) conductors = •
Sizing grounded (neutral) conductor = √

Given loads:

General lighting and receptacle load
- 2800 sq. ft. dwelling unit
- 2 small appliance circuits
- 1 laundry circuit

120 V, single-phase loads
- 1400 VA microwave
- 1200 VA disposal
- 1400 VA compactor
- 1600 VA dishwasher
- 14 A A/C window unit

240 V, single-phase loads
- 5000 VA dryer
- 10 kW heating unit (downstairs)
- 10 kW heating unit (upstairs)
- 16 A A/C central unit (downstairs)
- 16 A A/C central unit (upstairs)
- 6000 VA water heater
- 10,000 VA oven
- 9000 VA cooktop
- 1200 VA blower motor
- 1000 VA water pump

Column 1
Calculating general lighting and receptacle load

- **Step 1:** General lighting and receptacle load
  Table 220.12
  2800 sq. ft. x 3 VA = _____ VA [A]

- **Step 2:** Small appliance and laundry load
  220.52(A); (B)
  1500 VA x 2 = _____ VA [B]
  1500 VA x 1 = _____ VA [C]
  Total load = _____ VA [D]

- **Step 3:** Applying demand factors
  Demand load 1; Table 220.42
  First 3000 VA x 100% = 3,000 VA
  Next 9900 VA x 35% = _____ VA [E]
  Total load = _____ VA [F] • √

Column 2
Calculating cooking equipment load

- **Step 1:** Applying demand factors for phases
  Demand load 2; Table 220.55, Col. C
  9 kW and 10 kW = _____ VA [G] •

- **Step 2:** Applying demand factors for neutral
  220.61
  11,000 VA x 70% = _____ VA [H] √

Column 2
Calculating dryer load

- **Step 1:** Applying demand factors for phases
  Demand load 3; Table 220.54
  5000 VA x 100% = _____ VA [I] •

- **Step 2:** Applying demand factors for neutral
  220.61
  5000 VA x 70% = _____ VA [J] √

Column 2
Calculating fixed appliance load

- **Step 1:** Applying demand factors for phases
  Demand load 4; 220.53
  1400 VA x 75% = _____ VA [K]
  1200 VA x 75% = _____ VA [L]
  1400 VA x 75% = _____ VA [M]
  1600 VA x 75% = _____ VA [N]
  1200 VA x 75% = _____ VA [O]
  1000 VA x 75% = _____ VA [P]
  6000 VA x 75% = _____ VA [Q]
  Total load = _____ VA [R] •

- **Step 2:** Applying demand factors for neutral
  Demand load 4; 220.61
  1400 VA x 75% = _____ VA [S]
  1200 VA x 75% = _____ VA [T]
  1400 VA x 75% = _____ VA [U]
  1600 VA x 75% = _____ VA [V]
  Total load = _____ VA [W] √

Column 3
Largest load between heating and A/C load

- **Step 1:** Selecting largest load
  Demand load 5; 220.60
  Heating unit
  20,000 VA x 100% = _____ VA [X] •

Column 4
Calculating largest motor load

- **Step 1:** Selecting largest motor load for phases
  220.14(C); 430.24
  1680 VA x 25% = _____ VA [Y] •

- **Step 2:** Selecting largest motor load for neutral
  220.14(C); 430.24
  1680 VA x 25% = _____ VA [Z] √

Calculating ungrounded (phase) conductors

- General lighting load = _____ VA [AA] •
- Cooking equipment load = _____ VA [BB] •
- Dryer load = _____ VA [CC] •
- Appliance load = _____ VA [DD] •
- Heating load = _____ VA [EE] •
- Largest motor load = _____ VA [FF] •
- Six total loads = _____ VA [GG]

Calculating grounded (neutral) conductor

- General lighting load = _____ VA [HH] √
- Cooking equipment load = _____ VA [II] √
- Dryer load = _____ VA [JJ] √
- Appliance load = _____ VA [KK] √
- Largest motor load = _____ VA [LL] √
- Five total loads = _____ VA [MM]

Finding amps for ungrounded (phase) conductors - Phases A and B

I = VA ÷ V
I = _____ VA [NN] ÷ 240 V
I = _____ A [OO]

Finding amps for grounded (neutral) conductor

I = VA ÷ V
I = _____ VA [PP] ÷ 240 V
I = _____ A [QQ]

Finding conductor size

Table 310.16 and 310.15(B)(6)

Ungrounded (phase) conductors - Phases A and B
_____ AWG THWN copper [RR]

310.15(B)(6) allows _____ AWG THWN copper [SS]

Grounded (neutral) conductor
_____ AWG THWN copper [TT]

Design Example Problem 9-5. What is the load in VA for a residential dwelling unit with the following loads?

Given Loads:	120 V, single-phase loads	240 V, single-phase loads
General lighting and receptacle load • 3000 sq. ft. dwelling unit • 2 small appliance circuits • 1 laundry circuit	• 2600 VA water pump • 1200 VA disposal • 1400 VA compactor • 1600 VA dishwasher • 1200 VA microwave	• 6000 VA A/C unit • 10,000 VA heating unit • 6000 VA water heater • 12,000 VA oven • 16,000 VA cooktop • 5000 VA dryer • 800 VA blower motor

Column 1
Calculating general lighting and receptacle load

		Phase A	Phase B	Neutral

Step 1: General lighting and receptacle load
Table 220.12
3000 sq. ft. x 3 VA = 9,000 VA

Step 2: Small appliance and laundry load
220.52(A); (B)
1500 VA x 2 = 3,000 VA
1500 VA x 1 = 1,500 VA

Step 3: Applying demand factors
Table 220.42
General lighting load = 9,000 VA
Small appliance load = 3,000 VA
Laundry load = 1,500 VA
Total load = 13,500 VA

First 3000 VA x 100% = 3,000 VA
Next 10,500 VA x 35% = 3,675 VA
Total load = 6,675 VA •√   6,675 VA   6,675 VA   6,675 VA

Column 2
Calculating cooking equipment load

Step 1: Applying demand factors
Table 220.55, Note 2; 220.61
Total kW rating
12 kW + 16 kW = 28 kW
28 kW - 12 kW = 16 kW
16 kW x 5% = 80% (round up)
Table 220.55, Col. C
Total kVA rating
11 kVA x 180% x 1,000 = 19,800 VA •   19,800 VA   19,800 VA
19,800 VA x 70% = 13,860 VA √   13,860 VA

Calculating fixed appliance load

Step 1: Applying demand factors
220.53
			Phase A	Phase B	Neutral
Water heater	6,000 VA x 75%	= 4,500 VA •	4,500 VA	4,500 VA	
Water pump	2,600 VA x 75%	= 1,950 VA√	1,950 VA		1,950 VA
Disposal	1,200 VA x 75%	= 900 VA √		900 VA	900 VA
Compactor	1,400 VA x 75%	= 1,050 VA √		1,050 VA	1,050 VA
Dishwasher	1,600 VA x 75%	= 1,200 VA √	1,200 VA		1,200 VA
Microwave	1,200 VA x 75%	= 900 VA √		900 VA	900 VA
Blower motor	800 VA x 75%	= 600 VA •	600 VA	600 VA	

Calculating dryer load

Step 1: Applying demand factors
220.54; Table 220.54; 220.61
5000 VA x 100% = 5,000 VA •   5,000 VA   5,000 VA
5000 VA x 70% = 3,500 VA √   3,500 VA

Column 3
Largest load between heating or A/C

Step 1: Selecting the largest load
220.60
Heating load
10,000 VA x 100% = 10,000 VA •   10,000 VA   10,000 VA

Column 4
Calculating largest motor load

Step 1: Selecting largest motor load
220.14(C); 220.50; 430.24
Water pump
2600 VA x 25% = 650 VA √   650 VA   650 VA
Total VA   49,725 VA   50,075 VA   30,685 VA

Residential Calculations - Single-Family Dwellings

Design Exercise Problem 9-5. What is the load in VA for a residential dwelling unit with the following loads?

Given Loads:

General lighting and receptacle load
- 2800 sq. ft. dwelling unit
- 2 small appliance circuits
- 1 laundry circuit

120 V, single-phase loads

- 1400 VA microwave
- 1200 VA disposal
- 1400 VA compactor
- 1600 VA dishwasher

240 V, single-phase loads

- 6000 VA A/C unit
- 20,000 VA heating unit
- 6000 VA water heater
- 10,000 VA oven
- 9000 VA cooktop
- 5000 VA dryer
- 1000 VA blower motor
- 1000 VA water pump

Column 1
Calculating general lighting and receptacle load

		Phase A	Phase B	Neutral

Step 1: General lighting and receptacle load
Table 220.12
_____ [A] sq. ft. x _____ [B] VA = _____ [C] VA

Step 2: Small appliance and laundry load
220.52(A); (B)
_____ [D] VA x 2 = _____ [E] VA
_____ [F] VA x 1 = _____ [G] VA

Step 3: Applying demand factors
Table 220.42
General lighting load = _____ [H] VA
Small appliance load = _____ [I] VA
Laundry load = _____ [J] VA
Total load = _____ [K] VA

First _____ [L] VA x 100% = _____ [M] VA
Next _____ [N] VA x _____ [O] % = _____ [P] VA
Total load = _____ [Q] VA •√ _____ [R] VA _____ [S] VA _____ [T] VA

Column 2
Calculating cooking equipment load

Step 1: Applying demand factors
9 KW and 10 kW
_____ [X] VA x _____ [Y] % = _____ [U] VA _____ [V] VA _____ [W] VA
= _____ [Z] VA _____ [AA] VA

Calculating fixed appliance load

Step 1: Applying demand factors
1,400 VA x _____ [BB] % = _____ [CC] VA _____ [DD] VA _____ [EE] VA
1,200 VA x _____ [FF] % = _____ [GG] VA _____ [HH] VA _____ [II] VA
1,400 VA x _____ [JJ] % = _____ [KK] VA _____ [LL] VA _____ [MM] VA
1,600 VA x _____ [NN] % = _____ [OO] VA _____ [PP] VA _____ [QQ] VA
1,200 VA x _____ [RR] % = _____ [SS] VA _____ [TT] VA _____ [UU] VA
1,000 VA x _____ [VV] % = _____ [WW] VA _____ [XX] VA _____ [YY] VA
6,000 VA x _____ [ZZ] % = _____ [AAA] VA _____ [BBB] VA _____ [CCC] VA

Calculating dryer load

Step 1: Applying demand factors
_____ [DDD] VA x 100% = _____ [EEE] VA • _____ [FFF] VA _____ [GGG] VA
_____ [HHH] VA ÷ _____ [III] % = _____ [JJJ] VA√ _____ [KKK] VA

Column 3
Largest load between heating or A/C

Step 1: Selecting the largest load
Heating load
_____ [LLL] VA x 100% = _____ [MMM] VA _____ [NNN] VA _____ [OOO] VA

Column 4
Calculating largest motor load

Step 1: Selecting largest motor load
_____ [PPP] VA x _____ [QQQ] % = _____ [RRR] VA•√ _____ [SSS] VA _____ [TTT] VA
Total VA = _____ [UUU] VA _____ [VVV] VA _____ [WWW] VA

APPLYING STANDARD CALCULATION DIVIDING AMPS ON LINE AND NEUTRAL

When applying the standard calculation for dwelling units, the load in amps should be balanced when determining the ungrounded (phase) conductors and grounded (neutral) conductor. **(See Design Problem 9-6)**

(1) What size service-entrance conductors are required for the service?

    Step 1: Finding conductor amperage
    Largest phase (line) conductor
    208.65 A is larger than 207.19 A

    Step 2: Selecting conductor size
    Table 310.16; 75°C Column; **110.14(C)(1)(b)**
    4/0 AWG cu. = 230 A
    208.65 A is less than 230 A

Solution: The size THWN conductor required is 4/0 AWG copper.

(2) What size overcurrent protection device is required in the panelboard for the service-entrance conductors?

    Step 1: Finding conductor size
    Table 310.16; 75°C Column; **110.14(C)(1)(b)**
    4/0 AWG cu. = 230 A

    Step 2: Selecting OCPD size
    240.4(B); 240.6(A)
    230 A requires 250 A OCPD

Solution: The size OCPD required is 250 amps.

(3) What size copper grounding electrode conductor is required to ground the service equipment to a metal water pipe?

    Step 1: Sizing GEC to metal water pipe
    250.52(A)(1); Table 250.66
    4/0 AWG THWN cu. requires 2 AWG cu.

Solution: The size grounding electrode conductor is 2 AWG copper.

APPLYING THE OPTIONAL CALCULATION 220.82

The optional calculation for dwelling units provides a easier method for calculating the load per **220.82(B) and (C)**. The loads are separated into two columns. The first column of loads consists of all loads except heating and A/C loads, which are described as general loads. Heating and A/C loads are the second column of loads. The elements of the service shall be determined by using the total loads of these two columns.

The optional calculation has percentages applied that are derived from the demand factors. The total kVA of a dwelling unit shall be used in determining the demand factors to be applied. The percentages in **220.82(B) and (C)** shall be used to calculate the load in dwelling units. Separate units located in a multifamily dwelling shall be permitted to have the optional calculation applied per **220.82(B)** and **(C)**. The optional calculation method shall only be applied where an ampacity of at least 100 amps is applied to the service. This method shall be permitted be applied to newly constructed or older existing dwelling units.

GENERAL LOADS
220.82(B)

General loads are the first columns of loads to be calculated. General lighting and general-purpose receptacle loads shall be computed at 3 VA per sq. ft. per **220.82(B)(1)**. Small appliance loads shall calculated at 1500 VA per **220.82(B)(2)** and **220.52(A)** and **(B)**. These loads are then applied to the other loads in the column.

Heating and A/C loads shall not be applied to the general loads in the column. Heating and A/C loads shall not be applied to general loads per **220.82(B)(3)**. Special appliance loads shall be added to general loads per **220.82(B)(3)**. Section **220.82(B)(4)** includes motor loads in the general loads column.

Section **220.50** shall not require the largest motor load at 25 or 125 percent to be added to general loads. The nameplate ratings in the general columns of loads shall be calculated at 100 percent. The first 10,000 VA of the general loads shall be calculated at 100 percent of the demand factor, and the remaining VA shall be calculated at 40 percent of the demand factor allowed per **220.82(B)**. **(See Design Problem 9-7)**

HEAT OR A/C LOADS
TABLE 220.82(C)(1) thru (C)(4)

The second column of loads consist of heating, A/C, or heat pumps. The loads are determined by the following steps:

(1) Three or fewer units shall be calculated at 65 percent of the total kW rating of the heating load.

(2) Four or more units shall be calculated at 40 percent of the total kW rating of the heating load.

(3) The A/C load shall be calculated at 100 percent of its kVA rating.

(4) The total VA rating of the heating unit shall be compared to the A/C load, and the smaller load is dropped per **220.82(C)**.

(5) Heat pumps that operate with the heating unit shall be calculated at 100 percent and added to the heating load. **(See Design Problem 9-8(a) and (b))**

Design Example Problem 9-6. What is the load in amps for a residential dwelling unit with the following loads?

Given Loads:	120 V, single-phase loads	240 V, single-phase loads
General lighting and receptacle load • 3000 sq. ft. dwelling unit • 2 small appliance circuits • 1 laundry circuit **Note:** Amperage is evenly balanced by 240 volts.	• 2600 VA water pump • 1200 VA disposal • 1400 VA compactor • 1600 VA dishwasher • 1200 VA microwave	• 6000 VA A/C unit • 10,000 VA heating unit • 6000 VA water heater • 12,000 VA oven • 16,000 VA cooktop • 5000 VA dryer • 800 VA blower motor

Column 1
Calculating general lighting and receptacle load

		Phase A	Phase B	Neutral

Step 1: General lighting and receptacle load
Table 220.12
3000 sq. ft. x 3 VA = 9,000 VA

Step 2: Small appliance and laundry load
220.52(A); (B)
1500 VA x 2 = 3,000 VA
1500 VA x 1 = 1,500 VA

Step 3: Applying demand factors
Table 220.42
General lighting load = 9,000 VA
Small appliance load = 3,000 VA
Laundry load = 1,500 VA
Total load = 13,500 VA

First 3000 VA x 100% = 3,000 VA
Next 10,500 VA x 35% = 3,675 VA
Total load = 6,675 VA •√
6675 VA ÷ 240 V = 27.81 A √   27.81 A   27.81 A   27.81 A

Column 2
Calculating cooking equipment load

Step 1: Applying demand factors
Table 220.55, Note 2; 220.61
Total kW rating
12 kW + 16 kW = 28 kW
28 kW - 12 kW = 16 kW
16 kW x 5% = 80% (round up)
Table 220.19, Col. C
Total kVA rating
11 kVA x 180% x 1,000 = 19,800 VA •
19,800 VA ÷ 240 V = 82.50 A •   82.50 A   82.50 A
13,860 VA ÷ 240 V = 57.75 A √   57.75 A

Calculating fixed appliance load

Step 1: Applying demand factors
220.53
Water heater   6000 VA x 75% ÷ 240 V =   18.75 A •   18.75 A   18.75 A
Water pump   2600 VA x 75% ÷ 240 V =   8.13 A √   8.13 A   8.13 A
Disposal   1200 VA x 75% ÷ 240 V =   3.75 A √   3.75 A   3.75 A
Compactor   1400 VA x 75% ÷ 240 V =   4.38 A √   4.38 A   4.38 A
Dishwasher   1600 VA x 75% ÷ 240 V =   5.00 A √   5.00 A   5.00 A
Microwave   1200 VA x 75% ÷ 240 V =   3.75 A √   3.75 A   3.75 A
Blower motor   800 VA x 75% ÷ 240 V =   2.50 A •   2.50 A   2.50 A

Calculating dryer load

Step 1: Applying demand factors
220.54; Table 220.54; 220.61
5000 VA x 100% = 5,000 VA •
5000 VA ÷ 240 V = 20.83 A   20.83 A   20.83 A
5000 VA x 70% = 3,500 VA √
3500 VA ÷ 240 V = 14.58 A √   14.58 A

Column 3
Largest load between heating or A/C

Step 1: Selecting the largest load
220.51
Heating load
10,000 VA x 100% = 10,000 VA •
10,000 VA ÷ 240 V = 41.67 A •   41.67 A   41.67 A

Column 4
Calculating largest motor load

Step 1: Selecting largest motor load
220.50; 430.24
Water pump
2600 VA x 25% = 650 VA √
650 VA ÷ 240 V = 2.71 A √   2.71 A   2.71 A
Total VA   207.19 A   208.65 A   127.86 A

Residential Calculations - Single-Family Dwellings

Design Exercise Problem 9-6. What is the load in amps for a residential dwelling unit with the following loads?

Given Loads:

General lighting and receptacle load
- 2800 sq. ft. dwelling unit
- 2 small appliance circuits
- 1 laundry circuit

120 V, single-phase loads
- 1400 VA microwave
- 1200 VA disposal
- 1400 VA compactor
- 1600 VA dishwasher

240 V, single-phase loads
- 6000 VA A/C unit
- 20,000 VA heating unit
- 6000 VA water heater
- 10,000 VA oven
- 9000 VA cooktop
- 5000 VA dryer
- 1200 VA blower motor
- 1000 VA water pump

Note: Amperage is evenly balanced by 240 volts.

Column 1
Calculating general lighting and receptacle load

                                                                                Phase A   Phase B   Neutral

Step 1: General lighting and receptacle load
    Table 220.12
    _____[A] sq. ft. x _____[B] VA = _____[C] VA

Step 2: Small appliance and laundry load
    220.52(A); (B)
    _____[D] VA x 2 = _____[E] VA
    _____[F] VA x 1 = _____[G] VA

Step 3: Applying demand factors
    Table 220.42
    General lighting load = _____[H] VA
    Small appliance load = _____[I] VA
    Laundry load = _____[J] VA
    Total load = _____[K] VA

    First _____[L] VA x 100% = _____[M] VA
    Next _____[N] VA x _____[O] % = _____[P] VA
    Total load = _____[Q] VA
    _____[R] VA ÷ _____[S] V = _____[T] A √   _____[U] A   _____[V] A   _____[W] A

Column 2
Calculating cooking equipment load

Step 1: Applying demand factors
    9 KW and 10 kW
    _____[Y] VA ÷ _____[Z] V = _____[X] VA
    = _____[AA] A • _____[BB] A • _____[CC] A
    _____[DD] VA x _____[EE] % = _____[FF] VA
    _____[GG] VA ÷ _____[HH] V = _____[II] A √   _____[JJ] A

Calculating fixed appliance load

Step 1: Applying demand factors
    1,400 VA x 75% ÷ _____[KK] V = _____[LL] A √   _____[MM] A   _____[NN] A
    1,200 VA x 75% ÷ _____[OO] V = _____[PP] A √   _____[QQ] A   _____[RR] A
    1,400 VA x 75% ÷ _____[SS] V = _____[TT] A √   _____[UU] A   _____[VV] A
    1,600 VA x 75% ÷ _____[WW] V = _____[XX] A √   _____[YY] A   _____[ZZ] A
    1,200 VA x 75% ÷ _____[AAA] V = _____[BBB] A • _____[CCC] A   _____[DDD] A
    1,000 VA x 75% ÷ _____[EEE] V = _____[FFF] A • _____[GGG] A   _____[HHH] A
    6,000 VA x 75% ÷ _____[III] V = _____[JJJ] A • _____[KKK] A   _____[LLL] A

Calculating dryer load

Step 1: Applying demand factors
    _____[MMM] VA x 100% = _____[NNN] VA
    _____[OOO] VA ÷ _____[PPP] V = _____[QQQ] A • _____[RRR] A   _____[SSS] A
    _____[TTT] VA x _____[UUU] % = _____[VVV] VA
    _____[WWW] VA ÷ _____[XXX] V = _____[YYY] A √   _____[ZZZ] A

Column 3
Largest load between heating or A/C

Step 1: Selecting the largest load
    Heating load
    _____[AAAA] VA x 100% = _____[BBBB] VA
    _____[CCCC] VA ÷ _____[DDDD] V = _____[EEEE] A • _____[FFFF] A   _____[GGGG] A

Column 4
Calculating largest motor load

Step 1: Selecting largest motor load
    _____[HHHH] VA x _____[IIII] % = _____[JJJJ] VA
    _____[KKKK] VA ÷ _____[LLLL] V = _____[MMMM] A √   _____[NNNN] A   _____[OOOO] A
    Total VA =   _____[PPPP] A   _____[QQQQ] A   _____[RRRR] A

Design Example Problem 9-7. What is the load in VA and amps for a residential dwelling unit with the following loads? (See standard calculation in Design Example Problem 9-2 for sizing the grounded (neutral) conductor)

Given Loads:

General lighting and receptacle load
- 3000 sq. ft. dwelling unit
- 2 small appliance circuits
- 1 laundry circuit

120 V, single-phase loads
- 2600 VA water pump
- 1200 VA disposal
- 1400 VA compactor
- 1600 VA dishwasher
- 1200 VA microwave

240 V, single-phase loads
- 6000 VA A/C unit
- 10,000 VA heating unit
- 6000 VA water heater
- 12,000 VA oven
- 16,000 VA cooktop
- 5000 VA dryer
- 800 VA blower motor

Sizing ungrounded (phase) conductors = •

Column 1
Other loads

Step 1: General lighting load
220.82(B)(1)
3000 sq. ft. x 3 VA = 9,000 VA

Step 2: Small appliance and laundry load
220.82(B)(2); 220.52(A); (B)
1500 VA x 2 = 3,000 VA
1500 VA x 1 = 1,500 VA

Step 3: Appliance load
220.82(B)(3); (B)(4)
Water heater = 6,000 VA
Oven = 12,000 VA
Cooktop = 16,000 VA
Water pump = 2,600 VA
Disposal = 1,200 VA
Compactor = 1,400 VA
Dishwasher = 1,600 VA
Microwave = 1,200 VA
Dryer = 5,000 VA
Blower motor = 800 VA
Total load = 61,300 VA

Step 4: Applying demand load
220.82(B)
First 10,000 VA x 100% = 10,000 VA
Next 51,300 VA x 40% = 20,520 VA
Total load = 30,520 VA •

Column 2
Largest load between heating and A/C load

Step 5: Selecting largest load
220.82(C)(1); (C)(2); (C)(4)
Heating load = 10,000 VA x 1 x 65% = 6,500 VA •
A/C load = 6000 VA x 1 x 100% = 6,000 VA
Total load = 6,500 VA •

Totaling Column 1 and 2
220.82(B); (C)

Col. 1 ld. = 30,520 VA •
Col. 2 ld. = 6,500 VA •
Total load = 37,020 VA

Finding amps for ungrounded (phase) conductors - Phases A and B

I = VA ÷ VA
I = 37,020 VA ÷ 240 V
I = 154 A

Finding maximum conductor size

Table 310.16
Ungrounded (phase) conductors - Phases A and B
2/0 AWG THWN copper
Grounded (neutral) conductor per **250.24(C)(1)**
4 AWG THWN copper

Note: Section **310.15(B)(6)** allows 1/0 AWG THWN copper conductors (min. size) to be used for the phases.

Residential Calculations - Single-Family Dwellings

Design Exercise Problem 9-7. What is the load in VA and amps for a residential dwelling unit with the following loads? (See Standard Calculation in Design Exercise Problem 9-2 for sizing the grounded (neutral) conductor)

Given loads:

General lighting and receptacle load

- 2800 sq. ft. dwelling unit
- 2 small appliance circuits
- 1 laundry circuit

120 V, single-phase loads

- 1400 VA microwave
- 1200 VA disposal
- 1400 VA compactor
- 1600 VA dishwasher

240 V, single-phase loads

- 5000 VA dryer
- 6000 VA A/C unit
- 20,000 VA heating unit
- 6000 VA water heater
- 10,000 VA oven
- 9000 VA cooktop
- 1200 VA blower motor
- 1000 VA water pump

Sizing ungrounded (phase) conductors = •

Column 1
Other loads

Step 1: General lighting load
Table 220.82(B)(1)
2800 sq. ft. x 3 VA = _____ VA [A]

Step 2: Small appliance and laundry load
220.82(B)(2); 220.52(A); (B)
1500 VA x 2 = _____ VA [B]
1500 VA x 1 = _____ VA [C]

Step 3: Appliance load
220.82(B)(3); (B)(4)
Cooktop load = _____ VA [D]
Oven load = _____ VA [E]
Dryer load = _____ VA [F]
Water heater load = _____ VA [G]
Disposal load = _____ VA [H]
Compactor load = _____ VA [I]
Dishwasher load = _____ VA [J]
Microwave load = _____ VA [K]
Blower motor load = _____ VA [L]
Water pump load = _____ VA [M]
Total load = _____ VA [N]

Step 4: Applying demand load
220.82(B)
First 10,000 VA x 100% = _____ VA [O]
Next 40,700 VA x 40% = _____ VA [P]
Total load = _____ VA [Q] •

Column 2
Largest load between heating and A/C load

Step 5: Selecting lagest load
220.82(C)(1); (C)(2); (C)(4)
Heating load = 20,000 VA x 1 x 65% = _____ VA [R] •
A/C load = 6000 VA x 1 x 100% = _____ VA [S]
Total load = _____ VA [T] •

Totaling Column 1 and 2
220.82(B); (C)
Col. 1 ld. = _____ VA [U] •
Col. 2 ld. = _____ VA [V] •
Total load = _____ VA [W]

Finding amps for ungrounded (phase) conductors - Phases A and B

I = VA ÷ V
I = _____ VA [X] ÷ 240 V
I = _____ A [Y]

Finding conductor size

Table 310.16
Ungrounded (phase) conductors - Phases A and B
_____ AWG THWN [Z] copper

Grounded (neutral) conductor per 250.24(C)(1)
_____ AWG THWN [AA] copper

Design Example Problem 9-8(a): Using the optional calculation method, what is the largest load between a 25 kW heating unit and a 42.5 amp A/C unit supplied by a 120 / 240 volt, single-phase system?

Step 1: Selecting largest load
220.82(C)(1) thru (C)(4)
Heating unit
25 kW x 1000 x 65% = 16,250 VA
A/C unit
42.5 A x 240 V x 100% = 10,200 VA

Solution: The heating load is 16,250 VA.

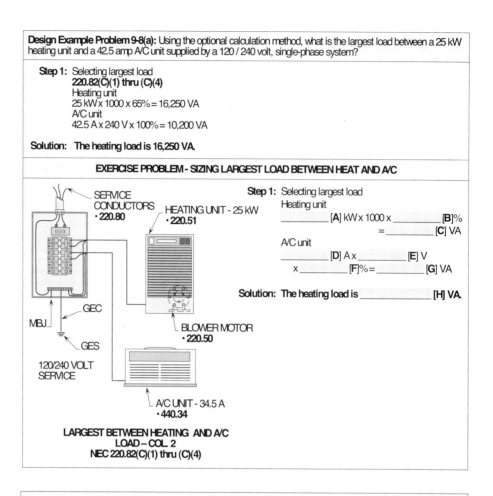

EXERCISE PROBLEM - SIZING LARGEST LOAD BETWEEN HEAT AND A/C

Step 1: Selecting largest load
Heating unit
_____ [A] kW x 1000 x _____ [B]%
= _____ [C] VA
A/C unit
_____ [D] A x _____ [E] V
x _____ [F]% = _____ [G] VA

Solution: The heating load is _____ [H] VA.

LARGEST BETWEEN HEATING AND A/C
LOAD – COL. 2
NEC 220.82(C)(1) thru (C)(4)

Design Example Problem 9-8(b): Using the optional calculation method, what is the largest load between 4 and 5000 VA floor heating units and a 42.5 amp A/C unit supplied by a 120 / 240 volt, single-phase system?

Step 1: Selecting largest load
220.82(C)(1) thru (C)(4)
Heating unit
4 - 5000 VA x 40% = 8000 VA
A/C unit
42.5 A x 240 V x 100% = 10,200 VA

Solution: The A/C load is 10,200 VA.

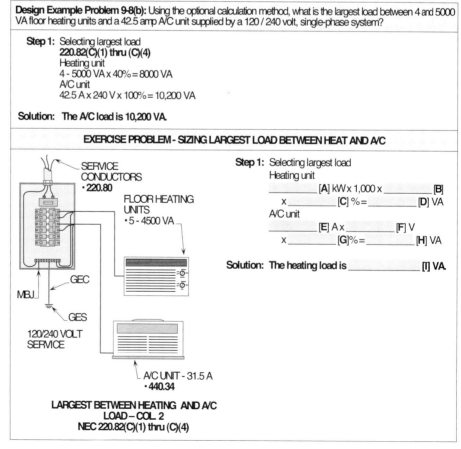

EXERCISE PROBLEM - SIZING LARGEST LOAD BETWEEN HEAT AND A/C

Step 1: Selecting largest load
Heating unit
_____ [A] kW x 1,000 x _____ [B]
x _____ [C] % = _____ [D] VA
A/C unit
_____ [E] A x _____ [F] V
x _____ [G]% = _____ [H] VA

Solution: The heating load is _____ [I] VA.

LARGEST BETWEEN HEATING AND A/C
LOAD – COL. 2
NEC 220.82(C)(1) thru (C)(4)

APPLYING THE OPTIONAL CALCULATION FOR EXISTING UNITS
220.83

Existing dwelling units per **220.83** shall be permitted to use the optional calculation to size the components for the service equipment to determine if additional loads can be added to the service. A 120/240 volt or 120/208 volt, three-wire, single-phase system shall be required to supply the service. Two columns are used in separating the loads in the dwelling unit. General lighting and general-purpose receptacle loads shall be calculated at 3 VA per sq. ft. per **220.83(A)(1)**. All other loads shall be calculated at 100 percent of their nameplate ratings. Heating and A/C loads shall be calculated at 100 percent, and the smaller load is dropped per **220.83(B)**.

OTHER LOADS
220.83

Existing loads are the first column to be selected. This column shall consist of adding 3 VA per sq. ft. to the small appliance loads, plus the nameplate ratings of each special appliance load, to determine the existing load in a dwelling unit. The demand factors in **220.83** shall be used in reducing the VA rating of the existing load, and this value is then added to the second column, which contains the appliance load to be added.

OTHER LOADS
ADDED APPLIANCE LOAD
220.83

Added appliance loads are the second column to be applied. This column shall be determined by adding the total VA rating of the load at 100 percent. The existing loads (after load) shall be added to the added load to determine the total VA rating of the dwelling unit. The added load can consist of an added appliance load. **(See Design Problem 9-9)**

OTHER LOADS
ADDED A/C LOAD

Added A/C load are the second column to be applied. This column shall be determined by adding the total VA rating of the load at 100 percent. The existing loads shall be added to the added loads to determine the total VA rating of the dwelling unit. The added loads usually consist of a heating unit, an air-conditioner, a dryer, or some other type of special appliance load. **(See Design Problem 9-10)**

    (1) What size THWN copper conductors are required to supply power to the A/C unit?

        Step 1:   Finding amperage
                    Text
                    I = VA ÷ V
                    I = 5040 VA ÷ 240 V
                    I = 21 A

        Step 2:   Calculating amperage
                    Table 220.3; 440.32
                    21 A x 125% = 26.25 A

Step 3: Selecting conductor size
Table 310.16; 75°C Column
10 AWG THWN cu. = 35 A
26.25 A is less than 35 A

Solution: The size THWN conductors are required to be 10 AWG copper.

(2) What size overcurrent protection device is required for the branch-circuit to the A/C unit? (Use the maximum rating)

Step 1: Calculating amperage
Table 240.4(G); 440.22(A)
21 A x 225% = 47.25 A

Step 2: Selecting OCPD size
Table 240.4(G); 240.6(A)
47.25 A requires 45 A OCPD

Solution: The size overcurrent protection device required is 45 amps.

OTHER LOADS
ADDED 120 VOLT, A/C WINDOW UNITS

Added A/C window unit loads are the second column to be applied. This column is determined by adding the total VA rating of the load at 100 percent. The existing loads (after loads) are added to the added load to determine the total VA rating of the dwelling unit. The added load can consist of added 120 volt A/C window units. **(See Design Problem 9-11)**

FEEDER TO MOBILE HOME - STANDARD CALCULATION
550.18; 550.53

Service calculations for mobile homes are performed at the factory. However, it is the responsibility of the test taker to size and select the proper size feeder-circuit to supply power to the mobile home. Such feeder-circuit shall be a four-wire circuit with all conductors insulated. After applying the standard calculation, the size of the conductors shall be permitted to be selected from **Table 310.16** or **310.15(B)(6)**. The service equipment on a pole or pedestal shall be rated at least 100 amps. **(See Design Problem 9-12)**

Residential Calculations - Single-Family Dwellings

Design Example Problem 9-9. Can a 5500 VA water heater be added to the existing dwelling unit without upgrading the service elements? (service is rated 100 amps)

Given Loads:

General lighting and receptacle load
- 2000 sq. ft. dwelling unit
- 2 small appliance circuits
- 1 laundry circuit

120 V, single-phase loads
- 900 VA disposal
- 1200 VA compactor
- 1400 VA dishwasher

240 V, single-phase loads
- 5000 VA dryer
- 11,000 VA range
- 5500 VA water heater (added load)

Column 1
Calculating general lighting and receptacle load

Step 1: General lighting and receptacle load
220.83(B)
2000 sq. ft. x 3 VA = 6,000 VA

Step 2: Small appliance and laundry load
1500 VA x 2 = 3,000 VA
1500 VA x 1 = 1,500 VA

Step 3: Existing load
220.83(B)
Range load = 11,000 VA
Dryer load = 5,000 VA
Disposal load = 900 VA
Compactor load = 1,200 VA
Dishwasher load = 1,400 VA
Total load = 30,000 VA

Step 4: Applying demand factor
Table 220.83
First 8000 VA x 100% = 8,000 VA
Next 22,000 VA x 40% = 8,800 VA
Total load = 16,800 VA •

Column 2
Added load

Step 5: Calculating added load
220.83(B)
Fixed appliance load
5500 VA x 100% = 5,500 VA •

Totalling Columns 1 and 2
220.83(B)
Col. 1 load = 16,800 VA •
Col. 2 load = 5,500 VA •
Total load = 22,300 VA

Finding amps for ungrounded (phase) conductors - Phases A and B

I = VA ÷ V
I = 22,300 VA ÷ 240 V
I = 93 A

Existing service = 100 A
New calculated = 93 A

93 A is less than 100 A
New water heater load can be added.

EXERCISE PROBLEM - ADDING APPLIANCE LOAD TO EXISTING DWELLING UNIT

Design Exercise Problem 9-9. Can a 240 volt, single-phase, 6000 VA water heater be added to the existing dwelling unit without upgrading the service elements? (service is rated 100 amps)

Given Loads:

General lighting and receptacle load
- 1800 sq. ft. dwelling unit
- 2 small appliance circuits
- 1 laundry circuit

120 V, single-phase loads
- 1200 VA disposal
- 1400 VA compactor
- 1600 VA dishwasher

240 V, single-phase loads
- 5000 VA dryer
- 12,000 VA range
- 6000 VA water heater (added load)

Column 1
Calculating general lighting and receptacle load

Step 1: General lighting and receptacle load
220.83(B)
____[A] sq. ft. x ____[B] VA = ____[C] VA

Step 2: Small appliance and laundry load
____[D] VA x 2 = ____[E] VA
____[F] VA x 1 = ____[G] VA

Step 3: Existing load
220.83(B)
Range load = ____[H] VA
Dryer load = ____[I] VA
Disposal load = ____[J] VA
Compactor load = ____[K] VA
Dishwasher load = ____[L] VA
Total load = ____[M] VA

Step 4: Applying demand factor
220.83(B)
First ____[N] VA x ____[O]% = ____[P] VA
Next ____[Q] VA x ____[R]% = ____[S] VA
Total load = ____[T] VA •

Column 2
Added load

Step 5: Calculating added load
220.83(B)
Fixed appliance load
____[U] VA x 100% = ____[V] VA •

Totalling Columns 1 and 2
220.83(B)
Col. 1 load = ____[W] VA •
Col. 2 load = ____[X] VA •
Total load = ____[Y] VA

Finding amps for ungrounded (phase) conductors - Phases A and B

I = VA ÷ V
I = ____[Z] VA ÷ ____[AA] V
I = ____[BB] A

Existing service = ____[CC] A
New calculated = ____[DD] A

97 A is less than 100 A
New water heater load can be added.

EXERCISE PROBLEM - ADDING A/C LOAD TO EXISTING DWELLING UNIT

Design Exercise Problem 9-10. Can a 5450 VA A/C load be added to the existing dwelling unit without upgrading the service elements? (service is rated 100 amps)

Given Loads:

General lighting and receptacle load
- 1800 sq. ft. dwelling unit
- 2 small appliance circuits
- 1 laundry circuit

120 V, single-phase loads
- 1200 VA disposal
- 1400 VA compactor
- 1600 VA dishwasher

240 V, single-phase loads
- 5000 VA dryer
- 12,000 VA range
- 5450 VA A/C unit (added load)

Column 1
Calculating general lighting and receptacle load

Step 1: General lighting and receptacle load
220.83(B)
   [A] 1800 sq. ft. x **[B]** 3 VA = **[C]** 5,400 VA

Step 2: Small appliance and laundry load
   [D] 1500 VA x 2 = **[E]** 3,000 VA
   [F] 1500 VA x 1 = **[G]** 1,500 VA

Step 3: Existing load
220.83(B)
   Range load   = **[H]** 12,000 VA
   Dryer load   = **[I]** 5,000 VA
   Disposal load   = **[J]** 1,200 VA
   Compactor load   = **[K]** 1,400 VA
   Dishwasher load = **[L]** 1,600 VA
   Total load   = **[M]** 31,100 VA

Step 4: Applying demand factor
220.83(B)
   First **[N]** 8,000 VA x **[O]** 100% = **[P]** 8,000 VA
   Next **[Q]** 23,100 VA x **[R]** 40% = **[S]** 9,240 VA
   Total load = **[T]** 17,240 VA •

Column 2
Added load

Step 5: Calculating added load
220.83(B)
Fixed appliance load
   [U] 5,450 VA x 100% = **[V]** 5,450 VA •

Totalling Columns 1 and 2
220.83(B)
   Col. 1 load = **[W]** 17,240 VA •
   Col. 2 load = **[X]** 5,450 VA •
   Total load = **[Y]** 22,690 VA

Finding amps for ungrounded (phase) conductors - Phases A and B

I = VA ÷ V
I = **[Z]** 22,690 VA ÷ **[AA]** 240 V
I = **[BB]** 95 A

Existing service   = **[CC]** 100 A
New calculated   = **[DD]** 95 A

95 A is less than 100 A
New A/C load can be added.

Residential Calculations - Single-Family Dwellings

Design Example Problem 9-11. Can a 240 volt, single-phase, 1680 VA A/C window unit be added to the existing dwelling unit without upgrading the service elements? (service is rated 100 amps)

Given Loads:

General lighting and receptacle load
- 2000 sq. ft. dwelling unit
- 2 small appliance circuits
- 1 laundry circuit

120 V, single-phase loads
- 900 VA disposal
- 1200 VA compactor
- 1400 VA dishwasher
- 1680 VA A/C window unit (added load)

240 V, single-phase loads
- 5000 VA dryer
- 11,000 VA range

Column 1
Calculating general lighting and receptacle load

Step 1: General lighting and receptacle load
220.83(B)
2000 sq. ft. x 3 VA = 6,000 VA

Step 2: Small appliance and laundry load
1500 VA x 2 = 3,000 VA
1500 VA x 1 = 1,500 VA

Step 3: Existing load
220.83(B)
Range load = 11,000 VA
Dryer load = 5,000 VA
Disposal load = 900 VA
Compactor load = 1,200 VA
Dishwasher load = 1,400 VA
Total load = 30,000 VA

Step 4: Applying demand factor
220.83(B)
First 8000 VA x 100% = 8,000 VA
Next 22,000 VA x 40% = 8,800 VA
Total load = 16,800 VA •

Column 2
Added load

Step 5: Calculating added load
220.83(B)
Fixed appliance load
1680 VA x 100% = 1,680 VA •

Totalling Columns 1 and 2
220.83(B)
Col. 1 load = 16,800 VA •
Col. 2 load = 1,680 VA •
Total load = 18,480 VA

Finding amps for ungrounded (phase) conductors - Phases A and B

I = VA ÷ V
I = 18,480 VA ÷ 240 V
I = 77 A
Existing service = 100 A
New calculated = 77 A

77 A is less than 100 A
New A/C load can be added.

EXERCISE PROBLEM - ADDING A/C WINDOW UNIT LOAD TO EXISTING DWELLING UNIT

Design Exercise Problem 9-11. Can a 2160 VA A/C window unit be added to the existing dwelling unit without upgrading the service elements? (service is rated 100 amps)

Given Loads:

General lighting and receptacle load
- 1800 sq. ft. dwelling unit
- 2 small appliance circuits
- 1 laundry circuit

120 V, single-phase loads
- 1200 VA disposal
- 1400 VA compactor
- 1600 VA dishwasher
- 1680 VA A/C window unit (added load)

240 V, single-phase loads
- 5000 VA dryer
- 12,000 VA range

Column 1
Calculating general lighting and receptacle load

Step 1: General lighting and receptacle load
220.83(B)
_____[A] sq. ft. x _____[B] VA = _____[C] VA

Step 2: Small appliance and laundry load
_____[D] VA x 2 = _____[E] VA
_____[F] VA x 1 = _____[G] VA

Step 3: Existing load
220.83(B)
Range load = _____[H] VA
Dryer load = _____[I] VA
Disposal load = _____[J] VA
Compactor load = _____[K] VA
Dishwasher load = _____[L] VA
Total load = _____[M] VA

Step 4: Applying demand factor
220.83(B)
First _____[N] VA x _____[O]% = _____[P] VA
Next _____[Q] VA x _____[R]% = _____[S] VA
Total load = _____[T] VA •

Column 2
Added load

Step 5: Calculating added load
Added A/C window unit
_____[U] VA x 100% = _____[V] VA •

Totalling Columns 1 and 2
220.83(B)
Col. 1 load = _____[W] VA •
Col. 2 load = _____[X] VA •
Total load = _____[Y] VA

Finding amps for ungrounded (phase) conductors - Phases A and B

I = VA ÷ V
I = _____[Z] VA ÷ _____[AA] V
I = _____[BB] A
Existing service = _____[CC] A
New calculated = _____[DD] A

79 A is less than 100 A
New A/C load can be added.

9-39

Design Example Problem 9-12. What is the load in VA and amps for a mobile home with the following loads?

Given Loads:
- 800 sq. ft. dwelling unit (mobile home)
- 2 small appliance circuits
- 1 laundry circuit
- 8,500 VA range — 240 volt, single-phase
- 6,000 VA water heater — 240 volt, single-phase
- 540 VA disposal — 120 volt, single-phase
- 800 VA dishwasher — 120 volt, single-phase
- 5,500 VA heating unit — 240 volt, single-phase

Sizing ungrounded (phase) conductors = •
Sizing grounded (neutral) conductor = √

Calculating general lighting and receptacle load

Step 1: General lighting and receptacle load
550.18(A)(1)
800 sq. ft. x 3 VA = 2,400 VA

Step 2: Small appliance and laundry load
550.18(A)(2); (A)(3)
1,500 VA x 2 = 3,000 VA
1,500 VA x 1 = 1,500 VA
Total load = 6,900 VA

Step 3: Applying demand factors
550.18(A)(5)
First 3,000 VA x 100% = 3,000 VA
Next 3,900 VA x 35% = 1,365 VA
Total load = 4,365 VA •√

Calculating special appliance loads

Step 1: Applying demand factors for phases
550.18(B)(2); (B)(3); (B)(4)
Water heater = 6,000 VA
Dishwasher = 800 VA √
Disposal = 540 VA √
Heating = 5,500 VA
Largest motor = 135 VA √
(540 VA x 25% = 135 VA)
Total load = 12,975 VA •

Calculating range load

Step 1: 550.18(B)(5); 220.61(B)(1)
8,500 VA x 80% = 6,800 VA •
6,800 VA x 70% = 4,760 VA √

Calculating ungrounded (phase) conductors

General lighting load = 4,365 VA •
Special appliance load = 12,975 VA •
Range load = 6,800 VA •
Three total loads = 24,140 VA

Finding amps for ungrounded (phase) conductors - Phases A and B

$I = VA \div V$
$I = 24,140 \text{ VA} \div 240 \text{ V}$
$I = 101 \text{ A}$

Calculating grounded (neutral) conductor

General lighting load = 4,365 VA √
Dishwasher load = 800 VA √
Disposal load = 540 VA √
Largest motor = 135 VA √
Range = 4,760 VA √
Five total loads = 10,600 VA

Finding amps for grounded (neutral) conductor

$I = VA \div V$
$I = 10,600 \text{ VA} \div 240 \text{ V}$
$I = 44 \text{ A}$

Finding conductors based on max. size

Table 310.16
Ungrounded (phase) conductors - Phases A and B
101 A requires 2 AWG THWN copper
Grounded (neutral) conductor
44 A requires 8 AWG THWN copper

310.15(B)(6) allows min. size
Ungrounded (phase) conductors - Phases A and B
101 A requires 3 AWG THWN copper
Grounded (neutral) conductor
44 A requires 8 AWG THWN copper

Residential Calculations - Single-Family Dwellings

Design Exercise Problem 9-12: What is the load in VA and amps for a mobile home with the following loads?

Given loads:
- 800 sq. ft. dwelling unit (mobile home)
- 2 small appliance circuits
- 1 laundry circuit
- 8500 VA range   240 volt, single-phase
- 5000 VA water heater   240 volt, single-phase
- 540 VA disposal   120 volt, single-phase
- 1000 VA dishwasher   120 volt, single-phase
- 6000 VA heating unit   240 volt, single-phase

Sizing ungrounded (phase) conductors = •
Sizing grounded (neutral) conductor = √

Calculating general lighting and receptacle load

Step 1: General lighting and receptacle load
550.18(A)(1)
800 sq. ft. x 3 VA = _____ VA [A]

Step 2: Small appliance and laundry load
550.18(A)(2); (A)(3)
1500 VA x 2 = _____ VA [B]
1500 VA x 1 = _____ VA [C]
Total load = _____ VA [D]

Step 3: Applying demand factor
550.18(A)(5)
First 3000 VA x 100% = 3,000 VA
Next 3900 VA x 35% = _____ VA [E]
Total load = _____ VA [F] •√

Calculating special appliance loads

Step 1: Applying demand factors for phases
550.18(B)(2); (B)(3); (B)(4)
Water heater = _____ VA [G]
Dishwasher = _____ VA [H] √
Disposal = _____ VA [I] √
Heating = _____ VA [J]
Total load = _____ VA [K] •

Calculating range load

Step 1: 550.18(B)(5); 220.61(B)(1)
8500 VA x 80% = _____ VA [L] •
6800 VA x 70% = _____ VA [M] √

Calculating ungrounded (phase) conductors

- General lighting load = _____ VA [N] •
- Special appliance load = _____ VA [O] •
- Range load = _____ VA [P] •
- Largest motor load (540 VA x 25%) = _____ VA [Q] •
- Four total loads = _____ VA [R]

Finding amps for ungrounded (phase) conductors - Phases A and B

I = VA ÷ V
I = _____ VA [S] ÷ 240 V
I = _____ A [T]

Calculating grounded (neutral) conductor

- General lighting load = _____ VA [U] √
- Dishwasher load = _____ VA [V] √
- Disposal load = _____ VA [W] √
- Largest motor = _____ VA [X] √
- Range = _____ VA [Y] √
- Five total loads = _____ VA [Z]

Finding amps for grounded (neutral) conductor

I = VA ÷ V
I = _____ VA [AA] ÷ 240
I = _____ A [BB]

Finding conductors based on max. size Table 310.16

Ungrounded (phase) conductors - Phases A and B
_____ AWG THWN [CC] copper conductors
Grounded (neutral) conductor
_____ AWG THWN [DD] copper conductors

310.15(B)(6) allows min. size

Ungrounded (phase) conductors - Phases A and B
_____ AWG THWN [EE] copper conductors
Grounded (neutral) conductor
_____ AWG THWN [FF] copper conductors

Name _____   Date _____

Chapter 9
Residential Calculations – Single-Family Dwellings

Section   Answer

1. The total VA for the general lighting and receptacle loads shall be derived by multiplying the square footage by 3 VA and applying demand factors.

    (a) True   **(b)** False

2. All small appliance branch-circuit loads shall be calculated at 1500 VA.

    (a) True   **(b)** False

3. For five or fewer dryers, the total VA shall be calculated at 100 percent with no demand factors applied.

    (a) True   **(b)** False

4. Three or fewer fixed appliances shall be calculated at 75 percent of their total VA.

    (a) True   **(b)** False

5. The heating and A/C loads shall be calculated at 125 percent, and the larger of the two loads is dropped.

    (a) True   **(b)** False

6. All laundry branch-circuit loads shall be calculated at _____ VA.

    (a) 1000   **(b)** 1200   **(c)** 1500   **(d)** 1800

7. Dryer equipment loads of four or fewer shall be calculated at _____ percent of the nameplate rating.

    (a) 65   **(b)** 75   **(c)** 85   **(d)** 100

8. The demand load for a household dryer shall be calculated at _____ kVA or the nameplate rating, whichever is greater.

    (a) 4   **(b)** 5   **(c)** 6   **(d)** 8

9. When there are four or more fixed appliances, a demand factor of _____ percent shall be permitted to be applied to the total VA.

    (a) 75   **(b)** 80   **(c)** 90   **(d)** 95

10. The motor's total full-load current rating shall be calculated at _____ percent when computing the largest motor load.

    (a) 80   **(b)** 100   **(c)** 125   **(d)** 150

11. What is the general lighting and receptacle loads and small appliance plus laundry loads for a 3,200 sq. ft. dwelling unit? (Calculate the load in VA)

    (a) 6675 VA   **(c)** 6885 VA   **(b)** 6785 VA   **(d)** 6975 VA

Section	Answer
_____	_____

12. What is the demand load in VA for two pieces of cooking equipment with a 3.5 kW rating?

    (a) 4550 VA   **(c)** 4725 VA   **(b)** 4650 VA   **(d)** 4800 VA

13. What is the demand load in VA for a range with a 18 kW rating?

    (a) 8450 VA   **(c)** 10,100 VA   **(b)** 9560 VA   **(d)** 10,400 VA

14. What is the demand load in VA for three pieces of cooking equipment with a 10 kW, 12 kW, and 18 kW rating?

    (a) 14,800 VA   **(c)** 15,400 VA   **(b)** 15,200 VA   **(d)** 16,200 VA

15. What is the branch-circuit demand load in VA for a 10 kW cooktop, 6 kW oven, and 4 kW oven installed in a dwelling unit?

    (a) 10,800 VA   **(c)** 11,600 VA   **(b)** 11,200 VA   **(d)** 12,400 VA

16. What is the demand load in VA for a 4.5 kW dryer?

    (a) 4500 VA   **(c)** 5000 VA   **(b)** 4800 VA   **(d)** 5200 VA

17. What is the demand load in VA for a 5.5 kW dryer?

    (a) 5000 VA   **(c)** 5400 VA   **(b)** 5200 VA   **(d)** 5500 VA

18. What is the fixed appliance load in VA for the ungrounded (phase) conductors with the following loads?

    • 6000 VA water heater   240 volt, single-phase
    • 2400 VA water pump   240 volt, single-phase
    • 1200 VA disposal   120 volt, single-phase
    • 1400 VA compactor   120 volt, single-phase
    • 1400 VA dishwasher   120 volt, single-phase
    • 800 VA microwave   120 volt, single-phase

    (a) 9900 VA   **(c)** 10,800 VA   **(b)** 10,200 VA   **(d)** 11,500 VA

19. What is the fixed appliance load in VA for the grounded (neutral) conductor with the following loads?

    • 6000 VA water heater   240 volt, single-phase
    • 1200 VA water pump   240 volt, single-phase
    • 1000 VA disposal   120 volt, single-phase
    • 1600 VA dishwasher   120 volt, single-phase
    • 800 VA blower motor   120 volt, single-phase

    (a) 2425 VA   **(c)** 2645 VA   **(b)** 2550 VA   **(d)** 2680 VA

	Section	Answer

20. What is the load in VA for a 1 HP, 240 volt, single-phase motor (all motors calculated at 100%) that is to be used for the largest motor load?

    (a) 420 VA   **(c)** 480 VA   **(b)** 460 VA   **(d)** 500 VA

Residential Calculations – Multifamily Dwellings

10

Residential calculations are the most difficult to perform due to the rules and regulations of the NEC being more restrictive than those for commercial and industrial facilities. The standard or optional calculation shall be permitted to be used to calculate the loads to size and select the elements of the feeder or service. The optional calculation seems to be the favorite of most testing agencies. This is true because once a load is calculated, this calculation produces smaller VA or amps than the longer, complicated standard calculation. The procedure for laying out residential calculations will be different in some ways from those used for commercial and industrial.

Quick Reference

APPLYING THE STANDARD CALCULATION	10-1
DEMAND FACTORS	10-2
GENERAL LIGHTING, RECEPTACLE LOADS AND SMALL APPLIANCE PLUS LAUNDRY LOADS	10-2
COOKING EQUIPMENT LOADS	10-4
FIXED APPLIANCE LOAD	10-8
DRYER LOAD	10-9
LARGEST LOAD BETWEEN HEAT AND A/C	10-13
LARGEST MOTOR LOAD	10-13
APPLYING THE STANDARD CALCULATION	
APPLYING THE OPTIONAL CALCULATION	10-23
PROBLEMS AND EXERCISES	10-26
TEST QUESTIONS	10-47

APPLYING THE STANDARD CALCULATION PART II TO ARTICLE 220

Residential occupancies are known in the industry as dwelling units. This chapter mainly deals with computing loads for multifamily dwelling units.

When using the standard calculation for calculating loads for a multifamily occupancy, all loads are divided into three groups and four columns. The groups are as follows:

Group 1:
    General lighting and receptacle loads
Group 2:
    Small appliance loads
Group 3:
    Special appliance loads

GENERAL LIGHTING LOAD
220.12; TABLE 220.12

Master Test Tip 1: The total VA for general lighting and receptacle loads shall be derived by multiplying the square footage by 3 VA and multiplying by the number of units and applying demand factors, where appropriate.

The general lighting load for multifamily dwelling units with the same square footage or different square footage shall be determined by multiplying the square footage by 3 VA per sq. ft. per **Table 220.12**. The required square footage per unit load VA (volt-amps) is found in **Table 220.12**. All lighting loads and general-purpose receptacles loads in a dwelling are determined by the general lighting load. All lighting loads per **210.70(A)** and general-purpose receptacle loads per **Article 100** and **210.52(A)** shall be calculated per **Table 220.12**.

SMALL APPLIANCE AND LAUNDRY LOAD
220.52(A); (B)

Master Test Tip 2: The small appliance and laundry circuits shall be permitted to be added to the VA per sq. ft. calculation, and demand factors applied.

All small appliance circuits shall be located in multifamily dwelling units per **210.52(B)(1)** and **210.11(C)(1)**. At least two small appliance circuits shall be required to supply receptacle outlets located in the kitchen, breakfast room, pantry and dining room per **220.52(A)**. A laundry room receptacle outlet shall be required per **220.52(B)**. All small appliance and laundry loads shall be calculated at 1500 VA to determine the size feeder conductors and elements to size the service.

DEMAND FACTORS
ARTICLE 220, PART II

Master Test Tip 3: Cooking equipment loads shall be permitted to have demand factors applied based upon the number of units.

The following four loads are separated into two columns of loads, and demand factors are applied accordingly:

Column 1:
   General lighting and receptacle loads and small appliance loads   **Table 220.11**

Column 2:
   Cooking equipment loads   **220.55; Table 220.55**
   Fixed appliance loads   **220.53**
   Dryer loads   **220.54; Table 220.54**

Master Test Tip 4: Where there are four or more fixed appliances, a demand factor of 75 percent shall be permitted to be applied to the total VA.

General lighting and receptacle loads and the small appliance circuits plus laundry circuit shall be permitted to have demand factors applied.

The total number of fixed appliance loads shall be permitted to have demand factors applied. The total number of ranges and dryers in multifamily dwelling units shall be permitted to be reduced by a percentage.

GENERAL LIGHTING AND RECEPTACLE LOADS AND SMALL APPLIANCE PLUS LAUNDRY LOADS
COLUMN 1 - TABLE 220.42; 220.52(A); (B)

Master Test Tip 5: For four or fewer dryers, the total VA shall be calculated at 100 percent with no demand factor applied.

The general lighting load for multifamily dwelling units shall be calculated by multiplying the square footage by 3 VA per sq. ft. per **Table 220.12**. The required square footage per unit load (volt-amps) is found in **Table 220.12**. All small appliance and laundry loads shall be calculated at 1500 VA per **220.52(A)** and **(B)**. A demand factor shall be permitted per **Table 220.42**. The demand factors for the general lighting and receptacle loads per **Table 220.42** for dwelling units are as follows:

   Volt-amps
      0 - 3000 at 100%
      3001 - 120,000 at 35%
      120,001 + at 25%

See Figure 10-1 for determining the general-purpose lighting and receptacle loads, including the small appliance and laundry loads in multifamily dwelling units with same square footage.

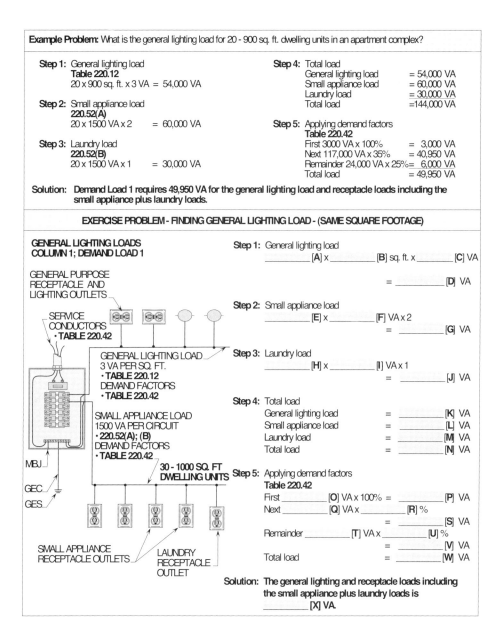

Figure 10-1. This figure illustrates an example and exercise problem for determining the general-purpose lighting and receptacle loads including the small appliance plus laundry loads in multifamily dwelling units with the same square footage.

Multifamily dwelling units in apartment complexes have varying square footages in most cases. Therefore, each dwelling unit with different square footage shall be multiplied by 3 VA per square foot, and the total added together to obtain the VA rating for the apartment complex.

See Figure 10-2 for determining the general-purpose lighting and receptacle loads including the small appliance plus laundry loads in multifamily dwelling units with different areas of square footage.

Figure 10-2. This figure illustrates an example and exercise problem for determining the general-purpose lighting and receptacle loads including the small appliance plus laundry loads in multi-family dwelling units with different square footages.

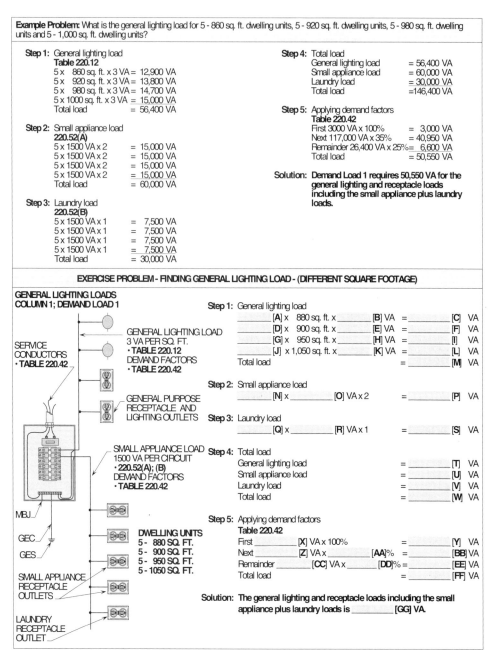

COOKING EQUIPMENT LOADS
COLUMN 2 - 220.55; TABLE 220.55

The cooking equipment loads and Demand load 2 are separated from Group 3 and placed in Column 2, and demand factors applied accordingly.

The demand factors listed in **Table 220.55** apply to the demand loads for cooking equipment. The Footnotes are based on the kW rating and number of units. Ranges, wall-mounted ovens, and counter-mounted cooktops are units of cooking equipment per **Table 220.55**.

DEMAND LOAD 2
TABLE 220.55, COLUMN A

The nameplate (kW) wattage rating of a range shall be calculated by the number of cooking units times the percentage factor found in **Table 220.55**. This calculation is used to obtain the maximum demand load. **(See Figure 10-3)**

Residential Calculations - Multifamily Dwellings

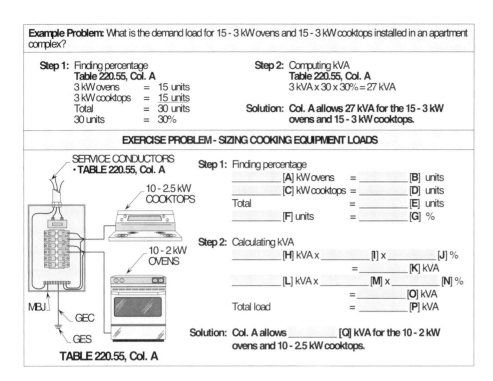

Figure 10-3. This figure illustrates an example and exercise problem for demand loads that shall be permitted to be applied per Table 220.55, Col. A for multifamily dwelling units.

DEMAND LOAD 2
TABLE 220.55, COLUMN B

The nameplate (kW) wattage rating of a range shall be calculated by the number of cooking units times the percentage factor applied in **Table 220.55**. This calculation derives the maximum demand load. **(See Figure 10-4)**

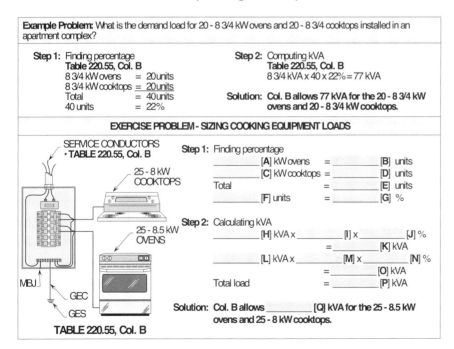

Figure 10-4. This figure illustrates an example and exercise problem for demand loads that shall be permitted to be applied per Table 220.55, Col. B for multifamily dwelling units.

DEMAND LOAD 2
TABLE 220.55, COLUMNS A AND B

Where the kW rating of each unit does not fall in **Columns A** or **B** but falls in both **Columns A** and **B**, each Column shall be used to determine demand factors based upon the number of units. **(See Figure 10-5)**

Figure 10-5. This figure illustrates an example and exercise problem for demand loads that shall be permitted to be applied per **Table 220.55, Columns A and B** for multifamily dwelling units.

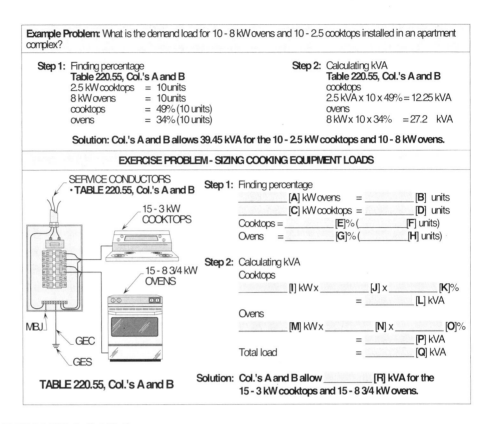

DEMAND LOAD 2
TABLE 220.55, COLUMN C

The demand load of cooking equipment is calculated in kW per **Table 220.55**. The maximum demand for the size and number of ranges is already calculated for selecting the elements of the feeder or service. Therefore, cooking equipment does not need to be multiplied by the kW rating of the unit by a percentage until **Column A** or **B** is utilized. **(See Figure 10-6)**

Figure 10-6. This figure illustrates an example and exercise problem for demand loads that shall be permitted to be applied per **Table 220.55, Col. C** for multifamily dwelling units.

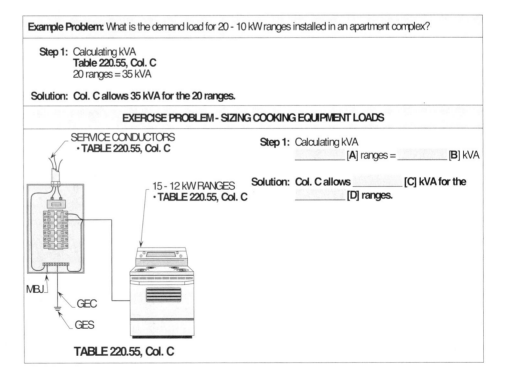

Cooking equipment shall be calculated at 15 kW plus 1 kW for each range when there are 26 through 40 units involved. **(See Figure 10-7)**

Cooking equipment shall be calculated at 25 kW plus 3/4 kW for each range when there are 41 and over units involved. **(See Figure 10-8)**

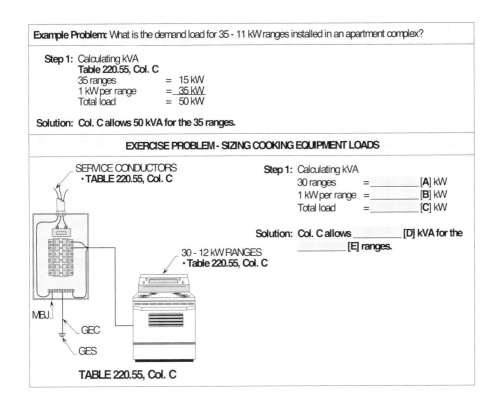

Figure 10-7. This figure illustrates an example and exercise problem for demand loads that shall be permitted to be applied per **Table 220.55, Col. C** for multifamily dwelling units.

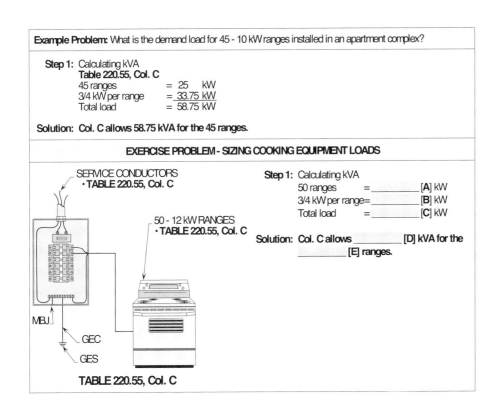

Figure 10-8. This figure illustrates an example and exercise problem for demand loads that shall be permitted to be applied per **Table 220.55, Col. C** for multifamily dwelling units.

DEMAND LOAD 2
TABLE 220.55, NOTE 1

For cooking equipment rated over 12 kW to 27 kW in **Column C, Note 1** shall be increased 5 percent for each kW over 12 kW to calculate the demand load. **(See Figure 10-9)**

Figure 10-9. This figure illustrates an example and exercise problem for demand loads that shall be permitted to be applied per **Table 220.55, Note 1** for multifamily dwelling units.

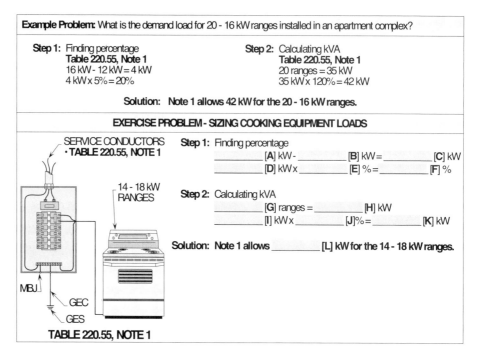

DEMAND LOAD 2
TABLE 220.55, NOTE 2

Cooking equipment of unequal values rated over 12 kW to 27 kW in **Column C, Note 2** shall be calculated by adding the kW ratings of all units and dividing by the number of units, all ranges below 12 kW shall be calculated at 12 kW. When an average rating is found, the number of units shall be increased by 5 percent for each kW exceeding 12 kW to derive the allowable kW. **(See Figure 10-10)**

DEMAND LOAD 2 - THREE-PHASE SYSTEM
TABLE 220.55, NOTE C

The load for cooking equipment when using three-phase systems shall be found by dividing the number of ranges by the three phases. The ranges per phase shall be multiplied by 2 (phases) to determine the number of ranges to be used per **Table 220.55, Col. A**. The kW rating for ranges shall be divided by 2 (phases) and multiplied by the three phases to determine the service load to be applied. **(See Figure 10-11)**

FIXED APPLIANCE LOAD
COLUMN 2 - 220.53

Cooking equipment loads, dryer equipment loads, air-conditioning loads and heating equipment loads shall not be applied to the fixed appliance load per **220.53**. The fixed appliance load for three or less fixed appliances shall be determined by adding wattage (volt-amps) values by the nameplate ratings for each appliance. However, these fixed appliances loads shall be permitted to have demand factors applied if there are four or more. All other fixed appliances of four or more grouped into a special appliance load are found by adding wattage ratings from appliance nameplates and multiplying the total wattage (volt-amps) by 75 percent to obtain demand load. **(See Figure 10-12)**

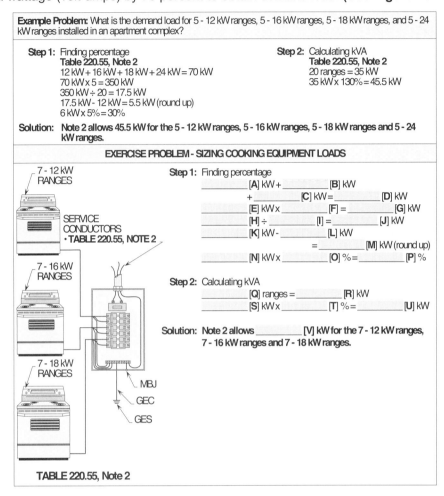

Figure 10-10. This figure illustrates an example and exercise problem for demand loads that shall be permitted to be applied per **Table 220.55, Note 2** for multifamily dwelling units.

DRYER LOAD
COLUMN 2 - 220.54; TABLE 220.54

The demand load for household dryers shall be calculated at 5 kVA or the nameplate rating, whichever is greater. Dryer equipment of four or fewer dryers shall be calculated at 100 percent of the nameplate rating. Dryer equipment of five or more dryers shall be permitted to have a percentage applied based on the number of units per **Table 220.54**. **(See Figure 10-13)**

The method for calculating the percent for 12 to 22 dryers is % = 47 minus (number of dryers minus 11). The method for calculating the percent for 24 to 42 dryers is % = 35 minus (0.5 x number of dryers minus 23). **(See Figure 10-14)**

DRYER LOAD
COLUMN 2 — 220.54; TABLE 220.54

The load for dryers, when using three-phase systems, shall be found by dividing the number of dryers by the three-phases. The dryers per phase shall be multiplied by 2 (phases) to determine the number of dryers to be used per **Table 220.54**. The kW rating for dryers shall be divided by 2 (phases) and multiplied by the three phases to determine the service load to be applied. **(See Figure 10-15)**

Figure 10-11. This figure illustrates an example and exercise problem for demand loads that shall be permitted to be applied per **Table 220.55, Note C** for multifamily dwelling units supplied by a three-phase service.

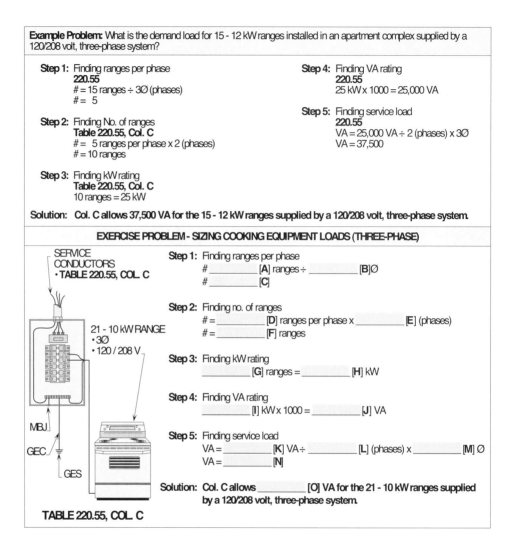

Example Problem: What is the demand load for 15 - 12 kW ranges installed in an apartment complex supplied by a 120/208 volt, three-phase system?

Step 1: Finding ranges per phase
220.55
= 15 ranges ÷ 3Ø (phases)
= 5

Step 2: Finding No. of ranges
Table 220.55, Col. C
= 5 ranges per phase x 2 (phases)
= 10 ranges

Step 3: Finding kW rating
Table 220.55, Col. C
10 ranges = 25 kW

Step 4: Finding VA rating
220.55
25 kW x 1000 = 25,000 VA

Step 5: Finding service load
220.55
VA = 25,000 VA ÷ 2 (phases) x 3Ø
VA = 37,500

Solution: Col. C allows 37,500 VA for the 15 - 12 kW ranges supplied by a 120/208 volt, three-phase system.

EXERCISE PROBLEM - SIZING COOKING EQUIPMENT LOADS (THREE-PHASE)

SERVICE CONDUCTORS
• TABLE 220.55, COL. C

21 - 10 kW RANGE
• 3Ø
• 120 / 208 V

MBJ
GEC
GES

TABLE 220.55, COL. C

Step 1: Finding ranges per phase
_____ [A] ranges ÷ _____ [B]Ø
_____ [C]

Step 2: Finding no. of ranges
= _____ [D] ranges per phase x _____ [E] (phases)
= _____ [F] ranges

Step 3: Finding kW rating
_____ [G] ranges = _____ [H] kW

Step 4: Finding VA rating
_____ [I] kW x 1000 = _____ [J] VA

Step 5: Finding service load
VA = _____ [K] VA ÷ _____ [L] (phases) x _____ [M] Ø
VA = _____ [N]

Solution: Col. C allows _____ [O] VA for the 21 - 10 kW ranges supplied by a 120/208 volt, three-phase system.

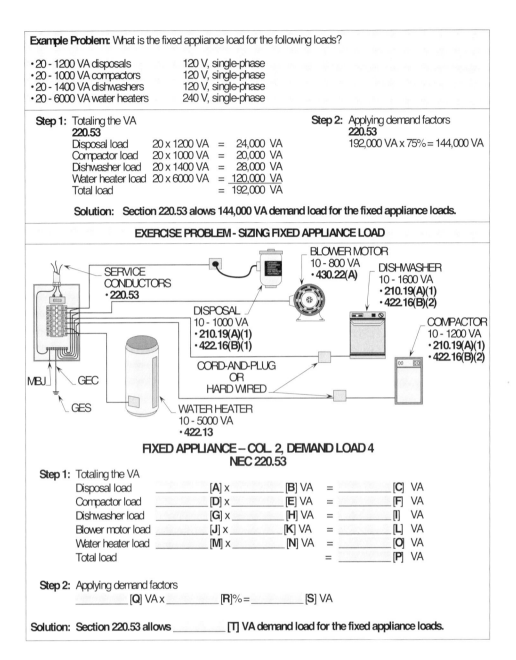

Figure 10-12. This figure illustrates an example and exercise problem for demand factors that shall be permitted to be applied per 220.53 for multifamily dwelling units.

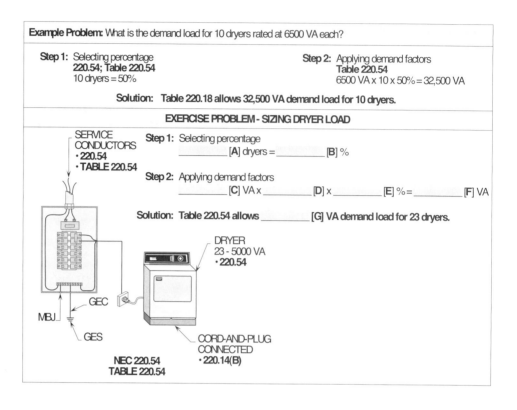

Figure 10-13. This figure illustrates an example and exercise problem for demand factors that shall be permitted to be applied per **220.54** and **Table 220.54** for multifamily dwelling units.

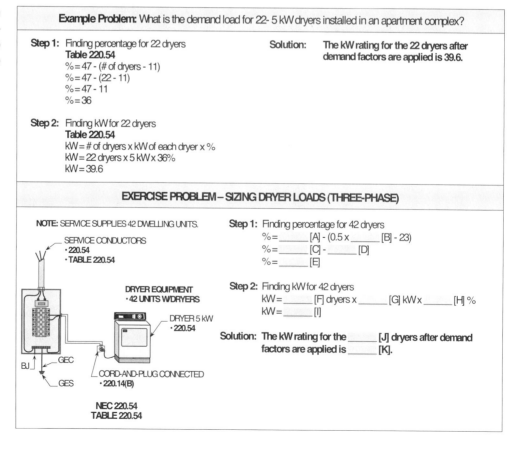

Figure 10-14. This figure illustrates an example and exercise problem for demand factors that shall be permitted to be applied per **220.54** and **Table 220.54** for multifamily dwelling units.

Residential Calculations - Multifamily Dwellings

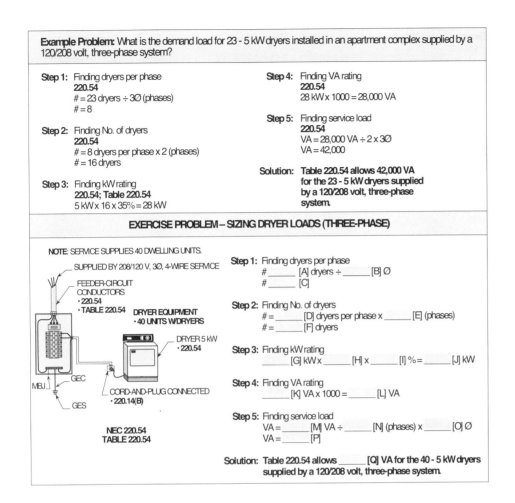

Figure 10-15. This figure illustrates an example and exercise problem for demand loads that shall be permitted to be applied per **Table 220.54** for three-phase multifamily dwelling units.

Note, for determining percentages (%) in (I) of step 3 in exercise problem 10-15, see steps 1 and 2 in exercise problem 10-14 on page 10-12.

LARGEST LOAD BETWEEN HEAT AND A/C
COLUMN 3 - 220.51

Section **220.51** shall be applied when determining the largest load between heat and A/C. The heating and A/C loads shall be calculated at 100 percent, and the smaller of the two loads is dropped. **(See Figure 10-16)**

LARGEST MOTOR LOAD
COLUMN 4 - 220.50

The motor's total full-load current rating shall be calculated at 25 or 125 percent per **220.50** which is required per **430.24** and **430.25**, for calculating the load of one or more motors with other loads. **(See Figure 10-17)**

The largest motor load shall always be added into the calculation, and it really doesn't matter what size the largest motor is. The NEC doesn't distinguish between the size of the motor, but requires one to be selected and calculated based upon HP and voltage.

Figure 10-16. This figure illustrates an example and exercise problem for demand factors that shall be permitted to be applied per **220.60** for multifamily dwelling units.

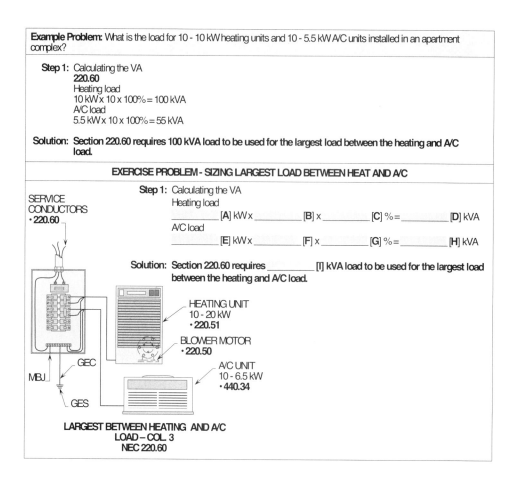

Figure 10-17. This figure illustrates an example and exercise problem for demand factors that shall be permitted to be applied per **220.14(C)** and **220.50** when finding the VA for a single-phase, 230 volt motor for multifamily dwelling units.

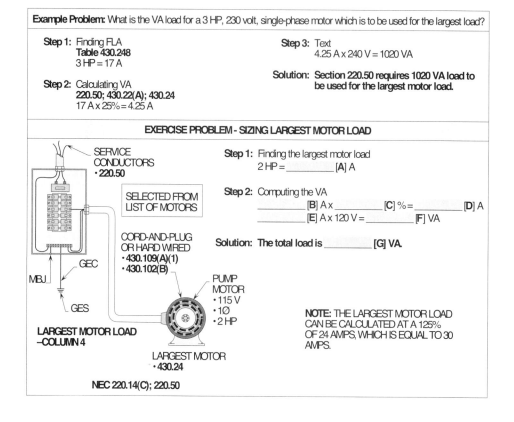

APPLYING THE STANDARD CALCULATION MULTIFAMILY

When applying the standard calculation for multifamily dwelling units, the loads shall be computed the same as using the standard calculation for single-family dwellings. The only difference is to calculate the loads of each unit and multiply the total number of dwelling units and pieces of electrical equipment together to derive the total VA or amps to size the elements of the service or feeder-circuit.

APPLYING THE STANDARD CALCULATION MULTIFAMILY - SAME SQUARE FOOTAGE

When applying the standard calculation for multifamily dwelling units with same square footage, the loads are separated into four columns. The first column of loads consists of the general lighting and receptacle and small appliance plus laundry loads. The second column of loads consists of special appliance loads and fixed appliance loads. The third column of loads consists of largest load between heating or A/C load. The fourth column of loads consists of the largest motor load. The elements of the service are determined by using the total load of these four columns. **(See Design Problem 10-1)**

Design Problem: Use Design Example Problem 10-1 to size the following components and elements of multifamily dwellings with the same square footage:

(1) What size OCPD is required to supply the water heater?

    Step 1: Finding amperage
    Text
    $I = VA \div V$
    $I = 6000 \text{ VA} \div 240 \text{ V}$
    $I = 25 \text{ A}$

    Step 2: Calculating amperage
    422.13
    25 A x 125% = 31.25 A

    Step 3: Finding size OCPD
    240.4(B); 240.6(A)
    31.25 A is less than 35 A

Solution: The size OCPD is 35 amps.

(2) How many 15 amp two-wire circuits are required for the general lighting and receptacle loads for each dwelling unit?

    Step 1: Finding total VA
    Table 220.12
    VA = 1000 sq. ft. x 3 VA
    VA = 3000

    Step 2: Finding # of outlets
    210.11(A)
    # = VA ÷ CB x 120 V
    # = 3000 VA ÷ 15 A x 120 V (1800 VA)
    # = 2

Solution: The number of outlets is 2 for each dwelling unit.

(3) What size rigid metal conduit is required for each parallel run?

> **Step 1:** Finding sq. in. area
> **Table 5, Ch. 9**
> 1000 KCMIL THWN   = 1.3478 sq. in.
> 1/0 AWG THWN   = .1855 sq. in.
>
> **Step 2:** Calculating sq. in. area
> **Table 5, Ch. 9**
> 1.3478 sq. in. x 2   = 2.6956 sq. in.
> .1855 sq. in. x 1   = .1855 sq. in.
> Total   = 2.8811 sq. in.
>
> **Step 3:** Selecting conduit size
> **Table 4, Ch. 9**
> 2.8811 sq. in. requires 3 in. (78)
>
> **Solution: The size rigid metal conduit is 3 in. (78).**

APPLYING THE STANDARD CALCULATION MULTIFAMILY - DIFFERENT SQUARE FOOTAGES

When applying the standard calculation for multifamily dwelling units with different square footages, the loads are calculated the same as for multifamily dwelling units with the same square footage. **(See Design Problem 10-2)**

Design Problem: Use Design Example Problem 10-2 to size the following components and elements of multifamily dwellings with different square footages:

(1) What size THWN copper conductors, paralleled six times per phase, is required for Phases A and B?

> **Step 1:** Finding amperage
> Text
> I = 3745 A ÷ 6 (No. runs per phase)
> I = 624 A
>
> **Step 2:** Selecting size conductors
> **Table 310.16, 75°C Column; 110.14(C)(1)(b)**
> 1500 KCMIL = 625 A
> 624 A is less than 625 A
>
> **Solution: The size THWN conductors required are 1500 KCMIL copper.**

(2) What size THWN copper conductors, paralleled six times per phase, are required for the grounded (neutral) conductors?

> **Step 1:** Finding amperage
> Text
> I = 498 A ÷ 6 (No. runs per phase)
> I = 83 A
>
> **Step 2:** Selecting size conductors
> **Table 310.16, 75°C Column; 310.4; 110.14(C)(1)(b)**
> 1/0 AWG THWN = 150 A
> 83 A is less than 150 A
>
> **Solution: The size THWN conductors required are 1/0 AWG copper.**

APPLYING THE STANDARD CALCULATION
MULTIFAMILY - SERVICE WITH UNITS HAVING GAS HEAT

When applying the standard calculation for multifamily dwelling units having gas heat, all gas-related loads shall not be applied to the total load when calculating the amperage for the service. **(See Design Problem 10-3)**

Design Problem: Use Design Example Problem 10-3 to size the following components and elements of multifamily dwellings having gas heat:

(1) What size THWN copper conductors, paralleled three times per phase, are required for Phases A and B?

    Step 1: Finding amperage
        Text
        I = 895 A ÷ 3 (No. runs per phase)
        I = 298 A

    Step 2: Selecting size conductors
        Table 310.16, 75°C Column; 110.14(C)(1)(b)
        350 KCMIL = 310 A
        298 A is less than 310 A

    Solution: The size THWN conductors required are 350 KCMIL copper.

(2) What size THWN copper conductors, paralleled three times per phase, is required for the grounded (neutral) conductors?

    Step 1: Finding amperage
        Text
        I = 359 A ÷ 3 (No. runs per phase)
        I = 120 A

    Step 2: Selecting size conductors
        Table 310.16, 75°C Column; 310.4; 110.14(C)(1)(b)
        1/0 AWG THWN = 150 A
        120 A is less than 150 A

    Solution: The size THWN conductors required are 1/0 AWG copper.

(3) What size rigid metal conduit is required for each parallel run?

    Step 1: Finding sq. in. area
        Table 5, Ch. 9
        350 KCMIL THWN   = .5242 sq. in.
        1/0 AWG THWN   = .1855 sq. in.

    Step 2: Calculating sq. in. area
        Table 5, Ch. 9
        .5242 sq. in. x 2   = 1.0484 sq. in.
        .1855 sq. in. x 1   = .1855 sq. in.
        Total =   1.2339 sq. in.

    Step 3: Selecting conduit size
        Table 4, Ch. 9
        1.2339 sq. in. requires 2 in. (53)

    Solution: The size rigid metal conduit is 2 in. (53).

APPLYING THE STANDARD CALCULATION MULTIFAMILY - SERVICE WITH 120 VOLT, A/C WINDOW UNITS

When applying the standard calculation for multifamily dwelling units with 120 volt A/C window units, all 120 volt A/C window units shall be applied to the total load (if largest load between heat and A/C) when calculating the amperage for the service. The A/C window unit shall also be used as the largest motor load for the ungrounded (phase) conductors and grounded (neutral) conductors. **(See Design Problem 10-4)**

Design Problem: Use Design Example Problem 10-4 to size the following components and elements of multifamily dwellings having 120 volt, A/C window units:

(1) What size THWN copper conductors, paralleled three times per phase, are required for Phases A and B?

> **Step 1:** Finding amperage
> Text
> I = 884 A ÷ 3
> I = 295 A
>
> **Step 2:** Selecting size conductors
> **Table 310.16, 75°C Col.; 110.14(C)(1)(b)**
> 350 KCMIL = 310 A
> 295 A is less than 310 A
>
> **Solution: The size THWN conductors required are 350 KCMIL copper.**

(2) What size THWN copper conductors, paralleled three times per phase, are required for the grounded (neutral) conductors?

> **Step 1:** Finding amperage
> Text
> 392 A ÷ 3
> I = 131 A
>
> **Step 2:** Selecting size conductors
> **Table 310.16, 75°C Col.; 310.4; 110.14(C)(1)(b)**
> 1/0 AWG THWN = 150 A
> 131 A is less than 150 A
>
> **Solution: The size THWN conductors required are 1/0 AWG copper.**

(3) What size CB is required for the service equipment?

> **Step 1:** Finding amperage
> **Table 310.16**
> 350 KCMIL = 310
>
> **Step 2:** Calculating amperage
> **240.4(C)**
> 310 A x 3 = 930 A
>
> **Step 3:** Selecting OCPD
> **240.6(A)**
> 930 A requires 900 A OCPD
>
> **Solution: The size OCPD required is 900 amps.**

(4) How many 15 amp, two-wire circuits are required for each general lighting and receptacle loads for all dwelling units?

Step 1: Finding total VA
Table 220.12
VA = 720 sq. ft. x 3 VA x 10
VA = 21,600 VA

Step 2: Finding # of outlets
210.11(A)
= VA ÷ CB x 120 V
= 21,600 ÷ 15 A x 120 V
= 12

Solution: The number of outlets is 12 for all dwelling units.

APPLYING THE STANDARD CALCULATION MULTIFAMILY - SERVICE WITH HEAT PUMPS

When applying the standard calculation for multifamily dwelling units with heat pumps, the electric heating and heat pump shall be applied to the total load (if largest load between heat and A/C) when calculating the amperage for the service. The heat pump shall also be used as the largest motor load for the ungrounded (phase) conductors. **(See Design Problem 10-5)**

Design Problem: Use Design Example Problem 10-5 to size the following components and elements of multifamily dwellings having heat pumps:

(1) What size THWN copper conductors, paralleled three times per phase, are required for Phases A and B?

Step 1: Finding amperage
Text
I = 888 A ÷ 3
I = 296 A

Step 2: Selecting size conductors
Table 310.16, 75°C Column; 110.14(C)(1)(b)
350 KCMIL = 310 A
296 A is less than 310 A

Solution: The size THWN conductors required are 350 KCMIL copper.

(2) What size THWN copper conductors, paralleled three times per phase, are required for the grounded (neutral) conductors?

Step 1: Finding amperage
Text
I = 215 A ÷ 3
I = 72 A

Step 2: Selecting size conductors
Table 310.16, 75°C Col.; 310.4; 110.14(C)(1)(b)
1/0 AWG THWN = 150 A
72 A is less than 150 A

Solution: The size THWN conductors required are 1/0 AWG copper.

(3) What size rigid metal conduit is required for each parallel run?

    Step 1: Finding sq. in. area
    Table 5, Ch. 9
    350 KCMIL THWN   = .5242 sq. in.
    1/0 AWG THWN   = .1855 sq. in.

    Step 2: Calculating sq. in. area
    Table 5, Ch. 9
    .5242 sq. in. x 2 = 1.0484 sq. in.
    .1855 sq. in. x 1 =   .1855 sq. in.
    Total   = 1.2339 sq. in.

    Step 3: Selecting conduit size
    Table 4, Ch. 9
    1.2399 sq. in. requires 2 in. (53)

Solution: The size rigid metal conduit is 2 in. (53).

(4) What size branch-circuit conductors, using nonmetallic-sheathed cable, are required to supply the 12,000 VA range?

    Step 1: Finding kW rating
    Table 220.55, Col. C
    12,000 VA = 8 kW

    Step 2: Finding amperage
    Text
    I = kW x 1000 ÷ V
    I = 8 kW x 1000 ÷ 240 V
    I = 33 A

    Step 3: Selecting conductors
    334.80; Table 310.16, 60°C Col.
    33 A requires 8-2 AWG w/ground

Solution: The size conductors are required to be 8-2 AWG w/ground.

(5) What size branch-circuit conductors, using nonmetallic-sheathed cable, are required to supply the 1200 VA disposal?

    Step 1: Finding amperage
    Text
    I = VA ÷ V
    I = 1200 VA ÷ 120 V
    I = 10 A

    Step 2: Calculating amperage
    430.6(A)(1); Table 220.3; 430.22(A)
    10 A x 125% = 12.5 A

    Step 3: Selecting conductors
    334.80; Table 310.16, 60°C Col.
    12.5 A requires 14-2 AWG w/ground

Solution: The size conductors are required to be 14-2 AWG w/ground.

APPLYING THE STANDARD CALCULATION MULTIFAMILY - DIVIDING VOLT-AMPS ON LINE AND NEUTRAL

When applying the standard calculation for multifamily dwelling units, the load in VA should be balanced when determining the ungrounded (phase) conductors and grounded (neutral) conductors. **(See Design Problem 10-6)**

Design Problem: Use Design Example Problem 10-6 to size the following components and elements of multifamily dwellings when dividing volt-amps on line and neutral:

(1) What is the load in VA for the 15 - 12,000 VA ranges supplied by a 120/208 volt service?

   Step 1: Finding ranges per phase
   220.55
   # = 15 ranges ÷ 3Ø
   # = 5

   Step 2: Finding ranges per phase
   220.55
   # = 5 ranges per phase x 2
   # = 10 ranges

   Step 3: Finding kW rating
   Table 220.55, Col. C
   10 ranges = 25 kW

   Step 4: Finding VA rating
   220.55
   25 kW x 1000 = 25,000 VA

   Step 5: Finding service load in VA
   220.55
   25,000 VA ÷ 2 phases x 3Ø = 37,500 VA

   Solution: The service load is 37,500 VA for the ranges.

(2) What size branch-circuit conductors, using nonmetallic-sheathed cable, is required to supply the dishwasher?

   Step 1: Finding amperage
   Text
   I = VA ÷ V
   I = 1000 VA ÷ 120 V
   I = 8.3 A

   Step 2: Selecting conductors
   334.80; Table 310.16, 60°C Col.
   8.3 A requires 14-2 AWG w/ground

   Solution: The size conductors are required to be 14-2 AWG w/ground.

(3) Can the dishwasher be cord-and-plug connected with a 42 in. SO cord supplied by a 15 amp branch-circuit?

Step 1: Finding amperage
210.23(A)
15 A x 80% = 12 A
8.3 A is less than 12 A

Step 2: Selecting SO cord
422.16(B)(2)(2)
3' to 4' cord allowed

Solution: Yes, the dishwasher can be cord-and-plug connected.

APPLYING THE STANDARD CALCULATION MULTIFAMILY - DIVIDING AMPS ON LINE AND NEUTRAL

When applying the standard calculation for multifamily dwelling units, the load in amps should be balanced when determining the ungrounded (phase) conductors and grounded (neutral) conductors. **(See Design Problem 10-7)**

Design Problem: Use Design Example Problem 10-7 to size the following components and elements of multifamily dwellings when dividing amps on line and neutral:

(1) What is the load in amps for the 15 - 12,000 VA ranges supplied by a 120/208 volt service?

Step 1: Finding ranges per phase
220.55
= 15 ranges ÷ 3Ø
= 5

Step 2: Finding ranges per phase
220.55
= 5 ranges per phase x 2
= 10 ranges

Step 3: Finding kW rating
Table 220.55, Col. C
10 ranges = 25 kW

Step 4: Finding VA rating
220.55
25 kW x 1000 = 25,000 VA

Step 5: Finding service load in VA
220.55
25,000 ÷ 2 phases x 3Ø = 37,500 VA

Step 6: Finding service load in amps
Text (using V x $\sqrt{3}$)
I = VA ÷ V x $\sqrt{3}$
I = 37,500 VA ÷ 208 V x 1.732
I = 104.2

Solution: The service load is 104.2 amps for the ranges.

(2) What size THWN copper conductors, paralleled three times per phase, are required for Phases A and B?

Step 1: Finding amperage
Text
I = 659 ÷ 3
I = 220 A

Step 2: Selecting size conductors
Table 310.16; 75°C Col.; 310.4; 110.14(C)(1)(b)
4/0 AWG THWN = 230 A
220 A is less than 230 A

Solution: The size THWN copper conductors are 4/0 AWG.

(3) What size THWN copper conductors, paralleled three times per phase, are required for grounded (neutral) conductors?

Step 1: Finding amperage
Text
I = 298 ÷ 3
I = 99 A

Step 2: Selecting size conductors
Table 310.16; 75°C Col.; 310.4; 110.14(C)(1)(b)
1/0 AWG THWN = 150 A
99 A is less than 150 A

Solution: The size THWN copper conductors are 1/0 AWG.

(4) What size rigid metal conduit is required for each parallel run?

Step 1: Finding sq. in. area
Table 5, Ch. 9
4/0 AWG THWN   = .3237 sq. in.
1/0 AWG THWN   = .1855 sq. in.

Step 2: Calculating sq. in. area
Table 5, Ch. 9
.3237 sq. in. x 3   = .9711 sq. in.
.1855 sq. in. x 1   = .1855 sq. in.
Total   = 1.1566 sq. in.

Step 3: Selecting conduit size
Table 4, Ch. 9
1.1566 sq. in. requires 2 in. (53)

Solution: The size rigid metal conduit is 2 in. (53).

APPLYING THE OPTIONAL CALCULATION
220.84

When applying the optional calculation for multifamily dwelling units, the loads are calculated the same as when using the optional calculation for single-family dwellings. The only difference would be to compute the total number of dwelling units and pieces of electrical equipment together and apply a percentage based upon the number to derive the total VA or amps to size the elements of the service or feeder-circuit.

APPLYING THE OPTIONAL CALCULATION MULTIFAMILY - SAME SQUARE FOOTAGE

The optional calculation is the simplest method to use in calculating the load to select the service elements. Each load shall be figured at 100 percent of its nameplate rating. The small appliance loads shall be computed at 1500 VA each, and the square footage of each dwelling unit (same square footage) shall be multiplied by 3 VA. The larger load between the heating and A/C is selected, and the smaller load dropped. Demand factors, based on the number of dwelling units, per **Table 220.84** shall be applied to the total load to determine the service load. **(See Design Problem 10-8)**

APPLYING THE OPTIONAL CALCULATION MULTIFAMILY - DIFFERENT SQUARE FOOTAGE

When applying the optional calculation for multifamily dwelling units with different square footages, the loads are calculated the same as for multifamily dwelling units with the same square footage. **(See Design Problem 10-9)**

APPLYING THE OPTIONAL CALCULATION MULTIFAMILY - SERVICE WITH UNITS AND HOUSE LOADS

When applying the optional calculation for multifamily dwelling units and house loads, the house loads are calculated separate from the dwelling unit loads and added together to determine the service load.

Design Problem: Use Design Example Problem 10-9 to size the house loads and service ampacity for the 30 multifamily dwelling units with the following house loads:

Laundry loads (Continuous and noncontinuous operation)	65,520 VA, 208 V, single-phase
Outside lighting loads (electric discharge)	50 units, 180 VA each, 120 V, single-phase
Outside receptacle loads (noncontinuous operation)	30 duplex receptacles, 120 V, single-phase
Outside sign loads (Electric discharge)	1800 VA, 120 V, single-phase

(1) What is the ampacity of the house loads (based on OCPD)?

    Step 1: Finding total house loads
    220.84(B)
    Laundry loads
    210.52(F), Ex. 1; 230.42(A)(1)
    65,520 VA x 100%   = 65,520 VA
    Outside lighting loads
    220.4(B); 230.42(A)(2)
    50 units x 180 VA x 125%   = 11,250 VA
    Outside receptacle loads
    220.14(I); Table 220.13
    30 receptacles x 180 VA x 100%   = 5,400 VA
    Outside sign loads
    600.5(A); 230.42(A)(2); 220.14(F)
    1800 VA x 125%   = 2,250 VA
    Total load   = 84,420 VA

Step 2: Finding ampacity
Text
$I = VA \div V \times \sqrt{3}$
$I = 84{,}420 \text{ VA} \div 208 \text{ V} \times 1.732$
$I = 234.5 \text{ A}$

Solution: The ampacity is 234.5 amps for the house loads.

(2) What is the service ampacity for Phases A, B and C and the grounded (neutral) conductors?

Step 1: Finding ampacity for phases
230.42(A)(1)
Apartment loads	= 1961 A
House loads	= 234.5 A
Total loads	= 2195.5 A

Solution: The ampacity for the phases is 2195.5 amps.

Step 1: Finding ampacity for grounded (neutral) conductors (house loads)
220.61
Outside lighting loads	= 9000 VA
Outside receptacle loads	= 5400 VA
Outside sign load	= 1800 VA
Total loads	= 16,200 VA

Step 2: Calculating ampacity for grounded (neutral) conductors (house loads)
220.61
$16{,}200 \div 240 \text{ V} = 67.5 \text{ A}$

Step 3: Finding ampacity for grounded (neutral) conductors (total)
220.61
Apartment load	= 498 A
House loads	= 67.5 A
Total loads	= 565.5 A

Solution: The ampacity for the grounded (neutral) conductors is 565.5 amps.

APPLYING THE OPTIONAL CALCULATION
MULTIFAMILY - DIVIDING VOLT-AMPS ON LINE AND NEUTRAL

When applying the optional calculation for multifamily dwelling units, the load in VA should be balanced when determining the ungrounded (phase) conductors and grounded (neutral) conductors. **(See Design Problem 10-10)**

Design Example Problem 10-1. What is the load in VA and amps for 25 multifamily dwelling units with the following loads? Note: Parallel service conductors, 6 times per phase.

Given loads:	120 V, single-phase loads	240 V, single-phase loads
• 25 - 1000 sq. ft. dwelling units • 2 small appliance circuits per unit • 1 laundry circuit per unit	• 25 - 1000 VA dishwashers • 25 - 1200 VA disposals	• 25 - 12,000 VA ranges • 25 - 6000 VA water heaters • 25 - 20,000 VA heating units • dryer facilities furnished by apartment complex

Sizing ungrounded (phase) conductors = •
Sizing grounded (neutral) conductors = √

Column 1
Calculating general lighting and receptacle load

Step 1: General lighting and receptacle load
Table 220.12
1000 sq. ft. x 3 VA x 25   =   75,000 VA

Step 2: Small appliance and laundry load
220.52(A); (B)
1500 VA x 2 x 25   =   75,000 VA
1500 VA x 1 x 25   =   37,500 VA
Total load   =   187,500 VA

Step 3: Applying demand factors
Demand load 1; **Table 220.42**
First 3000 VA x 100%   =   3,000 VA
Next 117,000 VA x 35%   =   40,950 VA
Remaining 67,500 VA x 25% =   16,875 VA
Total load   =   60,825 VA • √

Column 2
Calculating cooking equipment load

Step 1: Applying demand factors for phases
Demand load 2; **Table 220.55, Col. C**
25 - 12,000 VA ranges   =   40,000 VA •

Step 2: Applying demand factors for neutral
220.61
40,000 VA x 70%   =   28,000 VA √

Column 2
Calculating fixed appliance load

Step 1: Applying demand factors for phases
Demand load 4; **220.53**
1000 VA x 25 x 75%   =   18,750 VA
1200 VA x 25 x 75%   =   22,500 VA
6000 VA x 25 x 75%   =   112,500 VA
Total load   =   153,750 VA •

Step 2: Applying demand factors for neutral
Demand load 4; **220.61; 220.53**
1000 VA x 25 x 75%   =   18,750 VA
1200 VA x 25 x 75%   =   22,500 VA
Total load   =   41,250 VA √

Column 3
Largest load between heating and A/C load

Step 1: Selecting largest load
Demand load 5; **220.60**
Heating unit
20,000 VA x 25 x 100%   =   500,000 VA •

Column 4
Calculating largest motor load

Step 1: Selecting largest motor load for phases
220.14(C); 220.50; 430.24
1200 VA x 25%   =   300 VA •

Step 2: Selecting largest motor load for neutral
220.14(C); 220.50; 430.24
1200 VA x 25%   =   300 VA √

Calculating ungrounded (phase) conductors

General lighting load   =   60,825 VA •
Cooking equipment load   =   40,000 VA •
Appliance load   =   153,750 VA •
Heating load   =   500,000 VA •
Largest motor load   =   300 VA •
Five total loads   =   754,875 VA

Calculating grounded (neutral) conductors

General lighting load   =   60,825 VA √
Cooking equipment load   =   28,000 VA √
Appliance load   =   41,250 VA √
Largest motor load   =   300 VA √
Four total loads   =   130,375 VA

Finding amps for ungrounded (phase) conductors - Phases A and B

I = VA ÷ V
I = 754,875 VA ÷ 240 V
I = 3145 A

Finding amps for grounded (neutral) conductors

I = VA ÷ V
I = 130,375 VA ÷ 240 V
I = 543 A

Finding conductors

Ungrounded (phase) conductors - Phases A and B
I = 3145 ÷ 6 (No. run per phase)
I = 524 A

Grounded (neutral) conductors
220.61

543 A
First 200 A x 100%   =   200 A
Next 343 A x 70%   =   240 A
Total load   =   440 A
310.4
I = 440 A ÷ 6 (No. run per phase)
I = 73 A (1/0 AWG per **310.4**)

Sizing conductors (6 per phase)
Table 310.16; 310.4

Ungrounded (phase) conductors - Phases A and B
1000 KCMIL THWN copper
Grounded (neutral) conductors
1/0 AWG THWN copper

Residential Calculations - Multifamily Dwellings

Design Exercise Problem 10-1. What is the load in VA and amps for 20 multifamily dwelling units with the following loads? Note: Parallel service conductors, 6 times per phase.

Given loads:	120 V, single-phase loads	240 V, single-phase loads
• 20 - 1200 sq. ft. dwelling units • 2 small appliance circuits per unit • 1 laundry circuit per unit	• 20 - 1600 VA dishwashers • 20 - 1200 VA disposals	• 20 - 10,000 VA ranges • 20 - 5000 VA water heaters • 20 - 20,000 VA heating units • Dryer facilities furnished by apartment complex

Sizing ungrounded (phase) conductors = •
Sizing grounded (neutral) conductors = √

Column 1
Calculating general lighting and receptacle load

Step 1: General lighting and receptacle load
Table 220.12
1200 sq. ft. x 3 VA x 20   = _____ VA [A]

Step 2: Small appliance and laundry load
220.52(A); (B)
1500 VA x 2 x 20   = _____ VA [B]
1500 VA x 1 x 20   = _____ VA [C]
Total load   = _____ VA [D]

Step 3: Applying demand factors
Demand load 1; **Table 220.42**
First 3000 VA x 100%   =   3,000   VA
Next 117,000 VA x 35%   = _____ VA [E]
Remaining 42,000 VA x 25% = _____ VA [F]
Total load   = _____ VA [G] • √

Column 2
Calculating cooking equipment load

Step 1: Applying demand factors for phases
Demand load 2; **Table 220.55, Col. C**
20 - 10,000 VA ranges   = _____ VA [H] •

Step 2: Applying demand factors for neutral
220.61
35,000 VA x 70%   = _____ VA [I] √

Column 2
Calculating fixed appliance load

Step 1: Applying demand factors for phases
Demand load 4; **220.53**
1600 VA x 20 x 75%   = _____ VA [J]
1200 VA x 20 x 75%   = _____ VA [K]
5000 VA x 20 x 75%   = _____ VA [L]
Total load   = _____ VA [M] •

Step 2: Applying demand factors for neutral
Demand load 4; **220.61; 220.53**
1600 VA x 20 x 75%   = _____ VA [N]
1200 VA x 20 x 75%   = _____ VA [O]
Total load   = _____ VA [P] √

Column 3
Largest load between heating and A/C load

Step 1: Selecting largest load
Demand load 5; **220.53**
Heating unit
20,000 VA x 20 x 100%   = _____ VA [Q]

Column 4
Calculating largest motor load

Step 1: Selecting largest motor load for phases
220.14(C); 50; 430.24
1200 VA x 25%   = _____ VA [R] •

Step 2: Selecting largest motor load for neutral
220.14; 430.24
1200 VA x 25%   = _____ VA [S] √

Calculating ungrounded (phase) conductors

• General lighting load   = _____ VA [T] •
• Cooking equipment load   = _____ VA [U] •
• Appliance load   = _____ VA [V] •
• Heating load   = _____ VA [W] •
• Largest motor load   = _____ VA [X] •
Five total loads   = _____ VA [Y]

Calculating grounded (neutral) conductors

• General lighting load   = _____ VA [Z] √
• Cooking equipment load   = _____ VA [AA] √
• Appliance load   = _____ VA [BB] √
• Largest motor load   = _____ VA [CC] √
Four total loads   = _____ VA [DD]

Finding amps for ungrounded (phase) conductors - Phases A and B

I = VA ÷ V
I = _____ VA [EE] ÷ 240 V
I = _____ A [FF]

Finding amps for grounded (neutral) conductors

I = VA ÷ V
I = _____ VA [GG] ÷ 240 V
I = _____ A [HH]

Finding conductors

Ungrounded (phase) conductors - Phases A and B
I = _____ A [II] ÷ 6 (No. run per phases)
I = _____ A [JJ]

Grounded (neutral) conductors
220.61

_____ A [KK]
First 200 A x 100%   =   200   A
Next 305 A x 70%   = _____ A [LL]
Total load   = _____ A [MM]

310.4

I = _____ A [NN] ÷ 6 (No. run per phases)
I = _____ A [OO]

Sizing conductors (6 per phase)
Table 310.16; 310.4

Ungrounded (phase) conductors - Phases A and B
_____ KCMIL [PP] THWN copper conductors
Grounded (neutral) conductors
_____ AWG [QQ] copper conductor

10-27

Design Example Problem 10-2. What is the load in VA and amps for 30 multifamily dwelling units with the following loads? Note: Parallel service conductors, 6 times per phase.

Given loads:
- 10 - 820 sq. ft. dwelling units
- 10 - 960 sq. ft. dwelling units
- 10 - 1050 sq. ft. dwelling units
- 2 small appliance circuits per unit
- 1 laundry circuit per unit

120 V, single-phase loads
- 30 - 1000 VA dishwashers
- 30 - 1200 VA disposals

240 V, single-phase loads
- 30 - 12,000 VA ranges
- 30 - 6000 VA water heaters
- 30 - 20,000 VA heating units
- dryer facilities furnished by apartment complex

Sizing ungrounded (phase) conductors = •
Sizing grounded (neutral) conductors = √

Column 1
Calculating general lighting and receptacle load

Step 1: General lighting and receptacle load
Table 220.12
- 820 sq. ft. x 3 VA x 10 = 24,600 VA
- 960 sq. ft. x 3 VA x 10 = 28,800 VA
- 1050 sq. ft. x 3 VA x 10 = 31,500 VA
- Total load = 84,900 VA

Step 2: Small appliance and laundry load
220.52(A); (B)
- 1500 VA x 2 x 30 = 90,000 VA
- 1500 VA x 1 x 30 = 45,000 VA
- Total load = 219,900 VA

Step 3: Applying demand factors
Demand load 1; **Table 220.42**
- First 3000 VA x 100% = 3,000 VA
- Next 117,000 VA x 35% = 40,950 VA
- Remaining 99,900 VA x 25% = 24,975 VA
- Total load = 68,925 VA • √

Column 2
Calculating cooking equipment load

Step 1: Applying demand factors for phases
Demand load 2; **Table 220.55, Col. C**
- 30 ranges = 15 kW
- 1 kW per range = 30 kW
- Total load = 45 kW
- 45 kW x 1000 = 45,000 VA •

Step 2: Applying demand factors for neutral
220.61
- 45,000 VA x 70% = 31,500 VA √

Column 2
Calculating fixed appliance load

Step 1: Applying demand factors for phases
Demand load 4; **220.53**
- 1000 VA x 30 x 75% = 22,500 VA
- 1200 VA x 30 x 75% = 27,000 VA
- 6000 VA x 30 x 75% = 135,000 VA
- Total load = 184,500 VA •

Step 2: Applying demand factors for neutral
Demand load 4; **220.61; 220.53**
- 1000 VA x 30 x 75% = 22,500 VA
- 1200 VA x 30 x 75% = 27,000 VA
- Total load = 49,500 VA √

Column 3
Largest load between heating and A/C load

Step 1: Selecting largest load
Demand load 5; **220.60**
Heating unit
- 20,000 VA x 30 x 100% = 600,000 VA •

Column 4
Calculating largest motor load

Step 1: Selecting largest motor load for phases
220.14(C); 220.50; 430.24
- 1200 VA x 25% = 300 VA •

Step 2: Selecting largest motor load for neutral
220.14(C); 220.50; 430.24
- 1200 VA x 25% = 300 VA √

Calculating ungrounded (phase) conductors

- General lighting load = 68,925 VA •
- Cooking equipment load = 45,000 VA •
- Appliance load = 184,500 VA •
- Heating load = 600,000 VA •
- Largest motor load = 300 VA •
- Five total loads = 898,725 VA

Calculating grounded (neutral) conductors

- General lighting load = 68,925 VA √
- Cooking equipment load = 31,500 VA √
- Appliance load = 49,500 VA √
- Largest motor load = 300 VA √
- Four total loads = 150,225 VA

Finding amps for ungrounded (phase) conductors - Phases A and B

I = VA ÷ V
I = 898,725 VA ÷ 240 V
I = 3745 A

Finding amps for grounded (neutral) conductors

I = VA ÷ V
I = 150,225 VA ÷ 240 V
I = 626 A

Finding conductors

Ungrounded (phase) conductors - Phases A and B
I = 3745 ÷ 6 (No. run per phase)
I = 624 A

Grounded (neutral) conductors
220.61

626 A
- First 200 A x 100% = 200 A
- Next 426 A x 70% = 298 A
- Total load = 498 A

310.4
I = 498 A ÷ 6 (No. run per phase)
I = 83 A (1/0 AWG per **310.4**)

Sizing conductors (6 per phase)
Table 310.16; 310.4

Ungrounded (phase) conductors - Phases A and B
1500 KCMIL THWN copper
Grounded (neutral) conductors
1/0 AWG THWN copper

Residential Calculations - Multifamily Dwellings

Design Exercise Problem 10-2. What is the load in VA and amps for 15 multifamily dwelling units with the following loads? Note: Parallel service conductors, 6 times per phase.

Given loads:	120 V, single-phase loads	240 V, single-phase loads
• 5 - 1050 sq. ft. dwelling units • 5 - 1200 sq. ft. dwelling units • 5 - 1280 sq. ft. dwelling units • 2 small appliance circuits per unit • 1 laundry circuit per unit	• 15 - 1600 VA dishwashers • 15 - 1200 VA disposals	• 15 - 10,000 VA ranges • 15 - 5000 VA water heaters • 15 - 20,000 VA heating units • Dryer facilities furnished by apartment complex

Sizing ungrounded (phase) conductors = •
Sizing grounded (neutral) conductors = √

Column 1
Calculating general lighting and receptacle load

Step 1: General lighting and receptacle load
Table 220.12
- 1050 sq. ft. x 3 VA x 5 = _____ VA [A]
- 1200 sq. ft. x 3 VA x 5 = _____ VA [B]
- 1280 sq. ft. x 3 VA x 5 = _____ VA [C]
- Total load = _____ VA [D]

Step 2: Small appliance and laundry load
220.52(A); (B)
- 1500 VA x 2 x 15 = _____ VA [E]
- 1500 VA x 1 x 15 = _____ VA [F]
- Total load = _____ VA [G]

Step 3: Applying demand factors
Demand load 1; Table 220.42
- First 3000 VA x 100% = 3,000 VA
- Next 117,000 VA x 35% = _____ VA [H]
- Remaining 450 VA x 25% = _____ VA [I]
- Total load = _____ VA [J] • √

Column 2
Calculating cooking equipment load

Step 1: Applying demand factors for phases
Demand load 2; Table 220.55, Col. C
- 15 - 10,000 VA ranges = _____ VA [K] •

Step 2: Applying demand factors for neutral
220.61
- 30,000 VA x 70% = _____ VA [L] √

Column 2
Calculating fixed appliance load

Step 1: Applying demand factors for phases
Demand load 4; 220.53
- 1600 VA x 15 x 75% = _____ VA [M]
- 1200 VA x 15 x 75% = _____ VA [N]
- 5000 VA x 15 x 75% = _____ VA [O]
- Total load = _____ VA [P]

Step 2: Applying demand factors for neutral
Demand load 4; 220.61; 220.53
- 1600 VA x 15 x 75% = _____ VA [Q]
- 1200 VA x 15 x 75% = _____ VA [R]
- Total load = _____ VA [S] √

Column 3
Largest load between heating and A/C load

Step 1: Selecting largest load
Demand load 5; 220.60
Heating unit
- 20,000 VA x 15 x 100% = _____ VA [T] •

Column 4
Calculating largest motor load

Step 1: Selecting largest motor load for phases
220.14(C); 50; 430.24
- 1200 VA x 25% = _____ VA [U] •

Step 2: Selecting largest motor load for neutral
220.14; 430.24
- 1200 VA x 25% = _____ VA [V] √

Calculating ungrounded (phase) conductors
- General lighting load = _____ VA [W] •
- Cooking equipment load = _____ VA [X] •
- Appliance load = _____ VA [Y] •
- Heating load = _____ VA [Z] •
- Largest motor load = _____ VA [AA] •
- Five total loads = _____ VA [BB]

Calculating grounded (neutral) conductors
- General lighting load = _____ VA [CC] √
- Cooking equipment load = _____ VA [DD] √
- Appliance load = _____ VA [EE] √
- Largest motor load = _____ VA [FF] √
- Four total loads = _____ VA [GG]

Finding amps for ungrounded (phase) conductors - Phases A and B
I = VA ÷ V
I = _____ VA [HH] ÷ 240 V
I = _____ A [II]

Finding amps for grounded (neutral) conductors
I = VA ÷ V
I = _____ VA [JJ] ÷ 240 V
I = _____ A [KK]

Finding conductors

Ungrounded (phase) conductors - Phases A and B
I = _____ A [LL] ÷ 6 (No. run per phases)
I = _____ A [MM]

Grounded (neutral) conductors
220.61
_____ A [NN]
- First 200 A x 100% = 200 A
- Next 204 A x 70% = _____ A [OO]
- Total load = _____ A [PP]

Sizing conductors (6 per phase)
310.4
I = _____ A [QQ] ÷ 6 (No. run per phases)
I = _____ A [RR]

Table 310.16; 310.4

Ungrounded (phase) conductors - Phases A and B
_____ KCMIL [SS] THWN copper conductors
Grounded (neutral) conductors
_____ AWG [TT] copper conductor

10-29

Design Example Problem 10-3. What is the load in VA and amps for 25 multifamily dwelling units with the following loads? Note: Parallel service conductors, 3 times per phase.

Given loads:
- 25 - 1000 sq. ft. dwelling units
- 2 small appliance circuits per unit
- 1 laundry circuit per unit

120 V, single-phase loads
- 25 - 1000 VA dishwashers
- 25 - 1200 VA disposals

240 V, single-phase loads
- 25 - 12,000 VA ranges (gas)
- 25 - 6000 VA water heaters
- 25 - 20,000 VA heating units (gas)
- dryer facilities furnished by apartment complex

Sizing ungrounded (phase) conductors = •
Sizing grounded (neutral) conductors = √

Column 1
Calculating general lighting and receptacle load

Step 1: General lighting and receptacle load
Table 220.12
1000 sq. ft. x 3 VA x 25 = 75,000 VA

Step 2: Small appliance and laundry load
220.52(A); (B)
1500 VA x 2 x 25 = 75,000 VA
1500 VA x 1 x 25 = 37,500 VA
Total load = 187,500 VA

Step 3: Applying demand factors
Demand load 1; **Table 220.42**
First 3000 VA x 100% = 3,000 VA
Next 117,000 VA x 35% = 40,950 VA
Remaining 67,500 VA x 25% = 16,875 VA
Total load = 60,825 VA • √

Column 2
Calculating cooking equipment load

Step 1: Applying demand factors for phases
Demand load 2; **Table 220.55, Col. C**
(gas related load) = 0 VA •

Column 2
Calculating fixed appliance load

Step 1: Applying demand factors for phases
Demand load 4; **220.53**
1000 VA x 25 x 75% = 18,750 VA
1000 VA x 25 x 75% = 22,500 VA
6000 VA x 25 x 75% = 112,500 VA
Total load = 153,750 VA •

Step 2: Applying demand factors for neutral
Demand load 4; **220.61; 220.53**
1000 VA x 25 x 75% = 18,750 VA
1200 VA x 25 x 75% = 22,500 VA
Total load = 41,250 VA √

Column 3
Largest load between heating and A/C load

Step 1: Selecting largest load
Demand load 5; **220.60**
Heating unit
(gas related load) = 0 VA •

Column 4
Calculating largest motor load

Step 1: Selecting largest motor load for phases
220.14(C); 220.50; 430.24
1200 VA x 25% = 300 VA •

Step 2: Selecting largest motor load for neutral
220.14(C); 220.50; 430.24
1200 VA x 25% = 300 VA √

Calculating ungrounded (phase) conductors

General lighting load = 60,825 VA •
Cooking equipment load (gas) = 0 VA •
Appliance load = 153,750 VA •
Heating load (gas) = 0 VA •
Largest motor load = 300 VA •
Five total loads = 214,875 VA

Calculating grounded (neutral) conductors

General lighting load = 60,825 VA √
Cooking equipment load (gas) = 0 VA √
Appliance load = 41,250 VA √
Largest motor load = 300 VA √
Four total loads = 102,375 VA

Finding amps for ungrounded (phase) conductors - Phases A and B

I = VA ÷ V
I = 214,875 VA ÷ 240 V
I = 895 A

Finding amps for grounded (neutral) conductors

I = VA ÷ V
I = 102,375 VA ÷ 240 V
I = 427 A

Finding conductors

Ungrounded (phase) conductors - Phases A and B
I = 895 ÷ 3 (No. run per phase)
I = 298 A

Grounded (neutral) conductors
220.61
427 A
First 200 A x 100% = 200 A
Next 227 A x 70% = 159 A
Total load = 359 A
310.4
I = 359 A ÷ 3 (No. run per phase)
I = 120 A (1/0 AWG per **310.4**)

Sizing conductors (3 per phase)
Table 310.16; 310.4

Ungrounded (phase) conductors - Phases A and B
350 KCMIL THWN copper
Grounded (neutral) conductors
1/0 AWG THWN copper

Residential Calculations - Multifamily Dwellings

Design Exercise Problem 10-3. What is the load in VA and amps for 15 multifamily dwelling units with the following loads? Note: Parallel service conductors, 2 times per phase.

Given loads:
- 5 - 1050 sq. ft. dwelling units
- 5 - 1200 sq. ft. dwelling units
- 5 - 1280 sq. ft. dwelling units
- 2 small appliance circuits per unit
- 1 laundry circuit per unit

120 V, single-phase loads
- 15 - 1600 VA dishwashers
- 15 - 1200 VA disposals

240 V, single-phase loads
- 15 - 10,000 VA ranges (gas)
- 15 - 5000 VA water heaters
- 15 - 20,000 VA heating units (gas)
- Dryer facilities furnished by apartment complex

Sizing ungrounded (phase) conductors = •
Sizing grounded (neutral) conductors = √

Column 1
Calculating general lighting and receptacle load

Step 1: General lighting and receptacle load
Table 220.12
- 1050 sq. ft. x 3 VA x 5 = _____ VA [A]
- 1200 sq. ft. x 3 VA x 5 = _____ VA [B]
- 1280 sq. ft. x 3 VA x 5 = _____ VA [C]
- Total load = _____ VA [D]

Step 2: Small appliance and laundry load
220.52(A); (B)
- 1500 VA x 2 x 15 = _____ VA [E]
- 1500 VA x 1 x 15 = _____ VA [F]
- Total load = _____ VA [G]

Step 3: Applying demand factors
Demand load 1; Table 220.42
- First 3000 VA x 100% = 3,000 VA
- Next 117,000 VA x 35% = _____ VA [H]
- Remaining 450 VA x 25% = _____ VA [I]
- Total load = _____ VA [J] • √

Column 2
Calculating cooking equipment load

Step 1: Applying demand factors for phases
Demand load 2; Table 220.55, Col. C
(gas related load) = _____ VA [K] •

Column 2
Calculating fixed appliance load

Step 1: Applying demand factors for phases
Demand load 4; 220.53
- 1600 VA x 15 x 75% = _____ VA [L]
- 1200 VA x 15 x 75% = _____ VA [M]
- 5000 VA x 15 x 75% = _____ VA [N]
- Total load = _____ VA [O] •

Step 2: Applying demand factors for neutral
Demand load 4; 220.61; 220.53
- 1600 VA x 15 x 75% = _____ VA [P]
- 1200 VA x 15 x 75% = _____ VA [Q]
- Total load = _____ VA [R] √

Column 3
Largest load between heating and A/C load

Step 1: Selecting largest load
Demand load 5; 220.53
Heating unit
(gas related load) = _____ VA [S]

Column 4
Calculating largest motor load

Step 1: Selecting largest motor load for phases
220.14(C); 220.50; 430.24
- 1200 VA x 25% = _____ VA [T] •

Step 2: Selecting largest motor load for neutral
220.14(C); 220.50; 430.24
- 1200 VA x 25% = _____ VA [U] √

Calculating ungrounded (phase) conductors
- General lighting load = _____ VA [V] •
- Cooking equipment load (gas) = _____ VA [W] •
- Appliance load = _____ VA [X] •
- Heating load (gas) = _____ VA [Y] •
- Largest motor load = _____ VA [Z] •
- Five total loads = _____ VA [AA]

Calculating grounded (neutral) conductors
- General lighting load = _____ VA [BB] √
- Cooking equipment load (gas) = _____ VA [CC] √
- Appliance load = _____ VA [DD] √
- Largest motor load = _____ VA [EE] √
- Four total loads = _____ VA [FF]

Finding amps for ungrounded (phase) conductors - Phases A and B

I = VA ÷ V
I = _____ VA [GG] ÷ 240 V
I = _____ A [HH]

Finding amps for grounded (neutral) conductors

I = VA ÷ V
I = _____ VA [II] ÷ 240 V
I = _____ A [JJ]

Finding conductors

Ungrounded (phase) conductors - Phases A and B
I = _____ A [KK] ÷ 2 (No. run per phases)
I = _____ A [LL]

Grounded (neutral) conductors
220.61

_____ A [MM]
- First 200 A x 100% = 200 A
- Next 116 A x 70% = _____ A [NN]
- Total load = _____ A [OO]

310.4
I = _____ A [PP] ÷ 2 (No. run per phases)
I = _____ A [QQ]

Sizing conductors (2 per phase)
Table 310.16; 310.4

Ungrounded (phase) conductors - Phases A and B
_____ KCMIL [RR] THWN copper conductors
Grounded (neutral) conductors
_____ AWG [SS] copper conductor

10-31

Design Example Problem 10-4. What is the load in VA and amps for 10 multifamily dwelling units with the following loads? Note: Parallel service conductors, 3 times per phase.

Given loads:	120 V, single-phase loads	240 V, single-phase loads
• 10 - 720 sq. ft. dwelling units • 2 small appliance circuits per unit • 1 laundry circuit per unit	• 10 - 1000 VA dishwashers • 10 - 1200 VA disposals • 1,680 VA A/C window unit • 1,800 VA A/C window unit • 1,920 VA A/C window unit	• 10 - 12,000 VA ranges • 10 - 6000 VA water heaters • 10 - 10,000 VA heating units • dryer facilities furnished by apartment complex

Sizing ungrounded (phase) conductors = •
Sizing grounded (neutral) conductors = √

Column 1
Calculating general lighting and receptacle load

Step 1: General lighting and receptacle load
Table 220.12
720 sq. ft. x 3 VA x 10 = 21,600 VA

Step 2: Small appliance and laundry load
220.52(A); (B)
1500 VA x 2 x 10 = 30,000 VA
1500 VA x 1 x 10 = 15,000 VA
Total load = 66,600 VA

Step 3: Applying demand factors
Demand load 1; Table 220.42
First 3000 VA x 100% = 3,000 VA
Next 63,600 VA x 35% = 22,260 VA
Total load = 25,260 VA •√

Column 2
Calculating cooking equipment load

Step 1: Applying demand factors for phases
Demand load 2; Table 220.55, Col. C
10 - 12,000 VA ranges = 25,000 VA •

Step 2: Applying demand factors for phases
220.61
25,000 VA x 70% = 17,500 VA √

Column 2
Calculating fixed appliance load

Step 1: Applying demand factors for phases
Demand load 4; 220.53
1000 VA x 10 x 75% = 7,500 VA
1200 VA x 10 x 75% = 9,000 VA
6000 VA x 10 x 75% = 45,000 VA
Total load = 61,500 VA •

Step 2: Applying demand factors for neutral
Demand load 4; 220.61; 220.53
1000 VA x 10 x 75% = 7,500 VA
1200 VA x 10 x 75% = 9,000 VA
Total load = 16,500 VA √

Column 3
Largest load between heating and A/C load

Step 1: Selecting largest load
Demand load 5; 220.60
Heating unit
10,000 VA x 10 x 100% = 100,000 VA •
A/C unit
5400 x 10 x 100% = 54,000 VA √

Column 4
Calculating largest motor load

Step 1: Selecting largest motor load for phases
220.14(C); 220.50; 430.24
1920 VA x 25% = 480 VA •

Step 2: Selecting largest motor load for neutral
220.14(C); 220.50; 430.24
1920 VA x 25% = 480 VA √

Calculating ungrounded (phase) conductors

General lighting load = 25,260 VA •
Cooking equipment load = 25,000 VA •
Appliance load = 61,500 VA •
Heating load = 100,000 VA •
Largest motor load = 480 VA •
Five total loads = 212,240 VA

Calculating grounded (neutral) conductors

General lighting load = 25,260 VA √
Cooking equipment load = 17,500 VA √
Appliance load = 16,500 VA √
A/C load = 54,000 VA √
Largest motor load = 480 VA √
Five total loads = 113,740 VA

Finding amps for ungrounded (phase) conductors - Phases A and B

I = VA ÷ V
I = 212,240 VA ÷ 240 V
I = 884 A

Finding amps for grounded (neutral) conductors

I = VA ÷ V
I = 113,740 VA ÷ 240 V
I = 474 A

Finding conductors

Ungrounded (phase) conductors - Phases A and B
I = 884 ÷ 3 (No. run per phase)
I = 295 A

Grounded (neutral) conductors
220.61

474 A
First 200 A x 100% = 200 A
Next 274 A x 70% = 192 A
Total load = 392 A
310.4
I = 392 A ÷ 3 (No. run per phase)
I = 131 A (1/0 AWG per 310.4)

Sizing conductors (3 per phase)
Table 310.16; 310.4

Ungrounded (phase) conductors - Phases A and B
350 KCMIL THWN copper
Grounded (neutral) conductors
1/0 AWG THWN copper

Residential Calculations - Multifamily Dwellings

Design Exercise Problem 10-4. What is the load in VA and amps for 8 multifamily dwelling units with the following loads? Note: Parallel service conductors, 3 times per phase.

Given loads:
- 8 - 660 sq. ft. dwelling units
- 2 small appliance circuits per unit
- 1 laundry circuit per unit

120 V, single-phase loads
- 8 - 1600 VA dishwashers
- 8 - 1200 VA disposals
- 1440 VA A/C window unit
- 1680 VA A/C window unit
- 1800 VA A/C window unit

240 V, single-phase loads
- 8 - 10,000 VA ranges
- 8 - 5000 VA water heaters
- 8 - 10,000 VA heating units
- Dryer facilities furnished by apartment complex

Sizing ungrounded (phase) conductors = •
Sizing grounded (neutral) conductors = √

Column 1
Calculating general lighting and receptacle load

Step 1: General lighting and receptacle load
Table 220.12
660 sq. ft. x 3 VA x 8 = _____ VA [A]

Step 2: Small appliance and laundry load
220.52(A); (B)
1500 VA x 2 x 8 = _____ VA [B]
1500 VA x 1 x 8 = _____ VA [C]
Total load = _____ VA [D]

Step 3: Applying demand factors
Demand load 1; **Table 220.42**
First 3000 VA x 100% = 3,000 VA
Next 48,840 VA x 35% = _____ VA [E]
Total load = _____ VA [F] •√

Column 2
Calculating cooking equipment load

Step 1: Applying demand factors for phases
Demand load 2; **Table 220.55, Col. C**
8 - 10,000 VA ranges = _____ VA [G] •

Step 2: Applying demand factors for neutral
220.61
23,000 VA x 70% = _____ VA [H] √

Column 2
Calculating fixed appliance load

Step 1: Applying demand factors for phases
Demand load 4; **220.53**
1600 VA x 8 x 75% = _____ VA [I]
1200 VA x 8 x 75% = _____ VA [J]
5000 VA x 8 x 75% = _____ VA [K]
Total load = _____ VA [L] •

Step 2: Applying demand factors for neutral
Demand load 4; **220.61; 220.53**
1600 VA x 8 x 75% = _____ VA [M]
1200 VA x 8 x 75% = _____ VA [N]
Total load = _____ VA [O] √

Column 3
Largest load between heating and A/C load

Step 1: Selecting largest load
Demand load 5; **220.60**
Heating unit
10,000 VA x 8 x 100% = _____ VA [P] •
A/C unit
4920 x 8 x 100% = _____ VA [Q] √

Column 4
Calculating largest motor load

Step 1: Selecting largest motor load for phases
220.14(C); 50; 430.60
1800 VA x 25% = _____ VA [R] •

Step 2: Selecting largest motor load for neutral
220.14(C); 50; 430.24
1800 VA x 25% = _____ VA [S] √

Calculating ungrounded (phase) conductors

- General lighting load = _____ VA [T] •
- Cooking equipment load = _____ VA [U] •
- Appliance load = _____ VA [V] •
- Heating load = _____ VA [W] •
- Largest motor load = _____ VA [X] •
- Five total loads = _____ VA [Y]

Calculating grounded (neutral) conductors

- General lighting load = _____ VA [Z] √
- Cooking equipment load = _____ VA [AA] √
- Appliance load = _____ VA [BB] √
- A/C load = _____ VA [CC] √
- Largest motor load = _____ VA [DD] √
- Five total loads = _____ VA [EE]

Finding amps for ungrounded (phase) conductors - Phases A and B

I = VA ÷ V
I = _____ VA [FF] ÷ 240 V
I = _____ A [GG]

Finding amps for grounded (neutral) conductors

I = VA ÷ V
I = _____ VA [HH] ÷ 240 V
I = _____ A [II]

Finding conductors

Ungrounded (phase) conductors - Phases A and B
I = _____ A [JJ] ÷ 3 (No. run per phases)
I = _____ A [KK]

Grounded (neutral) conductors
220.61

_____ A [LL]
First 200 A x 100% = 200 A
Next 187A x 70% = _____ A [MM]
Total load = _____ A [NN]

310.4

I = _____ A [OO] ÷ 3 (No. run per phases)
I = _____ A [PP]

Sizing conductors (3 per phase)
Table 310.16; 310.4

Ungrounded (phase) conductors - Phases A and B
_____ KCMIL [QQ] THWN copper conductors
Grounded (neutral) conductors
_____ AWG [RR] copper conductor

Design Example Problem 10-5. What is the load in VA and amps for 8 multifamily dwelling units with the following loads? Note: Parallel service conductors, 3 times per phase.

Given loads:
- 4 - 1000 sq. ft. dwelling units
- 4 - 1200 sq. ft. dwelling units
- 2 small appliance circuits per unit
- 1 laundry circuit per unit

120 V, single-phase loads
- 8 - 1000 VA dishwashers
- 8 - 1200 VA disposals

240 V, single-phase loads
- 8 - 12,000 VA ranges
- 8 - 6000 VA water heaters
- 8 - 10,000 VA heating units
- 8 - 4500 VA heat pumps
- dryer facilities furnished by apartment complex

Sizing ungrounded (phase) conductors = •
Sizing grounded (neutral) conductors = √

Column 1
Calculating general lighting and receptacle load

Step 1: General lighting and receptacle load
Table 220.12
1000 sq. ft. x 3 VA x 4	=	12,000 VA
1200 sq. ft. x 3 VA x 4	=	14,400 VA

Step 2: Small appliance and laundry load
220.52(A); (B)
1500 VA x 2 x 8	=	24,000 VA
1500 VA x 1 x 8	=	12,000 VA
Total load	=	62,400 VA

Step 3: Applying demand factors
Demand load 1; **Table 220.42**
First 3000 VA x 100%	=	3,000 VA
Next 59,400 VA x 35%	=	20,790 VA
Total load	=	23,790 VA •√

Column 2
Calculating cooking equipment load

Step 1: Applying demand factors for phases
Demand load 2; **Table 220.55, Col. C**
8 - 12,000 VA ranges	=	23,000 VA •

Step 2: Applying demand factors for neutral
220.61
23,000 VA x 70%	=	16,100 VA √

Column 2
Calculating fixed appliance load

Step 1: Applying demand factors for phases
Demand load 4; **220.53**
1000 VA x 8 x 75%	=	6,000 VA
1200 VA x 8 x 75%	=	7,200 VA
6000 VA x 8 x 75%	=	36,000 VA
Total load	=	49,200 VA •

Step 2: Applying demand factors for neutral
Demand load 4; **220.61; 220.53**
1000 VA x 8 x 75%	=	6,000 VA
1200 VA x 8 x 75%	=	7,200 VA
Total load	=	13,200 VA √

Column 3
Largest load between heating and A/C load

Step 1: Selecting largest load
Demand load 5; **220.60**
Heating unit
14,500 VA x 8 x 100%	=	116,000 VA •

Column 4
Calculating largest motor load

Step 1: Selecting largest motor load for phases
220.14(C); 220.50; 430.24
4500 VA x 25%	=	1,125 VA •

Step 2: Selecting largest motor load for neutral
220.14(C); 220.50; 430.24
1200 VA x 25%	=	300 VA √

Calculating ungrounded (phase) conductors

General lighting load	=	23,790 VA •
Cooking equipment load	=	23,000 VA •
Appliance load	=	49,200 VA •
Heating load	=	116,000 VA •
Largest motor load	=	1,125 VA •
Five total loads	=	213,115 VA

Calculating grounded (neutral) conductors

General lighting load	=	23,790 VA √
Cooking equipment load	=	16,100 VA √
Appliance load	=	13,200 VA √
Largest motor load	=	300 VA √
Four total loads	=	53,390 VA

Finding amps for ungrounded (phase) conductors - Phases A and B

I = VA ÷ V
I = 213,115 VA ÷ 240 V
I = 888 A

Finding amps for grounded (neutral) conductors

I = VA ÷ V
I = 53,390 VA ÷ 240 V
I = 222 A

Finding conductors

Ungrounded (phase) conductors - Phases A and B
I = 888 ÷ 3 (No. run per phase)
I = 296 A

Grounded (neutral) conductors
220.61

222 A
First 200 A x 100%	=	200 A
Next 22 A x 70%	=	15 A
Total load	=	215 A

310.4
I = 215 A ÷ 3 (No. run per phase)
I = 72 A (1/0 AWG per **310.4**)

Sizing conductors (3 per phase)
Table 310.16; 310.4

Ungrounded (phase) conductors - Phases A and B
350 KCMIL THWN copper
Grounded (neutral) conductors
1/0 AWG THWN copper

Residential Calculations - Multifamily Dwellings

Design Exercise Problem 10-5. What is the load in VA and amps for 4 multifamily dwelling units with the following loads? Note: Parallel service conductors, 2 times per phase.

Given loads:
- 2 - 1200 sq. ft. dwelling units
- 2 - 1600 sq. ft. dwelling units
- 2 small appliance circuits per unit
- 1 laundry circuit per unit

120 V, single-phase loads
- 4 - 1600 VA dishwashers
- 4 - 1200 VA disposals

240 V, single-phase loads
- 4 - 10,000 VA ranges
- 4 - 5000 VA water heaters
- 4 - 10,000 VA heating units
- 4 - 5000 VA heat pumps
- Dryer facilities furnished by apartment complex

Sizing ungrounded (phase) conductors = •
Sizing grounded (neutral) conductors = √

Column 1
Calculating general lighting and receptacle load

- **Step 1:** General lighting and receptacle load
  Table 220.12
  1200 sq. ft. x 3 VA x 2 = _____ VA [A]
  1600 sq. ft. x 3 VA x 2 = _____ VA [B]

- **Step 2:** Small appliance and laundry load
  220.52(A); (B)
  1500 VA x 2 x 4 = _____ VA [C]
  1500 VA x 1 x 4 = _____ VA [D]
  Total load = _____ VA [E]

- **Step 3:** Applying demand factors
  Demand load 1; **Table 220.42**
  First 3000 VA x 100% = 3,000 VA
  Next 31,800 VA x 35% = _____ VA [F]
  Total load = _____ VA [G] • √

Column 2
Calculating cooking equipment load

- **Step 1:** Applying demand factors for phases
  Demand load 2; **Table 220.55, Col. C**
  4 - 10,000 VA ranges = _____ VA [H] •

- **Step 2:** Applying demand factors for neutral
  220.61
  17,000 VA x 70% = _____ VA [I] √

Column 2
Calculating fixed appliance load

- **Step 1:** Applying demand factors for phases
  Demand load 4; **220.53**
  1600 VA x 4 x 75% = _____ VA [J]
  1200 VA x 4 x 75% = _____ VA [K]
  5000 VA x 4 x 75% = _____ VA [L]
  Total load = _____ VA [M] •

- **Step 2:** Applying demand factors for neutral
  Demand load 4; **220.61; 220.53**
  1600 VA x 4 x 75% = _____ VA [N]
  1200 VA x 4 x 75% = _____ VA [O]
  Total load = _____ VA [P] √

Column 3
Largest load between heating and A/C load

- **Step 1:** Selecting largest load
  Demand load 5; **220.60**
  Heating unit
  15,000 VA x 4 x 100% = _____ VA [Q] •

Column 4
Calculating largest motor load

- **Step 1:** Selecting largest motor load for phases
  220.14(C); 220.50; 430.24
  5000 VA x 25% = _____ VA [R] •

- **Step 2:** Selecting largest motor load for neutral
  220.14(C); 50; 430.24
  1200 VA x 25% = _____ VA [S] √

Calculating ungrounded (phase) conductors
- General lighting load = _____ VA [T] •
- Cooking equipment load = _____ VA [U] •
- Appliance load = _____ VA [V] •
- Heating load = _____ VA [W] •
- Largest motor load = _____ VA [X] •
- Five total loads = _____ VA [Y]

Calculating grounded (neutral) conductors
- General lighting load = _____ VA [Z] √
- Cooking equipment load = _____ VA [AA] √
- Appliance load = _____ VA [BB] √
- Largest motor load = _____ VA [CC] √
- Four total loads = _____ VA [DD]

Finding amps for ungrounded (phase) conductors - Phases A and B

I = VA ÷ V
I = _____ VA [EE] ÷ 240 V
I = _____ A [FF]

Finding amps for grounded (neutral) conductors

I = VA ÷ V
I = _____ VA [GG] ÷ 240 V
I = _____ A [HH]

Finding conductors

Ungrounded (phase) conductors - Phases A and B
I = _____ A [II] ÷ 2 (No. run per phases)
I = _____ A [JJ]

310.4
I = _____ A [KK] ÷ 2 (No. run per phases)
I = _____ A [LL]

Sizing conductors (2 per phase)
Table 310.16; 310.4

Ungrounded (phase) conductors - Phases A and B
_____ KCMIL [MM] THWN copper conductors
Grounded (neutral) conductors
_____ AWG [NN] copper conductor

Design Example Problem 10-6. What is the load in VA for 15 multifamily dwelling units with the following loads?

Given Loads:	120 V, single-phase loads	208 V, three-phase loads
• 15 - 1000 sq. ft. dwelling units • 2 small appliance circuits per unit • 1 laundry circuit per unit	• 15 - 1000 VA dishwashers • 15 - 1200 VA disposals • 15 - 1920 VA bathroom heaters (ceiling mounted)	• 15 - 12,000 VA ranges • 15 - 5000 VA dryers • 15 - 6000 VA water heaters • 15 - 20,000 VA heating units • 15 - 5040 VA A/C units

Column 1
Calculating general lighting and receptacle load

			Phase A	Phase B	Phase C	Neutral

Step 1: General lighting and receptacle load
Table 220.12
1000 sq. ft. x 3 VA x 15 = 45,000 VA

Step 2: Small appliance and laundry load
220.52(A); (B)
1500 VA x 2 x 15 = 45,000 VA
1500 VA x 1 x 15 = 22,500 VA
Total load   = 112,500 VA

Step 3: Applying demand factors
Demand Load 1; **Table 220.42**
First 3000 VA x 100% = 3,000 VA
Next 109,500 VA x 35% = 38,325 VA
Total load   = 41,325 VA
                      = 41,325 VA ÷ 3 =   13,775 VA   13,775 VA   13,775 VA   13,775 VA

Column 2
Calculating cooking equipment load

Step 1: Applying demand factors
Demand Load 2; **Table 220.55, Col. C**
15 ranges ÷ 3Ø (phases) = 5
5 ranges per phase x 2   = 10
10 ranges   = 25 kW
25 kW x 1000   = 25,000 VA
25,000 VA ÷ 2 x 3Ø (phases) = 37,500 VA ÷ 3 = 12,500 VA   12,500 VA   12,500 VA

Step 2: Applying demand factors for neutral
Demand Load 2; **220.61**
12,500 VA x 70% = 8,750 VA   8,750 VA

Column 2
Calculating fixed appliance load

Step 1: Applying demand factors for phases and neutral
Demand Load 3; **220.53; 220.61**
1000 VA x 15 x 75% = 11,250 VA ÷ 3 = 3,750 VA   3,750 VA   3,750 VA   3,750 VA
1200 VA x 15 x 75% = 13,500 VA ÷ 3 = 4,500 VA   4,500 VA   4,500 VA   4,500 VA
1920 VA x 15 x 75% = 21,600 VA ÷ 3 = 7,200 VA   7,200 VA   7,200 VA   7,200 VA
6000 VA x 15 x 75% = 67,500 VA ÷ 3 = 22,500 VA 22,500 VA 22,500 VA

Column 2
Calculating dryer load

Step 1: Applying demand factors for phases
Demand Load 4; **220.54; Table 220.54**
15 dryers ÷ 3Ø (phases) = 5
5 dryers per phase x 2   = 10
5 kW x 10 x 50%   = 25 kW
25 kW x 1,000   = 25,000 VA
25,000 VA ÷ 2 x 3Ø (phases) = 37,500 VA ÷ 3 = 12,500 VA   12,500 VA   12,500 VA

Step 2: Applying demand factors for neutral
Demand Load 4; **220.22**
12,500 VA x 70% = 8,750 VA   8,750 VA

Column 3
Calculating largest load between heating and A/C load

Step 1: Selecting the largest load
Demand Load 5; **220.61**
Heating unit
20,000 x 15 x 100% = 300,000 VA ÷ 3 = 100,000 VA   100,000 VA   100,000 VA

Column 4
Calculating largest motor load

Step 1: Selecting largest motor load for phases
220.14(C); 220.50; 430.24
1920 VA x 25% = 480 VA   480 VA

Step 2: Selecting largest motor load for neutral
220.14(C); 220.50; 430.24
1920 VA x 25% = 480 VA   480 VA
Total load   =   177,205 VA   176,725 VA   176,725 VA   47,205 VA

Residential Calculations - Multifamily Dwellings

Design Exercise Problem 10-6. What is the load in VA for 10 multifamily dwelling units with the following loads?

Given Loads:	120 V, single-phase loads	208 V, three-phase loads
• 10 - 1200 sq. ft. dwelling units • 2 small appliance circuits per unit • 1 laundry circuit per unit	• 10 - 1600 VA dishwashers • 10 - 1200 VA disposals	• 10 - 10,000 VA ranges • 10 - 4500 VA dryers • 10 - 5000 VA water heaters • 10 - 20,000 VA heating units • 10 - 6500 VA A/C units

Column 1
Calculating general lighting and receptacle load   Phase A   Phase B   Phase C   Neutral

Step 1: General lighting and receptacle load
1200 sq. ft. x 3 VA x 10 = _____ [A] VA

Step 2: Small appliance and laundry load
1500 VA x 2 x 10 = _____ [B] VA
1500 VA x 1 x 10 = _____ [C] VA
Total load = _____ [D] VA

Step 3: Applying demand factors
First 3000 VA x 100% = _____ [E] VA
Next 78,000 VA x 35% = _____ [F] VA
Total load = _____ [G] VA ÷ 3 = _____ [H] VA _____ [I] VA _____ [J] VA _____ [K] VA

Column 2
Calculating cooking equipment load

Step 1: Applying demand factors
Demand Load 2; **Table 220.55, Col. C**
10 ranges ÷ 3Ø (phases) = _____ [L] kW
4 ranges per phase x 2 = _____ [M] kW
8 ranges = _____ [N] kW
23 kW x 1,000 = _____ [O] VA
23,000 VA ÷ 2 x 3Ø (phases) = _____ [P] VA ÷ 3 = _____ [Q] VA _____ [R] VA _____ [S] VA

Step 2: Applying demand factors for neutral
Demand Load 2; **220.61**
11,500 VA x 70% = _____ [T] VA _____ [U] VA

Column 2
Calculating fixed appliance load

Step 1: Applying demand factors for phases and neutral
Demand Load 3; **220.53; 220.61**
1200 VA x 10 x 75% = _____ [V] VA ÷ 3 = _____ [W] VA _____ [X] VA _____ [Y] VA _____ [Z] VA
1600 VA x 10 x 75% = _____ [AA] VA ÷ 3 = _____ [BB] VA _____ [CC] VA _____ [DD] VA _____ [EE] VA
5000 VA x 10 x 75% = _____ [FF] VA ÷ 3 = _____ [GG] VA _____ [HH] VA _____ [II] VA

Column 2
Calculating dryer load

Step 1: Applying demand factors for phases
Demand Load 4; **220.54; Table 220.54**
10 dryers ÷ 3Ø (phases) = _____ [JJ]
4 dryers per phase x 2 = _____ [KK]
5 kW x 8 x 60% = _____ [LL] kW
24 kW x 1,000 = _____ [MM] VA
24,000 VA ÷ 2 x 3Ø (phases) = _____ [NN] VA ÷ 3 = _____ [OO] VA _____ [PP] VA _____ [QQ] VA

Step 2: Applying demand factors for neutral
Demand Load 4; **220.61**
12,000 VA x 70% = _____ [RR] VA _____ [SS] VA

Column 3
Calculating largest load between heating and A/C load

Step 1: Selecting the largest load
Demand Load 5; **220.60**
Heating unit
20,000 x 10 x 100% = _____ [TT] VA ÷ 3 = _____ [UU] VA _____ [VV] VA _____ [WW] VA

Column 4
Calculating largest motor load

Step 1: Selecting largest motor load for phases
220.14(C); 220.50; 430.24
1200 VA x 25% = _____ [XX] VA _____ [YY] VA

Step 2: Selecting largest motor load for neutral
220.14(C); 220.50; 430.24
1200 VA x 25% = _____ [ZZ] VA _____ [AAA] VA
Total load = _____ [BBB] VA _____ [CCC] VA _____ [DDD] VA _____ [EEE] VA

10-37

Design Example Problem 10-7. What is the load in amps for 15 multifamily dwelling units with the following loads?

Given Loads:	120 V, single-phase loads	208 V, three-phase loads
• 15 - 1000 sq. ft. dwelling units • 2 small appliance circuits per unit • 1 laundry circuit per unit	• 15 - 1000 VA dishwashers • 15 - 1200 VA disposals • 15 - 1920 VA bathroom heaters (ceiling mounted)	• 15 - 12,000 VA ranges • 15 - 5000 VA dryers • 15 - 6000 VA water heaters • 15 - 20,000 VA heating units • 15 - 5040 VA A/C units

Column 1
Calculating general lighting and receptacle load

			Phase A	Phase B	Phase C	Neutral

Step 1: General lighting and receptacle load
Table 220.12
1000 sq. ft. x 3 VA x 15 = 45,000 VA

Step 2: Small appliance and laundry load
220.52(A); (B)
1500 VA x 2 x 15 = 45,000 VA
1500 VA x 1 x 15 = 22,500 VA
Total load = 112,500 VA

Step 3: Applying demand factors
Demand Load 1; **Table 220.42**
First 3000 VA x 100% = 3,000 VA
Next 109,500 VA x 35% = 38,325 VA
Total load = 41,325 VA
= 41,325 VA ÷ 120V ÷ 3 = 115 A | 115 A | 115 A | 115 A

Column 2
Calculating cooking equipment load

Step 1: Applying demand factors
Demand Load 2; **Table 220.55, Col. C**
15 ranges ÷ 3Ø (phases) = 5
5 ranges per phase x 2 = 10
10 ranges = 25 kW
25 kW x 1000 = 25,000 VA
25,000 VA ÷ 2 x 3Ø (phases) = 37,500 VA ÷ 360V ÷ 3 = 35 A | 35 A | 35 A

Step 2: Applying demand factors for neutral
Demand Load 2; **220.61**
35 A x 70% = 25 A | | | | 25 A

Column 2
Calculating fixed appliance load

Step 1: Applying demand factors for phases and neutral
Demand 3; **220.53**
1000 VA x 15 x 75% = 11,250 VA ÷ 120V ÷ 3 = 31 A | 31 A | 31 A | 31 A
1200 VA x 15 x 75% = 13,500 VA ÷ 120V ÷ 3 = 38 A | 38 A | 38 A | 38 A
1920 VA x 15 x 75% = 21,600 VA ÷ 120V ÷ 3 = 60 A | 60 A | 60 A | 60 A
6000 VA x 15 x 75% = 67,500 VA ÷ 360V ÷ 3 = 63 A | 63 A | 63 A

Column 2
Calculating dryer load

Step 1: Applying demand factors for phases
Demand Load 4; **220.54; Table 220.54**
15 dryers ÷ 3Ø (phases) = 5
5 dryers per phase x 2 = 10
5 kW x 10 x 50% = 25 kW
25 kW x 1,000 = 25,000 VA
25,000 VA ÷ 2 x 3Ø (phases) = 37,500 VA ÷ 360V ÷ 3 = 35 A | 35 A | 35 A

Step 2: Applying demand factors for neutral
Demand Load 4; **220.61**
35 A x 70% = 25 A | | | | 25 A

Column 3
Calculating largest load between heating and A/C load

Step 1: Selecting the largest load
Demand Load 5; **220.60**
Heating unit
20,000 x 15 x 100% = 300,000 VA ÷ 360V ÷ 3 = 278 A | 278 A | 278 A

Column 4
Calculating largest motor load

Step 1: Selecting largest motor load for phases
220.14(C); 220.50; 430.24
16 A x 25% = 4 A | 4 A

Step 2: Selecting largest motor load for neutral
220.14(C); 220.50; 430.24
16 A x 25% = 4 A | | | | 4 A
Total load = 659 A | 655 A | 655 A | 298 A

Residential Calculations - Multifamily Dwellings

Design Exercise Problem 10-7. What is the load in amps for 10 multifamily dwelling units with the following loads?

Given Loads:	120 V, single-phase loads	208 V, three-phase loads
• 10 - 1200 sq. ft. dwelling units • 2 small appliance circuits per unit • 1 laundry circuit per unit	• 10 - 1600 VA dishwashers • 10 - 1200 VA disposals	• 10 - 10,000 VA ranges • 10 - 4500 VA dryers • 10 - 5000 VA water heaters • 10 - 20,000 VA heating units • 10 - 6500 VA A/C units

Column 1
Calculating general lighting and receptacle load

                                                                      Phase A   **Phase B**   **Phase C**   **Neutral**

Step 1: General lighting and receptacle load
1200 sq. ft. x 3 VA x 10 = _____ [A] VA

Step 2: Small appliance and laundry load
1500 VA x 2 x 10 = _____ [B] VA
1500 VA x 1 x 10 = _____ [C] VA
Total load = _____ [D] VA

Step 3: Applying demand factors
First 3000 VA x 100% = _____ [E] VA
Next 78,000 VA x 35% = _____ [F] VA
Total load = _____ [G] VA ÷ 120V ÷ 3 = _____ [H] A _____ [I] A _____ [J] A _____ [K] A

Column 2
Calculating cooking equipment load

Step 1: Applying demand factors
Demand Load 2; **Table 220.55, Col. C**
10 ranges ÷ 3Ø (phases) = _____ [L] kW
4 ranges per phase x 2 = _____ [M] kW
8 ranges = _____ [N] kW
23 kW x 1,000 = _____ [O] VA
23,000 VA ÷ 2 x 3Ø (phases) = _____ [P] VA ÷ 360V ÷ 3 = _____ [Q] A _____ [R] A _____ [S] A

Step 2: Applying demand factors for neutral
Demand Load 2; **220.61**
32 A x 70% = _____ [T] A _____ [U] A

Column 2
Calculating fixed appliance load

Step 1: Applying demand factors for phases and neutral
Demand Load 3; **220.53**
1200 VA x 10 x 75% = _____ [V] VA ÷ 120V ÷ 3 = _____ [W] A _____ [X] A _____ [Y] A _____ [Z] A
1600 VA x 10 x 75% = _____ [AA] VA ÷ 120V ÷ 3 = _____ [BB] A _____ [CC] A _____ [DD] A _____ [EE] A
5000 VA x 10 x 75% = _____ [FF] VA ÷ 360V ÷ 3 = _____ [GG] A _____ [HH] A _____ [II] A

Column 2
Calculating dryer load

Step 1: Applying demand factors for phases
Demand Load 4; **220.154; Table 220.54**
10 dryers ÷ 3Ø (phases) = _____ [JJ]
4 dryers per phase x 2 = _____ [KK]
5 kW x 8 x 60% = _____ [LL]
24 kW x 1000 = _____ [MM]
24,000 VA ÷ 2 x 3Ø (phases) = _____ [NN] VA ÷ 360V ÷ 3 = _____ [OO] A _____ [PP] A _____ [QQ] A

Step 2: Applying demand factors for neutral
Demand Load 4; **220.61**
33 A x 70% = _____ [RR] VA _____ [SS] A

Column 3
Calculating largest load between heating and A/C load

Step 1: Selecting the largest load
Demand Load 5; **220.60**
Heating unit
20,000 x 10 x 100% = _____ [TT] VA ÷ 360V ÷ 3 = _____ [UU] A _____ [VV] A _____ [WW] A

Column 4
Calculating largest motor load

Step 1: Selecting largest motor load for phases
220.14(C); 220.50; 430.24
10 A x 25% = _____ [XX] A _____ [YY] A

Step 2: Selecting largest motor load for neutral
220.14(C); 220.50; 430.24
10 A x 25% = _____ [ZZ] A _____ [AAA] A
Total load = _____ [BBB] A _____ [CCC] A _____ [DDD] A _____ [EEE] A

Design Example Problem 10-8. What is the load in VA and amps for 25 multifamily dwelling units with the following loads? (See Standard Calculation in Example Problem 10-1 for sizing the grounded (neutral) conductors) Note: Parallel service conductors, 6 times per phase.

Given loads:
- 25 - 1000 sq. ft. dwelling units
- 2 small appliance circuits per unit
- 1 laundry circuit per unit

120 V, single-phase loads
- 25 - 1000 VA dishwashers
- 25 - 1200 VA disposals

240 V, single-phase loads
- 25 - 12,000 VA ranges
- 25 - 6000 VA water heaters
- 25 - 20,000 VA heating units

Sizing ungrounded (phase) conductors = •
Sizing grounded (neutral) conductors = √

Column 1
Calculating general lighting and receptacle load

Step 1: General lighting and receptacle load
220.84(C)(1)
1000 sq. ft. x 3 VA x 25 = 75,000 VA

Step 2: Small appliance and laundry load
220.84(C)(2)
1500 VA x 2 x 25 = 75,000 VA
1500 VA x 1 x 25 = 37,500 VA
Total load = 187,500 VA •

Column 2
Calculating cooking equipment load

Step 1: Applying demand factors for phases
Demand load 2; **220.84(C)(3)**
12,000 VA x 25 = 300,000 VA •

Calculating fixed appliance load

Step 1: Applying demand factors for phases
Demand load 4; **220.32(C)(3)**
1000 VA x 25 = 25,000 VA
1200 VA x 25 = 30,000 VA
6000 VA x 25 = 150,000 VA
Total load = 205,000 VA •

Column 3
Largest load between heat and A/C

Step 1: Selecting largest load
Demand load 5; **220.84(C)(5)**
Heating unit
20,000 VA x 25 = 500,000 VA •

Calculating ungrounded (phase) conductors

General lighting load = 187,500 VA •
Cooking load = 300,000 VA •
Appliance load = 205,000 VA •
Heating load = 500,000 VA •
Total load = 1,192,500 VA

Applying demand factors
Table 220.84

I = VA ÷ V
I = 1,192,500 VA ÷ 240 V
I = 4969 A

Finding amps for ungrounded (phase) conductors - Phases A and B
Table 220.84

I = A x %
I = 4,969 A x 35%
I = 1739 A

Finding ungrounded (phase) conductors

Phases A and B
I = 1739 A ÷ 6 (No. runs per phase)
I = 290 A

Table 310.16; 310.4
Ungrounded (phase) conductors - Phases A and B
350 KCMIL THWN copper
Grounded (neutral) conductors
1/0 AWG THWN copper

Design Exercise Problem 10-8. What is the load in VA and amps for 20 multifamily dwelling units with the following loads? (See Standard Calculation in Exercise Problem 10-1 for sizing the grounded (neutral) conductors) Note: Parallel service conductors, 6 times per phase.

Given loads:	120 V, single-phase loads	240 V, single-phase loads
• 20 - 1200 sq. ft. dwelling units • 2 small appliance circuits per unit • 1 laundry circuit per unit	• 20 - 1600 VA dishwashers • 20 - 1200 VA disposals	• 20 - 10,000 VA ranges • 20 - 5000 VA water heaters • 20 - 20,000 VA heating units

Sizing ungrounded (phase) conductors = •
Sizing grounded (neutral) conductors = √

Column 1
Calculating general lighting and receptacle load

Step 1: General lighting and receptacle load
220.84(C)(1)
1200 sq. ft. × 3 VA × 20 = __72,000__ VA **[A]**

Step 2: Small appliance and laundry load
220.84(C)(2)
1500 VA × 2 × 20 = __60,000__ VA **[B]**
1500 VA × 1 × 20 = __30,000__ VA **[C]**
Total load = __162,000__ VA **[D]** •

Column 2
Calculating cooking equipment load

Step 1: Applying demand factors for phases
Demand load 2; **220.84(C)(3)**
10,000 VA × 20 = __200,000__ VA **[E]** •

Column 2
Calculating fixed appliance load

Step 1: Applying demand factors for phases
Demand load 4; **220.84(C)(3)**
1600 VA × 20 = __32,000__ VA **[F]**
1200 VA × 20 = __24,000__ VA **[G]**
5000 VA × 20 = __100,000__ VA **[H]**
Total load = __156,000__ VA **[I]** •

Column 3
Largest load between heating and A/C load

Step 1: Selecting largest load
Demand load 5; **220.84(C)(5)**
Heating unit
20,000 VA × 20 × 100% = __400,000__ VA **[J]** •

Calculating ungrounded (phase) conductors

• General lighting load = __162,000__ VA **[K]** •
• Cooking equipment load = __200,000__ VA **[L]** •
• Appliance load = __156,000__ VA **[M]** •
• Heating load = __400,000__ VA **[N]** •
Total load = __918,000__ VA **[O]**

Applying demand factors
Table 220.84

I = VA ÷ V
I = __918,000__ VA **[P]** ÷ 240 V
I = __3825__ A **[Q]**

Finding amps for ungrounded (phase) conductors - Phases A and B
Table 220.84

I = A × %
I = __3825__ A **[R]** × 38%
I = __1453.5__ A **[S]**

Finding ungrounded (phase) conductors

Phases A and B
I = __1453.5__ A **[T]** ÷ 6 (No. run per phases)
I = __242.25__ A **[U]**

Table 310.16; 310.4
Ungrounded (phase) conductors - Phases A and B
__250__ KCMIL THWN **[V]** copper conductors
Grounded (neutral) conductors
_____ AWG THWN **[W]** copper conductors

Design Example Problem 10-9. What is the load in VA and amps for 30 multifamily dwelling units with the following loads? (See Standard Calculation in Design Example Problem 10-2 for sizing the grounded (neutral) conductors) Note: Parallel service conductors, 6 times per phase.

Given loads:	120 V, single-phase loads	240 V, single-phase loads
• 10 - 820 sq. ft. dwelling units • 10 - 960 sq. ft. dwelling units • 10 - 1050 sq. ft. dwelling units • 2 small appliance circuits per unit • 1 laundry circuit per unit	• 30 - 1000 VA dishwashers • 30 - 1200 VA disposals	• 30 - 12,000 VA ranges • 30 - 6000 VA water heaters • 30 - 20,000 VA heating units

Sizing ungrounded (phase) conductors = •
Sizing grounded (neutral) conductors = √

Column 1
Calculating general lighting and receptacle load

Step 1: General lighting and receptacle load
220.84(C)(1)
820 sq. ft. x 3 VA x 10   =   24,600 VA
960 sq. ft. x 3 VA x 10   =   28,800 VA
1050 sq. ft. x 3 VA x 10 =   31,500 VA

Step 2: Small appliance and laundry load
220.84(C)(2)
1500 VA x 2 x 30   =   90,000 VA
1500 VA x 1 x 30   =   45,000 VA
Total load   = 219,900 VA •

Column 2
Calculating cooking equipment load

Step 1: Applying demand factors for phases
Demand load 2; **220.84(C)(3)**
12,000 VA x 30   =   360,000 VA •

Calculating fixed appliance load

Step 1: Applying demand factors for phases
Demand load 4; **220.84(C)(3)**
1000 VA x 30   =   30,000 VA
1200 VA x 30   =   36,000 VA
6000 VA x 30   =   180,000 VA
Total load   =   246,000 VA •

Column 3
Largest load between heat and A/C

Step 1: Selecting largest load
Demand load 5; **220.84(C)(5)**
Heating unit
20,000 VA x 30   =   600,000 VA •

Calculating ungrounded (phase) conductors

General lighting load   =   219,900 VA •
Cooking load   =   360,000 VA •
Appliance load   =   246,000 VA •
Heating load   =   600,000 VA •
Total load   = 1,425,900 VA

Applying demand factors
Table 220.84

I = VA ÷ V
I = 1,425,900 VA ÷ 240 V
I = 5941 A

Finding amps for ungrounded (phase) conductors - Phases A and B
Table 220.84

I = A x %
I = 5941 A x 33%
I = 1961 A

Finding ungrounded (phase) conductors

Phases A and B
I = 1961 A ÷ 6 (No. runs per phase)
I = 327 A

Table 310.16; 310.4
Ungrounded (phase) conductors - Phases A and B
400 KCMIL THWN copper
Grounded (neutral) conductors
1/0 AWG THWN copper

Residential Calculations - Multifamily Dwellings

Design Exercise Problem 10-9. What is the load in VA and amps for 15 multifamily dwelling units with the following loads? (See Standard Calculation in Design Exercise Problem 10-2 for sizing the grounded (neutral) conductors) Note: Parallel service conductors, 6 times per phase.

Given loads:
- 5 - 1050 sq. ft. dwelling units
- 5 - 1200 sq. ft. dwelling units
- 5 - 1280 sq. ft. dwelling units
- 2 small appliance circuits per unit
- 1 laundry circuit per unit

120 V, single-phase loads
- 15 - 1000 VA dishwashers
- 15 - 1200 VA disposals

240 V, single-phase loads
- 15 - 10,000 VA ranges
- 15 - 5000 VA water heaters
- 15 - 20,000 VA heating units

Sizing ungrounded (phase) conductors = •
Sizing grounded (neutral) conductors = √

Column 1
Calculating general lighting and receptacle load

Step 1: General lighting and receptacle load
220.84(C)(1)
1050 sq. ft. x 3 VA x 5 = _____ VA [A]
1200 sq. ft. x 3 VA x 5 = _____ VA [B]
1280 sq. ft. x 3 VA x 5 = _____ VA [C]

Step 2: Small appliance and laundry load
220.84(C)(2)
1500 VA x 2 x 15 = _____ VA [D]
1500 VA x 1 x 15 = _____ VA [E]
Total load = _____ VA [F] •

Column 2
Calculating cooking equipment load

Step 1: Applying demand factors for phases
Demand load 2; **220.84(C)(3)**
10,000 VA x 15 = _____ VA [G] •

Column 2
Calculating fixed appliance load

Step 1: Applying demand factors for phases
Demand load 4; **220.84(C)(3)**
1600 VA x 15 = _____ VA [H]
1200 VA x 15 = _____ VA [I]
5000 VA x 15 = _____ VA [J]
Total load = _____ VA [K] •

Column 3
Largest load between heating and A/C load

Step 1: Selecting largest load
Demand load 5; **220.84(C)(5)**
Heating unit
20,000 VA x 15 x 100% = _____ VA [L] •

Calculating ungrounded (phase) conductors

- General lighting load = _____ VA [M] •
- Cooking equipment load = _____ VA [N] •
- Appliance load = _____ VA [O] •
- Heating load = _____ VA [P] •
Total load = _____ VA [Q]

Applying demand factors
Table 220.84

$I = VA \div V$
$I =$ _____ VA [R] ÷ 240 V
$I =$ _____ A [S]

Finding amps for ungrounded (phase) conductors - Phases A and B
Table 220.84

$I = A \times \%$
$I =$ _____ A [T] x 40%
$I =$ _____ A [U]

Finding ungrounded (phase) conductors

Phases A and B
$I =$ _____ A [V] ÷ 6 (No. run per phases)
$I =$ _____ A [W]

Table 310.16; 310.4
Ungrounded (phase) conductors - Phases A and B
_____ AWG THWN [X] copper conductors
Grounded (neutral) conductors
_____ AWG THWN [Y] copper conductors

10-43

Design Exercise Problem 10-10. What is the load in VA for 15 multifamily dwelling units with the following loads?

Given Loads:	120 V, single-phase loads	240 V, single-phase loads
• 15 - 880 sq. ft. dwelling units • 2 small appliance circuits per unit • 1 laundry circuit per unit	• 15 - 1400 VA dishwashers • 15 - 696 VA disposals • 15 - 765 VA compactors * 15 - 1000 VA microwaves • 15 - 920 VA exhaust fans	• 15 - 10,000 VA ranges • 15 - 5000 VA dryers • 15 - 6000 VA water heaters • 15 - 10,000 VA heating units • 15 - 5040 VA A/C units • 15 - 2400 VA water pumps • 15 - 800 VA blower motors

Column 1
Calculating general lighting and receptacle load (Standard Calculation)

Phase A   Phase B   Neutral

Step 1: General lighting and receptacle load
880 sq. ft. x 3 VA x 15 = _____ [A] VA ÷ 2 = _____ [B] VA _____ [C] VA

Step 2: Small appliance and laundry load
1500 VA x 2 x 15 = _____ [D] VA ÷ 2 = _____ [E] VA _____ [F] VA
1500 VA x 1 x 15 = _____ [G] VA ÷ 2 = _____ [H] VA _____ [I] VA _____ [J] VA

Column 2
Calculating cooking equipment load

Step 1: Applying demand factors for phases
10,000 VA x 15 = _____ [K] VA = _____ [L] VA _____ [M] VA _____ [N] VA

Column 2
Calculating fixed appliance load

Step 1: Applying demand factors for phases
1400 VA x 15 = _____ [O] VA ÷ 2 = _____ [P] VA _____ [Q] VA _____ [R] VA
696 VA x 15 = _____ [S] VA ÷ 2 = _____ [T] VA _____ [U] VA _____ [V] VA
765 VA x 15 = _____ [W] VA ÷ 2 = _____ [X] VA _____ [Y] VA _____ [Z] VA
1000 VA x 15 = _____ [AA] VA ÷ 2 = _____ [BB] VA _____ [CC] VA _____ [DD] VA
920 VA x 15 = _____ [EE] VA ÷ 2 = _____ [FF] VA _____ [GG] VA _____ [HH] VA
6000 VA x 15 = _____ [II] VA = _____ [JJ] VA _____ [KK] VA
2400 VA x 15 = _____ [LL] VA = _____ [MM] VA _____ [NN] VA
800 VA x 15 = _____ [OO] VA = _____ [PP] VA _____ [QQ] VA

Column 2
Calculating dryer load

Step 1: Applying demand factors for phases
5000 VA x 15 = _____ [RR] VA = _____ [SS] VA _____ [TT] VA _____ [UU] VA

Column 3
Calculating largest load between heat and A/C

Step 1: Selecting largest load   (plus largest motor)
Heating unit
10,000 VA x 15 = _____ [VV] VA = _____ [WW] VA _____ [XX] VA _____ [YY] VA
Total load = _____ [ZZ] VA _____ [AAA] VA _____ [BBB] VA

Note: Grounded (neutral) conductors are calculated based upon actual neutral load.

Residential Calculations - Multifamily Dwellings

Design Exercise Problem 10-10. What is the load in VA for 15 multifamily dwelling units with the following loads?

Given Loads:	120 V, single-phase loads	240 V, single-phase loads
• 15 - 880 sq. ft. dwelling units • 2 small appliance circuits per unit • 1 laundry circuit per unit	• 15 - 1400 VA dishwashers • 15 - 696 VA disposals • 15 - 765 VA compactors * 15 - 1000 VA microwaves • 15 - 920 VA exhaust fans	• 15 - 10,000 VA ranges • 15 - 5000 VA dryers • 15 - 6000 VA water heaters • 15 - 10,000 VA heating units • 15 - 5040 VA A/C units • 15 - 2400 VA water pumps • 15 - 800 VA blower motors

Column 1
Calculating general lighting and receptacle load   Phase A   Phase B   (Standard Calculation)
                                                                                                Neutral

Step 1: General lighting and receptacle load
        880 sq. ft. x 3 VA x 15   = _____ [A] VA ÷ 2 = _____ [B] VA _____ [C] VA

Step 2: Small appliance and laundry load
        1500 VA x 2 x 15   = _____ [D] VA ÷ 2 = _____ [E] VA _____ [F] VA
        1500 VA x 1 x 15   = _____ [G] VA ÷ 2 = _____ [H] VA _____ [I] VA   _____ [J] VA

Column 2
Calculating cooking equipment load

Step 1: Applying demand factors for phases
        10,000 VA x 15   = _____ [K] VA = _____ [L] VA _____ [M] VA _____ [N] VA

Column 2
Calculating fixed appliance load

Step 1: Applying demand factors for phases
        1400 VA x 15   = _____ [O] VA ÷ 2 = _____ [P] VA _____ [Q] VA _____ [R] VA
        696 VA x 15   = _____ [S] VA ÷ 2 = _____ [T] VA _____ [U] VA _____ [V] VA
        765 VA x 15   = _____ [W] VA ÷ 2 = _____ [X] VA _____ [Y] VA _____ [Z] VA
        1000 VA x 15   = _____ [AA] VA ÷ 2 = _____ [BB] VA _____ [CC] VA _____ [DD] VA
        920 VA x 15   = _____ [EE] VA ÷ 2 = _____ [FF] VA _____ [GG] VA _____ [HH] VA
        6000 VA x 15   = _____ [II] VA   = _____ [JJ] VA _____ [KK] VA
        2400 VA x 15   = _____ [LL] VA   = _____ [MM] VA _____ [NN] VA
        800 VA x 15   = _____ [OO] VA   = _____ [PP] VA _____ [QQ] VA

Column 2
Calculating dryer load

Step 1: Applying demand factors for phases
        5000 VA x 15   = _____ [RR] VA = _____ [SS] VA _____ [TT] VA _____ [UU] VA

Column 3
Calculating largest load between heat and A/C

Step 1: Selecting largest load   (plus largest
        Heating unit   motor)
        10,000 VA x 15   = _____ [VV] VA = _____ [WW] VA _____ [XX] VA _____ [YY] VA

        Total load   = _____ [ZZ] VA _____ [AAA] VA _____ [BBB] VA

Note: Grounded (neutral) conductors are calculated based upon actual neutral load.

Name _____   Date _____

Chapter 10
Residential Calculations – Multifamily Dwellings

Section   Answer

1. The general lighting load for a multifamily dwelling unit shall be determined by multiplying the square footage by 2 VA per **Table 220.12**.

    (a) True   **(b)** False

2. All small appliance and laundry loads shall be calculated at 1500 VA each to determine the size feeder conductors and elements to size the service.

    (a) True   **(b)** False

3. All fixed appliances for four or more grouped into a special appliance load shall be found by adding wattage ratings from appliance nameplates and multiplying the total wattage (volt-amps) by 65 percent to obtain demand load.

    (a) True   **(b)** False

4. The demand load for household dryers shall be figured at 3 kVA or the nameplate rating, whichever is greater.

    (a) True   **(b)** False

5. The heating and A/C loads shall be calculated, and the smaller of the two is dropped.

    (a) True   **(b)** False

6. Dryer equipment consisting of six units shall be calculated at _____ percent of the nameplate rating.

    (a) 65   **(b)** 75   **(c)** 90   **(d)** 100

7. The largest motor load shall be calculated at _____ percent, for calculating the load of one or more motors with other loads.

    (a) 80   **(b)** 100   **(c)** 125   **(d)** 150

8. All fixed appliances of four or more grouped into a special appliance load are found by adding wattage rating from appliance nameplates and multiplying the total wattage (volt-amps) by:

    (a) 65   **(b)** 75   **(c)** 80   **(d)** 100

9. The demand load for a 12 kW household range can be figured at a demand of:

    (a) 5 1/2 kVA   **(b)** 6 1/2 kVA   **(c)** 7 1/2 kVA   **(d)** 8 kVA

10. The branch-circuit heating and A/C load shall be calculated at:

    (a) 80 percent   **(b)** 100 percent   **(c)** 125 percent   **(d)** 150 percent

Section	Answer

_____ _____ **11.** What is the general lighting and general-purpose receptacle loads in VA for a 30 unit apartment complex with each dwelling unit having 690 sq. ft. of space?

        (a) 59,875 VA   **(b)** 60,925 VA   **(c)** 62,485 VA   **(d)** 63,225 VA

_____ _____ **12.** What is the general lighting and general-purpose receptacle loads in VA for 10 - 540 sq. ft. dwelling units, 10 - 680 sq. ft. dwelling units and 10 - 760 sq. ft. dwelling units?

        (a) 62,550 VA   **(b)** 64,225 VA   **(c)** 68,785 VA   **(d)** 72,350 VA

_____ _____ **13.** What is the demand load for 30 - 12 kW ranges installed in an apartment complex?

        (a) 45 kW   **(b)** 48 kW   **(c)** 50 kW   **(d)** 54 kW

_____ _____ **14.** What is the demand load for 5 - 3 kW ovens and 5 - 3 kW cooktops installed in an apartment complex?

        (a) 9.9 kW   **(b)** 11.4 kW   **(c)** 13.2 kW   **(d)** 14.7 kW

_____ _____ **15.** What is the demand load for 15 - 8 kW ovens and 15 - 3 kW cooktops installed in apartment complex?

        (a) 48.2 kW   **(b)** 56.4 kW   **(c)** 58.6 kW   **(d)** 62.5 kW

_____ _____ **16.** What is the demand load for 8 - 12 kW ranges, 8 - 14 kW ranges and 8 - 20 kW ranges installed in apartment complex?

        (a) 44.8 kW   **(b)** 46.8 kW   **(c)** 48.2 kW   **(d)** 52.1 kW

_____ _____ **17.** What is the demand load in VA for 10 - 11 kW ranges installed in an apartment complex supplied by a 120 / 208 volt, three-phase system?

        (a) 28,400 VA   **(b)** 30,500 VA   **(c)** 34,500 VA   **(d)** 38,200 VA

_____ _____ **18.** What is the fixed appliance load in VA for the ungrounded (phase) conductors with the following loads?

- 10 - 6000 VA water heaters   240 volt, single-phase
- 10 - 1200 VA disposals   120 volt, single-phase
- 10 - 1400 VA compactors   120 volt, single-phase
- 10 - 1600 VA dishwashers   120 volt, single-phase
- 10 - 800 VA microwaves   120 volt, single-phase

        (a) 74,100 VA   **(b)** 78,500 VA   **(c)** 80,400 VA   **(d)** 82,500 VA

_____ _____ **19.** What is the fixed appliance load in VA for the grounded (neutral) conductors with the following loads?

- 10 - 6000 VA water heaters   240 volt, single-phase
- 10 - 1200 VA disposals   120 volt, single-phase
- 10 - 1400 VA compactors   120 volt, single-phase
- 10 - 1600 VA dishwashers   120 volt, single-phase
- 10 - 800 VA microwaves   120 volt, single-phase

        (a) 37,500 VA   **(b)** 39,200 VA   **(c)** 42,225 VA   **(d)** 45,375 VA

Section **Answer**

20. What is the demand load in VA for 10 dryers rated at 5000 VA each?

    (a) 24,275 VA **(b)** 25,000 VA **(c)** 32,325 VA **(d)** 30,000 VA

Commercial Calculations

11

Those taking the master electrician examination must understand that commercial facilities such as offices, banks, stores and restaurants have diverse loads. These loads are classified as continuous or noncontinuous, or such loads may cycle on and off, permitting demand factors to be applied.

The procedure and manner in which the lighting, receptacle and equipment loads are used in the electrical system determines how they are classified.

Loads shall be calculated based upon the type of occupancy and the requirements of the equipment supplied. Either the standard or optional calculation is utilized to calculate the loads in VA or amps to size the service equipment and associated elements.

APPLYING THE STANDARD CALCULATION PART II TO ARTICLE 220

The standard calculation shall be permitted to be used to calculate the VA or amp rating to size and select the elements of the service equipment and associated components. The selection of loads to apply the standard calculation is arranged in a different manner for commercial loads than for the loads used in residential occupancies. There are seven loads utilized to determine the service load. Based upon conditions of use, demand factors shall be permitted to be applied to certain loads.

The loads are grouped into seven individual loads, and the proper NEC rule is applied to each of these loads based upon use. The loads are grouped and classified as follows:

(1) Lighting loads
- General lighting load per **220.12**

Quick Reference

APPLYING THE STANDARD CALCULATION	11-1
STANDARD CALCULATIONS	11-15
PARTS II AND III TO ARTICLE 220	11-15
POWER SOURCE - PARTS II AND III TO ARTICLE 220 (UNLISTED OCCUPANCY)	11-27
APPLYING THE OPTIONAL CALCULATION PART III TO ARTICLE 220	11-27
OPTIONAL CALCULATIONS FOR ADDITIONAL LOADS TO EXISTING INSTALLATIONS	11-29
TEST QUESTIONS	11-55

- Show window load per **220.43(A)**
- Lighting track load per **220.43(B)**
- Low-voltage lighting load per **Article 411**
- Outside lighting load per **230.42(A)(1)** and **(A)(2)**
- Outside sign lighting load per **220.14(F)** and **600.5(A)**

(2) Receptacle loads
- General-purpose receptacle load
(Noncontinuous per **220.14(I)** and **230.42(A)(1)** and **Table 220.44**)
- General-purpose receptacle load
(Continuous per **220.14(H), 230.42(A)(2)**)
- Multioutlet assembly load
(Used simultaneously per **220.14(H)(2)** and **Table 220.44**)
(Not used simultaneously per **220.14(H)(1)** and **Table 220.44**)

(3) Special appliance loads
- Noncontinuous load per **230.42(A)(1)**
- Continuous load per **230.42(A)(2)**
- Demand factors per various sections of the NEC

(4) Compressor loads
- Refrigeration per **440.34** and **230.42(A)(1)**
- Cooling per **440.34** and **230.42(A)(1)**

(5) Motor loads
- Single-phase per **430.25**
- Three-phase per **430.25**

(6) Heat or A/C loads
- Heating per **220.51**
- A/C per **440.34**
- Heat pump per **440.34**

(7) Largest motor load
- Taken from loads (4), (5), or (6) per **220.14(C)** and **220.50**

The seven loads shall be calculated at continuous operation or noncontinuous operation. Demand factors shall be permitted to be applied to the noncontinuous operated loads by specific sections of the NEC.

Generally, no demand factors shall be permitted to be applied to the loads for commercial occupancies, as the loads are usually used at continuous operation. However, **Table 220.42** permits the general lighting loads in hospitals, hotels, motels and warehouses to have demand factors applied because of load diversity.

LIGHTING LOADS
ARTICLE 220; ARTICLE 410; ARTICLE 411

Lighting loads are the first loads to be calculated. Six lighting loads are calculated to derive the total lighting load in commercial facilities. These six loads are as follows:

(1) General lighting load per **Table 220.12**
(2) Show window load per **220.43(A)**
(3) Lighting track load per **220.43(B)**
(4) Low-voltage lighting load per **Article 411**
(5) Outside lighting load per **230.42(A)(1)** and **(A)(2)**
(6) Sign lighting load per **220.14(F)** and **600.5(A)**

Each lighting load is calculated by the method in which it is used. Loads are calculated at continuous or noncontinuous operation, and other lighting loads shall be permitted to have demand factors applied where permitted by specific sections of the NEC.

GENERAL LIGHTING LOADS
220.12; TABLE 220.12

General lighting loads consist of lighting units installed inside the facility. This load shall be calculated by the VA rating times sq. ft. from **Table 220.12** based on the type of commercial occupancy.

The general lighting load found in listed occupancies shall be calculated according to the number of VA per sq. ft. and not based upon the number of outlets served. When fluorescent lighting is used, either the area load of VA per sq. ft. of the facility or the total connected load of each ballast is used, whichever is greater in rating.

Lighting loads for either the VA per sq. ft. or individual units shall be calculated at 100 percent for noncontinuous operation or 125 percent for continuous operation. **(See Figure 11-1)**

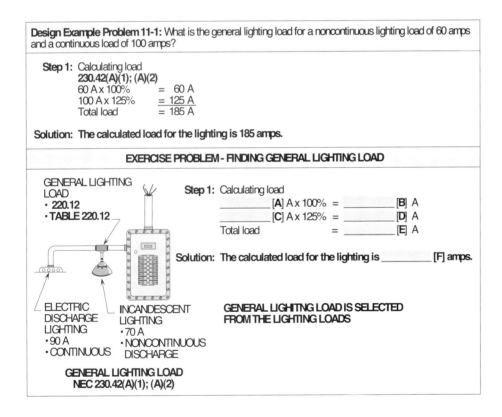

Figure 11-1. This figure illustrates an example and exercise problem for calculating the general lighting load.

LISTED OCCUPANCIES
220.12; TABLE 220.12

Table 220.12 shall be used to select the VA rating for listed occupancies. For example, a store building is figured at 3 VA per sq. ft., and an office building is figured at 3.5 VA per sq. ft. Other types of occupancies have different VA ratings per sq. ft. based on the type of occupancy involved. **(See Figure 11-2)**

Figure 11-2. This figure illustrates an example and exercise problem for computing the lighting load for a listed occupancy.

Note that the total VA in a test problem is only increased by 125 percent when the problem specifically states continuous use.

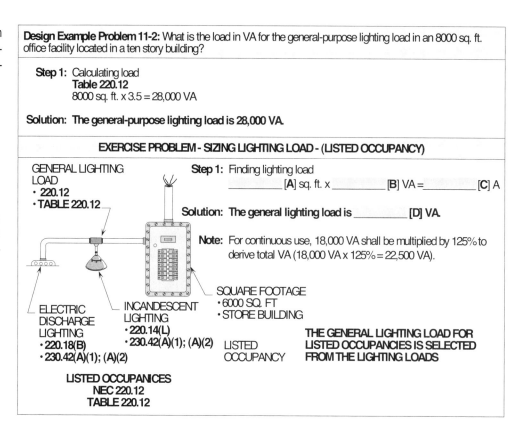

UNLISTED OCCUPANCIES
220.14(D); 220.18(B)

If an occupancy is not listed in **Table 220.12**, the general-purpose lighting load is computed in the following manner.

(1) Lamps for incandescent lighting per **220.14(D)**
(2) Lamps for recessed fixtures per **220.14(D)**
(3) Ballasts for electric discharge lighting per **220.18(B)**
(4) Track lighting units per **220.43(B)**
(5) Low-voltage systems per **Article 411**

An unlisted VA rating shall be calculated at 125 percent for sizing the OCPD and conductors for continuous operation and at 100 percent for noncontinuous operated loads per **230.42(A)(1)** and **(A)(2)**. **(See Figure 11-3)**

SHOW WINDOW LIGHTING LOAD
220.43(A)

The lighting load in VA for the show window shall be calculated by multiplying the linear feet of the show window by 200 VA per foot. Such lighting load shall be calculated at 100 percent for noncontinuous operation and 125 percent for continuous operation per **230.42(A)(1)** and **(A)(2)**. Conductors and OCPD's shall be increased to comply with **240.3** and **4**.

If the number of lighting outlets is known, the VA rating of each luminaire (lighting fixture) shall be multiplied by 125 percent for sizing the show window load. If the VA is not known, each outlet shall be calculated at 180 VA times 125 percent to obtain the lighting load in VA for the show window. Note that the greater between the 200 VA per linear foot or each individual calculation shall be used. **(See Figure 11-4)**

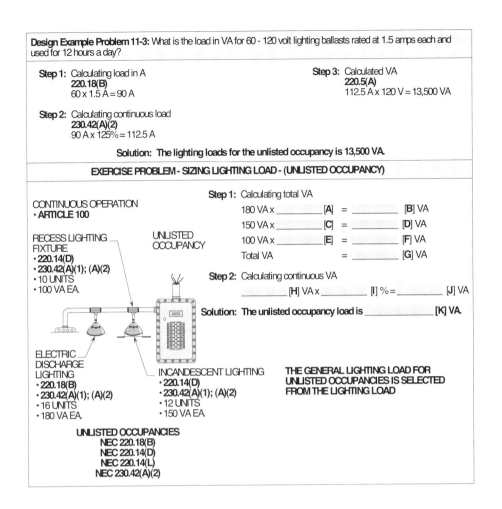

Figure 11-3. This figure illustrates an example and exercise problem for calculating the lighting load for an unlisted occupancy.

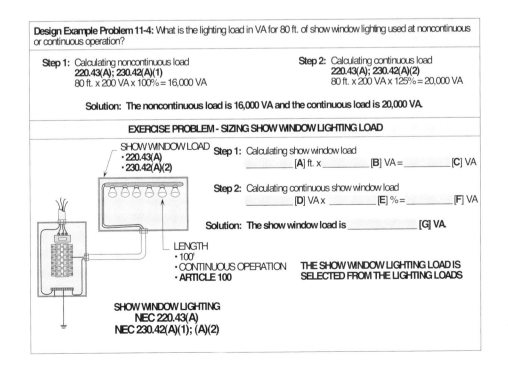

Figure 11-4. This figure illustrates an example and exercise problem for calculating the show window lighting load.

TRACK LIGHTING LOAD
220.43(B)

The lighting load in VA for lighting track shall be calculated by multiplying the lighting track by 150 VA and dividing by 2. Such VA rating shall be multiplied by 100 percent or 125 percent, based upon noncontinuous or continuous operation. **(See Figure 11-5)**

Design Example Problem 11-5. What is the load in VA and amps to size the elements for a 277/480 volt, three-phase, four-wire service supplying a 120,000 sq. ft. office with a 2,000 sq. ft. meeting hall and 1,000 sq. ft. storage space with the following loads? (All lighting is 277 volt) Note: 51 percent or more of the lighting load is electric discharge.

Sizing ungrounded (phase) conductors	= •
Sizing grounded (neutral) conductors	= √
Sizing load - Phases A and B	= *

120 V, single-phase loads

- 35 ft. of lighting track (noncontinuous operation)
- 15 - 180 VA ballast's outside lighting (continuous operation)
- 3600 VA sign lighting (continuous operation)
- 162 receptacles (noncontinuous operation)
- 108 receptacles (continuous operation)
- 6,000 VA isolation transformer for LVLS (continuous operation)
- 120 ft. multioutlet assembly (heavy-duty)

Note: All lighting is 277 volt.

480 V, three-phase loads

- 40 HP elevator (15 minute intermittent duty)
- 50 kW heating unit

208 V, three-phase loads

- 2 - 1225 VA copying machines
- 1 - 10,000 VA water heater
- 28 - 275 VA data processors
- 10 - 175 VA word processors
- 4 - 1200 VA printers

Calculating lighting loads

Step 1:	General lighting load			
	Table 220.12; 230.42(A)(2)			
	120,000 sq. ft. x 3.5 VA	=	420,000	VA √
	420,000 VA x 125%	=	525,000	VA •
	2000 sq. ft. x 1/2 VA	=	1,000	VA √
	1000 VA x 125%	=	1,250	VA •
	1000 sq. ft. x 1/4 VA	=	250	VA √
	250 VA x 125%	=	313	VA •
Step 2:	Track lighting load			
	220.43(B)			
	35' ÷ 2 x 150 VA	=	2,625	VA • √
Step 3:	Low-voltage lighting load			
	Art. 411; 230.42(A)(2)			
	6000 VA x 100%	=	6,000	VA √
	6000 VA x 125%	=	7,500	VA •
Step 4:	Outside lighting load			
	230.42(A)(2)			
	15 x 180 VA	=	2,700	VA √
	2700 VA x 125%	=	3,375	VA •
Step 5:	Sign lighting load			
	220.14(F); 600.5(A); 230.42(A)(2)			
	3600 VA x 100%	=	3,600	VA √
	3600 VA x 125%	=	4,500	VA •
	Total load	=	544,563	VA *

Calculating receptacle loads

Step 1:	Noncontinuous operation			
	220.14(I); 230.42(A)(1); 220.14(H)(1); (2)			
	162 x 180 VA	=	29,160	VA
	120' x 180 VA	=	21,600	VA
	Total load	=	50,760	VA
	Table 220.13			
	First 10,000 VA x 100%	=	10,000	VA
	Next 40,760 VA x 50%	=	20,380	VA
	Total load	=	30,380	VA •
Step 2:	Continuous operation			
	220.14(I); 230.42(A)(2)			
	108 x 180 VA	=	19,440	VA
	19,440 VA x 125%	=	24,300	VA •
	Total load	=	54,680	VA *

Calculating special loads

Step 1:	Copying machine load			
	230.42(A)(2)			
	1225 VA x 2	=	2,450	VA
	2450 VA x 125%	=	3,063	VA •
Step 2:	Water heater load			
	422.13; 230.42(A)(1)			
	10,000 VA x 100%	=	10,000	VA •
Step 3:	Data processor load			
	230.42(A)(2)			
	275 VA x 28	=	7,700	VA
	7700 VA x 125%	=	9,625	VA •
Step 4:	Word processor load			
	230.42(A)(2)			
	175 VA x 10	=	1,750	VA
	1750 VA x 125%	=	2,188	VA •
Step 5:	Printer load.			
	230.42(A)(2)			
	1200 VA x 4	=	4,800	VA
	4800 VA x 125%	=	6,000	VA •
	Total load	=	30,876	VA *

Calculating motor loads

Step 1:	40 HP elevators			
	430.24; 430.22(E); Table 430.22(E)			
	52 A x 831 V x 85%	=	36,730	VA • *

Calculating heating or A/C load

Step 1:	Heating load selected			
	220.60; 220.51			
	50,000 VA x 100%	=	50,000	VA • *

Calculating largest motor load

Step 1:	40 HP elevator			
	220.14(C); 220.50; 430.24			
	36,730 VA x 25%	=	9,183	VA • *
	Total load of facility	=	726,032	VA •

Finding amps for ungrounded (phase) conductors - Phases A, B and C

I = VA ÷ V x $\sqrt{3}$
I = 726,032 VA ÷ 831 V
I = 874 A

Calculating grounded (neutral) conductors

General lighting load			
(office building)	=	420,000	VA √
(halls)	=	1,000	VA √
(storage space)	=	250	VA
Track lighting load	=	2,625	VA √
Low-voltage lighting load	=	6,000	VA √
Outside lighting load	=	2,700	VA √
Sign lighting load	=	3,600	VA √
Total load	=	436,175	VA

Finding amps for grounded (neutral) conductors

I = VA ÷ V x $\sqrt{3}$
I = 436,175 VA ÷ 831 V
I = 525 A

Figure 11-5. This figure illustrates an example and exercise problem for calculating the track lighting load.

LOW-VOLTAGE LIGHTING LOAD
ARTICLE 411; 230.42(A)(1); (A)(2)

The lighting load in VA for low-voltage lighting systems shall be calculated by multiplying the full-load amps of the isolation transformer by 100 percent for noncontinuous operation and 125 percent for continuous operation. **(See Figure 11-6)**

OUTSIDE LIGHTING LOAD
220.18(B); 230.42(A)(1); (A)(2)

The lighting load in VA for outside lighting loads shall be calculated by multiplying the VA rating of each lighting unit by 100 percent for noncontinuous operation and 125 percent for continuous operation. **(See Figure 11-7)**

SIGN LIGHTING LOAD
220.14(F); 600.5(A); 230.42(A)(1); (A)(2)

The lighting loads for signs shall be calculated by the commercial occupancy or facility having ground floor footage accessible to pedestrians. Occupancies with grade level access for pedestrians shall have a minimum of 1200 VA provided for a sign lighting load. This VA rating shall be multiplied by 125 percent for signs operating for three hours or more and 100 percent for those operating less than three hours. **(See Figure 11-8)**

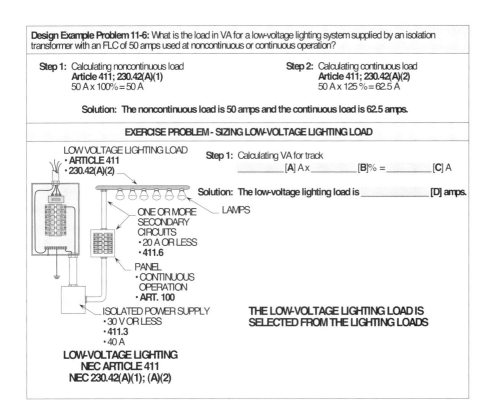

Figure 11-6. This figure illustrates an example and exercise problem for calculating the low-voltage lighting load.

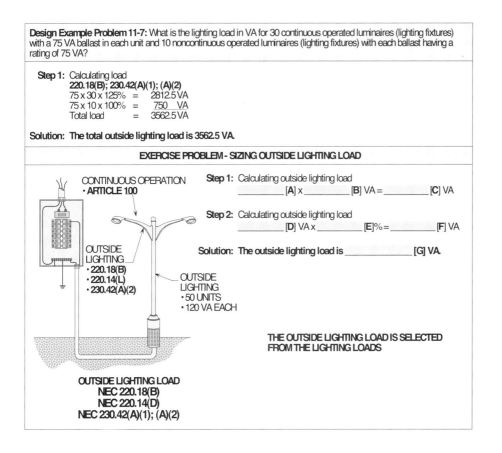

Figure 11-7. This figure illustrates an example and exercise problem for calculating the outside lighting load.

Figure 11-8. This figure illustrates an example and exercise problem for calculating the outside sign lighting load.

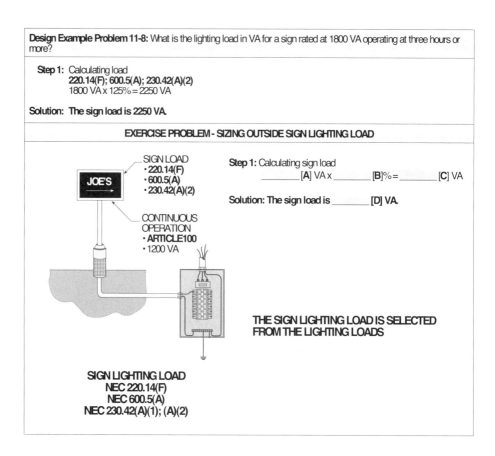

RECEPTACLE LOADS
220.14(I); TABLE 220.44

Receptacle loads are the second group of loads to be calculated. Such loads are divided into two subgroups as follows:

(1) General-purpose receptacle outlets
(2) Multioutlet assemblies

Each load in the subgroup must be calculated differently to derive the total VA. The load in VA for the general-purpose receptacle load shall be computed by multiplying the number of outlets times 180 VA each times 100 percent for noncontinuous operation and 125 percent for continuous operation. **(See Figure 11-9)**

APPLYING DEMAND FACTORS
220.14(I); TABLE 220.44

General-purpose receptacle outlets for cord-and-plug connected loads used at noncontinuous operation shall be calculated per **220.14(I)** and **Table 220.44**. Noncontinuous operated receptacles with a VA rating of 10,000 VA or less shall be calculated at 100 percent. If the VA rating of the receptacle load exceeds 10,000 VA, a demand factor of 50 percent shall be permitted to be applied to all VA exceeding 10,000 VA per **Table 220.44**. **(See Figure 11-10)**

Commercial Calculations

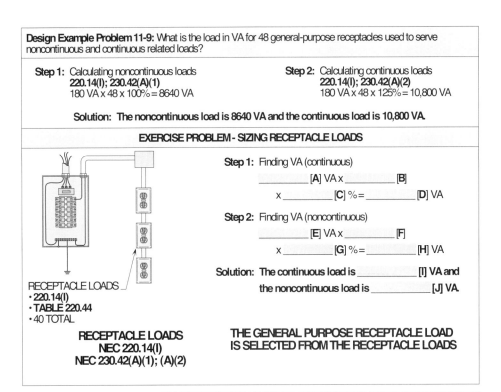

Figure 11-9. This figure illustrates an example and exercise problem for calculating the general-purpose outlet load.

Figure 11-10. This figure illustrates an example and exercise problem for calculating the general-purpose receptacle load (noncontinuous) when applying demand factors.

MULTIOUTLET ASSEMBLIES
220.14(H)(1); (H)(2)

The VA rating used to calculate the load for a multioutlet assembly shall be based on the use of such assembly. For connected loads not operating simultaneously, the VA rating shall be calculated by dividing the length of the assembly by 5 ft. and multiplying by 180 VA. For connected loads operating simultaneously, each foot of multioutlet assembly shall be multiplied by 180 VA. **(See Figure 11-11)**

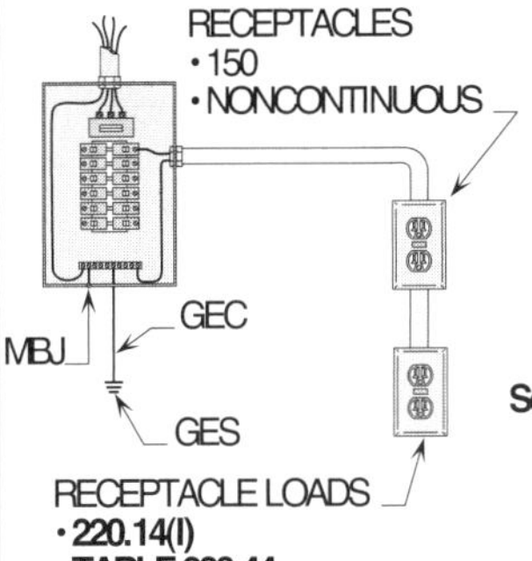

Figure 11-11. This figure illustrates an example exercise problem for computing the multioutlet assembly load.

Note that the multioutlet assembly load shall be permitted to be added to the noncontinuous receptacle load and demand factors applied per **Table 220.44**.

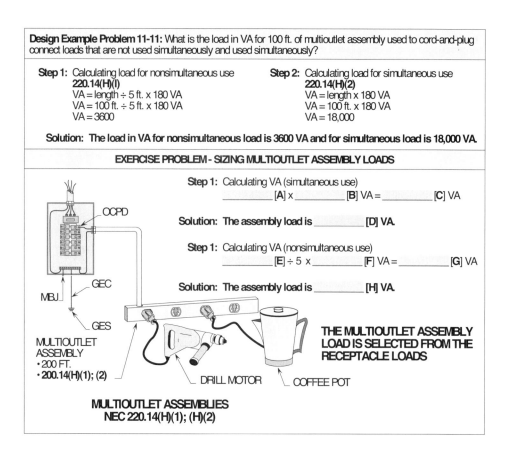

SPECIAL APPLIANCE LOADS
230.42(A)(1); (A)(2)

Special appliance loads are the third group of loads to be calculated. These loads, which include computers, processing machines, etc., are usually served by individual circuits.

CONTINUOUS AND NONCONTINUOUS OPERATION
230.42(A)(1); (A)(2)

The load in VA for special appliance loads shall be calculated by multiplying the VA rating of each load by 100 percent for noncontinuous operation and 125 percent for continuous operation. To determine classification, a special appliance load operating for less than three hours shall be classified as a noncontinuous operated load. However, a special appliance load operating for three hours or more shall be classified as a continuous operated load. **(See Figures 11-12(a) and (b))**

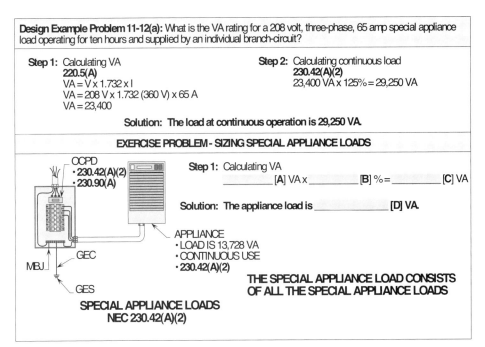

Figure 11-12(a). This figure illustrates an example and exercise problem for calculating the special appliance load at continuous operation.

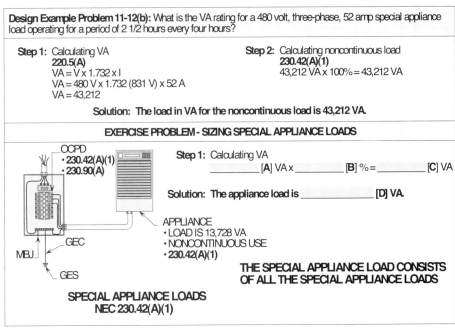

Figure 11-12(b). This figure illustrates an example and exercise problem for calculating the special appliance load at noncontinuous operation.

APPLYING DEMAND FACTORS
TABLE 220.56

Demand factors shall be permitted to be applied to cooking equipment in restaurants with three or more cooking units. The load in VA for cooking equipment shall be permitted to be calculated by applying the optional calculation listed in **Table 220.56**. **(See Figure 11-13)**

COMPRESSOR LOADS
440.34

Compressor loads are the fourth group of loads to be calculated. Special considerations shall be applied when calculating loads for hermetic sealed compressors supplying refrigerant and cooling related equipment. **(See Figure 11-14)**

Figure 11-13. This figure illustrates an example and exercise problem for calculating the special appliance load when demand factors are applied.

Note that if the heating load is 20 kW, the A/C load is 2.5 kVA, and an attic fan is 1.6 kVA, the A/C and attic fan loads shall be permitted to be dropped per **220.60**.

Figure 11-14. This figure illustrates an example and exercise problem for calculating the compressor load in VA.

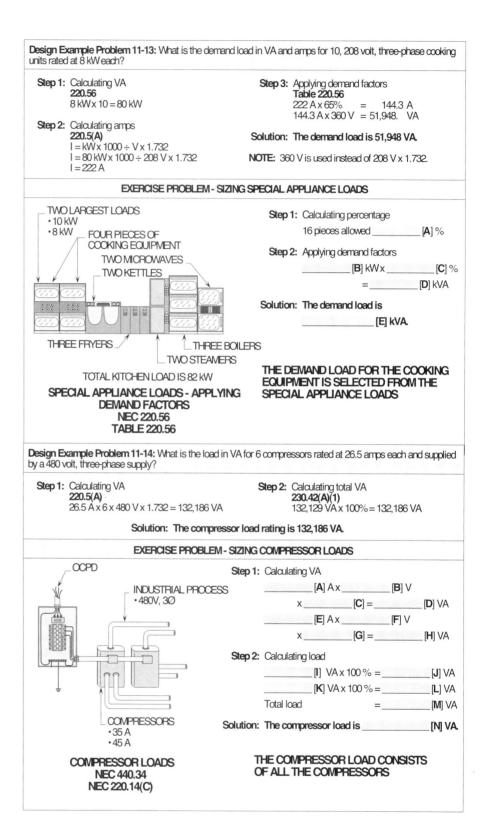

MOTOR LOADS
220.14(C); 220.50; 430.24

Motor loads are the fifth group of loads to be calculated. The VA rating of motors is converted from amperage to VA by multiplying the amperage from **Table 430.248** for single-phase or **Table 430.250** for three-phase by the supply voltage. **(See Figure 11-15)**

Commercial Calculations

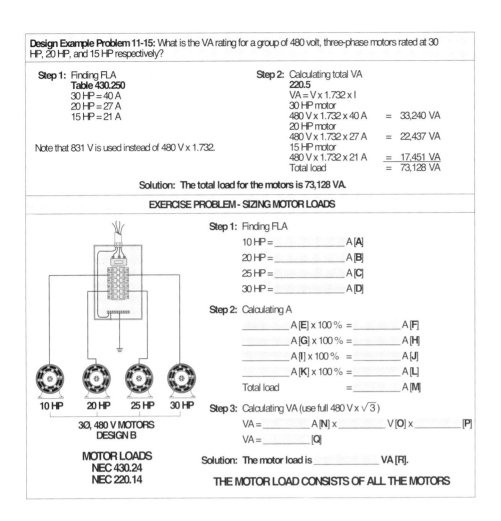

Figure 11-15. This figure illustrates an example and exercise problem for calculating the motor load in VA.

Note, in exercise problem 11-15, use the full square root of 3 (1.732) times 480 volts.

HEATING OR A/C LOADS
220.60

Heating or A/C loads is the forth group of loads to be calculated. The largest VA rating between the heating or A/C load shall be selected, and the smaller of the two loads is dropped. To determine the largest of the two loads, the VA rating of each load shall be computed at 100 percent, and the largest load of the two is selected. The load dropped is not used again in the calculation. **(See Figure 11-16)**

LARGEST MOTOR LOAD
220.14(C); 220.50; 430.24

The largest motor load in VA is the seventh of the loads to be calculated. The largest motor load is selected from one of the motor related loads. The VA rating of the largest motor shall be calculated by multiplying the amperage of the unit by the voltage times 25 or 125 percent. **(See Figure 11-17)**

Figure 11-16. This figure illustrates an example and exercise problem for calculating the largest load between the heat and A/C.

Note that a heat pump that operates with the heating unit shall be added to the VA of the heating at 100 percent of its VA rating.

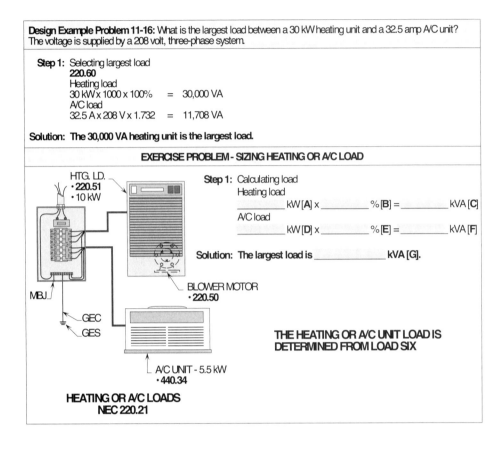

Figure 11-17. This figure illustrates an example and exercise problem for calculating the largest motor load in VA.

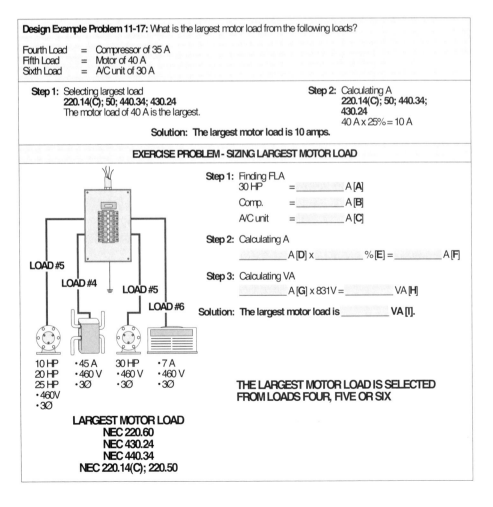

STANDARD CALCULATIONS
PARTS II AND III TO ARTICLE 220

The elements of electrical systems shall be permitted to be calculated by using the standard or optional calculation. The size of these elements are determined by which method testing agency requires the test taker to use to calculate these loads. The following computations are typical examples of how these loads are calculated, sized and selected. The step-by-step procedures are easy to follow and have condensed the more complicated rules pertaining to calculating loads into a compact listing, which provides easier understanding of how to perform calculations according to the provisions of the NEC. A broad assortment of basic code calculations have been selected to represent the main calculations that may appear on the master examination.

STORE BUILDING SUPPLIED BY 120/240 VOLT, SINGLE-PHASE POWER SOURCE
TABLE 220.12; (LISTED OCCUPANCY)

For a calculation on a test, it is the loads in a facility that determines the voltage to be selected to calculate and size the service or feeder-circuit elements. For example, 120/240 volt supply is usually used to supply a facility having an equal number of three-phase and single-phase related loads.

If a 120/240 volt supply is used on the master examination, the calculation will produce larger ratings to select the elements for service equipment and feeder-circuits. **(See Design Problem 11-1)**

Design Problem: Use Design Example Problem 11-1 to size the following components and elements of a store building supplied by a 120/240 volt, single-phase power source:

(1) What size THWN copper conductors, paralleled six times per phase and routed through an ambient temperature of 124°F, is required for Phases A and B?

    Step 1: Finding amperage
    Text
    I = 1719 A ÷ 6
    I = 287 A

    Step 2: Selecting size conductors
    Table 310.16, 75°C Column 3; 110.14(C)(1)(b)
    500 KCMIL = 380 A

    Step 3: Applying correction factors
    Table 310.16, Footnotes
    380 A x 76% = 289 A

  Solution: The size THWN conductors required are 500 KCMIL copper.

(2) What size OCPD is required for the switchgear?

    Step 1: Calculating amperage
    240.4(B)
    289 A x 6 = 1734 A

    Step 2: Selecting size OCPD
    240.6(A)
    1734 A requires 1600 A OCPD

  Solution: The size OCPD required is 1600 amps.

(3) What size switchgear is required for the service equipment?

    Step 1: Selecting size switchgear
            Chart on inside back cover
            1600 amp OCPD requires 1600 amp switchgear

    Solution: The size switchgear required is 1600 amps.

STORE BUILDING SUPPLIED BY 120/208, THREE-PHASE VOLT POWER SOURCE
TABLE 220.12; (LISTED OCCUPANCY)

A facility equipped with a large number of 120 volt and 208 volt three-phase loads is normally supplied by a four-wire, three-phase wye-connected system.

On a three-phase wye-connected system, three ungrounded (phase) conductors are obtained from ground; therefore, three ungrounded (phase) conductors may be routed with each grounded (neutral) conductor. This type of installation requires fewer grounded (neutral) conductors and therefore saves on the amount of copper or aluminum that would otherwise be needed if one grounded (neutral) conductor was pulled with each ungrounded (phase) conductor. **(See Design Problem 11-2)**

Design Problem: Use Design Example Problem 11-2 to size the following components and elements of a store building supplied by a 120/208 volt, three-phase power source:

(1) What size THWN copper conductors, paralleled four times per phase, are required for Phases A, B and C?

    Step 1: Finding amperage
            Text
            I = 1145 A ÷ 4
            I = 286 A
    Step 2: Selecting size conductors
            Table 310.16, 75°C Column; 110.14(C)(1)(b)
            350 KCMIL = 310 A
            286 A is less than 310 A

    Solution: The size THWN conductors required are 350 KCMIL copper.

(2) What size THWN copper conductors, paralleled four times per phase, are required for the grounded (neutral) conductor?

    Step 1: Finding amperage
            Text
            I = 742 A ÷ 4
            I = 186 A

    Step 2: Selecting size conductors
            Table 310.16, 75°C Column; 310.4; 110.14(C)(1)(b)
            3/0 AWG THWN = 200 A
            186 A is less than 200 A

    Solution: The size THWN conductors required are 3/0 AWG copper.

STORE BUILDING SUPPLIED BY 277/480 VOLT, THREE-PHASE POWER SOURCE
TABLE 220.12; (LISTED OCCUPANCY)

Smaller store buildings are usually supplied by 120/208 volt power systems. Loads such as lighting, receptacles and other related loads are served by the lower voltage of 120 volts. Equipment is normally served by the higher voltage rated at 208 volts phase-to-phase.

Larger office buildings utilize 277/480 volt, three-phase, four-wire systems. The higher 480 volt, three-phase voltage supplies the heavier equipment, and the 277 volt, single-phase voltage supplies lighting. Transformers are used to step down the 480 volts to 120/208 volts or 120/240 volts to supply the lower voltage loads and equipment. **(See Design Problem 11-3)**

Design Problem: Use Design Example Problem 11-3 to size the following components and elements of a store building supplied by a 277/480 volt, three-phase power source:

(1) What size THWN copper conductors, paralleled two times per phase, are required for Phases A, B and C?

   Step 1: Finding amperage
   Text
   I = 496 A ÷ 2
   I = 248 A

   Step 2: Selecting size conductors
   Table 310.16, 75°C Column; 110.14(C)(1)(b)
   250 KCMIL = 255 A
   248 A is less than 255 A

   Solution: The size THWN conductors required are 250 KCMIL copper.

(2) What size THWN copper conductors, paralleled two times per phase, is required for the grounded (neutral) conductors?

   Step 1: Finding amperage
   Text
   I = 229 A ÷ 2
   I = 115 A

   Step 2: Selecting size conductors
   Table 310.16, 75°C Column; 310.4; 110.14(C)(1)(b)
   1/0 AWG THWN = 150 A
   115 A is less than 150 A

   Solution: The size THWN conductors required are 1/0 AWG copper.

(3) What size rigid metal conduit is required for each parallel run?

   Step 1: Finding sq. in. area
   Table 5, Ch. 9
   250 KCMIL   = .397 sq. in.
   1/0 AWG THWN   = .1855 sq. in.

Step 2: Calculating sq. in. area
Table 5, Ch. 9
.397 sq. in. x 3   = 1.191 sq. in.
.1855 sq. in. x 1   = .1855 sq. in.
Total   = 1.3765 sq. in.

Step 3: Selecting conduit size
Table 4, Ch. 9
1.3765 sq. in. requires 2 1/2 in. (63)

Solution: The size rigid metal conduit is 2 1/2 in. (63).

STORE BUILDING SUPPLIED BY 120/240 VOLT, THREE-PHASE POWER SOURCE
TABLE 220.12; (LISTED OCCUPANCY)

Store buildings with a greater number of three-phase loads than single-phase loads may be supplied by a three-phase, four-wire, 120/240 volt service. The 120 volts may be obtained from two of the ungrounded (phase) conductors to ground. This phase-to-ground voltage is taken from the lighting and power transformer.

Transformers on a delta-connected system are connected in an open or closed delta configuration.

Closed delta systems require three transformers, while open delta systems require only two transformers. One of the advantages of closed delta systems is that if one transformer fails, the remaining two can be connected in an open delta configuration and continue to supply the load until a replacement transformer is purchased and installed. **(See Design Problem 11-4)**

Design Problem: Use Design Example Problem 11-4 to size the following components and elements of a store building supplied by a 120/240 volt, three-phase power source:

(1) What size THWN copper conductors, paralleled four times per phase, is required for Phases A, B and C?

Step 1: Finding amperage
Text
I = 1554 A ÷ 4
I = 389 A

Step 2: Selecting size conductors
Table 310.16, 75°C Column; 110.14(C)(1)(b)
600 KCMIL = 420 A
389 A is less than 420 A

Solution: The size THWN conductors required are 600 KCMIL copper.

(2) What size OCPD, using the minimum rating, is required for the walk-in cooler?

Step 1: Finding amperage
Text
I = VA ÷ V x 1.732
I = 9820 VA ÷ 416 V
I = 24 A

Step 2: Calculating amperage
Table 240.4(G); 440.22(A)
24 A x 175% = 42 A

Step 3: Selecting size OCPD
240.4(G); 240.6(A)
42 A requires 40 A OCPD

Solution: The size OCPD required is 40 amps.

(3) What size THWN copper conductors are required for the walk-in cooler?

Step 1: Finding amperage
Text
I = VA ÷ V x 1.732
I = 9820 V ÷ 416 V
I = 24 A

Step 2: Calculating amperage
440.32
24 A x 125% = 30 A

Step 3: Selecting size conductors
Table 310.16
30 A requires 10 AWG THWN cu.

Solution: The size THWN conductors are 10 AWG copper.

OFFICE BUILDING SUPPLIED BY 277/480 VOLT, THREE-PHASE POWER SOURCE
TABLE 220.12; (LISTED OCCUPANCY)

Office buildings consist mostly of lighting loads and receptacle loads supplying small sensitive electronic machines used for office-related work. Heavier loads such as equipment, heating, A/C and other three-phase loads are supplied by the higher voltage.

The advantage of 277/480 volt systems is that they allow OCPD's, conductors and other electrical elements to be smaller in size. Naturally, these smaller elements reduce the cost of installation. **(See Design Problem 11-5)**

Design Problem: Use Design Example Problem 11-5 to size the following components and elements of a office building supplied by a 277/480 volt, three-phase power source:

(1) How many 20 amp two-wire circuits are required for the general lighting load?

Step 1: Finding total VA
Table 220.12
VA = 525,000 VA + 1250 VA
VA = 526,250

Step 2: Finding # of outlets
210.11(A)
= VA ÷ CB x 277 V
= 526,250 VA ÷ 20 A x 277 V
= 95

Solution: The number of outlets is 95 for the general lighting load.

(2) How many 20 amp two-wire circuits are required for the general-purpose receptacle load?

Step 1: Finding total VA
Table 220.12
VA = 29,160

Step 2: Finding # of outlets
210.11(A)
= VA ÷ CB x 120 V
= 29,160 VA ÷ 20 A x 120 V
= 13

Solution: The number of outlets is 13 for the general-purpose receptacle load.

(3) What size transformer is required for the 120/208 volt loads?

Step 1: Finding total receptacle load
Noncontinuous operation = 30,380 VA
Continuous operation = 19,440 VA
Total load = 49,820 VA

Step 2: Special loads
Copying machines = 2,450 VA
Water heater = 10,000 VA
Data processors = 7,700 VA
Word processors = 1,750 VA
Printers = 4,800 VA
Total load = 26,700 VA

Step 3: Total loads
Receptacle loads = 49,820 VA
Special load = 26,700 VA
Total load = 76,520 VA

Step 4: Finding kVA
Text
kVA = VA ÷ 1000
kVA = 76,520 VA ÷ 1000
kVA = 76.52

Step 5: Selecting size transformer
Chart
76.52 kVA requires a 80 kVA transformer

Solution: The size transformer required 80 kVA.

(4) What size THWN copper conductors are required to supply the primary side of the transformer, if the load is continuous operation?

Step 1: Finding primary amperage
Text
A = kVA x 1000 ÷ V x $\sqrt{3}$
A = 80 kVA x 1000 ÷ 480 V x 1.732
A = 96.3

Step 2: Calculating amperage
230.42(A)(2)
96.3 A x 125% = 120.4 A

Step 3: Selecting size conductors
Table 310.16
1 AWG THWN cu. = 130 A
120.4 A is less than 130 A

Solution: The size THWN conductors are 1 AWG copper.

(5) What size OCPD is required for the primary side of the transformer?

Step 1: Calculating amperage
450.3(B); Table 450.3(B)
96.3 A x 125% = 120.4 A

Step 2: Selecting size OCPD
240.4(G); 450.3(B); Table 450.3(B); 240.6(A)
120.4 A requires 125 A OCPD

Solution: The size OCPD required is 125 amps.

SCHOOL BUILDING SUPPLIED BY 277/480 VOLT, THREE-PHASE POWER SOURCE
TABLE 220.12; (LISTED OCCUPANCY)

School buildings supplied by 277/480 volt services are calculated with the basic steps used to compute the load for any other commercial occupancy. **(See Design Problem 11-6)**

Design Problem: Use Design Example Problem 11-6 to size the following components and elements of a school building supplied by a 277/480 volt, three-phase power source:

(1) What size THWN copper conductors are required for Phases A, B and C?

Step 1: Finding amperage
Text
I = 284 A

Step 2: Selecting size conductors
Table 310.16, 75°C Column; 110.14(C)(1)(b)
300 KCMIL = 285 A
284 A is less than 285 A

Solution: The size THWN conductors required are 300 KCMIL copper.

(2) What size THWN copper conductors are required for the grounded (neutral) conductors?

Step 1: Finding amperage
Text
I = 94 A

Step 2: Selecting size conductors
Table 310.16, 75°C Column; 110.14(C)(1)(a)
3 AWG THWN = 100 A
94 A is less than 100 A

Solution: The size THWN conductors required are 3 AWG copper.

(3) What size rigid metal conduit is required?

> **Step 1:** Finding sq. in. area
> **Table 5, Ch. 9**
> 300 KCMIL THWN = .4608 sq. in.
> 3 AWG THWN = .0973 sq. in.
>
> **Step 2:** Calculating sq. in. area
> **Table 5, Ch. 9**
> .4608 sq. in. x 3 = 1.3824 sq. in.
> .0973 sq. in. x 1 = .0973 sq. in.
> Total = 1.4797 sq. in.
>
> **Step 3:** Selecting conduit size
> **Table 4, Ch. 9**
> 1.4797 sq. in. requires 2 1/2 in. (63)

Solution: The size rigid metal conduit is 2 1/2 in. (63).

RESTAURANT SUPPLIED BY 120/208 VOLT, THREE-PHASE POWER SOURCE
TABLE 220.12; (LISTED OCCUPANCY)

Small restaurants consist of lighting and receptacle loads, with the larger loads being the cooking equipment and other such pertinent apparatus.

Based upon the number of cooking units, demand factors shall be permitted to be applied to the total load per **220.56**. Larger restaurants are supplied with 277/480 volt services, with step down transformers being utilized to serve smaller 120 volt loads. **(See Design Problem 11-7)**

Design Problem: Use Design Example Problem 11-7 to size the following components and elements of a restaurant supplied by a 120/208 volt, three-phase power source:

(1) What is the demand load in VA and amps for the kitchen equipment?

> **Step 1:** Finding VA
> **220.56**
> Freezer = 2,704 VA
> Cooktop = 10,000 VA
> Ovens = 24,000 VA
> Range = 10,000 VA
> Refrigerator = 2,912 VA
> Ice cream box = 3,750 VA
> Boiler = 4,200 VA
> Deep fat fryer = 4,800 VA
> Walk-in cooler = 7,200 VA
> Water heater = 8,000 VA
> Total load = 77,566 VA
>
> **Step 2:** Applying demand factors
> **Table 220.56**
> 77,566 VA x 65% = 50,418 VA
>
> **Step 3:** Finding amperage
> Text
> I = VA ÷ V x $\sqrt{3}$
> I = 50,418 VA ÷ 360 V
> I = 140 A

Solution: The amperage is 140 amps.

HOSPITAL BUILDING SUPPLIED BY 277/480 VOLT, THREE-PHASE POWER SOURCE
TABLE 220.12; (LISTED OCCUPANCY)

The loads in hospitals are calculated by their conditions of use. They are either calculated at continuous operation, noncontinuous operation or demand factors shall be permitted to be applied for certain loads. The service voltage is determined by the size of the facility and related equipment. The procedure for calculating the load is to use the seven steps listed in this chapter for calculating the total load for a premises. **(See Design Problem 11-8)**

Design Problem: Use Design Example Problem 11-8 to size the following components and elements of a hospital supplied by a 277/480 volt, three-phase power source:

(1) What is the demand load in VA for the general lighting load?

    Step 1: Finding VA
        Table 220.12
        48,000 sq. ft. x 2 VA x 100% = 96,000 VA

    Step 2: Applying demand factors
        Table 220.42
        First 50,000 VA x 40%   = 20,000 VA
        Next 46,000 VA x 20%   = 9,200 VA
        Total load   = 29,200 VA

    Solution: The general lighting load is 29,200 VA.

(2) What is load in amps for the branch-circuit to the X-ray equipment with the short-time rating?

    Step 1: Finding amperage
        A = MA ÷ 1000 x Sec. V ÷ Pri. V
        A = 150 MA ÷ 1000 x 100,000 V ÷ 208 V
        A = 72

    Step 2: Applying demand factor
        660.6(A)
        72 A x 50% = 36 A

    Solution: The load is 36 amps.

(3) What size THWN copper conductors are required for the X-ray equipment with the short-time rating?

    Step 1: Selecting conductors
        Table 310.16
        36 A requires 8 AWG THWN cu.

    Solution: The size THWN conductors are 8 AWG copper.

(4) What is load in amps for the branch-circuit to the X-ray equipment with the longtime rating?

    Step 1: Finding amperage
        A = MA ÷ 1000 x Sec. V ÷ Pri. V
        A = 15 MA ÷ 1000 x 200,000 V ÷ 208 V
        A = 14

Step 2: Applying demand factor
660.6(A)
14 A x 100% = 14 A

Solution: The amperage is 14 amps.

(5) What size THWN copper conductors are required for the X-ray equipment with the longtime rating?

Step 1: Selecting conductors
Table 310.16
14 A requires 14 AWG THWN cu.

Solution: The size THWN conductors are 14 AWG copper.

(6) What is the total VA and amp ratings for the feeder and service conductors for the X-ray equipment?

Step 1: Finding amperage - short-time rating
A = MA ÷ 1000 x Sec. V ÷ Pri. V
A = 150 MA ÷ 1000 x 100,000 V ÷ 208 V
A = 72

Step 2: Finding amperage - longtime rating
A = MA ÷ 1000 x Sec. V ÷ Pri. V
A = 15 MA ÷ 1000 x 200,000 V ÷ 208 V
A = 14

Step 3: Finding VA
660.6(B)
Short-time rating
72 A x 2 x 50% x 208 V x 100% = 14,976 VA
Longtime rating
14 A x 2 x 208 V   =   5,824 VA
5824 VA x 20%   =   1,164.8 VA

Step 4: Finding total VA
660.6(B)
Short-time rating   = 14,976   VA
Longtime rating   =   1,164.8 VA
Total load   = 16,140.8 VA

Step 5: Finding amperage
Text
I = VA ÷ V
I = 16,140.8 VA ÷ 208 V
I = 77.6 A

Solution: The amperage is 77.6 amps.

(7) What size THWN copper conductors are required to supply power for the feeder to the X-ray equipment?

Step 1: Selecting size conductors
Table 310.16
77.6 A requires 4 AWG cu.

Solution: The size THWN conductors are 4 AWG copper.

Commercial Calculations

HOTELS AND MOTELS SUPPLIED BY 120/208 VOLT, THREE-PHASE POWER SOURCE
TABLE 220.12; (LISTED OCCUPANCY)

Hotels and motels differ from dwelling units in that guest rooms are not provided with permanent cooking facilities. **(See Design Problem 11-9)**

Design Problem: Use Design Example Problem 11-9 to size the following components and elements of a hotel or motel supplied by a 120/208 volt, three-phase power source:

(1) What size OCPD is required for the service supplied by copper busduct?

   Step 1: Selecting busduct
   Chart on inside back cover
   1948 A requires 2000 A busduct

   Step 2: Selecting size OCPD (based on busduct)
   240.4(B); 240.6(A); 368.10
   2000 A busduct requires 2000 A OCPD

   Solution: The size OCPD required is 2000 amps.

(2) What size switchgear is required for the service equipment?

   Step 1: Selecting switchgear
   Chart on inside back cover
   1948 A requires 2000 A switchgear

   Solution: The size switchgear required is 2000 amps.

(3) What size copper grounding electrode conductor is required?

   Step 1: Finding amps of busduct busbars
   4" x 1/2" x 1000 A = 2000 A busbars

   Step 2: Finding sq. mils of busbars
   Width   = 4" x 1000 = 4000 mils
   Thickness   = 1/2" x 1000 = 500 mils
   Area   = width x thickness (in mils)
   Area   = 4000 mils x 500 mils = 2,000,000 sq. mils

   Step 3: Finding CM of busbars
   CM = 2,000,000 sq. mils ÷ .7854
   CM = 2,546,473

   Step 4: Finding KCMIL
   KCMIL = CM ÷ 1000
   KCMIL = 2,546,473 ÷ 1000
   KCMIL = 2547 KCMIL

   Step 5: Selecting size conductor
   Table 250.66
   2547 KCMIL requires 3/0 AWG copper

   Solution: The size copper grounding electrode required is 3/0 AWG.

BANKS SUPPLIED BY 120/208 VOLT, THREE-PHASE POWER SOURCE TABLE 220.12; (LISTED OCCUPANCY)

The load in VA and amps for a bank is calculated using the same method used for office buildings. If the number of general-purpose receptacles is not known, an extra 1 VA per sq. ft. is calculated for the receptacle load. **(See Design Problem 11-10)**

Design Problem: Use Design Example Problem 11-10 to size the following components and elements of a bank supplied by a 120/208 volt, three-phase power source:

(1) What size THWN copper conductors, paralleled three times per phase, are required for Phases A, B and C?

    Step 1: Finding amperage
    Text
    I = 686 A
    I = 686 A ÷ 3 (No. run per phase)
    I = 229 A

    Step 2: Selecting size conductors
    Table 310.16, 75°C Column; 110.14(C)(1)(b)
    4/0 AWG THWN = 230 A
    229 A is less than 230 A

    Solution: The size THWN conductors required are 4/0 AWG copper.

(2) What size THWN copper conductors are required for the grounded (neutral) conductors?

    Step 1: Finding amperage
    Text
    I = 278 A
    I = 278 A ÷ 3 (No. run per phase)
    I = 93 A

    Step 2: Selecting size conductors
    Table 310.16, 75°C Column; 310.4; 110.14(C)(1)(b)
    1/0 AWG THWN = 150 A
    93 A is less than 150 A

    Solution: The size THWN conductors required are 1/0 AWG copper.

(3) What size rigid metal conduit is required for each parallel run?

    Step 1: Finding sq. in. area
    Table 5, Ch. 9
    4/0 AWG THWN   = .3237 sq. in.
    1/0 AWG THWN   = .1855 sq. in.

    Step 2: Calculating sq. in. area
    Table 5, Ch. 9
    .3237 sq. in. x 3   = .9711 sq. in.
    .1855 sq. in. x 1   = .1855 sq. in.
    Total   = 1.1566 sq. in.

    Step 3: Selecting conduit size
    Table 4, Ch. 9
    1.1566 sq. in. requires 2 in. (53)

    Solution: The size rigid metal conduit is 2 in. (53).

WELDING SHOPS SUPPLIED BY 120/208 VOLT POWER SOURCE
PARTS II AND III TO ARTICLE 220 (UNLISTED OCCUPANCY)

Unlisted occupancies are those not appearing in **Table 220.12**, therefore, their VA ratings shall be calculated from other sections than those listed in **Table 220.12**. The load for unlisted occupancies is calculated by applying the same step procedures listed in this chapter. The only difference is the lighting is not obtained per **Table 220.12** but from the VA or amp rating of each lighting unit. **(See Design Problem 11-11)**

Design Problem: Use Design Example Problem 11-11 to size the following components and elements of a welding shop supplied by a 120/208 volt, three-phase power source:

(1) What is the load in VA for the resistance welders?

    Step 1: Finding VA
    630.31(A); (B)
    11,000 VA x 55%   = 6,050 VA
    9000 VA x 55% x 60%   = 2,970 VA
    Total load   = 9,020 VA

    Solution: The load for the resistance welders is 9020 VA.

(2) What is the load in VA for the motor-generator arc welders?

    Step 1: Finding VA
    630.11(A); (B)
    12,000 VA x 91%   = 10,920 VA
    10,000 VA x 91%   = 9,100 VA
    Total load   = 20,020 VA

    Solution: The load for the motor-generator arc welders is 20,020 VA.

(3) What is the load in VA for the AC transformer and DC rectifier welders?

    Step 1: Finding VA
    630.11(A); (B)
    12,000 VA x 95%   = 11,400 VA
    9000 VA x 95%   = 8,550 VA
    Total load   = 19,950 VA

    Solution: The load for the AC transformer and DC rectifier welders is 19,950 VA.

APPLYING THE OPTIONAL CALCULATION
PART III TO ARTICLE 220

The load in VA and amps shall be permitted to be calculated by the optional calculation instead of the standard calculation. The optional calculation is based upon specific use of the electrical system or type occupancy and its use and operation.

KITCHEN EQUIPMENT
220.56; TABLE 220.56

Table 220.56 in the NEC shall be permitted to be used for load calculation for commercial electrical cooking equipment, such as dishwashers, booster heaters, water heaters and other kitchen equipment. The demand factors shown in that Table are applicable to all equipment that is thermostatically controlled or is only intermittently used as part of the kitchen equipment. In no way do the demand factors apply to the electric heating, ventilating or air-conditioning equipment. In calculating the demand, the demand load shall never be less than the sum of the two largest kitchen equipment loads.

Note: The answers to exercise problem 11-12, [A] through [O] are found in answer key on page 19.

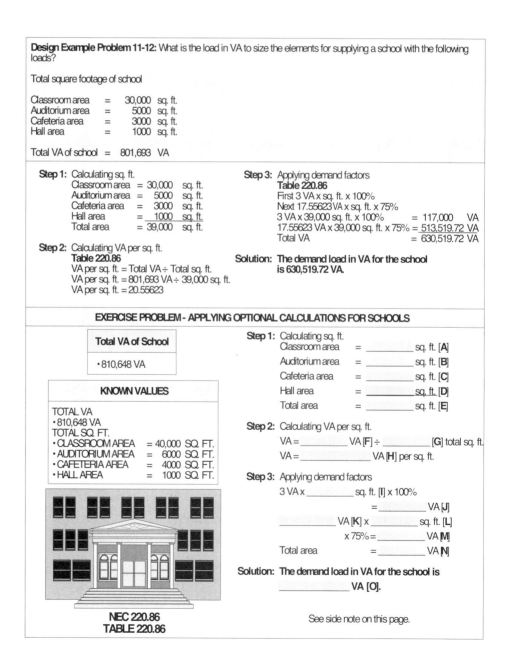

SCHOOLS
220.86; TABLE 220.86

Table 220.86 shall be permitted to be used to calculate the service or feeder loads for schools if they are equipped with electric space heating or air-conditioning or both. The demand factors in **Table 220.86** apply to both interior and exterior lighting, power, water heating, cooking or other loads, and the larger of the space heating load or the air-conditioning load.

When using the optional calculation, the grounded (neutral) conductor of the service or feeder loads shall be calculated as required per **220.61**. Feeders within the building or structure where the load is calculated by this optional method shall be permitted to use the reduced ampacity as connected, but the ampacity of any feeder need not be larger than the individual ampacity for the entire building. Portable classrooms or buildings shall not be included in this Section. **(See Design Problem 11-12)**

RESTAURANTS
220.88; TABLE 220.88

When calculating the service or feeder load for a new restaurant, and the feeder carries the entire load, **Table 220.36** shall be permitted to be used to size the elements necessary to supply the load. Overload protection shall be in accordance with **230.90(A) with exceptions** and **240.4(A) through (G)**. Also, feeder or subfeeder conductors do not have to be larger than service conductors, regardless of calculations. **(See Design Problem 11-13)**

OPTIONAL CALCULATIONS FOR ADDITIONAL LOADS TO EXISTING INSTALLATIONS
220.87

When additional loads are added to existing facilities having feeders and a service as originally calculated, the maximum kVA calculations in determining the load on the existing feeders and service shall be permitted to be used if the following conditions are complied with:

(1) If the maximum data of the demand in kVA is available for a minimum of one year, such as demand meter ratings.

(2) If the demand ratings for that period of one year at 125 percent and the addition of the new load does not exceed the rating of the service. Where demand meters are used, in most cases the load as calculated will probably be less than the demand meter indications.

(3) If the overcurrent protection meets **230.90** and **240.4** for feeders or a service.

APPLYING Ex. TO 220.87
220.87, Ex.

If the maximum demand data for a one year period is not available, the calculated loads shall be permitted to be based on the maximum demand (measure of average power demand over a 15 minute period) continuously recorded over a minimum 30 day period using a recording ammeter or power meter connected to the highest loaded ungrounded phase of the feeder or service, based on the initial loading at the start of the recording. **(See Design Problem 11-14)**

Note, the answers to exercise problem 11-13, [A] through [J] are found in answer key on page 19.

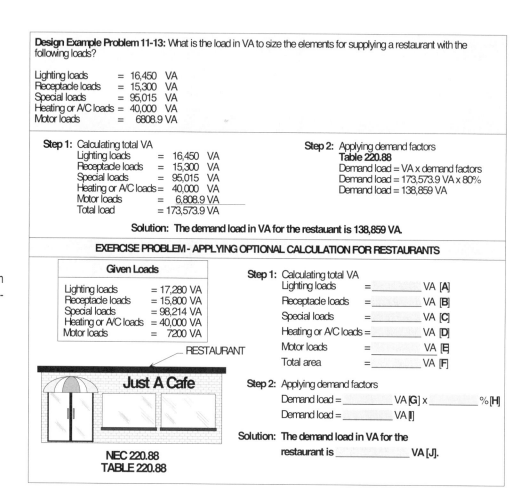

Design Example Problem 11-13: What is the load in VA to size the elements for supplying a restaurant with the following loads?

Lighting loads = 16,450 VA
Receptacle loads = 15,300 VA
Special loads = 95,015 VA
Heating or A/C loads = 40,000 VA
Motor loads = 6808.9 VA

Step 1: Calculating total VA
Lighting loads = 16,450 VA
Receptacle loads = 15,300 VA
Special loads = 95,015 VA
Heating or A/C loads = 40,000 VA
Motor loads = 6,808.9 VA
Total load = 173,573.9 VA

Step 2: Applying demand factors
Table 220.88
Demand load = VA x demand factors
Demand load = 173,573.9 VA x 80%
Demand load = 138,859 VA

Solution: The demand load in VA for the restaurant is 138,859 VA.

EXERCISE PROBLEM - APPLYING OPTIONAL CALCULATION FOR RESTAURANTS

Given Loads
Lighting loads = 17,280 VA
Receptacle loads = 15,800 VA
Special loads = 98,214 VA
Heating or A/C loads = 40,000 VA
Motor loads = 7200 VA

RESTAURANT — Just A Cafe

NEC 220.88
TABLE 220.88

Step 1: Calculating total VA
Lighting loads = _____ VA [A]
Receptacle loads = _____ VA [B]
Special loads = _____ VA [C]
Heating or A/C loads = _____ VA [D]
Motor loads = _____ VA [E]
Total area = _____ VA [F]

Step 2: Applying demand factors
Demand load = _____ VA [G] x _____ % [H]
Demand load = _____ VA [I]

Solution: The demand load in VA for the restaurant is _____ VA [J].

Commercial Calculations

Design Example Problem 11-14: Can a load of 15.1 kVA be added to an existing service with 400 KCMIL THW copper conductors having a maximum demand of 78.4 kVA?

Step 1: Finding demand
220.87
Maximum demand = 78.4 kVA

Step 2: Calculating existing demand
220.87
78.4 kVA x 125% = 98 kVA

Step 3: Calculating total kVA
230.42(A)(1)
98 kV + 15.1 kVA = 113.1 kVA

Step 4: Calculating amperage
Table 310.16
400 KCMIL THW copper = 335 A
113.1 x 1000 = 113,100 VA
113,100 VA ÷ 208 V x 1.732 = 314 A
314 A is less than 335 A

Solution: The 15.1 kVA load can be applied to the existing service without upgrading the elements.

EXERCISE PROBLEM - APPLYING ADDITIONAL LOADS TO EXISTING INSTALLATIONS

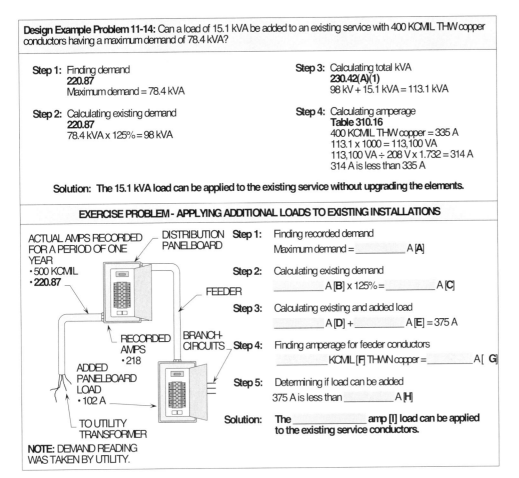

ACTUAL AMPS RECORDED FOR A PERIOD OF ONE YEAR
• 500 KCMIL
• 220.87

RECORDED AMPS
• 218

ADDED PANELBOARD LOAD
• 102 A

TO UTILITY TRANSFORMER

NOTE: DEMAND READING WAS TAKEN BY UTILITY.

DISTRIBUTION PANELBOARD

FEEDER

BRANCH-CIRCUITS

Step 1: Finding recorded demand
Maximum demand = _____ A [A]

Step 2: Calculating existing demand
_____ A [B] x 125% = _____ A [C]

Step 3: Calculating existing and added load
_____ A [D] + _____ A [E] = 375 A

Step 4: Finding amperage for feeder conductors
_____ KCMIL [F] THWN copper = _____ A [G]

Step 5: Determining if load can be added
375 A is less than _____ A [H]

Solution: The _____ amp [I] load can be applied to the existing service conductors.

Note, the answers to exercise problem 11-14, [A] through [I] are found in answer key on page 19.

Design Example Problem 11-1. What is the load in VA and amps to calculate and size the elements for a 120/240 volt, single-phase service supplying a 60,000 sq. ft. store including a 40,000 sq. ft. warehouse space with the following loads? Note: 51 percent or more of the lighting load is electric discharge.

120 V, single-phase loads
- 120 linear feet of show window (noncontinuous operation)
- 150 ft. of lighting track (noncontinuous operation)
- 50 - 180 VA ballasts outside lighting (continuous operation)
- 4,200 VA sign lighting (continuous operation)
- 82 receptacles (noncontinuous operation)
- 30 receptacles (continuous operation)
- 120 ft. multioutlet assembly (heavy-duty)

240 V, single-phase loads
- 12,000 VA water heater
- 7860 VA freezer
- 7240 VA ice cream box
- 9820 VA walk-in cooler
- 50,000 VA heating unit
- 26,400 VA A/C unit
- 1 - 1/2 HP exhaust fan
- 1 - 2 HP water pump

Sizing ungrounded (phase) conductors = •
Sizing grounded (neutral) conductors = √
Sizing load - Phases A and B = *

Calculating lighting load

Step 1: General lighting load
Table 220.12; 230.42(A)(2)
60,000 sq. ft. x 3 VA = 180,000 VA √
180,000 VA x 125% = 225,000 VA •
40,000 sq. ft. x 1/4 VA = 10,000 VA √
10,000 VA x 125% = 12,500 VA •

Step 2: Show window load
220.43(A)
120' x 200 VA = 24,000 VA •√

Step 3: Track lighting load
220.43(B)
150' ÷ 2 x 150 VA = 11,250 VA •√

Step 4: Outside lighting load
230.42(A)(2)
50 x 180 VA = 9,000 VA √
9000 VA x 125% = 11,250 VA •

Step 5: Sign lighting load
220.14(F); 600.5(A); 230.42(A)(2)
4200 VA x 100% = 4,200 VA √
4200 VA x 125% = 5,250 VA •
Total load = 289,250 VA *

Calculating receptacle loads

Step 1: Noncontinuous operation
220.14(I); 230.42(A)(1); 220.14(H)(1); (2)
82 x 180 VA = 14,760 VA
120' x 180 VA = 21,600 VA
Total load = 36,360 VA
Table 220.44
First 10,000 VA x 100% = 10,000 VA
Next 26,360 VA x 50% = 13,180 VA
Total load = 23,180 VA •√

Step 2: Continuous operation
220.14(I); 230.42(A)(2)
30 x 180 VA = 5,400 VA √
5400 VA x 125% = 6,750 VA •
Total load = 29,930 VA *

Calculating special loads

Step 1: 12,000 VA x 100% = 12,000 VA •*

Calculating compressor loads

Step 1: Freezer load
230.42(A)(1)
7860 VA x 100% = 7,860 VA •

Step 2: Ice cream boxes
230.42(A)(1)
7240 VA x 100% = 7,240 VA •

Step 3: Walk-in cooler
230.42(A)(1)
9820 VA x 100% = 9,820 VA •
Total load = 24,920 VA *

Calculating motor loads

Step 1: Water pump load
430.24; Table 430.248
12 A x 240 V x 100% = 2,880 VA •

Step 2: Exhaust fan load
430.24; Table 430.248
4.9 A x 240 V x 100% = 1,176 VA •
Total load = 4,056 VA *

Calculating heating or A/C load

Step 1: Heating load selected
220.60
50,000 VA x 100% = 50,000 VA •*
26,400 VA x 100% = 26,400 VA

Calculating largest motor load

Step 1: Walk-in cooler
220.14(C); 220.50; 430.24
9820 VA x 25% = 2,455 VA •*

Calculating ungrounded (phase) conductors

Lighting loads = 289,250 VA •
Receptacle loads = 29,930 VA •
Special loads = 12,000 VA •
Compressor loads = 24,920 VA •
Motor loads = 4,056 VA •
Heating or A/C loads = 50,000 VA •
Largest motor load = 2,455 VA •
Total load = 412,611 VA

Calculating grounded (neutral) conductors

Lighting load = 238,450 VA √
Receptacle load = 28,580 VA √
Total load = 267,030 VA

Finding amps for ungrounded (phase) conductors - Phases A and B

I = VA ÷ V
I = 412,611 VA ÷ 240 V
I = 1719 VA

Finding amps for grounded (neutral) conductors

I = VA ÷ V
I = 267,030 VA ÷ 240 V
I = 1113 A

Commercial Calculations

Design Exercise Problem 11-1. What is the load in VA and amps to calculate and size the elements for a 120/240 volt, single-phase service supplying a 50,000 sq. ft. store including a 25,000 sq. ft. of warehouse space with the following loads?

120 V, single-phase loads
- 100 linear feet of show window (noncontinuous operation)
- 100' of lighting track (noncontinuous operation)
- 40 - 180 VA ballasts outside lighting (continuous operation)
- 4200 VA sign lighting (continuous operation)
- 72 receptacles (noncontinuous operation)
- 34 receptacles (continuous operation)
- 120' multioutlet assembly (heavy-duty)

240 V, single-phase loads
- 12,000 VA water heater
- 7875 VA freezer
- 6325 VA ice cream boxes
- 9420 VA walk-in cooler
- 60,000 VA heating unit
- 24,640 VA A/C unit
- 1 - 1/2 HP exhaust fan
- 1 - 2 HP water pump

Sizing ungrounded (phase) conductors = •
Sizing grounded (neutral) conductors = √
Sizing load - Phases A and B = *

Calculating lighting load

Step 1: General lighting load
Table 220.12; 230.42(A)(2)
50,000 sq. ft. x 3 VA = _____ VA [A] √
150,000 VA x 125% = _____ VA [B] •
25,000 sq. ft. x 1/4 VA = _____ VA [C] √
6250 VA x 125% = _____ VA [D] •

Step 2: Show window load
220.43(A)
100' x 200 VA = _____ VA [E] • √

Step 3: Track lighting load
220.43(B)
100' ÷ 2 x 150 VA = _____ VA [F] • √

Step 4: Outside lighting load
230.42(A)(2)
40 x 180 VA = _____ VA [G] √
7200 VA x 125% = _____ VA [H] •

Step 5: Sign lighting load
220.14(F); 600.5(A); 230.42(A)(2)
4200 VA x 100% = _____ VA [I] √
4200 VA x 125% = _____ VA [J] •
Total load = _____ VA [K] *

Calculating receptacle loads

Step 1: Noncontinuous operation
220.14(I); 230.42(A)(1); 220.14(H)(1); (2)
72 x 180 VA = _____ VA [L]
120' x 180 VA = _____ VA [M]
Total load = _____ VA [N]
Table 220.44
First 10,000 VA x 100% = _____ VA [O]
Next 24,560 VA x 50% = _____ VA [P]
Total load = _____ VA [Q] • √

Step 2: Continuous operation
220.14(I); 230.42(A)(2)
34 x 180 VA = _____ VA [R] √
6120 VA x 125% = _____ VA [S] •
Total load = _____ VA [T] *

Calculating special loads

Step 1: Water heater load
12,000 VA x 100% = _____ VA [U] • *

Calculating compressor loads

Step 1: Freezer load
230.42(A)(1)
7875 VA x 100% = _____ VA [V] •

Step 2: Ice cream boxes
6325 VA x 100% = _____ VA [W] •

Step 3: Walk-in cooler
9420 VA x 100% = _____ VA [X] •
Total load = _____ VA [Y] *

Calculating motor loads

Step 1: Water pump load
430.24; Table 430.248
12 A x 240 V x 100% = _____ VA [Z] •

Step 2: Exhaust fan load
4.9 A x 240 V x 100% = _____ VA [AA] •
Total load = _____ VA [BB] *

Calculating heating or A/C load

Step 1: Heating load selected
220.60
60,000 VA x 100% = _____ VA [CC] • *
24,640 VA x 100% = _____ VA [DD]

Calculating largest motor load

Step 1: Walk-in cooler
220.14(C); 220.50; 430.24
9420 VA x 25% = _____ VA [EE] • *

Calculating ungrounded (phase) conductors

Lighting loads = _____ VA [FF] •
Receptacle loads = _____ VA [GG] •
Special loads = _____ VA [HH] •
Compressor loads = _____ VA [II] •
Motor loads = _____ VA [JJ] •
Heating or A/C load = _____ VA [KK] •
Largest motor load = _____ VA [LL] •
Total load for facility = _____ VA [MM]

Finding amps for ungrounded (phase) conductors - Phases A and B

I = VA ÷ V
I = _____ VA [NN] ÷ _____ V [OO]
I = _____ A [PP]

Calculating grounded (neutral) conductors

Lighting load = _____ VA [QQ] √
Receptacle load = _____ VA [RR] √
Total load = _____ VA [SS]

Finding amps for grounded (neutral) conductors

I = VA ÷ V
I = _____ VA [TT] ÷ _____ V [UU]
I = _____ A [VV]

11-33

Design Example Problem 11-2. What is the load in VA and amps to calculate and size the elements for a 120/208 volt, three-phase service supplying a 60,000 sq. ft. store including a 40,000 sq. ft. warehouse space with the following loads? Note: 51 percent or more of the lighting is electric discharge.

120 V, single-phase loads
- 120 linear feet of show window (noncontinuous operation)
- 150 ft. of lighting track (noncontinuous operation)
- 50 - 180 VA ballast's outside lighting (continuous operation)
- 4200 VA sign lighting (continuous operation)
- 82 receptacles (noncontinuous operation)
- 30 receptacles (continuous operation)
- 120 ft. multioutlet assembly (heavy-duty)

208 V, three-phase loads
- 12,000 VA water heater
- 7860 VA freezer
- 7240 VA ice cream box
- 9820 VA walk-in cooler
- 50,000 VA heating unit
- 26,400 VA A/C unit
- 1 - 1/2 HP exhaust fan
- 1 - 2 HP water pump

Sizing ungrounded (phase) conductors = •
Sizing grounded (neutral) conductors = √
Sizing load - Phases A, B and C = *

Calculating lighting load

Step 1: General lighting load
Table 220.12; 230.42(A)(2)
- 60,000 sq. ft. x 3 VA = 180,000 VA √
- 180,000 VA x 125% = 225,000 VA •
- 40,000 sq. ft. x 1/4 VA = 10,000 VA √
- 10,000 VA x 125% = 12,500 VA •

Step 2: Show window load
220.43(A)
- 120' x 200 VA = 24,000 VA • √

Step 3: Track lighting load
220.43(B)
- 150' ÷ 2 x 150 VA = 11,250 VA • √

Step 4: Outside lighting load
230.42(A)(2)
- 50 x 180 VA = 9,000 VA √
- 9000 VA x 125% = 11,250 VA •

Step 5: Sign lighting load
220.14(F); 600.5(A); 230.42(A)(2)
- 4200 VA x 100% = 4,200 VA √
- 4200 VA x 125% = 5,250 VA •
- Total load = 289,250 VA *

Calculating receptacle loads

Step 1: Noncontinuous operation
220.14(I); 230.42(A)(1); 220.14(H)(1); (2)
- 82 x 180 VA = 14,760 VA
- 120' x 180 VA = 21,600 VA
- Total load = 36,360 VA
Table 220.44
- First 10,000 VA x 100% = 10,000 VA
- Next 26,360 VA x 50% = 13,180 VA
- Total load = 23,180 VA • √

Step 2: Continuous operation
220.14(I); 230.42(A)(2)
- 30 x 180 VA = 5,400 VA √
- 5400 VA x 125% = 6,750 VA •
- Total load = 29,930 VA *

Calculating special load

Step 1: 12,000 VA x 100% = 12,000 VA • *

Calculating compressor loads

Step 1: Freezer load
230.42(A)(1)
- 7860 VA x 100% = 7,860 VA •

Step 2: Ice cream boxes
230.42(A)(1)
- 7240 VA x 100% = 7,240 VA •

Step 3: Walk-in cooler
230.42(A)(1)
- 9820 VA x 100% = 9,820 VA •
- Total load = 24,920 VA *

Calculating motor loads

Step 1: Water pump load
430.24; Table 430.250
- 7.5 A x 360 V x 100% = 2,700 VA •

Step 2: Exhaust fan load
430.24; Table 430.250
- 2.4 A x 360 V x 100% = 864 VA •
- Total load = 3,564 VA *

Calculating heating or A/C load

Step 1: Heating load selected
220.60
- 50,000 VA x 100% = 50,000 VA • *
- 26,400 VA x 100% = 26,400 VA

Calculating largest motor load

Step 1: Walk-in cooler
220.14(C); 220.50; 430.24
- 9820 VA x 25% = 2,455 VA • *

Calculating ungrounded (phase) conductors

- Lighting loads = 289,250 VA •
- Receptacle loads = 29,930 VA •
- Special loads = 12,000 VA •
- Compressor loads = 24,920 VA •
- Motor loads = 3,564 VA •
- Heating or A/C load = 50,000 VA •
- Largest motor load = 2,455 VA •
- Total load = 412,119 VA

Calculating grounded (neutral) conductors

- Lighting load = 238,450 VA √
- Receptacle load = 28,580 VA √
- Total load = 267,030 VA

Finding amps for ungrounded (phase) conductors - Phases A, B and C

I = VA ÷ V
I = 412,119 VA ÷ 360 V
I = 1145 A

Finding amps for grounded (neutral) conductors

I = VA ÷ V
I = 267,030 VA ÷ 360 V
I = 742 A

Note that the 360 volt is used (208 V x 1.732) instead of 360.256 volt.

Commercial Calculations

Design Exercise Problem 11-2. What is the load in VA and amps to calculate and size the elements for a 120/208 volt, three-phase, four-wire service supplying a 50,000 sq. ft. store including a 25,000 sq. ft. of warehouse space with the following loads?

120 V, single-phase loads
- 100 linear feet of show window (noncontinuous operation)
- 100' of lighting track (noncontinuous operation)
- 40 - 180 VA ballasts outside lighting (continuous operation)
- 4200 VA sign lighting (continuous operation)
- 72 receptacles (noncontinuous operation)
- 34 receptacles (continuous operation)
- 120' multioutlet assembly (heavy-duty)

208 V, three-phase loads
- 12,000 VA water heater
- 7875 VA freezer
- 6325 VA ice cream boxes
- 9420 VA walk-in cooler
- 60,000 VA heating unit
- 24,640 VA A/C unit
- 1 - 1/2 HP exhaust fan
- 1 - 2 HP water pump

Sizing ungrounded (phase) conductors = •
Sizing grounded (neutral) conductors = √
Sizing load - Phases A, B and C = *

Note that 360 volt is used (208 V x 1.732) instead of 360.256 volt.

Calculating lighting load

Step 1: General lighting load
Table 220.12; 230.42(A)(2)
50,000 sq. ft. x 3 VA = _____ VA [A] √
150,000 VA x 125% = _____ VA [B] •
25,000 sq. ft. x 1/4 VA = _____ VA [C] √
6250 VA x 125% = _____ VA [D] •

Step 2: Show window load
220.43(A)
100' x 200 VA = _____ VA [E] • √

Step 3: Track lighting load
220.43(B)
100' ÷ 2 x 150 VA = _____ VA [F] • √

Step 4: Outside lighting load
230.42(A)(2)
40 x 180 VA = _____ VA [G] √
7200 VA x 125% = _____ VA [H] •

Step 5: Sign lighting load
220.14(F); 600.5(A); 230.42(A)(2)
4200 VA x 100% = _____ VA [I] √
4200 VA x 125% = _____ VA [J] •
Total load = _____ VA [K] *

Calculating receptacle load

Step 1: Noncontinuous operation
220.14(I); 230.42(A)(1); 220.14(H)(1); (2)
72 x 180 VA = _____ VA [L]
120' x 180 VA = _____ VA [M]
Total load = _____ VA [N]
Table 220.44
First 10,000 VA x 100% = _____ VA [O]
Next 24,560 VA x 50% = _____ VA [P]
Total load = _____ VA [Q] • √

Step 2: Continuous operation
220.14(I); 230.42(A)(2)
34 x 180 VA = _____ VA [R] √
6120 VA x 125% = _____ VA [S] •
Total load = _____ VA [T] *

Calculating special loads

Step 1: Water heater load
12,000 VA x 100% = _____ VA [U] • *

Calculating compressor loads

Step 1: Freezer load
230.42(A)(1)
7875 VA x 100% = _____ VA [V] •

Step 2: Ice cream boxes
6325 VA x 100% = _____ VA [W] •

Step 3: Walk-in cooler
9420 VA x 100% = _____ VA [X] •
Total load = _____ VA [Y] *

Calculating motor loads

Step 1: Water pump load
430.24; Table 430.250
7.5 A x 360 V x 100% = _____ VA [Z] •

Step 2: Exhaust fan load
2.4 A x 360 V x 100% = _____ VA [AA] •
Total load = _____ VA [BB] *

Calculating heating or A/C load

Step 1: Heating load selected
220.60
60,000 VA x 100% = _____ VA [CC] • *
24,640 VA x 100% = _____ VA [DD]

Calculating largest motor load

Step 1: Walk-in cooler
220.14(C); 220.50; 430.24
9420 VA x 25% = _____ VA [EE] • *

Calculating ungrounded (phase) conductors

Lighting loads = _____ VA [FF] •
Receptacle loads = _____ VA [GG] •
Special loads = _____ VA [HH] •
Compressor loads = _____ VA [II] •
Motor loads = _____ VA [JJ] •
Heating or A/C load = _____ VA [KK] •
Largest motor load = _____ VA [LL] •
Total load for facility = _____ VA [MM]

Finding amps for ungrounded (phase) conductors - Phases A, B and C

I = VA ÷ V
I = _____ VA [NN] ÷ _____ V [OO]
I = _____ A [PP]

Calculating grounded (neutral) conductors

Lighting load = _____ VA [QQ] √
Receptacle load = _____ VA [RR] √
Total load = _____ VA [SS]

Finding amps for grounded (neutral) conductors

I = VA ÷ V
I = _____ VA [TT] ÷ _____ V [UU]
I = _____ A [VV]

Design Example Problem 11-3. What is the load in VA and amps to calculate and size the elements for a 277/480 volt, three-phase service supplying a 60,000 sq. ft. store including a 40,000 sq. ft. warehouse space with the following loads? Note: 51 percent or more of the lighting load is electric discharge.

120 V, single-phase loads
- 120 linear feet of show window (noncontinuous operation)
- 150 ft. of lighting track (noncontinuous operation)
- 50 - 180 VA ballasts outside lighting (continuous operation)
- 4200 VA sign lighting (continuous operation)
- 82 receptacles (noncontinuous operation)
- 30 receptacles (continuous operation)
- 120 ft. multioutlet assembly (heavy-duty)
- office area is 277 V lighting
- warehouse is 277 V lighting

480 V, three-phase loads
- 12,000 VA water heater
- 7860 VA freezer
- 7240 VA ice cream box
- 9820 VA walk-in cooler
- 50,000 VA heating unit
- 26,400 VA A/C unit
- 1 - 1/2 HP exhaust fan
- 1 - 2 HP water pump

Sizing ungrounded (phase) conductors = •
Sizing grounded (neutral) conductors = √
Sizing load - Phases A, B and C = *

Calculating lighting load

Step 1: General lighting load
Table 220.12; 230.42(A)(2)
60,000 sq. ft. x 3 VA = 180,000 VA √
180,000 VA x 125% = 225,000 VA •
40,000 sq. ft. x 1/4 VA = 10,000 VA √
10,000 VA x 125% = 12,500 VA •

Step 2: Show window load
220.43(A)
120' x 200 VA = 24,000 VA •

Step 3: Track lighting load
220.43(B)
150' ÷ 2 x 150 VA = 11,250 VA •

Step 4: Outside lighting load
230.42(A)(2)
50 x 180 VA = 9,000 VA
9000 VA x 125% = 11,250 VA •

Step 5: Sign lighting load
220.14(F); 600.5(A); 230.42(A)(2)
4200 VA x 100% = 4,200 VA
4200 VA x 125% = 5,250 VA •
Total load = 289,250 VA *

Calculating receptacle loads

Step 1: Noncontinuous operation
220.14(I); 230.42(A)(1); 220.14(H)(1); (2)
82 x 180 VA = 14,760 VA
120' x 180 VA = 21,600 VA
Total load = 36,360 VA
Table 220.44
First 10,000 VA x 100% = 10,000 VA
Next 26,360 VA x 50% = 13,180 VA
Total load = 23,180 VA •

Step 2: Continuous operation
220.14(I); 230.42(A)(2)
30 x 180 VA = 5,400 VA
5400 VA x 125% = 6,750 VA •
Total load = 29,930 VA *

Calculating special loads

Step 1: 12,000 VA x 100% = 12,000 VA • *

Calculating compressor loads

Step 1: Freezer load
230.42(A)(1)
7860 VA x 100% = 7,860 VA •

Step 2: Ice cream boxes
230.42(A)(1)
7240 VA x 100% = 7,240 VA •

Step 3: Walk-in cooler
230.42(A)(1)
9820 VA x 100% = 9,820 VA •
Total load = 24,920 VA *

Calculating motor loads

Step 1: Water pump load
430.24; Table 430.250
3.4 A x 831 V x 150% = 2,825 VA •

Step 2: Exhaust fan load
430.24; Table 430.250
1.1 A x 831 V x 100% = 914 VA •
Total load = 3,739 VA *

Calculating heating or A/C load

Step 1: Heating load selected
220.60
50,000 VA x 100% = 50,000 VA • *
26,400 VA x 100% = 26,400 VA

Calculating largest motor load

Step 1: Walk-in cooler
220.14(C); 220.50; 430.24
9820 VA x 25% = 2,455 VA • *

Calculating ungrounded (phase) conductors

Lighting loads = 289,250 VA •
Receptacle loads = 29,930 VA •
Special loads = 12,000 VA •
Compressor loads = 24,920 VA •
Motor loads = 3,739 VA •
Heating or A/C load = 50,000 VA •
Largest motor load = 2,455 VA •
Total load = 412,294 VA

Calculating grounded (neutral) conductors

Lighting load = 190,000 VA √
Total load = 190,000 VA

Finding amps for ungrounded (phase) conductors - Phases A, B and C

I = VA ÷ V
I = 412,294 VA ÷ 831 V
I = 496 A

Finding amps for grounded (neutral) conductors

I = VA ÷ V
I = 190,000 VA ÷ 831 V
I = 229 A

Note that the 831 volt is used (480 V x 1.732) instead of 831.36 volt.

Commercial Calculations

Design Exercise Problem 11-3. What is the load in VA and amps to calculate and size the elements for a 277/480 volt, three-phase, four-wire service supplying a 50,000 sq. ft. store including a 25,000 sq. ft. of warehouse space having 277 volt lighting with the following loads?

120 V, single-phase loads
- 100 linear feet of show window (noncontinuous operation)
- 100' of lighting track (noncontinuous operation)
- 40 - 180 VA ballasts outside lighting (continuous operation)
- 4200 VA sign lighting (continuous operation)
- 72 receptacles (noncontinuous operation)
- 34 receptacles (continuous operation)
- 120' multioutlet assembly (heavy-duty)
- store lighting is 277 V
- warehouse lighting is 277 V

480 V, three-phase loads
- 12,000 VA water heater
- 7875 VA freezer
- 6325 VA ice cream boxes
- 9420 VA walk-in cooler
- 60,000 VA heating unit
- 24,640 VA A/C unit
- 1 - 1/2 HP exhaust fan
- 1 - 2 HP water pump

Sizing ungrounded (phase) conductors = •
Sizing grounded (neutral) conductors = √
Sizing load - Phases A, B and C = *

Note that 831 volts is used (480 V x 1.732) instead of 831.36 volts.

Calculating lighting load

Step 1: General lighting load
Table 220.12; 230.42(A)(2)
50,000 sq. ft. x 3 VA = _____ VA [A] √
150,000 VA x 125% = _____ VA [B] •
25,000 sq. ft. x 1/4 VA = _____ VA [C] √
6250 VA x 125% = _____ VA [D] •

Step 2: Show window load
220.43(A)
100' x 200 VA = _____ VA [E] •

Step 3: Track lighting load
220.43(B)
100' ÷ 2 x 150 VA = _____ VA [F] •

Step 4: Outside lighting load
230.42(A)(2)
40 x 180 VA = _____ VA [G]
7200 VA x 125% = _____ VA [H] •

Step 5: Sign lighting load
220.14(F); 600.5(A); 230.42(A)(2)
4200 VA x 100% = _____ VA [I]
4200 VA x 125% = _____ VA [J] •
Total load = _____ VA [K] *

Calculating receptacle loads

Step 1: Noncontinuous operation
220.14(I); 230.42(A)(1); 220.14(H)(1); (2)
72 x 180 VA = _____ VA [L]
120' x 180 VA = _____ VA [M]
Total load = _____ VA [N]
Table 220.44
First 10,000 VA x 100% = _____ VA [O]
Next 24,560 VA x 50% = _____ VA [P]
Total load = _____ VA [Q] •

Step 2: Continuous operation
220.14(I); 230.42(A)(2)
34 x 180 VA = _____ VA [R]
6120 VA x 125% = _____ VA [S] •
Total load = _____ VA [T] *

Calculating special loads

Step 1: Water heater load
12,000 VA x 100% = _____ VA [U] •*

Calculating compressor loads

Step 1: Freezer load
230.42(A)(1)
7875 VA x 100% = _____ VA [V] •

Step 2: Ice cream boxes
6325 VA x 100% = _____ VA [W] •

Step 3: Walk-in cooler
9420 VA x 100% = _____ VA [X] •
Total load = _____ VA [Y] *

Calculating motor loads

Step 1: Water pump load
430.24; Table 430.250
3.4 A x 831 V x 100% = _____ VA [Z] •

Step 2: Exhaust fan load
1.1 A x 831 V x 100% = _____ VA [AA] •
Total load = _____ VA [BB] *

Calculating heating or A/C load

Step 1: Heating load selected
220.60
60,000 VA x 100% = _____ VA [CC] •*
24,640 VA x 100% = _____ VA [DD]

Calculating largest motor load

Step 1: Walk-in cooler
220.14(C); 220.50; 430.24
9420 VA x 25% = _____ VA [EE] •*

Calculating ungrounded (phase) conductors

Lighting loads = _____ VA [FF] •
Receptacle loads = _____ VA [GG] •
Special loads = _____ VA [HH] •
Compressor loads = _____ VA [II] •
Motor loads = _____ VA [JJ] •
Heating or A/C load = _____ VA [KK] •
Largest motor load = _____ VA [LL] •
Total load for facility = _____ VA [MM]

Finding amps for ungrounded (phase) conductors - Phases A, B and C

I = VA ÷ V
I = _____ VA [NN] ÷ _____ V [OO]
I = _____ A [PP]

Calculating grounded (neutral) conductors

Lighting load = _____ VA [QQ] √
Total load = _____ VA [RR]

Finding amps for grounded (neutral) conductors

I = VA ÷ V
I = _____ VA [SS] ÷ _____ V [TT]
I = _____ A [UU]

Design Example Problem 11-4. What is the load in VA and amps to calculate and size the elements for a 120 / 240 volt, three-phase, four-wire service supplying a 60,000 sq. ft. store including a 40,000 sq. ft. warehouse space with the following loads? Note: 51 percent or more of the lighting load is electric discharge.

120 V, single-phase loads
- 120 linear feet of show window (noncontinuous operation)
- 150 ft. of lighting track (noncontinuous operation)
- 50 - 180 VA ballasts outside lighting (continuous operation)
- 4200 VA sign lighting (continuous operation)
- 82 receptacles (noncontinuous operation)
- 30 receptacles (continuous operation)
- 120 ft. multioutlet assembly (heavy-duty)

240 V, three-phase loads
- 12,000 VA water heater
- 7860 VA freezer
- 7240 VA ice cream box
- 9820 VA walk-in cooler
- 50,000 VA heating unit
- 26,400 VA A/C unit
- 1 - 1/2 HP exhaust fan
- 1 - 2 HP water pump

Sizing ungrounded (phase) conductors = •
Sizing grounded (neutral) conductors = √
Sizing load - Phases A, B and C = *

Calculating lighting loads

Step 1: General lighting load
Table 220.12; 230.42(A)(2)
60,000 sq. ft. x 3 VA = 180,000 VA √
180,000 VA x 125% = 225,000 VA •
40,000 sq. ft. x 1/4 VA = 10,000 VA √
10,000 VA x 125% = 12,500 VA •

Step 2: Show window load
220.43(A)
120' x 200 VA = 24,000 VA • √

Step 3: Track lighting load
220.43(B)
150' ÷ 2 x 150 VA = 11,250 VA • √

Step 4: Outside lighting load
230.42(A)(2)
50 x 180 VA = 9,000 VA √
9000 VA x 125% = 11,250 VA •

Step 5: Sign lighting load
220.14(F); 600.5(A); 230.42(A)(2)
4200 VA x 100% = 4,200 VA √
4200 VA x 125% = 5,250 VA •
Total load = 289,250 VA *

Calculating receptacle loads

Step 1: Noncontinuous operation
220.14(I); 230.42(A)(1); 220.14(H)(1); (2)
82 x 180 VA = 14,760 VA •
120' x 180 VA = 21,600 VA
Total load = 36,360 VA
Table 220.44
First 10,000 VA x 100% = 10,000 VA
Next 26,360 VA x 50% = 13,180 VA
Total load = 23,180 VA • √

Step 2: Continuous operation
220.14(I); 230.42(A)(2)
30 x 180 VA = 5,400 VA √
5400 VA x 125% = 6,750 VA •
Total load = 29,930 VA *

Calculating special loads

Step 1: 12,000 VA x 100% = 12,000 VA • *

Calculating compressor loads

Step 1: Freezer load
230.42(A)(1)
7860 VA x 100 % = 7,860 VA •

Step 2: Ice cream boxes
230.42(A)(1)
7240 VA x 100% = 7,240 VA •

Step 3: Walk-in cooler
230.42(A)(1)
9820 VA x 100% = 9,820 VA •
Total load = 24,920 VA *

Note that 416 volts is used for the 240 volt, 3-phase calculations.

Calculating motor loads

Step 1: Water pump load
430.24; Table 430.250
6.8 A x 416 V x 100% = 2,829 VA •

Step 2: Exhaust fan load
430.24; Table 430.250
2.2 A x 416 V x 100% = 915 VA •
Total load = 3,744 VA *

Calculating heating or A/C load

Step 1: Heating load selected
220.60
50,000 VA x 100% = 50,000 VA • *
26,400 VA x 100% = 26,400 VA

Calculating largest motor load

Step 1: Walk-in cooler
220.14(C); 220.50; 430.24
9820 VA x 25% = 2,455 VA • *

Single-phase loads

Lighting loads = 289,250 VA •
Receptacle loads = 29,930 VA •
Total load = 319,180 VA

Three-phase loads

Special loads = 12,000 VA •
Compressor loads = 24,920 VA •
Motor loads = 3,744 VA •
Heating or A/C load = 50,000 VA •
Largest motor load = 2,455 VA •
Total load = 93,119 VA

Calculating grounded (neutral) conductors

Lighting loads = 238,450 VA √
Receptacle loads = 28,580 VA √
Total load = 267,030 VA

Finding single-phase load

I = VA ÷ V
I = 319,180 VA ÷ 240 V
I = 1330 A

Finding three-phase load

I = VA ÷ V
I = 93,119 VA ÷ 416 V
I = 224 A

Finding amps for grounded (neutral) conductors

I = VA ÷ V
I = 267,030 VA ÷ 240 V
I = 1113 A

Finding ungrounded (phase) conductors - Phases A and C

Single-phase load = 1,330 A
Three-phase load = 224 A
Total load = 1,554 A

Calculating ungrounded (phase) conductor - Phase A

Three-phase load = 224 A

Commercial Calculations

Design Exercise Problem 11-4. What is the load in VA and amps to calculate and size the elements for a 120/240 volt, three-phase, four-wire service supplying a 50,000 sq. ft. store including a 25,000 sq. ft. of warehouse space with the following loads?

120 V, single-phase loads
- 100 linear feet of show window (noncontinuous operation)
- 100' of lighting track (noncontinuous operation)
- 40 - 180 VA ballasts outside lighting (continuous operation)
- 4200 VA sign lighting (continuous operation)
- 72 receptacles (noncontinuous operation)
- 34 receptacles (continuous operation)
- 120' multioutlet assembly (heavy-duty)

240 V, three-phase loads
- 12,000 VA water heater
- 7875 VA freezer
- 6325 VA ice cream boxes
- 9420 VA walk-in cooler
- 60,000 VA heating unit
- 24,640 VA A/C unit
- 1 - 1/2 HP exhaust fan
- 1 - 2 HP water pump

Sizing ungrounded (phase) conductors = •
Sizing grounded (neutral) conductors = √
Sizing load - Phases A and B = *

Note that 416 volts is used for the 240 volts, 3-phase calculations.

Calculating lighting load

Step 1: General lighting load
Table 220.12; 230.42(A)(2)
50,000 sq. ft. x 3 VA = _____ VA [A] √
150,000 VA x 125% = _____ VA [B] •
25,000 sq. ft. x 1/4 VA = _____ VA [C] √
6250 VA x 125% = _____ VA [D] •

Step 2: Show window load
220.43(A)
100' x 200 VA = _____ VA [E] •√

Step 3: Track lighting load
220.43(B)
100' ÷ 2 x 150 VA = _____ VA [F] •√

Step 4: Outside lighting load
230.42(A)(2)
40 x 180 VA = _____ VA [G] √
7200 VA x 125% = _____ VA [H] •

Step 5: Sign lighting load
220.14(F); 600.5(A); 230.42(A)(2)
4200 VA x 100% = _____ VA [I] √
4200 VA x 125% = _____ VA [J] •
Total load = _____ VA [K] *

Calculating receptacle loads

Step 1: Noncontinuous operation
220.3(I); 230.42(A)(1); 220.14(H)(1); (2)
72 x 180 VA = _____ VA [L]
120' x 180 VA = _____ VA [M]
Total load = _____ VA [N]
Table 220.44
First 10,000 VA x 100% = _____ VA [O]
Next 24,560 VA x 50% = _____ VA [P]
Total load = _____ VA [Q] •√

Step 2: Continuous operation
220.14(I); 230.42(A)(2)
34 x 180 VA = _____ VA [R] √
6120 VA x 125% = _____ VA [S] •
Total load = _____ VA [T] *

Calculating special loads

Step 1: Water heater load
12,000 VA x 100% = _____ VA [U] •*

Calculating compressor loads

Step 1: Freezer load
230.42(A)(1)
7875 VA x 100% = _____ VA [V] •

Step 2: Ice cream boxes
6325 VA x 100% = _____ VA [W] •

Step 3: Walk-in cooler
9420 VA x 100% = _____ VA [X] •
Total load = _____ VA [Y] *

Calculating motor loads

Step 1: Water pump load
430.24; Table 430.250
6.8 A x 416 V x 100% = _____ VA [Z] •

Step 2: Exhaust fan load
2.2 A x 416 V x 100% = _____ VA [AA] •
Total load = _____ VA [BB] *

Calculating heating or A/C load

Step 1: Heating load selected
220.60
60,000 VA x 100% = _____ VA [CC] •*
24,640 VA x 100% = _____ VA [DD]

Calculating largest motor load

Step 1: Walk-in cooler
220.14(C); 220.50; 430.24
9420 VA x 25% = _____ VA [EE] •*

Single-phase loads

Lighting loads = _____ VA [FF] •
Receptacle loads = _____ VA [GG] •
Total load = _____ VA [HH]

Three-phase loads

Special loads = _____ VA [II] •
Compressor loads = _____ VA [JJ] •
Motor loads = _____ VA [KK] •
Heating or A/C load = _____ VA [LL] •
Largest motor load = _____ VA [MM] •
Total load = _____ VA [NN]

Calculating single-phase load

I = _____ VA [OO] ÷ _____ V [PP]
I = _____ A [QQ]

Calculating three-phase load

I = _____ VA [RR] ÷ _____ V [SS]
I = _____ A [TT]

Calculating grounded (neutral) conductors

I = _____ VA [UU] ÷ _____ V [VV]
I = _____ A [WW]

Calculating ungrounded (phase) conductors - Phases A and C

Single-phase load = _____ A [XX]
Three-phase load = _____ A [YY]
Total load = _____ A [ZZ]

Calculating ungrounded (phase) conductor - Phase B

Three-phase load = _____ A [AAA]

11-39

Design Example Problem 11-5. What is the load in VA and amps to size the elements for a 277/480 volt, three-phase, four-wire service supplying a 120,000 sq. ft. office with a 2,000 sq. ft. meeting hall and 1,000 sq. ft. storage space with the following loads? (All lighting is 277 volt) Note: 51 percent or more of the lighting load is electric discharge.

120 V, single-phase loads

- 35 ft. of lighting track (noncontinuous operation)
- 15 - 180 VA ballasts outside lighting (continuous operation)
- 3600 VA sign lighting (continuous operation)
- 162 receptacles (noncontinuous operation)
- 108 receptacles (continuous operation)
- 6,000 VA isolation transformer for LVLS (continuous operation)
- 120 ft. multioutlet assembly (heavy-duty)

Note: All lighting is 277 volt.

480 V, three-phase loads

- 40 HP elevator (15 minute intermittent duty)
- 50 kW heating unit

208 V, three-phase loads

- 2 - 1225 VA copying machines
- 1 - 10,000 VA water heater
- 28 - 275 VA data processors
- 10 - 175 VA word processors
- 4 - 1200 VA printers

Sizing ungrounded (phase) conductors = •
Sizing grounded (neutral) conductors = √
Sizing load - Phases A and B = *

Calculating lighting loads

Step 1: General lighting load
Table 220.12; 230.42(A)(2)
120,000 sq. ft. x 3.5 VA = 420,000 VA √
420,000 VA x 125% = 525,000 VA •
2000 sq. ft. x 1/2 VA = 1,000 VA √
1000 VA x 125% = 1,250 VA •
1000 sq. ft. x 1/4 VA = 250 VA √
250 VA x 125% = 313 VA •

Step 2: Track lighting load
220.43(B)
35' ÷ 2 x 150 VA = 2,625 VA • √

Step 3: Low-voltage lighting load
Art. 411; 230.42(A)(2)
6000 VA x 100% = 6,000 VA √
6000 VA x 125% = 7,500 VA •

Step 4: Outside lighting load
230.42(A)(2)
15 x 180 VA = 2,700 VA √
2700 VA x 125% = 3,375 VA •

Step 5: Sign lighting load
220.14(F); 600.5(A); 230.42(A)(2)
3600 VA x 100% = 3,600 VA √
3600 VA x 125% = 4,500 VA •
Total load = 544,563 VA *

Calculating receptacle loads

Step 1: Noncontinuous operation
220.14(I); 230.42(A)(1); 220.14(H)(1); (2)
162 x 180 VA = 29,160 VA
120' x 180 VA = 21,600 VA
Total load = 50,760 VA
Table 220.13
First 10,000 VA x 100% = 10,000 VA
Next 40,760 VA x 50% = 20,380 VA
Total load = 30,380 VA •

Step 2: Continuous operation
220.14(I); 230.42(A)(2)
108 x 180 VA = 19,440 VA
19,440 VA x 125% = 24,300 VA •
Total load = 54,680 VA *

Calculating special loads

Step 1: Copying machine load
230.42(A)(2)
1225 VA x 2 = 2,450 VA
2450 VA x 125% = 3,063 VA •

Step 2: Water heater load
422.13; 230.42(A)(1)
10,000 VA x 100% = 10,000 VA •

Step 3: Data processor load
230.42(A)(2)
275 VA x 28 = 7,700 VA
7700 VA x 125% = 9,625 VA •

Step 4: Word processor load
230.42(A)(2)
175 VA x 10 = 1,750 VA
1750 VA x 125% = 2,188 VA •

Step 5: Printer load.
230.42(A)(2)
1200 VA x 4 = 4,800 VA
4800 VA x 125% = 6,000 VA •
Total load = 30,876 VA *

Calculating motor loads

Step 1: 40 HP elevators
430.24; 430.22(E); Table 430.22(E)
52 A x 831 V x 85% = 36,730 VA • *

Calculating heating or A/C load

Step 1: Heating load selected
220.60; 220.51
50,000 VA x 100% = 50,000 VA • *

Calculating largest motor load

Step 1: 40 HP elevator
220.14(C); 220.50; 430.24
36,730 VA x 25% = 9,183 VA • *
Total load of facility = 726,032 VA •

Finding amps for ungrounded (phase) conductors - Phases A, B and C

I = VA ÷ V x √3
I = 726,032 VA ÷ 831 V
I = 874 A

Calculating grounded (neutral) conductors

General lighting load
(office building) = 420,000 VA √
(halls) = 1,000 VA √
(storage space) = 250 VA
Track lighting load = 2,625 VA √
Low-voltage lighting load = 6,000 VA √
Outside lighting load = 2,700 VA √
Sign lighting load = 3,600 VA √
Total load = 436,175 VA

Finding amps for grounded (neutral) conductors

I = VA ÷ V x √3
I = 436,175 VA ÷ 831 V
I = 525 A

Commercial Calculations

Design Exercise Problem 11-5. What is the load in VA and amps to compute and size the elements for a 277/480 volt, three-phase, four-wire service supplying a 148,000 sq. ft. office facility including a 4200 sq. ft. hall area equipped with 277 volt lighting units with the following loads?

120 V, single-phase loads

- 64' of lighting track (noncontinuous operation)
- 18 - 180 VA ballasts outside lighting (continuous operation)
- 4800 VA sign lighting (continuous operation)
- 196 receptacles (noncontinuous operation)
- 112 receptacles (continuous operation)
- 6000 VA isolation transformer for LVLS (continuous operation)
- 176' multioutlet assembly (heavy duty)

480 V, three-phase loads

- 50 kW heating units
- 40 HP elevator (15 minute intermittent duty)

208 V, three-phase loads

- 4 - 1225 VA copying machine
- 1 - 9600 VA water heater
- 30 - 275 VA data processors
- 10 - 175 VA word processor
- 2 - 1000 VA printers

Sizing ungrounded (phase) conductors = •
Sizing grounded (neutral) conductors = √
Sizing load - Phases A and B = *

Calculating lighting load

Step 1: General lighting load
Table 220.3(A); 230.42(A)(2)
148,000 sq. ft. x 3.5 VA = _____ VA [A] √
518,000 VA x 125% = _____ VA [B] •
4200 sq. ft. x 1/2 VA = _____ VA [C] √
2100 VA x 125% = _____ VA [D] •

Step 2: Track lighting load
220.12(B)
64' ÷ 2 x 150 VA = _____ VA [E] •

Step 3: Low-voltage lighting load
Art. 411; 230.42(A)(2)
6000 VA x 100% = _____ VA [F]
6000 VA x 125% = _____ VA [G] •

Step 4: Outside lighting load
230.42(A)(2)
18 x 180 VA = _____ VA [H]
3240 VA x 125% = _____ VA [I] •

Step 5: Sign lighting load
220.3(B)(6); 600.5(A); 230.42(A)(2)
4800 VA x 100% = _____ VA [J]
4800 VA x 125% = _____ VA [K] •
Total load = _____ VA [L] *

Calculating receptacle loads

Step 1: Noncontinuous operation
220.3(B)(9); 230.42(A)(1); 220.3(B)(8)(1); (2)
196 x 180 VA = _____ VA [M]
176' x 180 VA = _____ VA [N]
Total load = _____ VA [O]
Table 220.13
First 10,000 VA x 100% = _____ VA [P]
Next 56,960 VA x 50% = _____ VA [Q]
Total load = _____ VA [R] •

Step 2: Continuous operation
220.3(B)(9); 230.42(A)(2)
112 x 180 VA = _____ VA [S]
20,160 VA x 125% = _____ VA [T] •
Total load = _____ VA [U] *

Calculating special loads

Step 1: Copying machine load
230.42(A)(2)
1225 VA x 4 = _____ VA [V]
4900 VA x 125% = _____ VA [W] •

Step 2: Water heater load
442.13; 230.42(A)(1)
9600 VA x 100% = _____ VA [X] •

Step 3: Data processor load
275 VA x 30 = _____ VA [Y]
8250 VA x 125% = _____ VA [Z] •

Step 4: Word processor load
230.42(A)(2)
175 VA x 10 = _____ VA [AA]
1750 VA x 125% = _____ VA [BB] •

Step 5: Printer load
230.42(A)(2)
1000 VA x 2 = _____ VA [CC]
2000 VA x 125% = _____ VA [DD] •
Total load = _____ VA [EE] *

Calculating motor loads

Step 1: 40 HP elevator
430.24; 430.22(E); Table 430.22(E)
52 A x 831 V x 85% = _____ VA [FF] • *

Calculating heating or A/C load

Step 1: Heating load selected
220.21; 220.15
50,000 VA x 100% = _____ VA [GG] • *

Calculating largest motor

Step 1: 40 HP elevator
220.14; 430.24
36,730 VA x 25% = _____ VA [HH] • *
Total load of facility = _____ VA [II] •

Finding amps for ungrounded (phase) conductors - Phases A, B and C

I = VA ÷ V x √3
I = _____ VA [JJ] ÷ _____ V [KK]
I = _____ A [LL]

Calculating grounded (neutral) conductors

General lighting load
(office building) = _____ VA [MM] √
(halls) = _____ VA [NN] √
Total load = _____ VA [OO]

Finding amps for grounded (neutral) conductors

I = VA ÷ V x √3
I = _____ VA [PP] ÷ _____ V [QQ]
I = _____ A [RR]

Design Example Problem 11-6. What is the load in VA and amps to size the elements for a 277/480 volt, three-phase, four-wire service supplying a 24,000 sq. ft. classroom, 5,200 sq. ft. auditorium area, 200 sq. ft. stairway area, 800 sq. ft. storage space and 1,000 sq. ft. assembly hall area with the following loads? (school building lighting is supplied by 277 volt luminaires (lighting fixtures)) Note: 51 percent or more of the lighting load is electric discharge.

120 V, single-phase loads	208 V, single-phase cooking equipment	Sizing ungrounded (phase) conductors = •
• 176 receptacles (noncontinuous duty)	• 3 - 11 kW ranges	Sizing grounded (neutral) conductors = √
• 44 receptacles (continuous duty)	• 3 - 12 kW ovens	Sizing load - Phases A and B = *
• 148 ft. multioutlet assembly (heavy-duty)	• 3 - 3 kW fryers	

208 V, single-phase cooking equipment	208 V, single-phase motor loads	480 V, three-phase motor loads
• 2 - 1.2 kW toasters	• 3 - 1 HP vent-hood fans	• 18 - 3/4 HP exhaust fans
• 4 - 1.8 kW refrigerators	• 3 - 3/4 HP grill-vent fans	
• 2 - 1.6 kW freezers		

Calculating lighting loads

Step 1: General lighting load
Table 220.3(A); 230.42(A)(2)
24,000 sq. ft. x 3 VA = 72,000 VA √
72,000 VA x 125% = 90,000 VA •
5200 sq. ft. x 1 = 5,200 VA √
5200 VA x 125% = 6,500 VA •
1000 sq. ft. x 1 = 1,000 VA √
1000 VA x 125% = 1,250 VA •
800 sq. ft. x 1/4 VA = 200 VA √
200 VA x 125% = 250 VA •
200 sq. ft. x 1/2 VA = 100 VA √
100 VA x 125% = 125 VA •
Total load = 98,125 VA *

Calculating receptacle loads

Step 1: Noncontinuous operation
220.3(B)(9); 230.42(A)(1); 220.3(B)(8)(1); (2)
176 x 180 VA = 31,680 VA
148 x 180 VA = 26,640 VA
Total load = 58,320 VA
Table 220.13
First 10,000 VA x 100% = 10,000 VA
Next 48,320 VA x 50% = 24,160 VA
Total load = 34,160 VA •

Step 2: Continuous operation
220.3(B)(9); 230.42(A)(2)
44 x 180 VA = 7,920 VA
7920 VA x 125% = 9,900 VA •
Total load = 44,060 VA *

Calculating special loads

Step 1: Kitchen equipment
220.20
Toasters
2 x 1.2 kW x 1000 = 2,400 VA
Refrigerators
4 x 1.8 kW x 1000 = 7,200 VA
Freezers
2 x 1.6 kW x 1000 = 3,200 VA
Ranges
3 x 11 kW x 1000 = 33,000 VA
Ovens
3 x 12 kW x 1000 = 36,000 VA
Fryers
3 x 3 kW x 1000 = 9,000 VA
Total load = 90,800 VA

Step 2: Applying demand factors
Table 220.20
90,800 VA x 65% = 59,020 VA •*

Calculating motor loads
Tables 430.148 and 430.150

Step 1: Exhaust fans
28.8 A x 100% x 480 V = 13,824 VA
13,824 VA x 1.732 = 23,943 VA •
(1.6 A x 18 = 28.8 A)

Step 2: Hood fans
26.4 A x 100% x 208 V = 5,491 VA •
(8.8 A x 3 = 26.4 A)

Step 3: Grill vent fans
22.8 A x 100% x 208 V = 4,742 VA •
(7.6 A x 3 = 22.8 A)
Total load = 34,176 VA *

Calculating largest motor load

Step 1: Hood fan
220.14; 430.24
8.8 A x 100% x 208 V = 1,830 VA
1830 VA x 25% = 458 VA •*
Total load of facility = 235,839 VA •

Finding amps for ungrounded (phase) conductors - Phases A, B and C

I = VA ÷ V x √3
I = 235,839 VA ÷ 831 V
I = 284 A

Calculating grounded (neutral) conductors

General lighting load = 78,500 VA √
Total load = 78,500 VA

Finding amps for grounded (neutral) conductors

I = VA ÷ V x √3
I = 78,500 VA ÷ 831 V
I = 94 A

Commercial Calculations

Design Exercise Problem 11-6. What is the load in VA and amps to compute and size the elements for a 277/480 volt, three-phase, four-wire service supplying a 24,000 sq. ft. classroom area, 4,800 sq. ft. auditorium area and 1250 sq. ft. assembly hall area with the following loads? (School building lighting is supplied by 277 volt luminaires (lighting fixtures))

120 V, single-phase loads
- 184 receptacles (noncontinuous duty)
- 62 receptacles (continuous duty)
- 240' multioutlet assembly (heavy-duty)

208 V, single-phase motor loads
- 4 - 1 HP hood fans
- 3 - 3/4 HP grill vent fans

480 V, three-phase motor loads
- 20 - 3/4 HP exhaust fans

208 V, single-phase cooking equipment
- 2 - 1.2 kW toaster
- 3 - 1.6 kW refrigerators
- 3 - 1.8 kW freezers

208 V, single-phase cooking equipment
- 3 - 11 kW ranges
- 3 - 10 kW ovens
- 4 - 3.8 kW fryers

Sizing ungrounded (phase) conductors = •
Sizing grounded (neutral) conductors = √
Sizing load - Phases A and B = *

Calculating lighting load

Step 1: General lighting load
Table 220.3(A); 230.42(A)(2)
24,000 sq. ft. x 3 VA = _____ VA [A] √
72,000 VA x 125% = _____ VA [B] •
4800 sq. ft. x 1 = _____ VA [C] √
4800 VA x 125% = _____ VA [D] •
1250 sq. ft. x 1 = _____ VA [E] √
1250 VA x 125% = _____ VA [F] •
Total load = _____ VA [G] *

Calculating receptacle loads

Step 1: Noncontinuous operation
220.3(B)(9); 230.42(A)(1); 220.3(B)(8)(1); (2)
184 x 180 VA = _____ VA [H]
240' x 180 VA = _____ VA [I]
Total load = _____ VA [J]

Table 220.13
First 10,000 VA x 100% = _____ VA [K]
Next 66,320 VA x 50% = _____ VA [L]
Total load = _____ VA [M] •

Step 2: Continuous operation
220.3(B)(9); 230.42(A)(2)
62 x 180 VA = _____ VA [N]
11,160 VA x 125% = _____ VA [O] •
Total load = _____ VA [P] *

Calculating special loads

Step 1: Kitchen equipment
220.20
Toasters
2 x 1.2 kW x 1000 = _____ VA [Q]
Refrigerators
3 x 1.6 kW x 1000 = _____ VA [R]
Freezers
3 x 1.8 kW x 1000 = _____ VA [S]
Ranges
3 x 11 kW x 1000 = _____ VA [T]
Ovens
3 x 10 kW x 1000 = _____ VA [U]
Fryers
4 x 3.8 kW x 1000 = _____ VA [V]
Total load = _____ VA [W]

Step 2: Applying demand factors
90,800 x 65% = _____ VA [X] • *

Calculating motor loads
Tables 430.148 and 430.150

Step 1: Exhaust fans
32 A x 100% x 480 V = _____ VA [Y]
15,360 VA x 1.732 = _____ VA [Z] •
(1.6 A x 20 = 32 A)

Step 2: Hood fans
35.2 A x 100% x 208 V = _____ VA [AA] •
(8.8 A x 4 = 35.2 A)

Step 3: Grill vent fans
22.8 A x 100% x 208 V = _____ VA [BB] •
(7.6 A x 3 = 22.8)
Total load = _____ VA [CC] *

Calculating largest motor load

Step 1: Hood fan
220.14; 430.24
8.8 A x 100% x 208 V = _____ VA [DD]
1830 VA x 25% = _____ VA [EE] • *
Total load of facility = _____ VA [FF]

Finding amps for ungrounded (phase) conductors - Phases A, B and C

I = VA ÷ V x √3
I = _____ VA [GG] ÷ _____ V [HH]
I = _____ A [II]

Calculating grounded (neutral) conductors

General lighting load = _____ VA [JJ]
Total load = _____ VA [KK]

Finding amps for grounded (neutral) conductors

I = VA ÷ V x √3
I = _____ VA [LL] ÷ _____ V [MM]
I = _____ A [NN]

Design Example Problem 11-7. What is the load in VA and amps to size the elements for a 120/208 volt, three-phase, four-wire service supplying a restaurant with an area of 8000 sq. ft. with the following loads?

120 V, single-phase loads

Lighting Load

- 40 ft. of lighting track (continuous operation)
- 12 - 180 VA ballasts outside lighting (continuous operation)
- 1800 VA sign lighting (continuous operation)

Receptacle Load

- 27 receptacles (noncontinuous operation)
- 34 receptacles (continuous operation)
- 20 ft. multioutlet assembly (heavy duty)

Sizing ungrounded (phase) conductors = •
Sizing grounded (neutral) conductors = √
Sizing load - Phases A, B and C = *

208 V, single-phase loads

Motor Loads

- 6950 VA vent hood fans
- 4758 VA grill vent fans

208 V, three-phase loads

Special Loads

- 2 - 20 kW heating units
- 2 - 6840 VA A/C units

208 V, three-phase loads

Kitchen Equipment

- 1 - 4200 VA boiler
- 2 - 2400 VA deep fat fryer
- 1 - 20 A walk-in cooler
- 1 - 8000 VA water heater

208 V, single-phase loads

- 1 - 13 A freezer
- 1 - 10,000 VA cooktop
- 2 - 12,000 VA ovens
- 1 - 10,000 VA range
- 1 - 14 A refrigerator
- 1 - 3750 VA ice cream box

Calculating lighting load

Step 1: General lighting load
Table 220.3(A); 230.42(A)(2)
8000 sq. ft. x 2 VA = 16,000 VA √
16,000 VA x 125% = 20,000 VA •

Step 2: Track lighting load
220.12(B)
40' ÷ 2 x 150 VA = 3,000 VA √
3000 VA x 125% = 3,750 VA •

Step 3: Outside lighting load
230.42(A)(2)
12 x 180 VA = 2,160 VA √
2160 VA x 125% = 2,700 VA •

Step 4: Sign lighting load
220.3(B)(6); 600.5(A); 230.42(A)(2)
1800 VA x 100% = 1,800 VA √
1800 VA x 125% = 2,250 VA •
Total load = 28,700 VA *

Calculating receptacle load

Step 1: Receptacle load (noncontinuous)
220.3(B)(9); 230.42(A)(1)
27 x 180 VA = 4,860 VA • √

Step 2: Receptacle load (continuous)
220.3(B)(9); 230.42(A)(2)
34 x 180 VA = 6,120 VA √
6120 VA x 125% = 7,650 VA •

Step 3: Multioutlet assembly
220.3(B)(8)(2)
20' x 180 VA = 3,600 VA • √
Total load = 16,110 VA *

Calculating special loads

Step 1: Kitchen equipment
220.20
Freezer = 2,704 VA
Cooktop = 10,000 VA
Ovens = 24,000 VA
Range = 10,000 VA
Refrigerator = 2,912 VA
Ice cream box = 3,750 VA
Boiler = 4,200 VA
Deep fat fryer = 4,800 VA

Walk-in cooler = 7,200 VA
Water heater = 8,000 VA
Total load = 77,566 VA

Step 2: Applying demand factors
Table 220.20
77,566 VA x 65% = 50,418 VA • *

Calculating motor loads

Step 1: Hood fans
430.22(A); 430.24; 430.25
6950 VA x 100% = 6,950 VA •

Step 2: Grill vent fans
4758 VA x 100% = 4,758 VA •
Total load = 11,708 VA *

Calculating heating or A/C load

Step 1: Heating load
220.21; 220.15
20 kW x 2 x 1000 = 40,000 VA • *

Calculating largest motor load

Step 1: Walk-in cooler
220.14; 430.24
20 A x 100% x 360 V = 7,200 VA
7200 VA x 25% = 1,800 VA • *
Total load of facility = 148,736 VA •

Finding amps for ungrounded (phase) conductors - Phases A, B and C

I = VA ÷ V x √3
I = 148,736 VA ÷ 360 V
I = 413 A

Calculating grounded (neutral) conductors
220.22; 230.42(A)(1)

Lighting load = 22,960 VA √
Receptacle load = 14,580 VA √
Total load = 37,540 VA

Finding amps for grounded (neutral) conductors

I = VA ÷ V x √3
I = 37,540 VA ÷ 360 V
I = 104 A

Commercial Calculations

Design Exercise Problem 11-7: What is the load in VA and amps to compute and size the elements for a 120/208 volt, three-phase, four-wire service supplying a restaurant with an area of 6800 sq. ft. with the following loads?

120 V, single-phase loads

Lighting load
- 35' lighting track (continuous)
- 18 - 180 VA ballasts outside lighting (continuous)
- 3600 VA sign lighting (continuous)

Receptacle load
- 30 receptacles (noncontinuous)
- 20 receptacles (continuous)
- 20' multioutlet assembly (heavy-duty)

Sizing ungrounded (phase) conductors = •
Sizing grounded (neutral) conductors = √
Sizing load - Phases A and B = *

208 V, three-phase loads

Special loads
- 2 - 20 kW heating units
- 2 - 8650 VA A/C units

208 V, single-phase loads

Motor loads
- 7322 VA hood fans
- 4742 VA grill vent fans

208 V, three-phase loads

Kitchen Equipment
- 1 - 4125 VA boiler
- 2 - 2840 VA deep fat fryer
- 1 - 20 A walk-in cooler
- 1 - 8500 VA water heater

208 V, single-phase loads
- 1 - 13 A freezer
- 1 - 10,000 VA cooktop
- 2 - 8000 VA ovens
- 1 - 11,000 VA range
- 1 - 14 A refrigerator
- 1 - 3425 VA ice cream box

Calculating lighting loads

Step 1: General lighting load
Table 220.3(A); 230.42(A)(2)
6,800 sq. ft. x 2 VA = _____ VA [A] √
13,600 VA x 125% = _____ VA [B] •

Step 2: Track lighting load
220.12(B)
35' ÷ 2 x 150 VA = _____ VA [C] √
2625 VA x 125% = _____ VA [D] •

Step 3: Outside lighting load
230.42(A)(2)
180 VA x 18 units = _____ VA [E] √
3240 VA x 125% = _____ VA [F] •

Step 4: Sign lighting load
220.3(B)(6); 600.5(A); 230.42(A)(2)
3600 VA x 100% = _____ VA [G] √
3600 VA x 125% = _____ VA [H] •
Total load = _____ VA [I] *

Calculating receptacle loads

Step 1: Receptacle load (noncontinuous)
220.3(B)(9); 230.42(A)(1)
30 x 180 VA = _____ VA [J] • √

Step 2: Receptacle load (continuous)
220.3(B)(9); 230.42(A)(2)
20 x 180 VA = _____ VA [K] √
3600 VA x 125% = _____ VA [L] •

Step 3: Multioutlet assembly
220.3(B)(8)(2)
20' x 180 VA = _____ VA [M] • √
Total load = _____ VA [N] *

Calculating special load

Step 1: Kitchen equipment
220.20
Boiler = _____ VA [O]
Deep fat fryer = _____ VA [P]
Walk-in cooler = _____ VA [Q]
Water heater = _____ VA [R]
Ice cream box = _____ VA [S]
Freezer = _____ VA [T]
Cooktop = _____ VA [U]
Ovens = _____ VA [V]
Range = _____ VA [W]
Refrigerator = _____ VA [X]
Total load = _____ VA [Y]

Step 2: Applying demand factor
Table 220.20
71,546 VA x 65% = _____ VA [Z] • *

Calculating motor load

Step 1: Hood fans
430.22(A); 430.24; 430.25
7322 VA x 100% = _____ VA [AA] •

Step 2: Grill vent fans
4742 VA x 100% = _____ VA [BB] •
Total load = _____ VA [CC] *

Calculating heating or A/C load

Step 1: Heating load
220.21; 220.15
20 kW x 2 x 1000 = _____ VA [DD] • *

Calculating largest motor load

Step 1: Walk-in cooler
220.14; 430.24
20 A x 100% x 360 V = _____ VA [EE]
7200 VA x 25% = _____ VA [FF] • *
Total load of facility = _____ VA [GG] •

Calculating grounded (neutral) conductors
220.22; 230.42(A)(1)
Lighting load = _____ VA [HH] √
Receptacle load = _____ VA [II] √
Total load = _____ VA [JJ]

Finding amps for ungrounded (phase) conductors - Phases A, B and C

$I = VA \div V \times \sqrt{3}$
$I = $ _____ VA [KK] ÷ _____ V [LL]
$I = $ _____ A [MM]

Finding amps for grounded (neutral) conductors

$I = VA \div V \times \sqrt{3}$
$I = $ _____ VA [NN] ÷ _____ V [OO]
$I = $ _____ A [PP]

11-45

Design Example Problem 11-8. What is the load in VA and amps to size the elements for a 277/480 volt, three-phase service supplying a hospital with the following loads? (Hospital lighting is supplied by 277 volt luminaires (lighting fixtures))

120 V, single-phase loads
- 40 linear feet of show window (noncontinuous operation)
- 80 ft. of lighting track (noncontinuous operation)
- 24 - 225 VA ballasts outside lighting (continuous operation)
- 4 - 1800 VA sign lighting (continuous operation)
- 192 receptacles (noncontinuous operation)
- 84 receptacles (continuous operation)
- 160 ft. multioutlet assembly (heavy duty)
- 7 - 1200 VA copying machines
- 4 - 1600 VA soft drink machines

480 V, three-phase loads
- 6 - 30,000 VA heating units
- 6 - 9360 VA A/C units
- 4 - 5280 VA water pumps
- 8 - 2 HP exhaust fans
- 1 - 10 HP sprinkler pump

277 V, three-phase loads
- 48,000 sq. ft. hospital
Emergency System
- 45,000 VA life support branch load
(already computed at continuous operation)
- 45,000 VA critical branch load
(already computed at continuous operation)

X-Ray Equipment
- Short-time rating
two each at 3/4 seconds
rating = 150 MA
primary = 208 V, 1Ø
secondary = 100,000 V, 1Ø

- Long-term rating
two each at 15 minutes
rating = 15 MA
primary = 208 V, 1Ø
secondary = 200,000 V, 1Ø

Sizing ungrounded (phase) conductors = •
Sizing grounded (neutral) conductors = √
Sizing load - Phases A and B = *

Calculating lighting load

Step 1: General lighting load
Table 220.3(A); 230.42(A)(2); Table 220.11
48,000 sq. ft. x 2 VA x 100% = 96,000 VA
First 50,000 VA x 40% = 20,000 VA
Next 46,000 VA x 20% = 9,200 VA
Total load = 29,200 VA √

Step 2: Show window load
220.12(A)
40' x 200 VA = 8,000 VA •

Step 3: Track lighting load
220.12(B)
80' ÷ 2 x 150 VA = 6,000 VA •

Step 4: Outside lighting load
230.42(A)(2)
24 x 225 VA = 5,400 VA
5400 VA x 125% = 6,750 VA •

Step 5: Sign lighting load
220.3(B)(6); 600.5(A); 230.42(A)(2)
4 x 1,800 VA x 100% = 7,200 VA
7200 VA x 125% = 9,000 VA •
Total load = 58,950 VA *

Calculating receptacle loads

Step 1: Noncontinuous operation
220.3(B)(9); 230.42(A)(1); 220.3(B)(8)(1); (2)
192 x 180 VA = 34,560 VA
160' x 180 VA = 28,800 VA
Total load = 63,360 VA
Table 220.13
First 10,000 VA x 100% = 10,000 VA
Next 53,360 VA x 50% = 26,680 VA
Total load = 36,680 VA •

Step 2: Continuous operation
220.3(B)(9); 230.42(A)(2)
84 x 180 VA = 15,120 VA
15,120 VA x 125% = 18,900 VA •
Total load = 55,580 VA *

Calculating special loads

Step 1: Copying machine load
230.42(A)(2)
1200 VA x 7 = 8,400 VA
8400 VA x 125% = 10,500 VA •

Step 2: Soft drink machine load
230.42(A)(2)
1600 VA x 4 = 6,400 VA
6400 VA x 125% = 8,000 VA •

Step 3: X-ray equipment (short-time rating)
660.6(B)
14,976 VA x 100% = 14,976 VA •

Step 4: X-ray equipment (long-time rating)
660.6(B)
5824 VA x 20% = 1,165 VA •

Step 5: Emergency system (life support)
230.42(A)(1); 517.30(D)
45,000 VA x 100% = 45,000 VA •

Step 6: Emergency system (critical)
230.42(A)(1); 517.30(D)
45,000 VA x 100% = 45,000 VA •
Total load = 124,641 VA *

Calculating motor loads

Step 1: Water pump load
Table 430.150; 430.24
5280 VA x 4 = 21,120 VA •

Step 2: Exhaust fan load
Table 430.150; 430.24
3.4 A x 8 x 831 V = 22,603 VA •

Step 3: Sprinkler pump load
Table 430.150; 430.24
14 A x 831 V = 11,634 VA •
Total load = 55,357 VA *

Calculating heating and A/C load

Step 1: Heating load selected
220.21; 220.15
30,000 VA x 6 x 100% = 180,000 VA • *

Calculating largest motor load

Step 1: Sprinkler pump load selected
220.14; 430.24
11,634 VA x 25% = 2,909 VA • *
Total load of facility = 477,437 VA •

Finding amps for ungrounded (phase) conductors - Phases A, B and C

I = VA ÷ V x √3
I = 477,437 VA ÷ 831 V
I = 575 A

Calculating grounded (neutral) conductors

General lighting load = 29,200 VA √

Finding amps for grounded (neutral) conductors

I = VA ÷ V x √3
I = 29,200 V ÷ 831 V

Commercial Calculations

Design Exercise Problem 11-8. What is the load in VA and amps to size the elements for a 277/480 volt, three-phase service supplying a hospital with the following loads? (Hospital lighting is supplied by 277 volt luminaires (lighting fixtures))

120 V, single-phase loads
- 50 linear feet of show window (noncontinuous operation)
- 100 ft. of lighting track (noncontinuous operation)
- 30 - 180 VA ballasts outside lighting (continuous operation)
- 6 - 1200 VA sign lighting (continuous operation)
- 218 receptacles (noncontinuous operation)
- 98 receptacles (continuous operation)
- 120 ft. multioutlet assembly (heavy duty)
- 8 - 1400 VA copying machines
- 8 - 1600 VA soft drink machines

480 V, three-phase loads
- 8 - 25,000 VA heating units
- 8 - 16,205 VA A/C units
- 6 - 11,634 VA water pumps
- 10 - 3 HP exhaust fans
- 1 - 7 1/2 HP sprinkler pump

277 V, three-phase loads
- 59,000 sq. ft. hospital
Emergency System
- 50,000 VA life support branch load (already computed at continuous operation)
- 50,000 VA critical branch load (already computed at continuous operation)

X-Ray Equipment
- Short-time rating
  two each at 1 second
  rating = 200 MA
  primary = 208 V, 1Ø
  secondary = 100,000 V, 1Ø

- Long-term rating
  two each at 10 minutes
  rating = 20 MA
  primary = 208 V, 1Ø
  secondary = 200,000 V, 1Ø

Sizing ungrounded (phase) conductors = •
Sizing grounded (neutral) conductors = √
Sizing load - Phases A and B = *

Calculating lighting load

Step 1: General lighting load
59,000 sq. ft. x 2 VA x 100% = _____ VA [A]
First 50,000 VA x 40% = _____ VA [B]
Next 68,000 VA x 20% = _____ VA [C]
Total load = _____ VA [D] •√

Step 2: Show window load
50' x 200 VA = _____ VA [E] •

Step 3: Track lighting load
100' ÷ 2 x 150 VA = _____ VA [F] •

Step 4: Outside lighting load
30 x 180 VA = _____ VA [G]
5400 VA x 125% = _____ VA [H] •

Step 5: Sign lighting load
6 x 1200 VA x 100% = _____ VA [I]
7200 VA x 125% = _____ VA [J] •
Total load = _____ VA [K] *

Calculating receptacle loads

Step 1: Noncontinuous operation
218 x 180 VA = _____ VA [L]
120' x 180 VA = _____ VA [M]
Total load = _____ VA [N]
Table 220-13
First 10,000 VA x 100% = _____ VA [O]
Next 50,840 VA x 50% = _____ VA [P]
Total load = _____ VA [Q] •

Step 2: Continuous operation
98 x 180 VA = _____ VA [R]
17,640 VA x 125% = _____ VA [S] •
Total load = _____ VA [T] *

Calculating special loads

Step 1: Copying machine load
1400 VA x 8 = _____ VA [U]
11,200 VA x 125% = _____ VA [V] •

Step 2: Soft drink machine load
1600 VA x 8 = _____ VA [W]
12,800 VA x 125% = _____ VA [X] •

Step 3: X-ray equipment (short-time rating)
19,968 VA x 100% = _____ VA [Y] •

Step 4: X-ray equipment (long-time rating)
1581 VA x 20% = _____ VA [Z] •

Step 5: Emergency system (life support)
50,000 VA x 100% = _____ VA [AA] •

Step 6: Emergency system (critical)
50,000 VA x 100% = _____ VA [BB] •
Total load = _____ VA [CC] *

Calculating motor loads

Step 1: Water pump load
11,634 VA x 6 = _____ VA [DD] •

Step 2: Exhaust fan load
4.8 A x 10 x 831 V = _____ VA [EE] •

Step 3: Sprinkler pump load
11 A x 831 V = _____ VA [FF] •
Total load = _____ VA [GG] *

Calculating heating or A/C load

Step 1: Heating load selected
25,000 VA x 8 x 100% = _____ VA [HH] •*

Calculating largest motor load

Step 1: Water pump load selected
11,634 VA x 25% = _____ VA [II] •*
Total load of facility = _____ VA [JJ] •

Finding amps for ungrounded (phase) conductors - Phases A, B and C

I = VA ÷ V x √3
I = _____ VA [KK] ÷ _____ V [LL]
I = _____ A [MM]

Calculating grounded (neutral) conductors

General lighting load = _____ VA [NN] √

Finding amps for grounded (neutral) conductors

I = VA ÷ V x √3
I = _____ VA [OO] ÷ _____ V [PP]
I = _____ A [QQ]

Design Example Problem 11-9. What is the load in VA and amps to size the elements for a 120/208 volt, three-phase service supplying a hotel or motel with the following loads?

120 V, single-phase loads

- 50 units - 340 sq. ft. each
- 20 linear feet of show window (noncontinuous operation)
- 30 ft. of lighting track (noncontinuous operation)
- 18 - 225 VA ballasts outside lighting (continuous operation)
- 4 - 1200 VA sign lighting (continuous operation)
- 62 receptacles (continuous operation)
- 4 - 1200 VA copying machines
- 6 - 1600 VA soft drink machines
- 6500 VA house loads (continuous operation)
- 50 - 1/4 HP exhaust fans (noncontinuous operation)

208 V, three-phase loads

- 2 - 15,000 VA heating units (office and lobby)
- 50 - 7 1/2 HP A/C and heat pump units
- 8 - 8 kW water heaters
- 26,400 VA house loads (continuous operation)
- 22,860 VA laundry facilities load (continuous operation)

Sizing ungrounded (phase) conductors	= •
Sizing grounded (neutral) conductors	= √
Sizing load - Phases A and B	= *

Calculating lighting load

Step 1: General lighting load
Table 220.3(A); Table 220.11
340 sq. ft. x 2 VA x 50 = 34,000 VA
Applying demand factors
Table 220.11
First 20,000 VA x 50% = 10,000 VA
Next 14,000 VA x 40% = 5,600 VA
Total load = 15,600 VA √

Step 2: Show window load
220.12(A)
20' x 200 VA = 4,000 VA √

Step 3: Track lighting load
220.12(B)
30 ÷ 2 x 150 VA = 2,250 VA √

Step 4: Outside lighting load
230.42(A)(2)
18 x 225 VA = 4,050 VA √
4050 VA x 125% = 5,063 VA •

Step 5: Sign lighting load
220.3(B)(6); 600.5(A); 230.42(A)(2)
4 x 1200 VA x 100% = 4,800 VA √
4800 VA x 125% = 6,000 VA •
Total load = 32,913 VA *

Calculating receptacle loads

Step 1: Continuous operation
220.3(B)(9); 230.42(A)(2)
62 x 180 VA = 11,160 VA √
11,160 VA x 125% = 13,950 VA •*

Calculating special loads

Step 1: Copying machine load
230.42(A)(2)
4 x 1200 VA = 4,800 VA √
4800 VA x 125% = 6,000 VA •

Step 2: Soft drink machine load
230.42(A)(2)
6 x 1600 VA = 9,600 VA √
9600 VA x 125% = 12,000 VA •

Step 3: House loads
230.42(A)(2)
6500 VA x 100% = 6,500 VA √
6500 VA x 125% = 8,125 VA •
26,400 VA x 100% = 26,400 VA
26,400 VA x 125% = 33,000 VA •

Step 4: Exhaust fan load
230.42(A)(1); Table 430.148; 430.24
50 x 5.8 A x 120 V = 34,800 VA √

Step 5: Water heater load
230.42(A)(1)
8 x 8000 V x 100% = 64,000 VA •

Step 6: Laundry facility load
230.42(A)(2)
22,860 VA x 100% = 22,860 VA
22,860 VA x 125% = 28,575 VA •
Total load = 186,500 VA *

Calculating heating or A/C load

Step 1: Heating loads selected
220.21; 220.15
15,000 VA x 2 x 100% = 30,000 VA •
8712 VA x 50 x 100% = 435,600 VA •
Total load = 465,600 VA *

Calculating largest motor load

Step 1: Heat pump selected
220.14; 430.24
8712 VA x 25% = 2,178 VA •*
1600 VA x 25% = 400 VA √
Total load of facility = 701,141 VA

Finding amps for ungrounded (phase) conductors - Phases A, B and C

I = VA ÷ V x √3
I = 701,141 VA ÷ 360 V
I = 1948 A

Calculating grounded (neutral) conductors

Lighting loads = 30,700 VA √
Receptacle loads = 11,160 VA √
Special loads = 55,700 VA √
Largest motor load = 400 VA √
Total load = 97,960 VA √

Finding amps for grounded (neutral) conductors

I = VA ÷ V x √3
I = 97,960 VA ÷ 360 V
I = 272 A

Commercial Calculations

Design Exercise Problem 11-9. What is the load in VA and amps to size the elements for a 120/208 volt, three-phase service supplying a hotel or motel with the following loads?

120 V, single-phase loads
- 40 units - 300 sq. ft. each
- 15 linear feet of show window (noncontinuous operation)
- 25 ft. of lighting track (noncontinuous operation)
- 15 - 180 VA ballasts outside lighting (continuous operation)
- 3 - 1800 VA sign lighting (continuous operation)
- 54 receptacles (continuous operation)
- 4 - 1050 VA copying machines
- 4 - 1475 VA soft drink machines
- 5800 VA house loads (continuous operation)
- 40 - 1/6 HP exhaust fans (noncontinuous operation)

208 V, three-phase loads
- 2 - 10,000 VA heating units (office and lobby)
- 40 - 5 HP A/C and heat pump units
- 6 - 8 kW water heaters
- 24,800 VA house loads (continuous operation)
- 23,960 VA laundry facilities load (continuous operation)

Sizing ungrounded (phase) conductors = •
Sizing grounded (neutral) conductors = √
Sizing load - Phases A and B = *

Calculating lighting load

Step 1: General lighting load
Table 220.3(A); Table 220.11
300 sq. ft. x 2 VA x 40 = _____ VA [A]
Applying demand factors
First 20,000 VA x 50% = _____ VA [B]
Next 4,000 VA x 40% = _____ VA [C]
Total load = _____ VA [D] √

Step 2: Show window load
15' x 200 VA = _____ VA [E] √

Step 3: Track lighting load
25 ÷ 2 x 150 VA = _____ VA [F] √

Step 4: Outside lighting load
15 x 180 VA = _____ VA [G] √
2700 VA x 125% = _____ VA [H] •

Step 5: Sign lighting load
3 x 1800 VA x 100% = _____ VA [I] √
5400 VA x 125% = _____ VA [J] •
Total load = _____ VA [K] *

Calculating receptacle loads

Step 1: Continuous operation
220.3(B)(9); 230.42(A)(2)
54 x 180 VA = _____ VA [L] √
9720 VA x 125% = _____ VA [M] • *

Calculating special loads

Step 1: Copying machine load
230.42(A)(2)
4 x 1050 VA = _____ VA [N] √
4200 VA x 125% = _____ VA [O] •

Step 2: Soft drink machine load
230.42(A)(2)
4 x 1475 VA = _____ VA [P] √
5900 VA x 125% = _____ VA [Q] •

Step 3: House loads
230.42(A)(2)
5800 VA x 100% = _____ VA [R] √
5800 VA x 125% = _____ VA [S] •
24,800 VA x 100% = _____ VA [T]
24,800 VA x 125% = _____ VA [U] •

Step 4: Exhaust fan load
230.42(A)(1); Table 430.148; 430.24
40 x 4.4 A x 120 V = _____ VA [V] •√

Step 5: Water heater load
230.42(A)(1)
6 x 8,000 VA x 100% = _____ VA [W] •

Step 6: Laundry facility load
230.42(A)(2)
23,960 VA x 100% = _____ VA [X]
23,960 VA x 125% = _____ VA [Y] •
Total load = _____ VA [Z] *

Calculating heating or A/C load

Step 1: Heating loads selected
220.21; 220.15
10,000 VA x 2 x 100% = _____ VA [AA] •
6012 VA x 40 x 100% = _____ VA [BB] •
Total load = _____ VA [CC] *

Calculating largest motor load

Step 1: Heat pump selected
220.14; 430.24
6012 VA x 25% = _____ VA [DD] • *
1475 VA x 25% = _____ VA [EE] √
Total load of facility = _____ VA [FF]

Finding amps for ungrounded (phase) conductors - Phases A, B and C

I = VA ÷ V x √3
I = _____ VA [GG] ÷ _____ V [HH]
I = _____ A [II]

Calculating grounded (neutral) conductors

Lighting loads = _____ VA [JJ] √
Receptacle loads = _____ VA [KK] √
Special loads = _____ VA [LL] √
Largest motor load = _____ VA [MM] √
Total load = _____ VA [NN] √

Finding amps for grounded (neutral) conductors

I = VA ÷ V x √3
I = _____ VA [OO] ÷ _____ V [PP]
I = _____ A [QQ]

Design Example Problem 11-10. What is the load in VA and amps to size the elements for a 120/208 volt, three-phase service supplying a bank with the following loads?

120 V, single-phase loads

- 5800 sq. ft. bank
- 40 linear feet of show window (noncontinuous operation)
- 50 ft. of lighting track (noncontinuous operation)
- 20 - 225 VA ballasts outside lighting (continuous operation)
- 2400 VA sign lighting (continuous operation)
- 79 receptacles (noncontinuous operation)
- 32 receptacles (continuous operation)
- 30 ft. multioutlet assembly (heavy duty)
- 14 - 1/4 HP fans
- 5280 miscellaneous loads (continuous operation)
- 18 - 1,400 personal computers (continuous operation)

208 V, three-phase loads

- 30,000 VA heating unit
- 5 HP A/C unit
- 8 kW water heater
- 2 - 5 HP pump motors
- 8 - 3 HP exhaust fans
- 16 kW range (noncontinuous operation)
- 26,580 VA miscellaneous loads (continuous operation)

Sizing ungrounded (phase) conductors = •
Sizing grounded (neutral) conductors = √
Sizing load - Phases A and B = *

Calculating lighting load

Step 1: General lighting load
Table 220.3(A); 230.42(A)(2)
5800 sq. ft. x 3.5 = 20,300 VA √
20,300 VA x 125% = 25,375 VA •

Step 2: Show window load
220.12(A)
40' x 200 VA = 8,000 VA √

Step 3: Track lighting load
220.12(B)
50 ÷ 2 x 150 VA = 3,750 VA √

Step 4: Outside lighting load
230.42(A)(2)
20 x 225 VA = 4,500 VA √
4500 VA x 125% = 5,625 VA •

Step 5: Sign lighting load
220.3(B)(6); 600.5(A); 230.42(A)(2)
2400 VA x 100% = 2,400 VA √
2400 VA x 125% = 3,000 VA •
Total load = 45,750 VA *

Calculating receptacle loads

Step 1: Noncontinuous loads
220.3(B)(9); 230.42(A)(1); 220.3(B)(8)(1); (2)
79 x 180 VA = 14,220 VA
30' x 180 VA = 5,400 VA
Total load = 19,620 VA
Table 220.13
First 10,000 VA x 100% = 10,000 VA
Next 9620 VA x 50% = 4,810 VA
Total load = 14,810 VA √

Step 2: Continuous operation
220.3(B)(9); 230.42(A)(2)
32 x 180 VA = 5,760 VA √
5760 VA x 125% = 7,200 VA •
Total load = 22,010 VA *

Calculating special loads

Step 1: Miscellaneous loads
230.42(A)(2)
5280 VA x 100% = 5,280 VA √
5280 VA x 125% = 6,600 VA •
26,580 VA x 100% = 26,580 VA
26,580 VA x 125% = 33,225 VA •

Step 2: Personal computers
230.42(A)(2)
18 x 1400 VA = 25,200 VA √
25,200 VA x 125% = 31,500 VA •

Step 3: Water heater load
230.42(A)(1)
8000 VA x 100% = 8,000 VA •

Step 4: Range load
230.42(A)(1)
16,000 VA x 100% = 16,000 VA •
Total load = 95,325 VA *

Calculating motor loads

Step 1: Fan load
430.22(A); 430.24; Table 430.148
14 x 5.8 A x 120 V = 9,744 VA √

Step 2: Pump motors load
430.22(A); 430.24; Table 430.150
2 x 16.7 A x 360 V = 12,024 VA •

Step 3: Exhaust fan load
430.22(A); 430.24; Table 430.150
8 x 10.6 A x 360 V = 30,528 VA •
Total load = 52,296 VA *

Calculating heating or A/C load

Step 1: Heating load selected
220.21; 220.15
30,000 VA x 100% = 30,000 VA • *

Calculating largest motor load

Step 1: Pump motor load selected
220.14; 430.24
16.7 A x 360 V x 25% = 1,503 VA • *
5.8 A x 120 V x 25% = 174 VA √
Total load of facility = 246,884 VA

Finding amps for ungrounded (phase) conductors - Phases A, B and C

$I = VA \div V \times \sqrt{3}$
$I = 246,884 \div 360\ V$
$I = 686\ A$

Calculating grounded (neutral) conductors

General lighting loads = 38,950 VA √
Receptacle loads = 20,570 VA √
Special loads = 30,480 VA √
Motor loads = 9,744 VA √
Largest motor load = 174 VA √
Total load = 99,918 VA √

Finding amps for grounded (neutral) conductors

$I = VA \div V \times \sqrt{3}$
$I = 99,918 \div 360\ V$
$I = 278\ A$

Commercial Calculations

Design Exercise Problem 11-10. What is the load in VA and amps to size the elements for a 120/208 volt, three-phase service supplying a bank with the following loads?

120 V, single-phase loads
- 8400 sq. ft. bank
- 60 linear feet of show window (noncontinuous operation)
- 100 ft. of lighting track (noncontinuous operation)
- 40 - 180 VA ballasts outside lighting (continuous operation)
- 1800 VA sign lighting (continuous operation)
- 2,400 VA sign lighting (continuous operation)
- 94 receptacles (noncontinuous operation)
- 48 receptacles (continuous operation)
- 60 ft. multioutlet assembly (heavy duty)
- 18 - 1/6 HP fans
- 6420 miscellaneous loads (continuous operation)
- 28 - 1200 personal computers (continuous operation)

208 V, three-phase loads
- 50,000 VA heating unit
- 10 HP A/C unit
- 8 kW water heater
- 2 - 7 1/2 HP pump motors
- 12 - 2 HP exhaust fans
- 14 kW range (noncontinuous operation)
- 28,940 VA miscellaneous loads (continuous operation)

Sizing ungrounded (phase) conductors = •
Sizing grounded (neutral) conductors = √
Sizing load - Phases A and B = *

Calculating lighting load

Step 1: General lighting load
8400 sq. ft. x 3.5 = _____ VA [A] √
29,400 VA x 125% = _____ VA [B] •

Step 2: Show window load
60' x 200 VA = _____ VA [C] •√

Step 3: Track lighting load
100' ÷ 2 x 150 VA = _____ VA [D] •√

Step 4: Outside lighting load
40 x 180 VA = _____ VA [E] √
7200 VA x 125% = _____ VA [F] •

Step 5: Sign lighting load
1800 VA x 100% = _____ VA [G] √
1800 VA x 125% = _____ VA [H] •
2400 VA x 100% = _____ VA [I] √
2400 VA x 125% = _____ VA [J] •
Total load = _____ VA [K] *

Calculating receptacle loads

Step 1: Noncontinuous loads
94 x 180 VA = _____ VA [L]
60' x 180 VA = _____ VA [M]
Total load = _____ VA [N]
First 10,000 VA x 100% = _____ VA [O] •
Next 17,720 VA x 50% = _____ VA [P]
Total load = _____ VA [Q] •√

Step 2: Continuous operation
48 x 180 VA = _____ VA [R] √
8640 VA X 125% = _____ VA [S] •
Total load = _____ VA [T] *

Calculating special loads

Step 1: Miscellaneous loads
6420 VA x 100% = _____ VA [U] √
6420 VA x 125% = _____ VA [V] •
28,940 VA x 100% = _____ VA [W]
28,940 VA x 125% = _____ VA [X] •

Step 2: Personal computers
28 x 1200 VA = _____ VA [Y] √
33,600 VA x 125% = _____ VA [Z]

Step 3: Water heater load
8000 VA x 100% = _____ VA [AA] •

Step 4: Range load
14,000 VA x 100% = _____ VA [BB] •
Total load = _____ VA [CC] *

Calculating motor loads

Step 1: Fans load
18 x 4.4 A x 120 V = _____ VA [DD] •√

Step 2: Pump motors load
2 x 24.2 A x 360 V = _____ VA [EE] •

Step 3: Exhaust fan load
12 x 7.5 A x 360 V = _____ VA [FF] •
Total load = _____ VA [GG] *

Calculating heating or A/C load

Step 1: Heating load selected
50,000 VA x 100% = _____ VA [HH] •*

Calculating largest motor load

Step 1: Pump motor load selected
24.2 A x 360 V x 25% = _____ VA [II] •*
4.4 A x 120 V x 25% = _____ VA [JJ] √
Total load of facility = _____ VA [KK]

Finding amps for ungrounded (phase) conductors - Phases A, B and C

I = VA ÷ V x √3
I = _____ VA [LL] ÷ _____ V [MM]
I = _____ A [NN]

Calculating grounded (neutral) conductors

General lighting loads = _____ VA [OO] √
Receptacle loads = _____ VA [PP] √
Special loads = _____ VA [QQ] √
Motor loads = _____ VA [RR] √
Largest motor load = _____ VA [SS] √
Total load = _____ VA [TT] √

Finding amps for grounded (neutral) conductors

I = VA ÷ V x √3
I = _____ VA [UU] ÷ _____ V [VV]
I = _____ A [WW]

Design Example Problem 11-11. What is the load in VA and amps to size the elements for a 120/208 volt, three-phase service supplying a welding shop with the following loads?

120 V, single-phase loads

- 8500 VA inside lighting loads (continuous operation)
- 6 - 180 VA outside lighting loads (continuous operation)
- 1200 VA sign lighting load (noncontinuous operation)
- 50 receptacles (continuous operation)

Note: A welding shop is not a listed occupancy per **Table 220.3(A)**.

208 V, three-phase loads

- 2 - 12 kW heating units
- 6000 VA A/C unit
- 7.5 HP air-compressor
- 1 1/2 HP grinder
- 2 welders - resistance (30% duty cycle)
- 11 kW
- 9 kW
- 2 welders - motor generator arc (80% duty cycle)
- 12 kW
- 10 kW
- 2 welders - AC transformer and DC rectifier (90% duty cycle)
- 12 kW
- 9 kW

Sizing ungrounded (phase) conductors = •
Sizing grounded (neutral) conductors = √
Sizing load - Phases A and B = *

Calculating lighting loads

Step 1: Inside lighting load
230.42(A)(2)
8500 VA x 100% = 8,500 VA √
8500 VA x 125% = 10,625 VA •

Step 2: Outside lighting load
230.42(A)(2)
6 x 180 VA = 1,080 VA √
1080 VA x 125% = 1,350 VA •

Step 3: Sign lighting load
220.3(B)(6); 600.5(A); 230.42(A)(1)
1200 VA x 100% = 1,200 VA • √
Total load = 13,175 VA *

Calculating receptacle load

Step 1: Continuous operation
220.3(B)(9); 230.42(A)(2)
50 x 180 VA = 9,000 VA √
9000 VA x 125% = 11,250 VA *

Calculating special loads

Step 1: Welders - resistance
630.31(A); (B)
11,000 VA x 55% = 6,050 VA •
9000 VA x 55% x 60% = 2,970 VA •

Step 2: Welders - motor-generator arc
630.11(A); (B)
12,000 VA x 91% = 10,920 VA •
10,000 VA x 91% = 9,100 VA •

Step 3: Welders - AC transformer and DC rectifier
630.11(A); (B)
12,000 VA x 95% = 11,400 VA •
9000 VA x 95% = 8,550 VA •
Total load = 48,990 VA *

Calculating compressor and motor loads

Step 1: Air compressor
430.24; 430.22(A); Table 430.150
24.2 A x 100% x 360 V = 8,712 VA •

Step 2: Grinders
6.6 A x 100% x 360 V = 2,376 VA •
Total load = 11,088 VA *

Calculating heating or A/C load

Step 1: Heating load
220.21; 220.15
24,000 VA x 100% = 24,000 VA • *

Calculating largest motor load

Step 1: Air compressor
220.14; 430.24
24.2 A x 360 V x 25% = 2,178 VA • *

Calculating ungrounded (phase) conductors

Lighting loads = 13,175 VA •
Receptacle loads = 11,250 VA •
Special loads = 48,990 VA •
Motor loads = 11,088 VA •
Heating load = 24,000 VA •
Largest motor load = 2,178 VA •
Total load = 110,681 VA

Calculating grounded (neutral) conductors

Lighting loads = 10,780 VA √
Receptacle loads = 9,000 VA √
Total loads = 19,780 VA

Finding amps for ungrounded (phase) conductors - Phases A, B and C

I = VA ÷ V x √3
I = 110,681 VA ÷ 360 V
I = 307 A

Finding amps for grounded (neutral) conductors

I = VA ÷ V x √3
I = 19,780 VA ÷ 360 V
I = 55 A

Commercial Calculations

Design Exercise Problem 11-11: What is the load in VA and amps to compute and size the elements for a 120/208 volt, three-phase service supplying a welding shop with the following loads?

120 V, single-phase loads

- 9000 VA inside lighting loads (continuous operation)
- 6 - 180 VA outside lighting loads (continuous operation)
- 1200 VA sign lighting loads (noncontinuous operation)
- 60 receptacles (continuous operation)

Note: A welding shop is not a listed occupancy per **Table 220.3(A)**.

Sizing ungrounded (phase) conductors	= •
Sizing grounded (neutral) conductors	= √
Sizing load - Phases A and B	= *

208 V, three-phase loads

- 2 - 10 kW heating units
- 5400 VA A/C unit
- 7.5 HP air-compressor
- 1 1/2 HP grinder
- Welders - resistance (50% duty cycle)
- 12 kW
- 8 kW
- Welders - motor-generator arc (90% duty cycle)
- 14 kW
- 12 kW
- Welders - AC transformer and DC rectifier (80% duty cycle)
- 13 kW
- 9 kW

Calculating lighting loads

Step 1: Inside lighting load
230.42(A)(2)
9000 VA x 100% = _____ VA [A] √
9000 VA x 125% = _____ VA [B] •

Step 2: Outside lighting load
230.42(A)(2)
6 x 180 VA = _____ VA [C] √
1080 VA x 125% = _____ VA [D] •

Step 3: Sign lighting load
220.3(B)(6); 600.5(A); 230.42(A)(1)
1200 VA x 100% = _____ VA [E] • √
Total load = _____ VA [F] *

Calculating receptacle load

Step 1: Continuous duty
220.3(B)(9); 230.42(A)(2)
60 x 180 VA = _____ VA [G] √
10,800 VA x 125% = _____ VA [H] • *

Calculating special loads

Step 1: Welders - resistance
630.31(A); (B)
12,000 VA x 71% = _____ VA [I] •
8000 VA x 71% x 60% = _____ VA [J] •

Step 2: Welders - motor-generator arc
630.11(A); (B)
14,000 VA x 96% = _____ VA [K] •
12,000 VA x 96% = _____ VA [L] •

Step 3: Welders - AC transformer and DC rectifier
630.11(A); (B)
13,000 VA x 89% = _____ VA [M] •
9000 VA x 89% = _____ VA [N] •
Total load = _____ VA [O] *

Calculating compressor and motor loads

Step 1: Air-compressor
430.24; 430.22(A); Table 430.150
24.2 A x 100% x 360 V = _____ VA [P] •

Step 2: Grinders
6.6 A x 100% x 360 V = _____ VA [Q] •
Total load = _____ VA [R] *

Calculating heating or A/C load

Step 1: Heating load
220.21; 220.15
20,000 VA x 100% = _____ VA [S] • *

Calculating largest motor load

Step 1: Air-compressor
220.14; 430.24
24.2 A x 360 V x 25% = _____ VA [T] • *

Calculating ungrounded (phase) conductors

Lighting loads = _____ VA [U] •
Receptacle loads = _____ VA [V] •
Special loads = _____ VA [W] •
Motor loads = _____ VA [X] •
Heating loads = _____ VA [Y] •
Largest motor load = _____ VA [Z] •
Total load = _____ VA [AA]

Finding amps for ungrounded (phase) conductors - Phases A, B and C

I = VA ÷ V x √3
I = _____ VA [BB] ÷ _____ V [CC]
I = _____ A [DD]

Calculating grounded (neutral) conductors

220.22
Lighting loads = _____ VA [EE] √
Receptacle loads = _____ VA [FF] √
Total load = _____ VA [GG]

Finding amps for grounded (neutral) conductors

I = VA ÷ V x √3
I = _____ VA [HH] ÷ _____ V [II]
I = _____ A [JJ]

Name _____   Date _____

Chapter 11
Commercial Calculations

Section   Answer

1. The general-purpose lighting load for an office building shall be calculated at 3.5 VA per sq. ft.

      (a) True   (b) False

2. The general-purpose lighting for an auditorium in a school building shall be calculated at 1/2 VA per sq. ft.

      (a) True   (b) False

3. The lighting load for a sign shall be calculated at a minimum of 1400 volt-amperes.

      (a) True   (b) False

4. The largest VA rating between the heating and A/C load shall be selected, and the smaller of the two loads is dropped.

      (a) True   (b) False

5. The VA rating of the largest motor shall be calculated by multiplying the amperage of the unit by the voltage times 125 percent.

      (a) True   (b) False

6. For show window lighting, a load of not less than _____ volt-amperes shall be included for each linear foot of show window.

      (a) 150   (b) 180   (c) 200   (d) 250

7. General-purpose receptacle outlets shall be calculated by the number of receptacles times _____ volt-amperes.

      (a) 150   (b) 180   (c) 200   (d) 250

8. Special appliance loads that operate for more than three hours shall be calculated by the amperage of the appliance times _____ percent.

      (a) 80   (b) 100   (c) 125   (d) 150

9. The demand load for 22 pieces of cooking equipment in a school shall be calculated by the equipment load times _____ percent.

      (a) 65   (b) 75   (c) 80   (d) 100

10. The volt-ampere rating for each heating and A/C load is calculated at _____ percent, and the largest load of the two is selected.

      (a) 80   (b) 100   (c) 125   (d) 150

11-55

Section	Answer

11. What is the load in VA for the general-purpose lighting load in an 40,000 sq. ft. store including 20,000 sq. ft. warehouse space? (Used at continuous operation)

(a) 142,860 VA (b) 156,250 VA (c) 158,940 VA (d) 164,550 VA

12. What is the load in VA for the general-purpose lighting load in an school building with the following area: (Compute at continuous operation)

- 20,000 sq. ft. classroom area
- 4000 sq. ft. auditorium area
- 1000 sq. ft. assembly hall area

(a) 81,250 VA (b) 84,960 VA (c) 88,125 VA (d) 89,895 VA

13. What is the load in VA for the following lighting fixtures:

- 12 - 100 VA each recess lighting fixtures
- 14 - 150 VA each incandescent lighting fixtures
- 20 - 180 VA each electric discharge lighting fixtures
- Compute at continuous operation

(a) 6840 VA (b) 7250 VA (c) 8115 VA (d) 8625 VA

14. What is the lighting load in VA for a 120 ft. of show window used at continuous operation?

(a) 28,250 VA (b) 30,000 VA (c) 31,400 VA (d) 34,680 VA

15. What is the lighting load in VA for a sign rated 2400 VA operating for three hours?

(a) 2400 VA (b) 2800 VA (c) 3000 VA (d) 3250 VA

16. What is the load in VA for 120 general-purpose receptacle outlets to cord-and-plug connect loads used at noncontinuous operation?

(a) 15,580 VA (b) 15,960 VA (c) 16,250 VA (d) 16,820 VA

17. What is the demand load in VA and amps for 20, 208 volt, three-phase cooking units rated 8 kW each used in a school building?

(a) 101,240 VA (b) 102,860 VA (c) 103,990 VA (d) 104,860 VA

18. What is the load in VA for a group of 240 volt, three-phase motors rated at 40 HP, 30 HP, and 20 HP?

(a) 84,485 VA (b) 88,240 VA (c) 95,525 VA (d) 99,008 VA

19. What is the load in VA for 4 compressors rated at 24.5 amps each and supplied by a 480 volt, three-phase supply?

(a) 81,438 VA (b) 84,256 VA (c) 85,881 VA (d) 88,245 VA

20. What is the largest load in VA between a 20 kW heating unit and a 6.5 kW A/C unit supplied by 208 volt, three-phase system?

(a) 6500 VA (b) 8125 VA (c) 20,000 VA (d) 25,000 VA

Test Questions

12

Chapter 12 is a compilation of questions based on the 2005 NEC. These questions are designed to reinforce as well as test the skills of each individual on questions that may appear on the Master's examination.

All questions are either illustrated and discussed in each chapter of this book, or they can be found in the 2005 NEC. For example, to properly answer a question, the user may need to refer to the NEC instead of the *Master Electrician's Study Guide* to obtain the requirements for determining his or her answer. This procedure will prevent the user from depending completely on the information in this book and help the user learn to use the NEC as well. Each test in Chapter 12 begins with a listing of NEC references that will substantiate the answer determined for that particular question.

Chapter 12 contains one thousand (1000) questions based on the nine chapters of the NEC. Students can use these tests after studying each chapter to verify how much of the material they have retained during each session.

ARTICLE 90 12-3	ARTICLE 427 12-56
ARTICLE 100 12-4	ARTICLE 430 12-56
ARTICLE 110 12-5	ARTICLE 440 12-62
ARTICLE 200 12-6	ARTICLE 450 12-63
ARTICLE 210 12-7	ARTICLE 460 12-65
ARTICLE 215 12-16	ARTICLE 500 12-65
ARTICLE 220 12-17	ARTICLE 501 12-66
ARTICLE 225 12-18	ARTICLE 502 12-68
ARTICLE 230 12-19	ARTICLE 503 12-68
ARTICLE 240 12-23	ARTICLE 511 12-68
ARTICLE 250 12-24	ARTICLE 513 12-69
ARTICLE 300 12-29	ARTICLE 514 12-69
ARTICLE 305 12-32	ARTICLE 515 12-70
ARTICLE 310 12-33	ARTICLE 517 12-70
ARTICLE 318 12-35	ARTICLE 518 12-71
ARTICLE 321 12-36	ARTICLE 520 12-72
ARTICLE 324 12-36	ARTICLE 525 12-72
ARTICLE 325 12-37	ARTICLE 530 12-73
ARTICLE 326 12-37	ARTICLE 545 12-74
ARTICLE 328 12-37	ARTICLE 547 12-74
ARTICLE 330 12-38	ARTICLE 550 12-74
ARTICLE 331 12-38	ARTICLE 551 12-75
ARTICLE 333 12-39	ARTICLE 553 12-75
ARTICLE 334 12-39	ARTICLE 600 12-75
ARTICLE 336 12-40	ARTICLE 604 12-77
ARTICLE 338 12-40	ARTICLE 605 12-77
ARTICLE 339 12-41	ARTICLE 610 12-77
ARTICLE 340 12-41	ARTICLE 620 12-78
ARTICLE 342 12-41	ARTICLE 625 12-78
ARTICLE 345 12-42	ARTICLE 630 12-79
ARTICLE 346 12-42	ARTICLE 645 12-79
ARTICLE 347 12-43	ARTICLE 650 12-80
ARTICLE 348 12-43	ARTICLE 660 12-80
ARTICLE 349 12-43	ARTICLE 665 12-80
ARTICLE 350 12-44	ARTICLE 668 12-81
ARTICLE 351 12-44	ARTICLE 669 12-81
ARTICLE 352 12-45	ARTICLE 670 12-81
ARTICLE 354 12-45	ARTICLE 675 12-81
ARTICLE 362 12-46	ARTICLE 680 12-81
ARTICLE 364 12-46	ARTICLE 690 12-84
ARTICLE 365 12-46	ARTICLE 700 12-84
ARTICLE 314 12-47	ARTICLE 701 12-86
ARTICLE 312 12-48	ARTICLE 705 12-87
ARTICLE 366 12-49	ARTICLE 710
ARTICLE 404 12-49	IS NOW
ARTICLE 408 12-50	ARTICLE 490 12-87
ARTICLE 400 12-51	ARTICLE 725 12-88
ARTICLE 402 12-51	ARTICLE 760 12-88
ARTICLE 410 12-52	ARTICLE 800 12-89
ARTICLE 411 12-54	ARTICLE 810 12-90
ARTICLE 422 12-54	ARTICLE 820 12-91
ARTICLE 424 12-55	CHAPTER 9 12-91
ARTICLE 426 12-56	TABLES 12-91
	APPENDIX C 12-91

CHAPTER 12
1000 TEST QUESTIONS

Section   Answer

Article 90
Introduction

1. Electrical equipment under the exclusive control of utility companies for the purpose of generating electricity is exempt from the NEC.

    A. True   B. False

2. If persons disagree with a section in the NEC, they can request a formal interpretation from the NFPA.

    A. True   B. False

3. A main purpose of the NEC is for the protection of persons from dangers that can occur from the use of electricity.

    A. True   B. False

4. The NEC covers all of the following electrical installations, except:

    A. Recreational vehicles   B. Floating buildings
    C. Carnivals   D. Public utilities

5. Which of the following statements about the authority having jurisdiction of enforcement of the NEC is not true?

    A. Uses the authority given for making interpretations of the NEC rules, etc.
    B. Uses the authority given for deciding the approval of equipment and materials
    C. Uses the authority given to grant special permission
    D. Uses the authority given to automatically waive specific requirements in the NEC

6. Chapters 1 through 4 of the NEC applies to _____.

    A. General   B. Special occupancies
    C. Special conditions   D. Communication systems

7. Chapter 8 of the NEC applies to _____.

    A. General   B. Special occupancies
    C. Special conditions   D. Communication systems

8. The _____ wiring of equipment shall not be required to be reinspected at the time of installation if the equipment is listed by a qualified testing laboratory.

    A. Associated   B. Internal
    C. External   D. None of the above

Stallcup's Master Electrician's Study Guide

Section   **Answer**

Article 100
Definitions

_____ _____ 9. *Accessible* (as applied to equipment) is defined as admitting close approach, not guarded by locked doors, elevation, or other effective means.

        A. True   B. False

_____ _____ 10. A bathroom can defined as an area with a toilet, a tub, or a shower.

        A. True   B. False

_____ _____ 11. A bonding jumper is defined as the permanent joining of metallic parts to form an electrically conductive path that will ensure electrical continuity and the capacity to conduct safely any current likely to be imposed.

        A. True   B. False

_____ _____ 12. A branch-circuit is defined as a branch-circuit that supplies a number of outlets for lighting and appliances.

        A. True   B. False

_____ _____ 13. A building is defined as a structure that stands alone or that is cut off from adjoining structures by fire walls with all openings therein protected by approved fire doors.

        A. True   B. False

_____ _____ 14. A continuous load is defined as a load where the maximum current is expected to continue for three hours or less.

        A. True   B. False

_____ _____ 15. A feeder is defined as the circuit conductors between the final overcurrent device protecting the circuit and the outlet(s).

        A. True   B. False

_____ _____ 16. *Grounded* is defined as a system or circuit conductor that is intentionally grounded.

        A. True   B. False

_____ _____ 17. *Grounding conductor* is defined as a conductor used to connect equipment or the grounded circuit of a wiring system to a grounding electrode or electrodes.

        A. True   B. False

_____ _____ 18. Where the NEC specifies that one equipment shall be "in sight from," "within sight from," or "within sight," or etc., of another equipment, the specified equipment is to be visible and not more than _____ ft. distant from the other.

_____ _____ 19. A location not normally subject to dampness or wetness is called a _____ location.

_____ _____ 20. The conductors from the service point to the service disconnecting means are called the _____ conductors.

_____ _____ 21. The _____ conductors are the overhead conductors between the last utility pole and the service-entrance conductors.

_____ _____ 22. The service _____ is where the premises wiring or equipment connects to the utility wiring.

Test Questions

Section   Answer

Article 110
Requirements for electrical installations

23. The conductors and equipment required or permitted by the NEC shall be acceptable only if _____ by the AHJ.

24. Electric equipment shall be installed in a neat and _____ manner.

25. Termination provisions of equipment for circuits rated 100 amperes or less, or marked _____ AWG through _____ AWG, shall be used only for conductors rated 60°C. (General rule)

26. Termination provisions of equipment for circuits rated over 100 amperes, or marked for conductors larger than _____ AWG, shall be used only with conductors rated 75°C.

27. For mounting of electrical equipment, the code specifically prohibits which of the following methods?

    A. Machine screws   B. Sheet metal screws
    C. Bolts of malleable iron   D. Wooden plugs driven into holes in masonry

28. Connections, by means of wire binding screws or studs and nuts having upturned lugs, shall be permitted for conductors of a maximum size _____ AWG.

    A. 12   B. 10   C. 8   D. 6

29. Where used on circuits of 100 amperes or less, the load on conductors shall be limited to ampacities rated _____ degrees C, unless the terminals are rated higher.

    A. 60   B. 75   C. 85   D. 90

30. The load on conductors used on circuits of over 100 amperes shall be limited to ampacities rated _____ degrees C, unless the terminals are rated higher.

    A. 60   B. 75   C. 85   D. 90

31. The work space in front of electric equipment shall be at least _____ inches wide.

    A. 24   B. 30   C. 36   D. 40

32. For electrical equipment operating with a nominal voltage of 150 volts-to-ground, the maximum working space clearance required in front of the equipment shall be _____ in. deep.

    A. 30   B. 36   C. 42   D. 48

33. Where electrical switchboards face each other and exposed live parts of electrical equipment operating at 277/480 volts are accessible on each side of the work space, the clear working space between shall be at least _____ inches.

    A. 36   B. 42   C. 48   D. 60

34. Where access is required to the back of a motor control center for working on de-energized parts, a minimum working space of _____ in. horizontally shall be required.

    A. 24   B. 30   C. 36   D. 42

35. For equipment rated over 1200 amps or more, containing overcurrent devices, switching devices, or control devices, there shall be one entrance not less than _____ in. wide and _____ ft. high at each end.

    A. 24; 6   B. 24; 6 1/2   C. 30; 6   D. 30; 6 1/2

Section Answer

36. Enclosures for electrical installations over 600 volts, a fence shall not be less than _____ ft. with _____ ft. of fence fabric utilizing three or more strands of barbed wire or equivalent is permitted.

   A. 6; 1   B. 6; 2   C. 7; 1   D. 7; 2

37. The minimum headroom about a 100 amp panelboard in a store building shall be at least _____ ft. to ensure safe standing room.

   A. 6 1/4   B. 6 1/2   C. 7   D. 8

Article 200
Use and identification of grounded conductors

38. The grounded conductor shall be permitted to be identified by notching the insulation.

   A. True   B. False

39. Terminals for attachment of grounded conductors are usually identified by a white finish.

   A. True   B. False

40. Devices with screw-shells shall have the grounded conductor connected the screw shell.

   A. True   B. False

41. Grounded conductors connected to leads shall not be required to be terminated to prevent reverse polarity.

   A. True   B. False

42. Grounded conductors at 600 volts can be used with ungrounded conductors rated 1000 volts or greater.

   A. True   B. False

43. Where one grounded and two ungrounded circuit conductors are run in a raceway, the grounded conductor shall have an outer finish of white or gray its entire length, if it is a maximum size _____ AWG or smaller.

   A. 10   B. 8   C. 6   D. 4

44. The maximum size grounded conductor that is required to be white or gray along its entire length is _____ AWG.

   A. 10   B. 8   C. 6   D. 4

45. Multiconductor flat cable _____ AWG or larger shall be permitted to employ an external ridge on the grounded conductor.

   A. 8   B. 6
   C. 4   D. 2

46. Where a conductor with black insulation is used as a grounded conductor in a raceway, it shall be permitted to be identified with white markings at its terminations if it is at least size _____ AWG or larger.

   A. 8   B. 6   C. 4   D. 2

Test Questions

Section   Answer

47. An insulated conductor intended for use as a grounded conductor, where contained in a flexible cord, shall be identified by a _____ outer finish.

    A. White
    B. Gray
    C. None of the above
    D. All of the above

48. A continuous white or gray color identifies the _____ conductor of a branch-circuit.

    A. Bonding   B. Isolated   C. Grounded   D. Grounding

49. The white conductor of a nonmetallic sheathed cable shall be permitted to be used in a switch loop, as an ungrounded conductor:

    A. For a single-phase switch loop only
    B. As the supply to the switched outlet
    C. As the return from the switched outlet
    D. Where reidentified as an ungrounded conductor

50. The identification of terminals to which a grounded conductor is to be connected shall be substantially _____ in color.

    A. Brass   B. White   C. Green   D. Silver

51. For screw-shell devices with attached leads, the conductor attached to the center terminal or contact point at the bottom of the screw shall, be _____.

    A. Gray   B. Green   C. White   D. Black

Article 210
Branch-circuits

52. Multiwire branch-circuits shall only supply phase-to-phase loads.

    A. True   B. False

53. Where a multiwire branch-circuit supplies more than one device on the same yoke, a means shall be provided to disconnect simultaneously all ungrounded conductors.

    A. True   B. False

54. A 6 AWG or smaller EGC of a branch-circuit shall be identified by a continuous green color or a continuous green color with one or more yellow stripes unless it's bare.

    A. True   B. False

55. A nongrounding type receptacle shall be permitted to be replaced with a grounding type receptacle where supplied through a ground-fault circuit-interrupter. (General rule)

    A. True   B. False

56. Multioutlet branch-circuits greater than 40 amperes shall be permitted to supply nonlighting outlet loads on industrial premises where conditions of maintenance and supervision ensure that only qualified persons service the equipment.

    A. True   B. False

57. In dwelling units, wet bar sinks shall have ground-fault circuit-interrupter protection for all 120 volt, 15 or 20 amp receptacles, where located within 6 ft. of the outside edge of the wet bar sink.

    A. True   B. False

Section	Answer

58. A 40 or 50 amp branch-circuit shall supply only nonlighting outlet loads.

    A. True   B. False

59. Each wall space 6 ft. or more wide shall be treated individually and separately from other wall spaces within the room.

    A. True   B. False

60. Receptacles that are an integral part of a lighting fixture, appliance, or cabinet shall not be counted as one of the required receptacle outlets in a dwelling unit.

    A. True   B. False

61. A maximum of 2 - 20 amp, 1500 VA small appliance circuits shall be required to supply receptacle outlets that are located in the kitchen, pantry, breakfast room, and dining room in a dwelling unit.

    A. True   B. False

62. Receptacle outlets shall be required to be installed so that there is no point on the wall greater than 6 ft. from a receptacle outlet in a dwelling unit.

    A. True   B. False

63. Sliding glass doors or panels shall be considered wall space when spacing receptacles in a dwelling unit. (**Note:** Wall space and not door openings)

    A. True   B. False

64. An outside outlet shall be permitted to be installed on the small appliance circuit in a dwelling unit.

    A. True   B. False

65. Receptacle outlets shall be permitted to be mounted not more than 18 in. below the countertop in a dwelling unit for the physically impaired.

    A. True   B. False

66. At least two wall receptacle outlets shall be installed in a bathroom adjacent to each basin location in a dwelling unit.

    A. True   B. False

67. At least one receptacle outlet shall be installed for the laundry in a dwelling unit.

    A. True   B. False

68. Hallways of 20 ft. or more in length shall have at least one receptacle outlet in a dwelling unit.

    A. True   B. False

69. At least one receptacle outlet shall be installed directly above a show window for each 12 linear ft.

    A. True   B. False

70. A three-phase, four-wire, _____ connected system used to supply power to nonlinear loads such as personal computers, electric discharge lighting, data processing, etc. shall be designed to allow for the possibility of high harmonic neutral currents.

Test Questions

Section   Answer

71. All 120 volt, single phase, 15 and 20 ampere branch-circuits supplying outlets installed in dwelling units _____ shall be protected by a listed AFCI, combination type installed to provide protection of the branch-circuit.

72. GFCI protection for personnel shall be provided for outlets that supply boat _____ installed in dwelling unit locations and supplied by 125 volt, 15 and 20 ampere branch-circuits.

73. Tap conductors supplying electric ranges, wall-mounted electric ovens, and counter-mounted electric cooking units from a _____ ampere branch-circuit shall have an ampacity of not less than 20 amperes.

74. Branch-circuit conductors supplying loads other than cooking appliances shall have an ampacity sufficient for the loads served and shall not be smaller than _____ AWG.

75. A heavy-duty lampholder shall have a rating of not less than _____ watts if of the admedium type.

76. The rating of any one cord-and-plug connected utilization equipment shall not exceed _____ percent of the branch-circuit amp rating.

77. The total rating of utilization equipment fastened-in-place shall not exceed _____ percent of the branch-circuit amp rating.

78. Appliance receptacle outlets installed in a dwelling unit for specific appliances shall be installed within _____ ft. of the intended location of the appliance.

79. Receptacle outlets in a dwelling unit shall be installed so that there is no point on the wall greater than _____ ft. from a receptacle.

80. Receptacle outlets in a dwelling unit shall not serve as one of the required outlets if located over _____ ft. _____ in. from the floor.

81. Receptacles installed in the kitchen of dwelling units to serve countertop surfaces shall be supplied by not less than _____ small appliance branch-circuits.

82. A receptacle outlet shall be installed at each wall counter space _____ in. or wider in a dwelling unit.

83. Receptacle outlets shall be installed for each peninsular countertop in a dwelling unit with a long dimension of _____ in. or greater and a short dimension of _____ or greater.

84. In dwelling units, bathroom receptacle outlets shall be supplied by at least one _____ amp branch-circuit.

85. Grade level access shall be considered _____ ft. _____ in. or less from finished grade if installed outside of a dwelling unit.

86. Receptacle outlets shall be located not more than _____ in. above the countertop in a dwelling unit.

87. At least one receptacle outlet shall be installed in bathrooms within _____ ft. of the outside edge of each basin.

88. A 15 or 20 amp receptacle outlet shall be installed at an _____ location for the servicing of air-conditioning equipment on rooftops.

89. A vehicle door in a _____ shall not be considered as an outdoor entrance or exit for a dwelling unit.

12-9

Section   Answer

90. Multiwire circuits feeding more than _____ device on the same yoke shall be disconnected simultaneously.

91. Voltage between conductors supplying terminals of luminaries (lighting fixtures) in dwelling units shall not exceed _____ volts.

92. For ranges of 8 3/4 kW or more rating, the minimum branch-circuit rating shall be _____ amperes.

    A. 40   B. 50   C. 60   D. 100

93. General purpose branch-circuits shall be rated in accordance with the:

    A. Voltage of the system
    B. Sizing of the conductor used
    C. Ampere rating or setting of the overcurrent device
    D. Temperature rating of the terminals on the equipment supplied

94. All of the following are ratings (per NEC) for general purpose branch-circuits, except _____ amperes.

    A. 15   B. 20   C. 25   D. 30

95. The following is a rating (per NEC) only of an individual branch-circuit of _____ amperes.

    A. 15   B. 20   C. 25   D. 30

96. The code specifically requires a means to disconnect simultaneously all ungrounded conductors at the panelboard where the circuit originated, for all _____ branch-circuits supplying more than one device.

    A. Multiwire   B. Individual   C. Multioutlet   D. General purpose

97. A metal junction box contains the following eight conductors for two multiwire branch-circuits:
    2 black
    2 blue
    2 red
    1 white
    1 white with a green stripe

The white conductor with a green stripe is:

    A. A grounded conductor
    B. A grounding conductor
    C. An insulated equipment grounding conductor
    D. Not permitted by the code for either a grounded or grounding conductor

98. A continuous white or gray color identifies the _____ conductor of a branch-circuit.

    A. Bonding   B. Isolated   C. Grounded   D. Grounding

99. Where installed in a raceway, a white 8 AWG insulated conductor, used as an equipment grounding conductor, is:

    A. A code violation, even it is marked with green tape
    B. Permitted if it is colored green for the entire exposed length
    C. Permitted if the white insulation is stripped from the entire exposed length
    D. Permitted if the conditions of maintenance ensure that only qualified persons will service the installation.

	Section	Answer

100. In dwelling units, the code requires that the voltage be limited to 120 volts between conductors for all branch-circuits that supply the terminals of:

    A. Receptacles   B. Luminaires (lighting fixtures)
    C. Equipment of 1 HP or less
    D. Cord-and-plug connected equipment of 2400 VA or less

101. Branch-circuits exceeding 120 volts between conductor but not more than 277 volts-to-ground shall be permitted to supply luminaires (lighting fixtures) equipped with _____ base screw shell lampholders.

    A. Mogul   B. Medium   C. Admedium   D. Intermediate

102. On a circuit that supplies electric discharge luminaires (lighting fixtures) at the top of poles 22 ft. in height, the maximum permitted voltage is _____ volts-to-ground.

    A. 120   B. 277   C. 300   D. 600

103. In an existing dwelling, built and wired before GFCI receptacles where required in kitchens, a receptacle serving the kitchen countertop surfaces is replaced. This replacement receptacle:

    A. Is not covered by the current edition of the code
    B. Is required to be ground-fault circuit-interrupter protected
    C. Is not required to be ground-fault circuit-interrupter protected
    D. Is required to be ground-fault circuit-interrupter protected only if within 6 ft. of the kitchen sink

104. Where connected to a branch-circuit having a rating in excess of _____ amperes, lamp holders shall be of the heavy-duty type.

    A. 15   B. 20   C. 25   D. 30

105. Ground-fault circuit-interrupter protection shall be required for receptacles in all of the following locations, except receptacles:

    A. In industrial facilities
    B. In an office building restroom with a toilet and sink
    C. On the roof of a bank building for servicing an air conditioning unit
    D. In the kitchen of a dwelling to serve countertops

106. GFCI-protection shall be required for all receptacles, without exception or qualification, in dwelling unit:

    A. Garages   B. Wet bar sinks
    C. Bathrooms   D. Unfinished basements

107. In a dwelling unit garage, one cord-and-plug connected freezer will occupy a dedicated space, but the receptacle for the freezer will be "readily accessible" after the freezer is installed. This receptacle is:

    A. Required to be a combination single receptacle with a switch
    B. Not required to be a single receptacle, but if a single receptacle is used, it shall be GFCI-protected
    C. Permitted to be a duplex receptacle, and if a duplex receptacle is used to connect another appliance, it shall be GFCI-protected
    D. Required to be installed on the same GFCI-protected circuit as the other readily accessible garage receptacles

Stallcup's Master Electrician's Study Guide

Section   **Answer**

108. Outdoor receptacles at a dwelling unit shall not be required to be GFCI-protected if they are supplied from a dedicated branch-circuit and installed:

    A. In a weatherproof box
    B. At a second floor balcony
    C. At least 6 ft. 6 in. above grade
    D. For electric snow-melting or deicing equipment that is GFCI protected

109. Which of the following outdoor receptacles at a dwelling unit shall not be required to be GFCI-protected? A receptacle installed:

    A. In an open carport
    B. On a second floor balcony
    C. On the wall at a height of 7 ft. above the ground level
    D. On a dedicated branch-circuit to supply snow-melting equipment at the roof

110. At a store, ground-fault circuit-interrupter protection for personnel shall be required for 20 amp receptacles installed at which of the following locations?

    A. Rooftop   B. Bathrooms   C. All of the above   D. None of the above

111. An autotransformer that supplies a 480 volt circuit from a 600 volt system shall not be required to have a grounded conductor that is electrically connected between the supplying system and the circuit supplied where the autotransformer:

    A. Has no exposed noncurrent-carrying metal parts
    B. Is connected in a three-phase open delta arrangement
    C. Is located at least 8 ft. above the floor or working surface
    D. Is in an industrial occupancy and only qualified persons will service the installation

112. Where the ampacity of multioutlet branch-circuit conductors does not correspond to the rating of a standard size overcurrent device, the next smaller size device shall be used if the branch-circuit supplies:

    A. A motor load
    B. A load of less than 800 amps
    C. Receptacles for cord-and-plug connected loads
    D. Fixed loads such as lighting units and appliances

113. The code requires that branch-circuit conductors shall have an ampacity of at least the rating of the branch-circuit if it pertains to conductors of:

    A. All branch-circuits
    B. Only individual branch-circuits
    C. Multioutlet branch-circuits supplying fixed lighting and appliance loads
    D. Multioutlet branch-circuits supplying receptacles for cord-and-plug connected per NEC Table loads

114. Where a 30 amp branch-circuit serves fixed lighting units, tap conductors to individual luminaires (light fixtures) shall be permitted to extend a maximum of ____ in. beyond any portion of the fixture.

    A. 6   B. 12   C. 18   D. 24

115. A mogul base lampholder installed on a 30 amp branch-circuit shall have a rating of at least ____ watts.

    A. 250   B. 300   C. 660   D. 750

Test Questions

Section   Answer

116. All of the following are code violations, except:

    A. A single 15 amp rated receptacle on a 20 amp general purpose branch-circuit
    B. A single 15 amp rated receptacle on a 20 amp individual branch-circuit
    C. A duplex 20 amp rated receptacle on a 15 amp multiwire branch-circuit
    D. A single 30 amp rated receptacle on a 40 amp individual branch-circuit

117. The rating of cord-and-plug connected utilization equipment shall not exceed _____ percent of the branch-circuit ampere rating.

    A. 50   B. 80   C. 100   D. 125

118. A 20 amp branch-circuit shall be permitted to supply lighting units together with utilization equipment in all of the following situations except:

    A. If the utilization equipment is fastened-in-place
    B. If the utilization equipment is cord-and-plug connected
    C. Where the 20 amp circuit is a small appliance branch-circuit
    D. Where the circuit supplies a motor-operated utilization equipment with a motor larger than 1/8 horsepower

119. A branch-circuit in a multifamily dwelling that supplies outdoor lighting in a common area shall:

    A. Not be supplied from a panelboard in an individual dwelling unit
    B. Be permitted to be supplied from a panelboard in the dwelling unit for the manager of the multifamily dwelling
    C. Be permitted to be supplied from a panelboard in an individual dwelling unit if the lighting is controlled by a photocell
    D. Be permitted to be supplied from a panelboard in an individual dwelling unit where the circuit is not over 20 amperes

120. A cord connector that is supported by a permanently installed cord pendant shall be considered a(n) _____ outlet.

    A. Power   B. Lighting   C. Individual   D. Receptacle

121. Receptacle outlets shall be spaced a maximum of 12 feet apart in all of the following rooms or areas of dwellings, except the _____.

    A. Kitchen   B. Hallway   C. Dining room   D. Living room

122. In the living room of a dwelling unit, a wall space at least _____ in. or more in width shall have a receptacle outlet installed.

    A. 12   B. 18   C. 24   D. 30

123. In the family room of a dwelling unit, which of the following is not included in the measurement of wall space for required spacing of receptacle outlets?

    A. An open railing next to a stairway
    B. A sliding panel (door opening) in an exterior door
    C. A fixed window panel next to an exterior door
    D. A free standing bar-type counter serving as a room divider

Stallcup's Master Electrician's Study Guide

Section   Answer

_____ _____ 124. In dwellings, a floor outlet shall be permitted to be counted as part of the required number of receptacle outlets where located a maximum of _____ in. from the wall.

   A. 6   B. 12   C. 18   D. 24

_____ _____ 125. In a dwelling unit, a receptacle outlet that is factory-installed in a listed baseboard electric heater shall be permitted to serve as the required outlet for the wall space utilized by the baseboard heater if:
_____

   A. The heater circuit is a maximum of 120 volts.
   B. The outlet is connected to the heater circuit.
   C. The outlet is not connected to the heater circuit.
   D. The baseboard heater is 6 ft. or more in length.

_____ _____ 126. For a commercial occupancy, the code requires receptacle outlets to be installed:

   A. Within 6 ft. of any point along the floor line
   B. Within 6 ft. of the intended location of equipment
   C. Within 12 ft. of the intended location of equipment
   D. Wherever flexible cords with attachment plugs are used

_____ _____ 127. For dwelling units, a branch-circuit supplying _____ receptacle outlet(s) in an attached garage may also supply outdoor receptacles.

   A. Outdoor   B. Bathroom   C. Laundry room   D. Kitchen small appliance

_____ _____ 128. In a dwelling unit, a separate circuit with no other outlets or loads on the same circuit shall be required for all of the following locations or loads except:

   A. Bathroom receptacles in two bathrooms
   B. Laundry room receptacles
   C. A refrigerator supplied by a small-appliance 20-amp circuit
   D. Central heating equipment other than fixed electric space-heating equipment.

_____ _____ 129. In a dwelling unit kitchen, the small appliance branch-circuits that supply countertop receptacles shall not be permitted to supply receptacle outlets:

   A. Outdoors   B. In a pantry
   C. In a dining room   D. For refrigeration equipment

_____ _____ 130. In a dwelling unit kitchens, the two or more small appliance branch-circuits installed to serve countertop surfaces shall be permitted to serve which of the following cord-and-plug connected appliances?

   A. Trash compactor
   B. Built-in dishwasher
   C. Electrically operated kitchen waste disposer
   D. Gas-fired range with supplemental electrical equipment

_____ _____ 131. In a dwelling unit, the receptacle outlet for the refrigerator shall be:

   A. Permitted to be supplied from at least an individual 15 amp branch-circuit
   B. Permitted to be supplied from a general purpose 15 amp branch-circuit
   C. Required to be supplied from an individual 20 amp branch-circuit
   D. Required to be supplied from (only) one of the two or more 20 amp small appliance branch-circuits

Test Questions

Section | Answer

132. In a dwelling unit, receptacle outlets installed for a specific appliance shall be installed within at least _____ ft. of the intended location of the appliance.

    A. 3   B. 6   C. 10   D. 12

133. In the kitchen of a dwelling unit, a receptacle outlet shall be installed at each wall counter space that is at least _____ in. or wider.

    A. 10   B. 12   C. 8   D. 24

134. In the kitchen of a dwelling unit, an island countertop space is 90 in. long with a width of 28 in., and has a 30 in. wide range installed in the center, evenly dividing the space on each side of the range. How many receptacle outlets is/are required for the island?

    A. 1   B. 2   C. 3   D. 4

135. In a dwelling unit kitchen, a peninsular countertop is 2 1/2 ft. wide and 7 ft. long measured from the connecting edge. At least _____ receptacle outlet(s) is/are required to be installed at this peninsular counter space.

    A. 0   B. 1   C. 2   D. 3

136. In the kitchen of a dwelling unit, receptacles installed below the countertop of a peninsular counter space shall not be located where the countertop extends at least _____ in. or more beyond its support base.

    A. 6   B. 8   C. 10   D. 12

137. In the kitchen of a dwelling unit, receptacle outlets for counter spaces shall be located a maximum of _____ in. above the countertop.

    A. 10   B. 12   C. 18   D. 20

138. In two dwelling unit bathrooms, the receptacle outlets shall be supplied by at least one:

    A. 15 ampere branch-circuit that supplies no other outlets
    B. 15 ampere branch-circuit that is permitted to also supply outdoor outlets
    C. 20 ampere branch-circuit that supplies no other outlets
    D. 20 ampere branch-circuit that is permitted to also supply outdoor outlets

139. The outdoor receptacle outlets for a dwelling unit that are required to be accessible at grade level shall be installed a maximum of _____ ft. above grade.

    A. 5 1/2   B. 6   C. 6 1/2   D. 8

140. In a dwelling unit, the receptacle outlet provided for the clothes washer shall be:

    A. Permitted to be supplied from an individual 15 amp branch-circuit
    B. Permitted to be supplied from a general purpose 15 amp branch-circuit
    D. Required to be supplied from a 20 amp branch-circuit that supplies only the outlet or outlets for the laundry

141. In dwelling units, hallways of _____ ft. or more in length shall have at least one receptacle outlet installed.

    A. 3   B. 6   C. 10   D. 20

142. For a store, at least one receptacle shall be installed above a show window for each _____ linear ft. or major fraction thereof of show window.

    A. 10   B. 12   C. 18   D. 25

12-15

Section **Answer**

143. Where an air conditioning unit is located on the rooftop of an office building, a receptacle outlet shall be located on the roof on the same level and within _____ ft. of the air conditioning unit.

    A. 6   B. 12   C. 25   D. 50

144. In a dwelling unit, a luminaire (light fixture) over a stairway shall be controlled by a wall switch at each floor level where the difference between floor levels is _____ risers or more.

    A. 3   B. 4   C. 5   D. 6

145. In which of the following rooms of a dwelling unit would the required lighting outlet be permitted to be a wall-switched controlled receptacle?

    A. Hallway   B. Kitchen   C. Bathroom   D. None of the above

146. In a dwelling unit, a lighting outlet being controlled by an occupancy sensor is:

    A. Prohibited by the code
    B. Permitted for all locations except at outdoor entrances
    C. Permitted for all rooms except rooms containing electrical equipment requiring servicing
    D. Permitted if a wall switch or manual override is also installed at the customary wall switch location

Article 215
Feeders

147. Feeders with a common neutral shall be allowed to supply two or three sets of three-wire feeders.

    A. True   B. False

148. Feeder-circuits supplying 15 and 20 amp receptacle branch-circuits shall not be permitted to be protected by a GFCI.

    A. True   B. False

149. Feeders shall be permitted to be derived from autotransformers.

    A. True   B. False

150. All feeder-circuit conductors, with a common neutral, pulled in metal raceways shall be grouped together.

    A. True   B. False

151. The feeder conductor ampacity shall not be less than that of the service-entrance conductors where the feeder conductors carry the total load supplied by service-entrance conductors with an ampacity of _____ amp. or less.

152. The rating of the overcurrent device for a feeder shall not be less than the noncontinuous load plus _____ percent of the continuous load.

153. Ground-fault protection of equipment shall be provided for a feeder disconnect rated _____ amps or more in a solidly grounded wye system with greater than 150 volts-to-ground, but not exceeding 600 volts phase-to-phase.

Article 220
Branch-circuit, Feeder, and Service Calculations

154. Neutral conductors shall be considered current-carrying conductors where _____ processing equipment is supplied.

155. Existing demand shall be calculated at _____ percent of the maximum demand and added to the new load according to the optional method.

156. Before the optional method for dwelling units can be applied, the ampacity of the service-entrance or feeder conductors shall be rated at least _____ amps.

157. For a branch-circuit that supplies only lighting in an office building, the conductors shall have an minimum allowable ampacity of _____ percent of the lighting load on the circuit.

    A. 80   B. 100   C. 115   D. 125

158. When calculating the general lighting load for an office building, a unit load of _____ volt-amperes shall be included for general purpose receptacle outlets when the number of outlets to be installed is unknown.

    A. 1/2   B. 1   C. 1 1/2   D. 3 1/2

159. When calculating the general lighting and receptacle load for a dwelling unit, a unit load of _____ volt-amperes shall be used.

    A. 1   B. 2   C. 3   D. 3 1/2

160. When calculating the general lighting load for an restaurant, a unit load of _____ volt-amperes shall be used.

    A. 1/4   B. 1/2   C. 1   D. 2

161. For the feeder calculation for a store, the load for each noncontinuous duty receptacle outlet shall be _____ volt-amperes.

    A. 150   B. 180   C. 200   D. 300

162. For calculating the load on a branch-circuit, an outlet for a heavy-duty lampholder not used for general illumination shall be _____ volt-amperes.

    A. 250   B. 600   C. 660   D. 750

163. For calculating the load on a branch-circuit, sign and outline lighting shall be calculated a minimum of _____ volt-amperes.

    A. 1000   B. 1200   C. 1500   D. 1800

164. When applying demand factors for nondwelling receptacle loads, the first 10 kVA is calculated at 100 percent and the remainder is calculated at _____ percent.

    A. 25   B. 50   C. 75   D. 85

165. The largest motor load shall be added to the other loads calculated per Article 220 at _____ percent or the motor's full load current rating.

    A. 25   B. 50   C. 60   D. 75

166. In a dwelling unit, at least _____ branch-circuits shall be provided for the small appliance circuits.

    A. One 15-ampere   B. Two 15-ampere
    C. One 20 ampere   D. Two 20-ampere

Section	Answer

167. For calculating the load on a feeder using the standard calculation method, a fixed electric space heating load shall be calculated at _____ percent of the total connected load.

    A. 80   B. 100   C. 125   D. 150

168. For fixed electric space-heating equipment, the ampacity of the branch-circuit conductors shall be at least _____ percent of the total load of the supplied.

    A. 80   B. 100   C. 125   D. 150

169. In a dwelling unit, each small appliance branch-circuit used for the kitchen countertop surface shall be calculated at _____ volt-amperes.

    A. 1000   B. 1500   C. 3000   D. 5000

170. For calculating the service conductor size for a dwelling unit, at least _____ volt-amperes shall be included for each two-wire laundry branch-circuit.

    A. 500   B. 1000   C. 1500   D. 3000

171. For calculating a feeder load for a two-family dwelling, the demand factor permitted to be applied to the nameplate rating load of four or more appliances would apply to all of the following except:

    A. Pool pumps   B. Dishwashers
    C. Clothes dryers   D. Kitchen waste disposals

172. Four or more fixed appliances fastened-in-place shall have a demand factor of _____ percent applied.

    A. 50   B. 60   C. 70   D. 75

173. Electric clothes dryers shall be _____ watts or the nameplate rating, whichever is larger, for each dryer served in a dwelling unit.

    A. 4000   B. 5000   C. 7500   D. 8000

174. Demand factors for five electric clothes dryers shall be calculated at _____ percent.

    A. 65   B. 70   C. 80   D. 100

175. A school with twelve pieces of electrical kitchen equipment can have a feeder demand factor of _____ percent applied.

    A. 65   B. 70   C. 80   D. 90

Article 225
Outside Branch-circuits and Feeders

176. Luminaires (lighting fixtures) rated at 277 volts can be installed outdoors if mounted at least 2 ft. from windows.

    A. True   B. False

177. An overhead span of conductors shall not be supported by trees.

    A. True   B. False

178. Festoon messenger wires shall not be supported by a fire escape on a building.

    A. True   B. False

	Test Questions	
	Section	Answer

179. Luminaires (lighting fixtures) rated at 277 volts shall not be used for outdoor stairstep illumination of commercial buildings.

    A. True   B. False

180. Open individual conductors shall not be smaller than 10 AWG copper for 600 volts, nominal, or less, for spans up to _____ ft. in length.

    A. 25   B. 40   C. 50   D. 75

181. Overhead conductors for festoon lighting shall not be smaller than _____ AWG.

    A. 14   B. 12   C. 10   D. 8

182. The conductors for festoon supports shall be supported by a messenger wire in spans exceeding _____ ft.

    A. 25   B. 40   C. 50   D. 75

183. Conductors on poles shall have a separation of not less than _____ ft. where not placed on racks or brackets.

    A. 1   B. 2   C. 3   D. 6

184. Outside conductors shall have a clearance from television antennas not less than _____ ft.

    A. 1   B. 2   C. 3   D. 6

Article 230
Branch-circuit, Feeder, and Service Calculations

185. A permanent plaque or directory shall be installed at each service disconnecting location where a building or structure is supplied by more than one service.

    A. True   B. False

186. Service conductors supplying a building or other structure shall be permitted to run through the interior of another building or other structure.

    A. True   B. False

187. Overhead service drop conductors shall not be smaller then 8 AWG copper.

    A. True   B. False

188. Service lateral conductors shall not be smaller than 6 AWG copper.

    A. True   B. False

189. Service lateral conductors exposed to physical damage can be protected by Schedule 40 rigid nonmetallic conduit.

    A. True   B. False

190. Each service disconnecting means shall be permanently marked to identify it as a service disconnecting means.

    A. True   B. False

191. A dwelling unit disconnect shall be rated at least 100 amp.

    A. True   B. False

Section   Answer

_____ _____ 192. Service-entrance conductors shall not be spliced by bolted connections.

      A. True   B. False

_____ _____ 193. Service-entrance conductors in a raceway under at least 2 ft. of concrete shall be considered outside the building.

      A. True   B. False

_____ _____ 194. Service-entrance conductors over 600 volts shall not be smaller than 8 AWG.

      A. True   B. False

_____ _____ 195. Individual meter socket enclosures shall not be considered service equipment.

      A. True   B. False

_____ _____ 196. Disconnects for fire pumps or emergency service shall be located at a remote point from the disconnect for the normal service.

      A. True   B. False

_____ _____ 197. Each service may have _____ disconnecting means to disconnect the service equipment.

_____ _____ 198. Two to six disconnecting means for each service shall be _____ for easy access by the user.

_____ _____ 199. The smallest copper conductor permitted for an underground service-entrance conductor is _____ AWG copper. (General rule)

_____ _____ 200. A(n) _____ service from the main service may be installed to supply a fire pump.

_____ _____ 201. Flexible metal conduit in _____ ft. or shorter lengths can be used in the service raceway run to the equipment.

_____ _____ 202. Each occupant of a multifamily building shall have _____ to the occupancy's service disconnecting means. (General rule)

_____ _____ 203. The ground-fault protection system shall be _____ tested when first installed on site.

_____ _____ 204. Each service disconnect shall be _____ marked to identify it as a service disconnect.

_____ _____ 205. Service conductors installed as open conductors shall have a clearance of not less than _____ ft. from windows that are designed to be opened.

_____ _____ 206. Overhead conductors that are installed to supply only limited loads of a single branch-circuit, such as controlled water heaters, shall not be smaller than _____ AWG hard-drawn copper.

_____ _____ 207. Conductors shall have a vertical clearance of not less than _____ ft. above the roof surface.

_____ _____ 208. Service drop conductors of 240 volts-to-ground crossing residential property or driveways shall have a clearance of _____ ft. from finished grade.

Test Questions

	Section	Answer

209. Service drop conductors crossing over public streets that are subject to truck traffic shall have a clearance of _____ ft. from finished grade.

210. Service-entrance cables shall be supported by straps within _____ in. of every service head and at intervals not exceeding _____ in.

211. Each service disconnect shall _____ disconnect all ungrounded conductors that it controls from the premises wiring system.

212. The service disconnecting means shall be plainly marked to identify whether it is in the _____ or _____ position.

213. Service conductors shall be permitted to pass through the interior of another building, such as through the crawlspace of a multifamily dwelling, where they are:

    A. Installed in rigid conduit
    B. Installed in a cable
    C. Installed in conduit properly supported to the lower edges of the floor joists
    D. Installed in a raceway that is encased in concrete or brick at least 2 in. thick

214. Overhead service conductors run above the top level of a window, shall:

    A. Be enclosed in a cable or raceway
    B. Have a maximum voltage of 150 volts-to-ground
    C. Have a clearance of at least 3 ft. from the window
    D. Be permitted to be less than 3 ft. from the window

215. Where service conductors pass over a flat roof, the vertical clearance required above the roof level shall be maintained for a distance of at least _____ ft. in all directions from the edge of the roof.

    A. 3   B. 4   C. 6   D. 8

216. Where a building with a flat roof is supplied by a 240 volt service drop passing 6 ft. horizontally above the roof overhang and terminated through the roof raceway, the service drop conductors shall have a clearance from the roof of at least _____ ft.

    A. 1 1/2   B. 3   C. 4   D. 8

217. Service drop conductors with a voltage of 240 volts-to-ground shall have a clearance of _____ ft. over a commercial driveway used by delivery trucks.

    A. 12   B. 15   C. 16   D. 18

218. The point of attachment of service drop conductors to a building shall be at least _____ ft. above finished grade.

    A. 10   B. 12   C. 16   D. 18

219. Service-entrance conductors for a one-family dwelling shall have a disconnecting means of _____ amp to disconnect the service equipment.

    A. 30   B. 60   C. 70   D. 100

220. For installations consisting of not more than two 2-wire branch-circuits, the service disconnecting means shall have a rating of not less than _____ amperes.

    A. 20   B. 30   C. 40   D. 50

Stallcup's Master Electrician's Study Guide

Section   **Answer**

_____ _____ 221. If the service disconnecting means is located inside the building, it shall be located _____ the point of entrance of the service conductors.

    A. At a point nearest   B. A maximum of 10 ft. from
    C. A maximum of 15 ft. from   D. A maximum of 25 ft. from

_____ _____ 222. The code requires the service disconnecting means to be installed at a(n):

    A. Accessible location outside the building
    B. Accessible location either outside the building or inside the nearest point of entrance of the service conductors
    C. Readily accessible location outside the building
    D. Readily accessible location either outside the building or inside the building nearest the point of entrance of the service conductors

_____ _____ 223. The service disconnecting means shall consist of not more than _____ switches or circuit breakers.

    A. 2   B. 4   C. 5   D. 6

_____ _____ 224. In a multiple occupancy building where electrical maintenance is not provided by the building management, the service disconnect for each occupancy shall be:

    A. Accessible to each occupant
    B. Located outside the building
    C. Located inside each occupancy
    D. Accessible only to building management personnel

_____ _____ 225. For installations consisting of not more than two two-wire branch-circuits, the rating of the service disconnecting means shall have a rating of at least _____ amperes.

    A. 20   B. 30   C. 60   D. 100

_____ _____ 226. Ground-fault protection of equipment shall be provided for solidly grounded wye electrical services of more than 150 volts-to-ground, but not exceeding _____ volts phase-to-phase.

    A. 240   B. 300   C. 480   D. 600

_____ _____ 227. On a 1600 amp, solidly grounded wye electrical service, the maximum setting of the ground-fault protection shall be _____ amperes.

    A. 800   B. 1000   C. 1200   D. 1600

_____ _____ 228. The disconnecting means for a circuit controlling the ground-fault protection of a service is not counted as one of the _____ disconnects.

    A. Six   B. Seven   C. Eight   D. None of the above

_____ _____ 229. For installations to supply only limited loads of a single branch-circuit, the service disconnecting means shall have a rating of not less than _____ amperes.

    A. 15   B. 20   C. 30   D. 40

_____ _____ 230. Service cables shall be supported by straps or other approved means within _____ in. of every service head.

    A. 6   B. 12   C. 18   D. 24

Article 240
Overcurrent Protection

231. OCPD's shall be permitted to be installed in clothes closets around combustible materials.

    A. True   B. False

232. All ungrounded conductors supplying a 240 volt branch-circuit may be opened at one time by single-pole CB's without using handle ties.

    A. True   B. False

233. Supplementary overcurrent protection devices shall not be required to be readily accessible.

    A. True   B. False

234. Overcurrent protection for conductors and equipment shall be provided to open the circuit if the current reaches a value that will cause an excessive or dangerous temperature in conductors or conductor insulation.

    A. True   B. False

235. Where a flexible cord is used as an extension cord, 16 AWG and larger conductors shall be protected by 30 ampere circuits.

    A. True   B. False

236. In grounded systems for single-phase circuits, individual single pole circuit breakers with identified handle ties shall be not be approved as the protection for each ungrounded conductor for line-to-line connected loads.

    A. True   B. False

237. Overcurrent devices shall be readily accessible. (General rule)

    A. True   B. False

238. A 3 amp fuse is a standard rating and can be used for motor control circuits.

    A. True   B. False

239. An 18 AWG fixture wire run 20 ft. that is connected to a 120 volt branch-circuit shall be protected by an _____ amp circuit. (General rule)

240. Overcurrent devices shall be located where they will not be exposed to _____ damage.

241. Plug fuses shall not be used in circuits exceeding _____ volts between conductors.

242. Where circuit breakers handles are operated vertically, the up position of the handle shall be the _____ position.

243. The general rule does not permit an OCPD to be placed in _____ with a grounded conductor.

244. An fixture wire run 75 ft. which is connected to a 120 volt branch-circuit shall be protected by an _____ amp circuit.

    A. 15   B. 20   C. 30   D. 40

12-23

Section	Answer

245. The requirements that feeder tap conductors be at least 6 AWG copper, contain no splices, and not penetrate walls, floors, or ceilings are some of the conditions for feeder taps:

    A. Not over 10 ft. long   B. Not over 25 ft. long
    C. Over 25 ft. long   D. Supplying a transformer

246. Tap conductors supplying a wall-mounted oven from a 50 ampere branch-circuit shall have an ampacity of at least _____ amperes.

    A. 20   B. 30   C. 40   D. 50

247. In a multifamily dwelling where electrical service and maintenance is not provided by the building manager, the branch-circuit overcurrent devices, where installed with proper clearance, would be permitted to be located in which of the following locations?

    A. A laundry room
    B. A clothes closet
    C. The bathroom of each dwelling
    D. In an electrical room kept locked where only the manager has the key

248. Disconnecting means shall be provided on the supply side of cartridge fuses in circuits of _____ where accessible to other than qualified persons.

    A. Over 50 volts-to-ground   B. Over 150 volts-to-ground
    C. Over 300 volts-to-ground   D. Any voltage

249. Where the interrupting capacity rating of _____ is other than 10,000 amperes, it shall be marked on the device.

    A. Plug fuses   B. Cartridge fuses
    C. Circuit breakers   D. Ground-fault sensors for equipment

250. Cartridge fuses rated at _____ volts, nominal or less, shall be permitted to be used for voltages at or below their ratings.

    A. 300   B. 600   C. 800   D. 1,000

Article 250
Grounding

251. AC systems of less than 50 volts shall be grounded where supplied by transformers that exceed 150 volts-to-ground.

    A. True   B. False

252. A neutral conductor shall be bonded to the generator frame where the generator is a component of a separately derived system.

    A. True   B. False

253. A grounding electrode shall be required for the grounded system in each building or structure when two or more buildings or structures are supplied from a common AC service.

    A. True   B. False

254. Any nonconductive paint, enamel, or similar coating shall be removed at threads, contact points, and contact surfaces.

    A. True   B. False

Test Questions

Section **Answer**

255. Aluminum wire can be installed when used an equipment bonding jumper.

    A. True   B. False

256. The interior metal water piping system shall be bonded to service equipment enclosure.

    A. True   B. False

257. A metal underground gas piping system can be used as a grounding electrode.

    A. True   B. False

258. The size of the grounding electrode conductor for a DC system shall not be smaller than 6 AWG copper where connected to a ground ring.

    A. True   B. False

259. The size of the grounding electrode conductor connected to a concrete-encased electrode for a DC system shall not be smaller than 4 AWG copper.

    A. True   B. False

260. The connection of a grounding electrode conductor to a grounding electrode is not required to be accessible where connected to building steel.

    A. True   B. False

261. Ground clamps shall be protected from physical damage.

    A. True   B. False

262. Where a submersible pump is used in a metal well casing, the well casing shall be _____ to the pump circuit equipment grounding conductor.

263. Lightning protection electrodes shall be _____ to the grounding electrode system of the service equipment.

264. The earth shall not be considered as an _____ ground-fault current path for grounded systems.

265. Class 1 control transformers rated at _____ VA or less shall not be required to have a separate grounding electrode conductor.

266. When rigid conduit is threaded _____ in threaded bosses on enclosures of electrical equipment, no further bonding is required.

267. Where only a _____ circuit is supplying another building with an equipment grounding conductor, a separate grounding electrode conductor is not required.

268. The grounding electrode conductor to a ground rod shall not be required to be larger than _____ AWG copper wire.

269. A ground rod using an grounding electrode that does not have a resistance of _____ ohms or less shall be augmented by one additional electrode.

270. Where multiple ground rods are installed, they shall not be less than _____ ft. apart.

271. Grounding electrode conductors run in the hollow spaces of a wall shall not be required to be enclosed in a raceway if installed in sizes smaller than _____ AWG.

12-25

Stallcup's Master Electrician's Study Guide

Section   Answer

_____ _____ 272. The connection of a grounding electrode conductor to a grounding electrode shall be made in a manner that will ensure a _____ and effective grounding path.

_____ _____ 273. Secondary circuits of current and potential instrument transformers shall be grounded where the primary windings are connected to circuits of _____ volts or more to ground.

_____ _____ 274. The grounding conductor for secondary circuits of instrument transformers and for instrument cases shall not be smaller than _____ AWG copper.

_____ _____ 275. _____ insulated office equipment does not require grounding with an equipment grounding conductor.

        A. Single   B. Partly   C. Double   D. None of the above

_____ _____ 276. A single structural member shall be considered a grounding electrode where _____ ft. or more is in direct contact with the earth.

        A. 5   B. 6   C. 10   D. 12

_____ _____ 277. The grounding electrode conductor shall be connected to the grounded service conductor:

    A. Within 5 ft. of the service equipment on the load side
    B. At any point on the load side of the service disconnecting means
    C. At a subpanel outside the building
    D. At any accessible point between the load end of the service drop or lateral to and including at the service disconnecting means

_____ _____ 278. Where a separately derived AC system provided by a transformer is required to be grounded, the connection of the grounding electrode conductor of the derived system shall be made:

    A. At the service panel
    B. At the overcurrent device on the load side
    C. At any point on the load side of the first disconnecting means
    D. At any point on the separately derived system from the transformer to the first disconnecting means or overcurrent device

_____ _____ 279. Where a separately derived AC system is located on the third floor of a shopping mall, the preferred or first choice for a grounding electrode is:

    A. A metal gas pipe
    B. A concrete-encased electrode
    C. A driven ground rod at the first floor of the building
    D. The nearest available effectively grounded structural member of the building

_____ _____ 280. Exposed metal parts of fastened-in-place equipment shall be grounded where located _____ from grounded metal objects and subject to contact by persons.

    A. 7 1/2 ft. vertically and 5 ft. horizontally
    B. 7 1/2 ft. vertically and 3 ft. horizontally
    C. 8 ft. vertically and 3 ft. horizontally
    D. 8 ft. vertically and 5 ft. horizontally

_____ _____ 281. The code permits frames of ranges and dryers in dwelling units to be grounded to the grounded conductor if the branch-circuits are _____.

    A. New   B. Existing   C. Commercial   D. Residential

	Section	Answer

282. An equipment bonding jumper installed outside of a raceway shall be permitted to be a maximum _____ ft. in length.

    A. 3   B. 4   C. 6   D. 8

283. The metal water piping system in the area served by a separately derived system is not effectively grounded, therefore, a ground rod is used as the grounding electrode for this system. The metal water piping in the area shall be:

    A. Bonded to a separate ground rod
    B. Bonded to the grounded conductor of the separately derived system
    C. Used as a supplemental grounding electrode for the separately derived system
    D. Isolated from any contact with the electrical system, since it is not effectively grounded

284. The connection of the grounding electrode conductor to a metal underground water pipe used as a grounding electrode shall be at any point:

    A. Outside the building
    B. On the water piping system within the building
    C. Within 5 ft. from where the water pipe enters the building
    D. Between the water pipe entrance to the building and the first insulating joint or device such as a water meter

285. Which one of the grounding electrodes listed below is required to be supplemented by an additional electrode.

    A. Ground rod   B. Ground ring
    C. Concrete-encased electrode   D. Metal underground water pipe

286. All of the following statements are true for a concrete-encased electrode, except:

    A. It may be used as the sole grounding electrode
    B. It may serve as the supplemental electrode to a metal underground waterpipe electrode
    C. If it is available on the premises, it is required to be made part of the grounding electrode system
    D. Where the concrete-encased electrode is a copper conductor, it shall be sized in accordance with **250.94**.

287. A metal underground water pipe may serve as a grounding electrode if it is in direct contact with the earth for at least _____ ft. or more.

    A. 6   B. 8   C. 10   D. 12

288. In an industrial building where the conditions of maintenance and supervision ensure that only qualified persons will service the installation:

    A. The resistance of a single-made electrode may exceed 25 ohms
    B. A metal underground water pipe shall be permitted to serve as the sole grounding electrode
    C. A metal underground gas piping system shall be permitted to serve as a grounding electrode
    D. The supplemental electrode shall be permitted to be bonded to a totally exposed interior metal water piping system at any convenient point.

Stallcup's Master Electrician's Study Guide

Section   Answer

289. For a bare copper conductor installed in a concrete footing as a concrete-encased grounding electrode, which of the following is not a correct code requirement? The bare copper conductor shall be:

A. At least size 2 AWG
B. At least 20 ft. in length
C. Located near the bottom of the footing
D. Encased by at least 2 in. of concrete

290. Where two made electrodes are installed, each of a different system, they shall be spaced at least _____ ft. apart.

A. 3   B. 6   C. 8   D. 10

291. Where a ground rod cannot be driven to the required depth, it shall be driven at an angle or buried in a trench at least _____ ft. deep.

A. 2   B. 2 1/2   C. 3   D. 3 1/2

292. A ground shall be installed so that at least _____ ft. of the length of the rod is in contact with the soil.

A. 6   B. 7 1/2   C. 8   D. 10

293. A single ground rod that has a resistance to ground of 50 ohms has:

A. Acceptable resistance as is
B. Too low a resistance and shall be replaced by a longer rod
C. Too high a resistance and shall be replaced by a rod with a larger diameter
D. Too high a resistance and shall be augmented by one additional grounding electrode

294. The minimum size grounding electrode conductor that shall be permitted to be run along the surface of the building construction without protection is _____ AWG.

A. 8   B. 6   C. 4   D. 2

295. Which of the following wiring methods is not permitted by the code to be used as an equipment grounding conductor?

A. Rigid metal conduit   B. Electrical metallic tubing
C. Armor of Type AC cable   D. Flexible metal conduit over 6 ft. in length

296. Bare aluminum grounding conductors that are used outside shall not be installed within _____ in. of the earth.

A. 6   B. 12   C. 18   D. 24

297. The size of the grounding electrode conductor connected to a concrete-encased electrode for a grounded AC system shall not be smaller than _____ AWG copper.

A. 6   B. 4   C. 2   D. 1

298. The minimum size copper grounding electrode conductor required for a 200 ampere service consisting of 3/0 THW copper service-entrance conductors is _____ AWG.

A. 8   B. 6   C. 4   D. 2

Test Questions

Section   Answer

299. A grounding electrode conductor that is the sole connection to a plate electrode shall not be required to be larger than _____ AWG.

    A. 8   B. 6   C. 4   D. 2

300. Which of the following are not allowed for the connection of grounding conductors and bonding jumpers?

    A. Exothermic welding   B. Listed pressure connectors
    C. Listed clamps   D. Sheet-metal screws

301. The terminal for the connection of the equipment grounding conductor shall be identified by _____.

    A. A green colored, not readily removable terminal screw with a hexagonal head
    B. A green colored, hexagonal, not readily removable terminal nut
    C. A green colored pressure wire connector
    D. All of the above

302. When bonding circuits over _____ volts-to-ground, two locknuts and a bushing attached to a box with cleanly punched holes will comply.

    A. 125   B. 250   C. 300   D. 480

303. When equipment bonding jumpers are installed on the outside of the raceway or enclosure, the bonding jumper shall not exceed _____ ft.

    A. 6   B. 8   C. 10   D. 10

304. The grounded conductor (existing branch-circuits) used to ground the frames of ranges shall not be smaller than _____ AWG copper.

    A. 6   B. 8   C. 10   D. 12

305. Separation from lightning protection conductors through the air is _____ ft..

    A. 5   B. 6   C. 8   D. 10

306. If two or more buildings are supplied by a grounded system from one main service, each building shall have separate _____.

    A. Phases   B. Neutrals
    C. Grounding electrode systems   D. Service drops

307. Equipment grounding conductors larger than _____ AWG shall be green, green with yellow stripes, or stripped bare.

    A. 6   B. 8   C. 10   D. 12

Article 300
Wiring methods

308. Aluminum conductors carrying over 50 amps shall be grouped in AC circuits to prevent inductive heating of metal enclosures and raceways.

    A. True   B. False

Stallcup's Master Electrician's Study Guide

Section   **Answer**

_____ _____ 309. Minimum burial depth for underground conduits is measured from the top of the conduit to the finished grade.

        A. True   B. False

_____ _____ 310. A single enclosure or raceway may be used for AC-DC system of 600 volts or less.

        A. True   B. False

_____ _____ 311. Phase conductors for AC lighting and power circuits may be run in separate metal raceway systems.

        A. True   B. False

_____ _____ 312. Grounding electrode conductors shall be grouped with phase conductors when installed in a raceway.

        A. True   B. False

_____ _____ 313. It is permissible to run 120 volt lighting conductors in the same raceway with 2300 volt power conductors when all conductors are insulated for the higher voltage.

        A. True   B. False

_____ _____ 314. Direct-buried conductors or cables shall be permitted to be spliced without the use of splice boxes.

        A. True   B. False

_____ _____ 315. Conductors of circuits rated over 600 volts can occupy the same raceway with conductors of circuits rated 600 volts or less.

        A. True   B. False

_____ _____ 316. Where a raceway is exposed to physical damage, Schedule 80 rigid nonmetallic conduit shall be permitted to be used.

        A. True   B. False

_____ _____ 317. Cable systems that are not run through the center of the framing members can be placed in a cut notch in the framing member and protected by a steel plate of _____ in.

_____ _____ 318. Raceways shall be provided with _____ joints where necessary to compensate for thermal expansion and contraction.

_____ _____ 319. No wiring systems of any type shall be installed in _____ used to transport dust, loose stock, or flammable vapors.

_____ _____ 320. For conductors rated 600 volt or less, conductors of different voltage ratings are permitted to occupy the same raceway where all conductors in the raceway:

        A. Are power conductors, with no control wires included
        B. Have an insulation rating of at least 600 volts
        C. Have an insulation rating equal to that of the lowest rated conductor in the raceway
        D. Have an insulation rating equal to at least the maximum circuit voltage applied to any conductor in the raceway

Test Questions

Section   Answer

321. In exposed locations where a cable is run through bored holes in studs, the edge of the hole shall be at least _____ in. from the nearest edge of the stud.

    A. 1   B. 1 1/8   C. 1 1/4   D. 1 1/5

322. For raceways installed in shallow groves, where nails or screws or likely to penetrate, physical protection such as steel plates is required for all of the following raceways except:

    A. Flexible metal conduit   B. Electrical metallic tubing
    C. Electrical nonmetallic tubing   D. Liquidtight flexible metal conduit

323. A fitting such as an insulating bushing or insulating material shall be provided to protect conductors where raceways containing ungrounded conductors of at least _____ AWG enter a cabinet.

    A. 8   B. 6   C. 4   D. 2

324. A 120/240 volt, 40 ampere branch-circuit from a house to an outbuilding is direct buried underground with Type UF cable. The cable shall be at least _____ in.

    A. 6   B. 12   C. 18   D. 24

325. A direct burial cable that is run underground and emerges on the outside wall of a building and then enters the building, shall be protected by an enclosure or raceway to:

    A. At least 3 ft. above grade
    B. At least 6 ft. above grade
    C. At least 8 ft. above grade
    D. The point where it enters the building

326. A direct burial UF cable for a 24 volt circuit is installed outside of an office building for landscape lighting. This cable shall be buried at least _____ in. deep.

    A. 6   B. 12   C. 18   D. 24

327. A rigid nonmetallic conduit is buried under a single family dwelling and contains a 120 volt, 30 ampere circuit for an irrigation pump for the yard of the dwelling. This conduit shall be buried at least _____ in. deep.

    A. 6   B. 12   C. 18   D. 24

328. Service lateral cables that are direct buried under a driveway shall be buried at least _____ in. deep.

    A. 12   B. 18   C. 24   D. 36

329. A direct buried service lateral run down a utility pole and then underground to emerge at a building shall be protected at the pole by an enclosure or raceway to a point at least _____ ft. above finished grade.

    A. 8   B. 10   C. 12   D. 15

330. Where direct buried conductors emerge from the ground and are subject to physical damage, which of the following conduits provide the protection required by the code?

    A. Intermediate metallic conduit
    B. Electrical metallic tubing
    C. Electrical nonmetallic tubing
    D. Schedule 40 rigid nonmetallic conduit

Section	Answer

331. Rigid metallic conduit shall be buried in the ground at a minimum of _____ in.

    A. 4   B. 6   C. 12   D. 18

332. In indoor wet locations, which of the following types of raceways shall not be required to be mounted so that there is at least 1/4 in. space between it and the mounting surface?

    A. Armored cables   B. Rigid metal conduits
    C. Flexible metal conduits   D. Rigid nonmetallic conduits

333. A cabinet installed in a wet location shall be mounted so there is an airspace of at least _____ in. between the cabinet and the supporting surface.

    A. 1/8   B. 1/4   C. 1/2   D. 3/4

334. At each lighting outlet having a dimension of 8 in. or more shall have at least _____ in. of free conductor left for the connection of luminaries (light fixtures).

    A. 4   B. 6   C. 8   D. 10

335. Where installed in vertical raceways, copper conductors of size 250 KCMIL shall be supported every _____ ft.

    A. 60   B. 80   C. 135   D. 180

336. Which of the following wiring methods shall be permitted to be installed in a plenum used for environmental air? (General rule)

    A. Type AC cable   B. Type MI cable
    C. Rigid nonmetallic conduit   D. Type MC cable with a nonmetallic cover

337. Where installed in an environmental air duct, liquidtight flexible metal conduit shall be limited to a maximum length of _____ ft.

    A. 3   B. 4   C. 6   D. 8

338. Conductors over 600 volts shall not be bent to a radius less than _____ times the overall diameter for nonshielded conductors.

    A. 4   B. 6   C. 8   D. 10

Article 590
Temporary wiring

339. For wiring over _____ volts, suitable fencing, barriers or other effective means shall be provided.

340. _____ wire branch-circuits shall not be laid on the floor or ground at construction sites.

341. All lamps used for temporary wiring at construction sites shall be protected by a(n) _____ over the lampholder.

342. Temporary electrical power that may be of a class less than would be required for a permanent installation shall be permitted for a maximum of _____ days for Christmas lighting.

    A. 30   B. 60   C. 90   D. 120

Test Questions

Section Answer

343. Temporary electrical power and lighting installations shall be permitted during emergencies for which of the following?

   A. Tests
   B. Experiments
   C. Developmental work
   D. All of the above

344. For temporary wiring on a construction site where the voltage does not exceed 150 volts-to-ground, which of the following is not a code violation?

   A. Feeders installed using nonmetallic sheathed cable
   B. Receptacles installed on branch-circuits that supply temporary lighting
   C. Branch-circuits run as open conductors to provide lighting throughout the period of construction
   D. Receptacles connected to the same ungrounded conductor of a multiwire circuit that supplies temporary lighting

345. Where a multiconductor cord or cable is used, a box shall not be required for splices for temporary wiring:

   A. On construction sites
   B. Indoors or in dry locations only
   C. For Christmas decorative lighting or displays
   D. Only in industrial establishments where conditions of maintenance ensure that qualified personnel are involved

346. Where a store is being remodeled, temporary power for hand tools used by personnel during construction is supplied from an existing receptacle outlet not affected by the remodeling. This receptacle:

   A. Shall be removed and the circuit deenergized during construction
   B. Is required to be GFCI-protected during the construction period
   C. Is permitted to be used for temporary power without GFCI-protection during construction if located indoors
   D. Is permitted to be used for temporary power as is, without GFCI protection since it is part of the permanent wiring

Article 310
Conductors for general wiring

347. The largest size solid conductor permitted to be installed in raceways is _____ AWG.

   A. 12   B. 10   C. 8   D. 6

348. The minimum size feeder conductors permitted to be installed in parallel is _____ AWG.

   A. 2   B. 1   C. 1/0   D. 2/0

349. Where conductors are run in parallel, several conditions are required. Which one of the following conditions is not one of those requirements? Paralleled conductors shall:

   A. Be run in metal raceways only
   B. Have the same insulation type
   C. Have the same conductor material
   D. Be the same size in circular mil area

350. Conductors in sizes smaller than _____ AWG shall be permitted to be run in parallel to supply control power.

   A. 8   B. 6   C. 4   D. 1/0

12-33

Stallcup's Master Electrician's Study Guide

Section   **Answer**

351. In an existing installation, where nonlinear loads are causing an increased load on neutral conductors, under an engineer's supervision, it shall be permitted to run with the neutral conductors in parallel for sizes as small as _____ AWG.

　　A. 8　　　B. 6　　　C. 4　　　D. 2

352. Which of the following conductors is permitted to be used in wet locations?

　　A. RH　　　B. XHH　　　C. THHN　　　D. THWN

353. Multiconductor flat cable _____ AWG or larger shall be permitted to employ an external ridge on the grounded conductor.

　　A. 8　　　B. 6　　　C. 4　　　D. 1/0

354. A black conductor shall be permitted to be used as an equipment grounding conductor where permanently and properly identified, if it is at least size _____ AWG.

　　A. 8　　　B. 6　　　C. 4　　　D. 2

355. The ampacity of a conductor is the current in amperes the conductor can carry continuously under the conditions of use without exceeding its _____ rating.

　　A. Voltage　　　B. Impedance　　　C. Temperature　　　D. Circuit interrupting

356. An derating factor of _____ percent shall be applied for THHN copper conductors running through an ambient temperature of 120°F.

　　A. 71　　　B. 76　　　C. 82　　　D. 87

357. Where multiconductor cables not installed in raceways are bundled longer than _____ in. without maintaining spacing, the allowable ampacity of each conductor shall be derated.

　　A. 18　　　B. 24　　　C. 36　　　D. 48

358. A derating factor of _____ percent shall be applied for 4 current-carrying conductors pulled through a raceway.

　　A. 45　　　B. 50　　　C. 70　　　D. 80

359. A derating factor of _____ percent shall be applied for 20 current-carrying conductors pulled through a raceway.

　　A. 45　　　B. 50　　　C. 70　　　D. 80

360. A derating factor shall not apply to conductors in nipples having a length not exceeding _____ in.

　　A. 6　　　B. 12　　　C. 18　　　D. 24

361. Where burial depths are deeper than shown in a specific underground ampacity Table, an ampacity derating factor of _____ percent per increased ft. of depth, beyond 25 percent for all values of RHO, can be utilized.

　　A. 2　　　B. 4　　　C. 6　　　D. 8

362. Where a 4 conductor UF cable leaves an outdoor trench and is extended up a pole in rigid metal conduit, ampacity derating shall not be required if the conduit has a maximum length of _____ ft.

    A. 6   B. 8   C. 10   D. 12

Article 392
Cable trays

363. For multiconductor cables installed in cable trays, the sum of the diameter of the cables shall not exceed the width of the cable tray for conductors _____ AWG and larger.

364. The maximum fill area for multiconductor cables in a 24 in. cable tray is _____ sq. in.

365. The maximum fill area for an 18 in. ladder type cable tray holding single conductor cables is _____ sq. in.

366. Cable tray systems shall be permitted to have mechanically _____ segments between cable tray runs.

367. Supports shall be provided to prevent _____ on cables where they enter raceways.

368. Steel cable tray systems shall be permitted to be used as _____ grounding conductors provided certain requirements are complied with.

369. In industrial establishments where only qualified persons will service the installation, single conductors shall be permitted to be installed in cable trays but shall be at least size _____ AWG.

370. Cable trays shall have side rails or equivalent _____ members.

371. In other than horizontal runs, the cables shall be fastened securely to _____ members of the cable trays.

372. For raceways terminating at the tray, a _____ cable tray clamp or adapter shall be used to securely fasten the raceway to the cable tray system.

373. Where any of the single conductors installed in ladder or ventilated through cable trays are 1/0 through 4/0 AWG, all single conductors shall be installed in a _____ layer.

Article 398
Open wiring on insulators

374. Open wiring on insulators shall be installed within _____ in. from a tap or splice.

    A. 3   B. 6   C. 12   D. 18

375. Open wiring on insulators shall be installed within _____ in. of a dead-end connection to a lampholder or receptacle.

    A. 6   B. 12   C. 18   D. 24

376. Open wiring on insulators shall be installed at intervals not exceeding _____ ft.

    A. 3   B. 3 1/2   C. 4 1/2   D. 6

377. Open conductors shall be separated at least _____ in. from metal raceways, piping, or other conducting material.

    A. 1   B. 2   C. 3   D. 6

378. Open conductors within _____ ft. from the floor shall be considered exposed to physical damage.

    A. 3   B. 6   C. 7   D. 10

Article 396
Messenger Supported Wiring

379. Messenger supported wiring is permitted only in _____ establishments where maintenance personnel service the system.

    A. Industrial   B. Commercial   C. Both A and B   D. Neither A or B

380. The ampacity of conductors supported by a messenger wiring system is determined per _____ of the NEC.

    A. 310.12   B. 310.13   C. 310.14   D. 310.15

381. Spices and _____ in messenger wire shall be made by approved methods.

    A. Devices   B. Taps   C. Both A and B   D. Neither A or B

Article 394
Concealed knob-and-tube wiring

382. Concealed knob-and-tube wiring can be installed, with special permission, in hollow spaces of walls and ceilings of dwelling units.

    A. True   B. False

383. Knob-and-tube wiring installed in inaccessible attics shall be run through bored holes in the joists, studs, or rafters, or run along the sides.

    A. True   B. False

384. Conductors of knob-and-tube wiring can be tied to solid knobs with any type of insulated tie.

    A. True   B. False

385. Concealed knob-and-tube wiring shall have a clearance of not less than _____ in. between conductors.

    A. 1   B. 2   C. 3   D. 6

Test Questions

Section   Answer

386. Concealed knob-and-tube wiring shall have a clearance of not less than _____ in. between the conductor and the surface over which it passes.

    A. 1   B. 2   C. 3   D. 6

387. Concealed knob-and-tube wiring passing through wood cross members shall be protected by noncombustible, nonabsorbent, insulating tubes extending not less than _____ in. beyond the wood member.

    A. 1   B. 3   C. 6   D. 12

388. Concealed knob-and-tube wiring run through bored holes in joists shall be above the floor at a height of not less than _____ ft.

    A. 6   B. 7   C. 8   D. 10

Article 326
Integrated gas spacer cable

389. A 250 KCMIL, integrated gas spacer cable has an ampacity of 168 amps.

    A. True   B. False

390. The maximum size integrated gas spacer cable is 4750 KCMIL.

    A. True   B. False

391. Integrated gas spacer cable can be used as service-entrance conductors when routed underground.

    A. True   B. False

Article 328
Medium-voltage cable

392. Medium-voltage cable is available in 2000 volts or less.

    A. True   B. False

393. The ampacity of medium-voltage cable is determined per **310.60** (general rule).

    A. True   B. False

394. Medium-voltage cable exposed to sunlight shall be identified for such use.

    A. True   B. False

Article 324
Flat conductor cable

395. A flat conductor cable shall not exceed _____ volts between ungrounded conductors.

396. General-purpose branch-circuits are limited to _____ amps or less for flat conductor cables.

397. Floor-mounted type FCC cable, cable connectors and insulating ends shall be covered with carpet squares not larger than _____ square.

Article 332
Mineral-insulated, metal-sheathed cable

398. Type MI cable shall be installed and supported so that the nearest outside surface of the cable or raceway is not less than _____ in. from the nearest edge of the framing member.

    A. 1 1/8   B. 1 1/4   C. 1 1/2   D. 1 3/4

399. Where Type MI cable is installed in notches of wood, the cable shall be protected by a steel plate at least _____ in. thickness.

    A. 1/16   B. 1/8   C. 1/4   D. 1/2

400. Type MI cable shall be securely supported at intervals not exceeding _____ ft.

    A. 3   B. 4   C. 6   D. 10

401. The radius of the inner edge of any bend for Type MI cable shall not be less than _____ times the diameter of the metallic sheath for cable not more than 3/4 in. in external diameter.

    A. 3   B. 5   C. 8   D. 10

Article 362
Electrical Nonmetallic Tubing

402. Which of the following raceways is defined by the code as a pliable raceway or one that can be bent by hand with a reasonable force, but without other assistance.

    A. Flexible metal conduit   B. Rigid nonmetallic conduit
    C. Electrical metallic conduit   D. Electrical nonmetallic conduit

403. Electrical nonmetallic tubing shall be permitted in a four story building concealed in walls that have a thermal barrier of material that has at least a _____ minute finish rating.

    A. 15   B. 30   C. 45   D. 60

404. The maximum size electrical nonmetallic tubing allowed to be used is _____ in.

    A. 1   B. 2   C. 3   D. 4

405. Electrical nonmetallic tubing shall not be more than _____ degrees of bends between pull points, conduit bodies, and boxes.

    A. 120   B. 250   C. 360   D. 420

406. Electrical nonmetallic tubing shall be securely fastened-in-place within _____ ft. of every outlet box.

    A. 1   B. 2   C. 3   D. 5

Test Questions

Section   Answer

Article 320
Armored cable

407. Which of the following wiring methods requires an insulating bushing to be installed that is visible for inspection where connected to a box or cabinet?

    A. AC cable
    B. MC cable
    C. Flexible metal conduit
    D. Electrical metallic tubing

408. Type AC cable shall be secured at intervals not exceeding _____ ft.

    A. 3   B. 4 1/2   C. 6   D. 10

409. Exposed runs of AC cable shall closely follow the building surface except for a maximum length of _____ ft. to serve luminaires (lighting fixtures) in accessible ceilings.

    A. 3   B. 4   C. 6   D. 8

410. Where AC cable is run across the face of rafters in an accessible attic that is not accessible by a permanent ladder or stairs, the cable shall be protected within _____ ft. of the nearest edge of the attic entrance.

    A. 1 1/2   B. 3   C. 4 1/4   D. 6

411. Type AC cable installed in thermal insulation shall have conductors rated at _____ °C.

    A. 60   B. 75   C. 80   D. 90

Article 330
Metal-clad cable

412. Type MC cable shall be secured at maximum intervals not exceeding _____ ft.

    A. 3   B. 4 1/2   C. 5   D. 6

413. MC cable shall be supported within 12 in. of boxes where it contains a maximum of _____ conductors sized a maximum of _____ AWG.

    A. 3, 12   B. 3, 10   C. 4, 12   D. 4, 10

414. Type MC cable shall be installed and supported so that the nearest outside surface of the cable or raceway is not less than _____ in. from the nearest edge of the framing member.

    A. 1 1/8   B. 1 1/4   C. 1 1/2   D. 1 3/4

415. Where Type MC cable is installed in notches of wood, the cable shall be protected by a steel plate at least _____ in. thickness.

    A. 1/16   B. 1/8   C. 1/4   D. 1/2

416. Where MC cable is run across the face of rafters in an accessible attic that is not accessible by a permanent ladder or stairs, the cable shall be protected within _____ ft. of the nearest edge of the attic entrance.

    A. 1 1/2   B. 3   C. 4 1/4   D. 6

Section   Answer

Article 334
Nonmetallic sheathed cable

417. Where NM cable is installed exposed and passes through a floor, the cable shall be protected by a conduit extending at least _____ in. above the floor.

   A. 6   B. 12   C. 18   D. 24

418. Where NM cable is run at angles with floor joists in an unfinished basement, the minimum size three conductor cable permitted to be secured to the lower edge of the joists is _____ AWG.

   A. 10   B. 8   C. 6   D. 4

419. Where NM cable is run across the face of the rafters in an attic accessible by a permanent pull-down ladder, the cable shall be protected by guard strips within _____ ft. of the attic entrance.

   A. 3   B. 6   C. 7   D. 8

420. Type NM cable shall be installed and supported so that the nearest outside surface of the cable or raceway is not less than _____ in. from the nearest edge of the framing member.

   A. 1 1/8   B. 1 1/4   C. 1 1/2   D. 1 3/4

421. Where Type NM cable is installed in notches of wood, the cable shall be protected by a steel plate at least _____ in. thickness.

   A. 1/16   B. 1/8   C. 1/4   D. 1/2

422. Type NM cable shall be secured in place at intervals not exceeding _____ ft.

   A. 3   B. 4 1/2   C. 6   D. 10

423. Where Type NM cable is run into a panelboard, it shall be secured in place within at least _____ in. of the panelboard.

   A. 6   B. 12   C. 36   D. 54

424. The insulation of the conductors of Type NM cable shall be rated at _____ degrees C, and the allowable ampacity of NM cable shall be that of _____ degrees C conductors.

   A. 60, 90   B. 75, 75   C. 90, 75   D. 90, 60

Article 338
Service-entrance and underground feeder cable

425. Which of the following is not a correct code provision for a Type SE service-entrance cable that consists of only three conductors? (General rule)

   A. All three conductors shall be insulated where the cable is used inside a building
   B. One conductor shall be permitted to be bare only if the cable is used to supply new installations of ranges or clothes dryers and originates in the service equipment
   C. One conductor shall be permitted to be bare where the cable is used as a feeder to supply only other buildings on the same premises
   D. If the cable is used inside a building, one of the three conductors shall be permitted to be bare only if it is used as an equipment grounding conductor

	Test Questions	
	Section	Answer

426. Multiconductor Type UF cable shall be permitted for all of the following uses, except:

    A. In corrosive locations   B. Installed in cable trays
    C. In interior wet locations   D. As service-entrance cables

427. The ampacity of UF cable shall be that of _____ °C conductors.

    A. 60   B. 75   C. 85   D. 90

Article 340
Underground feeder and branch-circuit cable

428. Underground feeder and branch-circuit cable (UF) is available in sizes 14 AWG copper through 250 KCMIL copper.

    A. True   B. False

429. Underground feeder cables shall not be used when exposed to sunlight unless identified for such use.

    A. True   B. False

430. Underground feeder cable can be used for service-entrance.

    A. True   B. False

Article 336
Power and control tray cable

431. Power and control tray cable (TC) can be used for direct burial in the earth when identified for such use.

    A. True   B. False

432. The ampacity of conductors in TC cable is selected per **402.5** for sizes smaller than 14 AWG.

    A. True   B. False

433. The maximum size conductor permitted in TC cable is _____ KCMIL copper.

Article 382
Nonmetallic Extensions

434. Splices or taps shall not be permitted to be made in _____ lengths of nonmetallic extensions.

435. Nonmetallic surface extensions shall be secured at intervals not exceeding _____ in.

436. One or more extensions shall be permitted to be run in any direction from an existing outlet, but not on the floor or within _____ in. from the floor.

12-41

Section   Answer

Article 342
Intermediate metal conduit

437. The maximum size intermediate metal conduit shall not be larger than _____ in.

    A. 2   B. 3   C. 4   D. 6

438. Where intermediate metal conduit is threaded in the field, a cutting die with a _____ in. taper per foot shall be used.

    A. 3/8   B. 1/2   C. 3/4   D. 1

439. Intermediate metal conduit shall be supported at least every _____ ft. (General rule)

    A. 3   B. 5   C. 10   D. 20

440. Intermediate metal conduit shall be securely fastened within _____ ft. of each outlet box.

    A. 1   B. 3   C. 5   D. 6

441. Where intermediate metal conduit is installed as a vertical riser from industrial machinery and made up with threaded couplings, supports shall be permitted to be a maximum of every _____ ft. if supported at the top and bottom of the riser.

    A. 5   B. 10   C. 15   D. 20

Article 344
Rigid metal conduit

442. The minimum size rigid metal conduit shall not be smaller than _____ in.

443. Where rigid metal conduit is threaded in the field, a standard conduit cutting die with a _____ in. taper per ft. shall be used.

444. Rigid metal conduit (1 1/2 in.) shall be supported at least every _____ ft.

445. The maximum size rigid metal conduit shall not be larger than _____ in.

    A. 2   B. 3   C. 4   D. 6

446. Rigid metal conduit (3/4 in.) shall be supported at least every _____ ft.

    A. 3   B. 5   C. 6   D. 10

447. Rigid metal conduit shall be securely fastened within _____ ft. of each outlet box.

    A. 1   B. 3   C. 6   D. 10

448. Where approved, rigid metal conduit installed as a through-the-roof mast shall not be required to be securely fastened within _____ ft. of the service head.

    A. 1 1/2 B. 3   C. 4 1/2 D. 5

Article 352
Rigid nonmetallic conduit

449. Rigid nonmetallic conduit shall not be permitted:

    A. In cinder fill
    B. For the support of luminaires (fixtures)
    C. To be concealed in floors or ceilings
    D. In locations subject to severe corrosive influences

450. Rigid nonmetallic conduit shall be securely fastened within _____ ft. of each box.

    A. 3   B. 5   C. 6   D. 7

451. Where 1 1/4 in. rigid nonmetallic conduit is run horizontally along a wall, it shall be supported at least every _____ ft.

    A. 3   B. 5   C. 6   D. 7

452. Expansion fittings shall be installed on runs of rigid nonmetallic conduit where the length change due to thermal expansion and contraction is expected to be at least _____ in.

    A. 1/4   B. 1/2   C. 3/4   D. 1

Article 358
Electrical metallic tubing

453. The maximum size electrical metallic tubing shall not be larger than _____ in.

    A. 3   B. 4   C. 5   D. 6

454. Electrical metallic tubing shall not have more than _____ degrees bends between pull points, conduit bodies, and boxes.

    A. 120   B. 250   C. 360   D. 420

455. Electrical metallic tubing shall be securely fastened in place at least every _____ ft.

    A. 3   B. 5   C. 6   D. 10

456. Electrical metallic tubing shall be securely fastened in place within _____ ft. of each outlet box.

    A. 3   B. 5   C. 6   D. 10

457. Electrical metallic tubing installed where structural members are spaced greater than 3 ft. apart shall be permitted to be supported a maximum of _____ ft. from boxes and cabinets if unbroken lengths are used.

    A. 4   B. 4 1/2   C. 5   D. 6

Article 360
Flexible metallic tubing

458. The maximum number of 12 AWG THWN conductors permitted in 3/8 in. flexible metallic tubing with outside fittings is two.

    A. True   B. False

Section   Answer

459. The maximum size flexible metallic tubing is 1 in.

   A. True   B. False

460. Flexible metallic tubing cannot be used in hoistways.

   A. True   B. False

Article 348
Flexible metal conduit

461. The maximum size flexible metal conduit shall not be larger than _____ in.
   A. 1   B. 2   C. 3   D. 4

462. Listed flexible metal conduit shall be permitted to be installed as a grounding means if the total length in any ground return path is _____ ft. or less.

   A. 3   B. 5   C. 6   D. 10

463. Flexible metal conduit shall be securely fastened in place by an approved means within _____ in. of each box.

   A. 6   B. 12   C. 18   D. 24

464. Flexible metal conduit shall be supported and secured at intervals not exceeding _____ ft.

   A. 4   B. 4 1/2   C. 5   D. 6

465. Flexible metal conduit (1/2 in. through 1 1/4 in.) shall not be required to be supported for lengths not exceeding _____ ft. at terminals where flexibility is required.

   A. 3   B. 4   C. 5   D. 6

Article 350
Liquidtight flexible metal conduit and liquidtight flexible nonmetallic conduit

466. The minimum size liquidtight flexible metal conduit shall not be smaller than _____ in. (General rule)

   A. 1/2   B. 3/4   C. 1   D. 1 1/4

467. Liquidtight flexible metal conduit shall be securely fastened in place by an approved means within _____ in. of each box.

   A. 6   B. 12   C. 18   D. 24

468. A 6 ft. run of 1 1/4 inch liquidtight flexible metal conduit shall be permitted to be used as a grounding means where the fittings are listed for grounding, if the branch-circuit is rated a maximum of _____ amperes.

   A. 20   B. 40   C. 50   D. 60

469. Liquidtight flexible nonmetallic conduit shall be securely fastened at intervals not greater than _____ ft.

   A. 3   B. 4 1/2   C. 5   D. 6

	Test Questions
Section	Answer

470. Liquidtight flexible nonmetallic conduit shall not be required to be "securely fastened" where lengths not exceeding _____ ft. at terminals where flexibility is necessary.

    A. 12   B. 24   C. 36   D. 48

Article 384, 386 and 388
Strut-type channel raceway, surface metal raceways and surface nonmetallic raceways

471. For conductors installed in surface metal raceways, ampacity derating factors do not apply where certain conditions are met. Which one of the following is not one of those conditions?
    A. The maximum number of conductors does not exceed 30
    B. The maximum cross-sectional area of the raceway is 4 sq. in.
    C. The surface metal raceway shall not pass transversely through walls, partitions or floors
    D. The sum of the conductor cross-sectional areas does not exceed 20 percent of the cross-sectional area of the raceway

472. Splices and taps shall not fill a surface metal raceway to more than _____ percent of its area at that point.

    A. 50   B. 65   C. 75   D. 80

473. An _____ grounding conductor shall be connected to surface metal raceway enclosures providing a transition from other wiring methods.

    A. Equipment   B. Grounded   C. Grounding electrode   D. Grounding

474. Surface mount strut-type channel raceway shall be secured to the mounting surface at intervals not exceeding _____ ft.

    A. 3   B. 5   C. 6   D. 10

475. Strut-type channel raceways shall be permitted to be suspension mounted in air at intervals not to exceed _____ ft.

    A. 5   B. 10   C. 15   D. 20

476. Each length of strut-type channel raceways shall be clearly and durably _____.

477. Surface nonmetallic raceway and associated fittings shall be _____.

Article 390
Underfloor raceways

478. Underfloor raceways shall have dead ends _____.

479. Underfloor raceways shall have a suitable _____ near or at each end to locate the last insertion.

12-45

Section	Answer

Article 376
Metal wireways and nonmetallic wireways

480. Wireways shall not contain more than _____ current-carrying conductors without derating the ampacities of the conductors.

    A. 20   B. 25   C. 30   D. 40

481. Conductors installed in wireways shall not contain more than _____ percent of the interior cross-sectional area.

    A. 10   B. 20   C. 30   D. 40

482. Conductors with splices shall not fill the wireway to more than _____ percent of its area to that point.

    A. 20   B. 40   C. 50   D. 75

Article 368
Busways

483. Lighting busway shall not be installed less than _____ ft. above the floor.

    A. 7   B. 8   C. 10   D. 12

484. Busways shall be supported at maximum intervals of _____ ft.

    A. 3   B. 5   C. 6   D. 10

485. Busways shall be permitted to pass vertically through floors if they are totally enclosed where passing through the floor and for a distance of at least _____ ft. above the floor.

    A. 3   B. 5   C. 6   D. 8

486. A cable assembly shall be permitted to be used as a branch from a busway where the length of the cable from the busway plug-in device to the tension take-up support device is a maximum of _____ ft.

    A. 1 1/2   B. 3   C. 5   D. 6

487. In industrial establishments, overcurrent protection shall not be required at the point where a busway is reduced in ampacity, if certain conditions are met. Which of the following is not one of those conditions?

    A. The smaller busway shall be a maximum of 50 ft. in length
    B. The smaller busway shall not be in contact with any combustible material
    C. The rating of the overcurrent device next back on the line shall not exceed 800 amperes
    D. The ampacity of the smaller busway shall be at least 1/3 the rating of the overcurrent device next back on the line

Article 370
Cablebus

488. Cablebus framework can be used as an equipment grounding conductor for feeders when bonded properly per **Article 250**.

    A. True   B. False

Test Questions

Section   Answer

489. Cablebus shall be supported at intervals not exceeding 15 ft. (General rule)

    A. True   B. False

490. Cablebus shall extend 6 ft. above the floor line that it passes through and be totally enclosed.

    A. True   B. False

Article 314
Outlet, device, pull and junction boxes, conduit bodies and fittings

491. Metal plugs used with nonmetallic boxes shall be recessed at least 1/8 in. from the outer surface.

    A. True   B. False

492. In straight pulls for junction boxes, the length of the box shall not be less than six times the trade diameter of the largest raceway.

    A. True   B. False

493. Where permanent barriers are installed in a junction box, each section shall be considered as a separate box.

    A. True   B. False

494. Conductors passing through the box unbroken and not pulled into a loop or spliced together with scotchlocks are counted as _____ conductor each.

495. Equipment grounding conductors passing through or spliced together in the box count as _____ conductor.

496. A receptacle or switch that is mounted on a strap or yoke is counted as _____ conductors.

497. Boxes that contain fixture studs or hickeys shall have _____ conductor added for each fitting.

498. Conduit bodies enclosing _____ AWG or smaller conductors shall have a cross-sectional area not less than twice the cross-sectional area of the largest conduit to which it is attached.

499. Spliced conductors within a box count as _____ each.

    A. One   B. Two   C. Three   D. Four

500. An isolated equipment grounding conductor run with an equipment grounding conductor shall count as _____ conductor.

    A. Zero   B. One   C. Two   D. Three

501. Boxes with one or more cable clamps that are installed to support cables shall have _____ conductor added toward the fill.

    A. One   B. Two   C. Three   D. Four

Section	Answer	

502. Where nonmetallic sheathed cable is installed in a 2 1/4 in. x 4 in. wall mounted nonmetallic box, securing the cable to the box is not required if the sheath extends inside the box at least 1/4 in. and is stapled a maximum of _____ in. from the box.

    A. 6   B. 8   C. 10   D. 12

503. Boxes mounted in walls of noncombustible material shall be installed so that the front edge of the box will be set back a maximum of _____ in.

    A. 1/4   B. 3/8   C. 1/2   D. 3/4

504. The front edge of a metal switch box that is mounted in a wall covered with wood paneling shall be _____ from the surface of the wall.

    A. Flush with or project out
    B. Permitted to be set back a maximum of 1/4 in.
    C. Permitted to be set back a maximum of 3/8 in.
    D. Permitted to be set back a maximum of 1/2 in.

505. Plasterboard that is broken around the edges of a switch box shall be repaired so there will be no gaps or open spaces greater than _____ in. at the edge of the box.

    A. 1/16   B. 1/8   C. 1/4   D. 3/8

506. A 100 cu. in. box that does not contain devices or support luminaires (fixtures) shall be permitted to be supported only by conduit threaded wrenchtight into hubs on the box, if the conduit is supported within _____ (in. or ft.) of the box and enters the box on at least two side(s) of the box.

    A. 6 in.   B. 1 ft.   C. 2 ft.   D. 3 ft.

507. Boxes intended to enclose flush devices shall have an internal depth of not less than _____ in.

    A. 3/4   B. 1/8   C. 15/16   D. 3/8

Article 312
Cabinets, cutout boxes, and meter socket enclosures

508. Cabinets installed in noncombustible material such as walls of concrete shall be so installed that the front edge of the cabinet will not set back of the finished surface more than _____ in.

    A. 1/8   B. 1/4   C. 1/2   D. 3/4

509. In a cabinet containing overcurrent devices, the conductors shall fill a maximum of _____ percent of the wiring space.

    A. 20   B. 40   C. 60   D. 75

510. Other than at points of support, there shall be an airspace of at least _____ in. between the base of the device and the wall of any metal cabinet in which the device is mounted.

    A. .0625   B. .0840   C. 1.4   D. 1.2

511. Where the voltage is 277 volts-to-ground, the airspace clearance between any metal part of a cabinet and the nearest exposed current-carrying live part of an electrical device in the cabinet shall be _____ in.

    A. 1   B. 3   C. 4   D. 6

Test Questions

Section   Answer

Article 366
Auxiliary gutters

512. Where auxiliary gutters are used at the electrical equipment, they shall be permitted to extend a maximum of _____ ft. from the equipment they supplement.

    A. 10   B. 20   C. 30   D. 40

513. Nonmetallic auxiliary gutters shall be supported at intervals of _____ ft..

    A. 3   B. 5   C. 8   D. 10

514. Sheet metal auxiliary gutters shall not contain more than _____ current-carrying conductors at any cross-section.

    A. 10   B. 20   C. 30   D. 50

515. The sum of cross-sectional areas of all contained conductors at any cross-section of the nonmetallic auxiliary gutter shall not exceed _____ percent of the interior cross-section area of the nonmetallic auxiliary gutter.

    A. 10   B. 20   C. 30   D. 40

516. Conductors, including splices and taps, shall not fill the gutter space to more than _____ percent of its area.

517. The cross-sectional area of an auxiliary gutter shall not be filled to more than _____ percent.

518. The current carried continuously in bare copper bars in sheet metal auxiliary gutters shall not exceed _____ amps/in. of cross section of the conductor.

Article 404
Switches

519. Switches shall be installed so that the center of the grip of the operating handle, when in its highest position, will not be more than _____ above the floor.

    A. 6 ft., 3 in.   B. 6 ft., 4 in.   C. 6 ft., 6 in.   D. 6 ft., 7 in.

520. Multiple switches installed in one enclosure without barriers between switches shall be arranged so that the voltage between adjacent switches does not exceed _____ volts.

    A. 120   B. 150   C. 240   D. 300

521. A knife switch rated 800 amps and 480 volts shall:

    A. Be used to interrupt inductive loads only
    B. Be used to interrupt noninductive loads only
    C. Have contacts of a renewable or quick-break type
    D. Be used as an isolating switch only and not opened under load

522. AC general use snap switches shall be permitted to supply motor loads not exceeding _____ percent of the ampere rating of the switch at its rated voltage.

    A. 50   B. 75   C. 80   D. 90

Section	Answer

523. AC-DC general use snap switches can supply inductive loads not exceeding _____ percent of the ampere rating of the switch at the applied voltage.

    A. 50   B. 75   C. 80   D. 90

Article 408
Switchboard and panelboards

524. Delta breakers shall be permitted to be installed in panelboards.

    A. True   B. False

525. Suitable protection shall be provided for switchboards where installed over a combustible floor.

    A. True   B. False

526. The _____ phase shall be that phase having the higher voltage-to-ground on three-phase, four-wire, delta-connected systems.

527. Not more than _____ overcurrent protection devices of a lighting and appliance branch-circuit panelboard shall be installed in any one cabinet.

528. Each lighting and appliance branch-circuit panelboard shall be individually protected on the supply side by not more than _____ main circuit breakers.

529. A power panelboard is one having _____ percent or fewer of its overcurrent devices protecting lighting or appliance branch-circuits.

    A. 10   B. 15   C. 20   D. 25

530. For other than a totally enclosed switchboard, a space not less than _____ ft. shall be provided between the top of the switchboard and any combustible ceiling.

    A. 1   B. 2   C. 3   D. 6

531. Where conduits enter a floor standing panelboard at the bottom, the conduits, including their fittings, shall rise a maximum of _____ in. above the bottom of the enclosure.

    A. 2   B. 3   C. 4   D. 6

532. Where conductors are run under the floor and enter a switchboard at the bottom, the wiring space between the bottom of the switchboard and the busbars shall be at least _____ in. if the busbars are bare.

    A. 6   B. 8   C. 10   D. 12

533. In an industrial plant, a panelboard designed for 24 overcurrent devices contains seven three-pole circuit breakers that feed three-phase motors and three single-pole, 20 amp circuit breakers that feed receptacles. This panelboard is:

    A. Being used as a power panel
    B. Being used as a split-bus panelboard
    C. Being used as a lighting and appliance branch-circuit panelboard
    D. In violation of the code because receptacle circuits shall not be permitted in a three-phase panelboard supplying motors

Test Questions

Section   Answer

534. A power panelboard is one having a maximum of _____ percent of its overcurrent devices rated 30 amps or less for which neutral connections are provided.

    A. 10   B. 20   C. 30   D. 50

535. A lighting and appliance branch-circuit panelboard in a store contains 10 two-pole breakers and 10 single-pole breakers. What is the maximum number of additional single-pole circuit breakers permitted to be installed in this panelboard?

    A. 4   B. 10   C. 12   D. 22

536. Panelboards equipped with snap switches rated at 30 amps or less shall have overcurrent protection of _____ amperes or less.

    A. 100   B. 150   C. 200   D. 225

Article 400
Flexible cords and cables

537. An SO cord containing an equipment grounding conductor and two current-carrying conductors of size 12 AWG has an allowable ampacity of _____ amperes.

    A. 15   B. 18   C. 20   D. 25

538. Flexible cords shall be permitted indoors where:

    A. Concealed behind a wall
    B. Run through a wall to serve nonstationary equipment
    C. Used for wiring of luminaires (fixtures)
    D. Run along the surface of a wall and secured at the required intervals

539. A three conductor flexible cord used in a luminaire (light fixture) has one conductor marked with ridges or groves to identify it as the _____ conductor.

    A. Grounded   B. Grounding   C. Ungrounded D. Switched

540. Multiconductor portable cables used to connect mobile equipment shall be _____ AWG copper or larger and shall employ flexible stranding.

    A. 8   B. 6   C. 4   D. 2

Article 402
Fixture wires

541. The allowable ampacity of 14 AWG fixture wires is _____ amperes.

    A. 14   B. 17   C. 20   D. 23

542. The allowable ampacity of 12 AWG fixture wires is _____ amperes.

    A. 17   B. 20   C. 23   D. 28

543. The minimum size fixture wires shall not be smaller than _____ AWG.

    A. 18   B. 16   C. 14   D. 12

Section   **Answer**

Article 410
Luminaires (lighting fixtures), lampholders, lamps, and receptacles

544. A recessed fluorescent luminaire (light fixture) shall be permitted to be installed in a clothes closet on the wall above the door where the clearance between the luminaire (light fixture) and the nearest point of storage space is at least 12 in.

    A. True   B. False

545. A fixture unit that weighs more than 6 lbs. or exceeds 16 in. in any dimension shall not be supported by the screw shell of a lampholder.

    A. True   B. False

546. Luminaires are permitted to be used as a raceway for circuit conductors. (General rule)

    A. True   B. False

547. Recessed portions of a luminaire (lighting fixture) enclosure shall be spaced at least 1/2 in. from combustible materials.

    A. True   B. False

548. Lighting track installed over a bathtub shall have a clearance of 3 ft. horizontally and _____ ft. vertically.

549. A recessed incandescent luminaire (lighting fixture) shall be permitted to be installed in a clothes closet on the wall above the door where the clearance between the luminaire (light fixture) and the nearest point of storage space is at least _____ in.

550. Luminaires (lighting fixtures) supported by metal poles shall be required to have an accessible handhole not less than _____ in. x _____ in.

551. Pendant conductors longer than _____ ft. shall be twisted together where not labeled in a listed assembly.

    A. 3   B. 5   C. 6   D. 10

552. Unless part of listed decorative lighting assemblies, pendant conductors shall not be smaller _____ AWG for mogul-base lampholders.

    A. 18   B. 16   C. 14   D. 12

553. Electric-discharge luminaires (lighting fixtures) mounted with mogul-base, screw-base, screw-shell lampholders shall be permitted to be connected to branch-circuits of _____ amperes or less by cords.

    A. 20   B. 30   C. 40   D. 50

554. A recessed luminaire (fixture) that is not identified for contact with insulation shall have all recessed parts spaced not less than _____ in. from combustible materials.

    A. 1/4   B. 1/2   C. 3/4   D. 1

555. Portable lampholders shall not be required to be grounded where supplied through an isolating transformer with an ungrounded secondary of not over _____ volts.

    A. 15   B. 20   C. 50   D. 100

Test Questions

Section   Answer

556. Electric discharge luminaires (lighting fixtures) shall be permitted to be connected by flexible cord if the cord is visible for _____ percent of its entire length.

      A. 25   B. 50   C. 75   D. 100

557. Electric discharge luminaires (light fixtures) installed _____ an outlet box shall have an opening with access to the box.

      A. under   B. over   C. on either side

558. Externally wired luminaires (fixtures) shall be permitted to be installed in _____.

      A. Show windows   B. Clothes closets
      C. Both A and B   D. Neither A or B

559. Electric-discharge lighting systems with an open-circuit voltage exceeding _____ volts shall be listed.

      A. 277   B. 480
      C. 600   D. 1000

560. Candelabra sockets can be wired with _____ AWG rubber-covered conductors.

      A. 12   B. 14   C. 16   D. 18

561. Locations where luminaires (light fixtures) are installed outdoors under a canopy, they are considered _____ locations.

      A. Dry   B. Wet   C. Damp   D. Outdoor

562. Where a bathroom ceiling is 9 1/2 feet high from the floor and 8 feet above the rim of the bathtub, which of the following luminaires (fixtures) shall not be permitted on the ceiling, 30 in. horizontally from the rim of the bathtub?

      A. A recessed fluorescent fixture
      B. A lighting track mounted fixture
      C. A surface mounted fluorescent fixture
      D. A surface mounted incandescent fixture

563. Metal poles that support luminaires (lighting fixtures) where supply conductors are installed within the pole shall not require a handhole for access to conductors if the pole has a hinged base and is a maximum of _____ ft. in height.

      A. 10   B. 15   C. 20   D. 25

564. A surface mounted incandescent luminaire (light fixture) shall be permitted to be installed in a clothes closet on the wall above the door where the clearance between the fixture and the nearest point of storage space is at least _____ in.

      A. 6   B. 10   C. 12   D. 18

565. When a light unit weighs a maximum of _____ pounds, an outlet box can be the only support for the luminaire (light fixture) if the outlet box meets code provisions for boxes.

      A. 20   B. 30   C. 40   D. 50

566. Tap conductors shall be permitted to be run from the luminaire (fixture) terminal connection to an outlet box placed at least _____ ft. from the luminaire (fixture).

      A. 1   C. 5
      B. 3   D. 6

12-53

Section   Answer

567. Thermal insulation shall not be installed within _____ in. of a recessed luminaire (lighting fixture) enclosure, wiring compartment, or ballast.

    A. 1   B. 2   C. 3   D. 6

568. An electric-discharge lighting system with equipment having exposed live parts and an open-circuit voltage exceeding _____ volts shall not be installed in a dwelling occupancy.

    A. 120   B. 240   C. 300   D. 600

569. Where a surface-mounted light unit, with a ballast, is installed on combustible low-density cellulose fiberboard, it shall be spaced at least _____ in. from the fiberboard.

    A. 1/2   B. 1   C. 1 1/2   D. 3

Article 411
Lighting systems operating at 30 volts or less

570. Lighting systems operating at 30 volts or less shall not be installed within _____ ft. of swimming pools.

    A. 5   B. 10   C. 15   D. 20

571. Bare conductors for lighting systems operating at 30 volts or less shall not be installed less than _____ ft. above the finished floor.

    A. 6   B. 6 1/2   C. 7   D. 8

572. Lighting systems operating at 30 volts or less shall be supplied from a maximum _____ amp circuit.

    A. 15   B. 20   C. 25   D. 30

Article 422
Appliances

573. A kitchen waste disposer shall be permitted to be connected with a flexible cord under certain conditions. Which of the following is not one of those conditions.

    A. The receptacle shall be accessible
    B. It shall be permitted only in dwelling units
    C. The cord shall be at least 18 in. long
    D. The rating of the disposer shall not exceed 80 percent of the branch-circuit rating

574. Built-in dishwashers shall be permitted to be cord-and-plug connected by a flexible cord with a length of _____ ft. to _____ ft.

    A. 2 to 3   B. 3 to 4   C. 4 to 5   D. 5 to 6

575. Screw shell lampholders shall not be used with infrared lamps over _____ watts rating.

    A. 150   B. 175   C. 225   D. 300

576. A plug and receptacle combination in the supply line to a counter-mounted cooking unit shall:

    A. Not be installed as the disconnecting means
    B. Be permitted as the disconnecting means if it is accessible
    C. Be permitted as the disconnecting means if it is readily accessible
    D. Be permitted as the disconnecting means if the cooking unit has a unit switch with a marked "off" position

577. Ceiling fans shall be permitted to be supported by outlet boxes identified for such use if the maximum weight of the ceiling fan is _____ pounds. (General Rule)

    A. 35   B. 40   C. 45   D. 50

578. The branch-circuit overcurrent device shall be permitted to serve as the disconnecting means for a permanently connected appliance not rated over _____ volt-amperes.

    A. 180   B. 200   C. 300   D. 480

579. A circuit breaker serving as the disconnecting means for a permanently connected motor driven appliance of more than _____ horsepower shall be located within sight from the motor controller.

    A. 1/8   B. 1/4   C. 1/2   D. 3/4

580. An electric heating appliance that has resistance-type heating elements rated at more than 48 amps shall have heating elements subdivided with each subdivided load protected at a maximum of _____ amps.

    A. 50   B. 60   C. 75   D. 90

Article 424
Fixed electric space heating equipment

581. Conductors supplying a heating unit shall be calculated at _____ percent times the heating elements load plus the blower motor.

    A. 110   B. 115   C. 120   D. 125

582. Resistance elements in heating units shall be protected at not more than _____ amps.

    A. 30   B. 40   C. 50   D. 60

583. A disconnecting means shall be installed in sight of the heating unit for a self-contained blower motor over _____ HP.

    A. 1/8   B. 1/4   C. 1/3   D. 1/2

584. Wiring located above heated ceilings shall be spaced not less than _____ in. above the heated ceiling.

    A. 1   B. 2   C. 3   D. 4

585. Heating elements shall be separated at least _____ in. from the edge of outlet boxes and junction boxes that are used for mounting surface lighting units.

    A. 4   B. 6   C. 8   D. 10

Stallcup's Master Electrician's Study Guide

Section   Answer

Article 426
Fixed outdoor electric deicing and snow-melting equipment

_____ _____ 586. Panels or units of embedded deicing and snow-melting equipment shall not exceed _____ VA per sq. ft. of a heated area.

    A. 80   B. 90   C. 120   D. 180

_____ _____ 587. Conductors supplying outdoor deicing and snow-melting equipment shall be at least _____ percent of the heating load.

    A. 50   B. 75   C. 100   D. 125

_____ _____ 588. Exposed outdoor electric deicing equipment operating above _____ °C shall be installed to protect against personnel contact.

    A. 30   B. 40   C. 50   D. 60

_____ _____ 589. The ampacity of branch-circuit conductors supplying fixed outdoor electric deicing and snow-melting equipment shall not be less than _____ percent of the total load of the heaters.

    A. 100   B. 115   C. 125   D. 150

_____ _____ 590. Embedded snow-melting cables shall be installed at least _____ in. from the top surface.

    A. 1   B. 1 1/2   C. 2   D. 3 1/2

_____ _____ 591. For fixed outdoor deicing equipment, the disconnecting means shall be permitted to be all of the following, except:

    A. A remote temperature controller
    B. The branch-circuit breaker where readily accessible
    C. A factory-installed attachment plug rated 20 amps and 150 volts-to-ground
    D. A temperature controlled switching device with an "off" position with a lockout provided

Article 427
Fixed electric heating equipment for pipelines and vessels

_____ _____ 592. Pipelines with impedance heating shall not operate at greater than _____ volts AC.

    A. 24   B. 30   C. 50   D. 120

_____ _____ 593. Secondaries of transformers supplying voltage for impedance heating of vessels shall be calculated at not less than _____ percent of the heating load.

    A. 80   B. 100   C. 125   D. 150

Article 430
Motors, motor circuits, and controllers

_____ _____ 594. The full-load current in amps for three-phase motors shall be selected from **Table 430.248** of the NEC.

    A. True   B. False

Test Questions

Section   Answer

595. Conductors for a motor used for periodic duty shall be required to be sized with a current-carrying capacity of 125 percent of the motor's full-load current in amps.

    A. True   B. False

596. Motors shall be located so that adequate ventilation is provided.

    A. True   B. False

597. The motor branch-circuit overcurrent device shall not be required to be capable of carrying the starting current of the motor in amps.

    A. True   B. False

598. Code letters shall be marked on the nameplate, and such letters shall be used for determining the locked-rotor current in amps of the motor.

    A. True   B. False

599. Time-delay fuses that are sized at 125 percent or less of the motor's FLC rating shall be permitted to provide overload protection for the motor.

    A. True   B. False

600. The maximum percentage that can be applied for a time-delay fuse is 400 percent.

    A. True   B. False

601. The FLA ratings of the remaining motors are added to the rating of the largest overcurrent protection device when sizing the OCPD for two or more motors on a feeder.

    A. True   B. False

602. The controller shall be permitted to be an attachment plug and receptacle which is acceptable for use with portable motors rated 3/4 horsepower or less.

    A. True   B. False

603. A molded case switch rated in amperes shall not be permitted as a controller for all motors.

    A. True   B. False

604. The controller shall be permitted to be a general-use switch rated for at least twice the motor's full-load current for stationary motors rated 2 horsepower or less.

    A. True   B. False

605. Branch-circuit conductors supplying multispeed motors shall be sized and selected from the largest amperage on the motor's nameplate based on the motor's RPM.

    A. True   B. False

606. Open motors that have commutators or collector rings shall be permitted to be located adjacent to combustible material.

    A True   B. False

607. Conductors supplying two or more motors on a feeder-circuit shall be sized by multiplying all the motors in the group by 125 percent.

    A. True   B. False

12-57

Stallcup's Master Electrician's Study Guide

Section   Answer

608. Feeder conductors supplying two or more motors can be routed to a gutter and spliced to the conductors feeding each motor.
    A. True   B. False

609. Resistor leads are the conductors between the resistor bank and drum controller.
    A. True   B. False

610. A Class 20 or Class 30 overload relay will provide a longer motor acceleration time.
    A. True   B. False

611. Adjustable speed drive systems shall protect against motor overtemperature conditions.
    A. True   B. False

612. Branch-circuit conductors supplying a single motor shall have an ampacity not less than _____ percent of the motor's full-load current rating.

613. The selection of conductors for wye start and delta run motors between the controller and motor shall be based on _____ percent of the motor's full-load current times 125 percent for continuous use.

614. The disconnecting means for power conversion equipment shall not be less than _____ percent of the input current rating of the conversion unit.

615. The full-load current rating of the largest motor shall be multiplied by _____ percent to select the size of conductors for a feeder supplying a group of two or more motors.

616. The motor branch-circuit overcurrent device shall be capable of carrying the _____ current of the motor.

617. When a motor won't start and run, a nontime-delay fuse not exceeding 600 amps in rating shall be permitted to be increased up to _____ percent of the full-load current.

618. When a motor won't start and run, an inverse-time CB, greater than 100 amps, shall be permitted to be increased up to _____ percent of the FLC.

619. Code letters are installed on motors by manufacturers for calculating the _____ in amps based upon the kVA per horsepower which is selected from the motors code letter.

620. An overcurrent protection device rated at _____ amps or less can protect a 120 volt or less branch-circuit supplying motors rated less than 1 horsepower.

621. The service factor or _____ rise of the motor shall be used when sizing and installing the minimum running overload protection for motors.

622. The frames of portable motors that operate over _____ volts to ground shall be guarded or grounded.

623. The motor controller for a torque motor shall have a continuous duty, full-load current rating not less than the _____ current rating of the motor in amps.

624. Control-circuit conductors of 18 AWG and 16 AWG that extend beyond the control equipment enclosure shall not be protected when fused at 15 or _____ amps.

Test Questions

Section   Answer

625. Control conductors routed to remote start-stop stations shall be permitted to be protected by a device set at _____ percent of the conductor's ampacity.

626. Control conductors inside a starter enclosure that run to the start-stop station on the cover of the enclosure can be protected at _____ times the control conductor's ampacity.

627. For general motor applications (other than for torque motors and AC adjustable voltage motors), where the current rating of a motor is used to determine the ampacity of conductors, which of the following values shall be used?

    A. The current rating marked on the nameplate of the motor
    B. The current values given in the Tables at the end of Article 430
    C. 125 percent of the rating of the overcurrent device
    D. 125 percent of the value of the overload protection required

628. Generally, separate motor overload protection for motors shall be based on the:

    A. Motor nameplate current rating
    B. Size of the conductors supplying the motor
    C. Rating of the branch-circuit overcurrent device
    D. Values given in the Tables at the end of Article 430

629. For torque motors, the locked-rotor current shall be used to determine all of the following, except the ampere rating of the:

    A. Overload protection
    B. Disconnecting means
    C. Ground-fault protection
    D. Branch-circuit conductors

630. The code does not specify the wire bending space within an enclosure for motor controllers where the maximum size wire used is _____ AWG.

    A. 12   B. 10   C. 8   D. 6

631. When an automatically started motor has a long accelerating time to arrive at normal running speed, motor overload protection shall be permitted to be shunted during the starting period if several conditions are met. Which of the following is not one of those conditions?

    A. Shunting will automatically be prevented in the event the motor fails to start
    B. If normal motor running condition is not reached, means shall be provided for shutdown and manual restarting
    C. The time period of overload protection shunting shall be less than the locked-rotor time rating of the protected motor
    D. Inverse time fuses or circuit breakers rated at a maximum of 400 percent of the motor's full-load current shall be operative during the starting period of the motor

632. A motor that is used for short-time duty shall be considered as being protected against overcurrent by the:

    A. Overload relay
    B. Branch-circuit overcurrent device
    C. Thermal protector integral with the motor
    D. Fuses that are used for overload protection

Stallcup's Master Electrician's Study Guide

Section   Answer

_____ _____ 633. A continuous duty motor shall be permitted to be protected solely by the branch-circuit protective device if the motor meets certain conditions. Which of the following is not one of those conditions? The motor shall be:

   A. Automatically started
   B. Rated a maximum of 1 horsepower
   C. Within sight of the controller location
   D. Portable and not permanently installed

_____ _____ 634. A motor _____ is where the motor operates in excess of its normal full-load current rating for a sufficient length of time so as to cause dangerous overheating.

   A. Overload   B. Short-circuit   C. Ground-fault   D. Shunt-circuit

_____ _____ 635. The size of branch-circuit conductors for a three-phase motor shall be based on the:

   A. Motor nameplate current rating
   B. Rating of the overload protective devices
   C. Current rating of the motor given in **Table 430.250**
   D. Maximum permitted ampere rating or setting of the branch-circuit protective device

_____ _____ 636. The maximum size or rating of a branch-circuit protective device for a motor circuit is determined by the:

   A. Starting current of the motor
   B. Rating of the overload protection devices
   C. Full-load current of the motor in amperes
   D. Size of the motor branch-circuit conductors

_____ _____ 637. Which of the following statements is true regarding motor circuits?

   A. Motor controllers are intended to open short circuits and ground faults
   B. The motor overload protective device is intended to open short-circuits and ground-faults.
   C. Branch-circuit conductors shall be sized according to the rating of the branch-circuit protective device
   D. The branch-circuit protective device shall protect the conductors, motor and motor control apparatus from overcurrent due to short-circuits or grounds

_____ _____ 638. An attachment plug and receptacle shall be permitted as the controller for a portable motor rated a maximum of _____ horsepower.

   A. 1/8   B. 1/4   C. 1/3   D. 1/2

_____ _____ 639. A motor control enclosure that provides a degree of protection against windblown dust where used outdoors is Type _____.

   A. 1   B. 3   C. 3R   D. 5

_____ _____ 640. For a motor, the disconnecting means shall be located in sight from and not more than _____ ft. from the motor. (General Rule)

   A. 6   B. 10   C. 25   D. 50

_____ _____ 641. The branch-circuit overcurrent device shall be permitted to serve as the disconnecting means for stationary motors of a maximum _____ horsepower.

   A. 1/8   B. 1/2   C. 3/4   D. 1

12-60

642. Which of the following switches is rated in horsepower?

    A. Transfer switch   B. Isolating switch
    C. Motor-circuit switch   D. Bypass isolation switch

643. The disconnecting means for motor circuits rated 600 volts or less shall have an ampere rating of at least _____ percent of the full-load current rating of the motor.

    A. 100   B. 115   C. 125   D. 150

644. Conductors for power conversion equipment shall be sized at _____ percent of the rated input of such equipment.

    A. 100   B. 115   C. 125   D. 150

645. For stationary motors rated at more than _____ HP AC, the disconnecting means shall be permitted to be a general-use or isolating switch where plainly marked "Do Not Operate Under Load."

    A. 40   B. 50   C. 75   D. 100

646. The disconnecting means shall open all _____ conductors and shall be designed so that no pole can be operated independently.

    A. Grounding   B. Grounded   C. Ungrounded   D. All of the above

647. Motor with a marked service factor not less than 1.15 shall have the minimum running overload protection sized at _____ percent.

    A. 115   B. 125   C. 130   D. 140

648. Motor marked with a temperature rise not over 40°C shall have the minimum running overload protection sized at _____ percent.

    A. 115   B. 125   C. 130   D. 140

649. Motors marked with a service factor not less than 1.15 shall have the maximum running overload protection sized at _____ percent.

    A. 115   B. 125   C. 130   D. 140

650. Motors marked with a temperature rise not over 40°C shall have the maximum running overload protection sized at _____ percent.

    A. 115   B. 125   C. 130   D. 140

651. An AC general-use snap switch shall be permitted to be installed as the controller for a stationary motor where the full-load current of the switch does not exceed _____ percent of the branch-circuit rating.

    A. 50   B. 75   C. 80   D. 90

652. Live parts of motors or controllers shall be guarded against accidental contact where operating at over _____ volts-to-ground.

    A. 50   B. 150   C. 250   D. 600

Article 440
Air-conditioning and refrigerating equipment

653. The OCPD shall be sized and selected by the information (BCSC) provided on the nameplate listing for air-conditioning and refrigeration equipment.

    A. True   B. False

654. A horsepower rated switch may be used as a disconnecting means for air-conditioners.

    A. True   B. False

655. Cord-and-plug connected room air-conditioners shall not be permitted to serve as the disconnecting means for hermetic motors.

    A. True   B. False

656. When sizing the conductors for a feeder supplying A/C units and motors, the FLC, in amps, of the largest motor shall be multiplied by _____ percent.

657. To size the disconnecting means, the full-load current rating on the nameplate or the nameplate branch-circuit selection current of the compressor, whichever is greater, shall be multiplied by _____ percent.

658. The disconnecting means for air-conditioning equipment shall be located within sight and within _____ ft. and shall be readily accessible to the user.

659. To size a CB for a feeder, the full-load current rating on the nameplate or the branch-circuit selection current rating of the largest compressor, whichever is greater, shall be multiplied by _____ percent, if there are two or more hermetic sealed motors installed on the same feeder-circuit.

660. When sizing conductors, two or more motor compressors can be connected to a feeder-circuit with the largest motor compressor calculated at _____ percent of its FLA and the remaining compressor loads added to this total at _____ percent of their FLA rating.

661. The overload relay for the motor compressor shall trip at not more than _____ percent of the full-load current rating.

    A. 100   B. 125   C. 130   D. 140

662. When attachment plugs and receptacles are used for circuit connections they shall be rated no higher than _____ amps, for 208 or 240 volt, single-phase branch-circuits.

    A. 15   B. 20   C. 25   D. 30

663. The full-load current rating of a room air-conditioner shall be marked on the nameplate and shall not operate at more than _____ amps on 250 volts, single-phase.

    A. 20   B. 30   C. 40   D. 50

664. A cord-and-plug shall be permitted to serve as the disconnecting means if the room air-conditioner's manual controller _____ ft. or less from the floor.

    A. 4   B. 5   C. 6   D. 7

665. Room air-conditioners installed with flexible cords shall be required to be a length that is limited to _____ ft. for 120 volt circuits.

    A. 6   B. 8   C. 10   D. 12

Test Questions

Section　　Answer

666. The OCPD for hermetic sealed compressors shall be selected at _____ percent (for minimum) of the compressor FLA rating or the branch-circuit selection current, whichever is greater.

　　　A. 125　　　B. 150　　　C. 175　　　D. 225

667. The rating of the overload protection device shall be determined by using the rating of the nameplate of the cord-and-plug connected equipment having single-phase, _____ volt or less hermetic sealed motors.

　　　A. 120　　　B. 250　　　C. 480　　　D. 600

668. Room air-conditioners shall be required to be grounded where operating over _____ volts-to-ground.

　　　A. 120　　　B. 150　　　C. 250　　　D. 300

669. A cord-and-plug shall be permitted to serve as the disconnecting means for the room air-conditioner if it operates at _____ volts or less.

　　　A. 120　　　B. 150　　　C. 250　　　D. 300

Article 450
Transformers and transformer vaults

670. Transformers shall be permitted to be connected in parallel and switched as a unit, provided that each transformer has overcurrent protection.

　　　A. True　　　B. False

671. Transformers enclosed by fences or guards shall not be required to be grounded.

　　　A. True　　　B. False

672. Doors to transformer vaults shall be kept locked at all times to prevent access of unqualified persons to the vault.

　　　A. True　　　B. False

673. Foreign piping or duct systems shall be permitted to be installed in a transformer vault.

　　　A. True　　　B. False

674. Storage of materials shall be permitted in a transformer vault. (General Rule)

　　　A. True　　　B. False

675. When the primary current of a transformer rated 600 volts or less is _____ amps or more and a 125 percent factor does not correspond to a standard size device, the next higher standard shall be permitted to be used.

676. When the primary current of a transformer rated 600 volts or less is less than _____ amps, the overcurrent protection device shall not exceed 300 percent of the primary current.

677. When the primary side of a transformer rated over 600 volts is individually protected by a circuit breaker, the setting shall not exceed _____ percent of the primary current.

Stallcup's Master Electrician's Study Guide

Section   Answer

678. Transformers rated 600 volts or less and _____ kVA or less shall be permitted to be installed in suspended ceilings of buildings.

679. Where practicable, vaults containing more than _____ kVA transformer capacity shall be provided with a drain.

680. Transformers shall be installed in vaults when rated over _____ volts.

681. When transformer vaults are protected with an automatic sprinkler system, the door leading to a vault may have a minimum fire rating of _____ hour(s).

682. Dry-type transformers not over 600 volts that are located on open walls or steel columns shall not be required to be _____ accessible.

683. If installing fuses, the overcurrent protection device for the primary side of a transformer rated over 600 volts shall be rated not greater than _____ percent of the rated primary current of the transformer. (Individual (any) OCP in primary)

684. The OCPD and conductors on the secondary side of a transformer, over 600 volts and in any location per Table, with a secondary voltage of 600 volts or less, shall be sized at _____ percent of the FLC rating. (Primary and secondary protection)

685. A transformer 600 volts or less, nominal, where the rated secondary current of a transformer is less than 9 amps but 2 amps or more, an overcurrent protection device rated or set at no more than _____ percent of secondary current may be used.

686. Transformers shall be elevated at least _____ ft. above the floor to prevent unauthorized personnel from contact.

    A. 6   B. 6 1/2   C. 8   D. 10

687. All indoor dry-type transformers of over _____ volts shall be required to be installed in a vault.

    A. 10,000   B. 15,000   C. 25,000   D. 35,000

688. The walls and roof for a transformer vault shall have a _____ in. thickness.

    A. 3   B. 4   C. 5   D. 6

689. The door sills for a transformer vault shall be at least _____ in. high.

    A. 1   B. 2   C. 3   D. 4

690. Where automatic closing fire dampers are installed in a vault, such dampers shall possess a standard fire rating or not less than _____ hours.

    A. 1   B. 1 1/2   C. 2   D. 3

691. For a transformer in a supervised location with a secondary voltage rated 600 volts or less, the OCPD on the secondary side shall be sized at not more than _____ percent of the FLC rating.

    A. 100   B. 125   C. 225   D. 250

692. Dry-type transformers rated 112 1/2 kVA or less shall be separated at least _____ in. from the combustible material where the voltage is 600 volts or less.

    A. 6   B. 12   C. 18   D. 24

	Test Questions
	Section Answer

693. Askarel-insulated transformers of over _____ kVA shall be furnished with a relief vent such as a chimney.

    A. 25   B. 35   C. 50   D. 75

694. The floor for a transformer vault shall be at least _____ in. thick.

    A. 2   B. 4   C. 6   D. 12

695. Dampers used to close ventilating openings in vaults and protect against fire shall be rated not less than _____ hour(s).

    A. 1   B. 1 1/2   C. 2   D. 3

696. Doors leading into a transformer vault shall be kept locked and access permitted only to _____.

    A. The owner   B. Any person working in the building
    C. Qualified persons   D. General public

697. A 112 1/2 kVA transformer that is completely enclosed and equipped with a 155°C rise insulation shall be permitted to be installed in a room _____.

    A. With concrete walls   B. Designed as a vault
    C. Of fire resistant construction   D. Built with tile blocks

698. Dry-type transformers 600 volts, nominal, or less and not exceeding _____ kVA shall be permitted in hollow spaces of buildings not permanently closed in by structure.

    A. 25   B. 100
    C. 150   D. 200

Article 460
Capacitors

699. The ampacity of capacitor circuit conductors shall not be less than _____ percent of the rated current of the capacitor.

700. The ampacity of capacitor circuit conductors shall be at least _____ of the ampacity of the motor's circuit conductors.

701. The disconnecting means for a capacitor circuit conductor shall be at least _____ percent of the capacitor's FLC rating.

Article 500
Hazardous (classified) locations

702. Explosionproof equipment shall be permitted to be installed in Class I, Division 1 locations.

    A. True   B. False

703. Purged and pressurized equipment shall be permitted to be installed in any hazardous (classified) location for which it is identified.

    A. True   B. False

Section	Answer

704. Nonincendive circuits shall be permitted for equipment in Class I, Division 1 locations.

    A. True   B. False

705. Class I, Division 1 locations are those locations in which volatile flammable liquids or flammable gases are handled, processed, or used. (Gases are not normally present)

    A. True   B. False

706. Class II, Division 1 locations are those locations in which combustible dust is in the air under normal operation conditions in quantities sufficient to produce explosive or ignitable mixtures.

    A. True   B. False

707. An atmosphere containing acetylene shall be classified as Group _____.

    A. A   B. B   C. C   D. D

708. An atmosphere containing ethylene, gases or vapors is classified as Group _____ hazardous location.

    A. A   B. B   C. C   D. D

709. An atmosphere containing combustible metal dusts is classified as Group _____.

    A. D   B. E   C. F   D. G

710. Equipment marked "Division _____" shall be suitable for both Division 1 and 2.

    A. 1   B. 2   C. 1 and 2   D. None of the above

711. Class III locations shall be those areas that are hazardous because of the presence of easily ignitable fibers or _____.

    A. Gases   B. Combustible dusts
    C. Flyings   D. Liquids

Article 501
Class I locations

712. Flexible fittings shall be approved for Class I, Division 1 locations.

    A. True   B. False

713. Seals shall be provided within 24 in. of each arcing device in hazardous locations.

    A. True   B. False

714. Self-sealed devices in hazardous locations shall not require additional seals.

    A. True   B. False

715. Portable lighting equipment meeting requirements for fixed lighting equipment shall be permitted for use in Class I, Division 2 locations.

    A. True   B. False

Test Questions

Section   Answer

716. All boxes and fittings shall be approved for Class I, Division 1 locations.

      A. True   B. False

717. Seals shall be provided in a conduit and cable system to minimize the passage of gases or vapors from one portion of the system to another portion.

      A. True   B. False

718. A multiwire branch-circuit shall be permitted in a Class I, Division 1 location.

      A. True   B. False

719. Squirrel-cage induction motors without arcing devices shall be explosionproof when installed in Class I, Division 2 locations.

      A. True   B. False

720. Each luminaire (lighting fixture) shall be protected against physical damage by a suitable guard or by location.

      A. True   B. False

721. Cables without gastight and vaportight sheaths shall be _____ at the boundary of Division 2 and nonhazardous locations.

722. Listed cartridge fuses installed in luminaires (lighting fixtures) shall be permitted to be installed in Class _____, Division 2 locations.

723. Threaded joint of rigid metal conduit shall be made up with at least _____ threads fully engaged.

      A. Four   B. Five   C. Six   D. Seven

724. Rigid nonmetallic conduit shall be permitted to be installed in Class I, Division 1 locations if encased in a concrete envelope of at least _____ in. and buried below the surface not less than _____ ft. of earth.

      A. 1; 1   B. 1; 2   C. 2; 2   D. 2; 3

725. In Class I, Division 2 locations, pendant luminaires (fixtures) shall be suspended by flexible hanger unless rigid stems not over _____ in. long are used.

      A. 6   B. 12   C. 18   D. 24

726. Explosionproof equipment or equipment with arcing or sparking devices shall have a seal placed within _____ in. of such equipment in Class I, Division 1 locations.

      A. 6   B. 12   C. 18   D. 24

727. The cross-sectional area of the conductors permitted in a seal shall not exceed _____ percent of the cross-sectional area of a conduit of the same trade size.

      A. 20   B. 25   C. 40   D. 50

Section   Answer

Article 502
Class II locations

728. A 4 ft. run of conduit between two enclosures (one with a switching device) located in a Class II, Division 1 location shall have a seal.

   A. True   B. False

729. In Class II, Division 2 locations, flexible metal conduit can be used in 6 ft. or shorter lengths without a bonding jumper for loads of _____ amps or less.

730. Rigid metal or _____ conduit can be used in Class II, Division 1 locations.

731. A flexible _____ or liquidtight flexible metal conduit can shall be permitted to be used in Class II, Division 1 locations.

732. Totally enclosed pipe _____ motors shall be permitted to be installed in Class II, Division 1 locations.

733. Type MC cable can be used for general wiring in Class _____, Division 1 locations.

Article 503
Class III locations

734. A totally enclosed pipe-ventilated motor shall not be permitted to be installed in a Class III, Division 2 location.

   A. True   B. False

735. Electric cranes with contacts isolated from all other systems shall be equipped with an alarm.

   A. True   B. False

Article 511
Commercial garages, repair and storage

736. Where Class I liquids are transferred, the entire area for each floor up to a level of 24 in. above the floor shall be a Class I, Division 2 location.

   A. True   B. False

737. Packing garages used for parking shall be permitted to be unclassified.

   A. True   B. False

738. Equipment that is less than _____ ft. above the floor level and that may produce arcs, sparks or particles of hot metal shall be of the totally enclosed type.

   A. 6   B. 12   C. 18   D. 24

739. Where a cord is suspended overhead for plug connections to vehicles, the cord shall be arranged so that the lowest point of sag is at least _____ in. above the floor.

   A. 6   B. 12   C. 18   D. 24

Article 513
Aircraft hangars

740. A seal shall be required at every point in an aircraft hangar where a conduit is routed from a hazardous area to a nonhazardous area.

    A. True   B. False

741. Flexible cord shall have a separate equipment grounding conductor when used to connect cord-and-plug portable equipment in aircraft hangars.

    A. True   B. False

742. The entire area of an aircraft hangar shall be classified as a Class I, Division 2 location up to a level _____ in. above the floor.

    A. 6   B. 12   C. 18   D. 24

743. Any equipment installed in aircraft hangars that is over _____ ft. above engines shall be permitted to be the general-purpose type.

    A. 3   B. 5   C. 10   D. 20

744. Aircraft energizers shall be so designed and mounted that all electric equipment and fixed wiring will be at least _____ in. above the floor level.

    A. 12   B. 18   C. 24   D. 30

Article 514
Gasoline dispensing and service stations

745. PVC installed beneath a gas island is located in a Class I, Division 1 location.

    A. True   B. False

746. RNC installed in the area beneath a gas island shall be buried at least 24 in.

    A. True   B. False

747. Emergency controls for attended self-service motor fuel dispensing facilities shall be installed at a location no more than 150 ft. from dispensers.

    A. True   B. False

748. Metallic portions of dispensing pumps and all noncurrent-carrying metal parts of electric equipment shall be grounded.

    A. True   B. False

749. The surrounding space out to a distance of _____ ft., measured from a point vertically below the edge of the dispenser enclosure, shall be Class I, Division 2 up to a height of _____ in. above grade.

    A. 10; 12   B. 20; 12   C. 20; 18   D. 20; 24

12-69

Section	Answer

750. Where rigid nonmetallic conduit is used for underground wiring in a service station, threaded rigid metal conduit shall be used for the last _____ ft. of the underground run to emergence.

    A. 1   B. 2   C. 3   D. 6

751. Emergency controls for unattended self-service stations shall be more than _____ ft. but less than _____ ft. from the dispensers.

    A. 10; 50   B. 20; 50   C. 10; 100   D. 20; 100

Article 515
Bulk storage plants

752. An indoor pit for a bulk storage tank that is located in the _____ or _____ ft. sphere around pumps, bleeders, etc. shall be considered as a Class I, Division 1 area.

    A. 3; 10   B. 5; 10   C. 3; 25   D. 5; 20

753. Storage and repair garages for tank vehicles shall be considered as Class I, Division 2 up to a height of _____ in. above the floor.

    A. 6   B. 12   C. 18   D. 24

754. Underground wiring routed to and around aboveground storage tanks shall be permitted to be installed with RMC where buried less than _____ ft. in the earth.

    A. 1   B. 2   C. 3   D. 6

Article 517
Health care facilities

755. The outer metal armor of listed Type AC cable is not an acceptable grounding return path when installed in patient care areas.

    A. True   B. False

756. Each patient bed location in general care areas shall be supplied by at least two branch-circuits. (General rule)

    A. True   B. False

757. Each patient bed location in critical care areas shall be supplied by at least four branch-circuits.

    A. True   B. False

758. Ground-fault interrupter protection shall be required for receptacles installed in those critical care areas where the toilet and basin are installed within the patient room.

    A. True   B. False

759. The life safety branch and critical branch of the emergency system shall be kept entirely independent of all other wiring and equipment.

    A. True   B. False

Test Questions

Section   Answer

760. Receptacles installed in psychiatric locations shall be listed _____ resistant.

761. Any room in which flammable _____ are stored shall be considered to be a Class I, Division 1 location from floor to ceiling.

762. If mounted _____ ft. above the floor, receptacles in flammable, anesthetizing locations shall not be required to be totally enclosed.

763. The _____ shield of an isolated power transformer is connected to the reference grounding point.

764. Individual branch-circuits shall not be required for portable x-ray equipment requiring a capacity of not over _____ amperes.

765. Conductors and overcurrent protection devices shall be rated at least _____ percent of the current rating of medical, therapy, or X-ray equipment.

766. Equipment grounding terminals in the normal service panelboard shall be bonded to equipment grounding terminals in the essential system's panelboard with not less than _____ AWG copper conductors.

767. Luminaires (light fixtures) more than _____ ft. above the floor in patient care areas shall not be required to be grounded by an insulated grounding conductor.

    A. 6   B. 6 1/2   C. 7   D. 7 1/2

768. The equipment grounding terminal buses of the normal and essential branch-circuit panelboards serving the same individual patient vicinity shall be bonded together with an insulated continuous copper conductor not smaller than _____ AWG.

    A. 12   B. 10   C. 8   D. 6

769. Each patient bed location in general care areas shall be provided with a minimum of _____ receptacles.

    A. Two   B. Four   C. Six   D. Eight

770. Each patient bed location in critical care areas shall be provided with a minimum of _____ receptacles.

    A. Two   B. Four   C. Six   D. Eight

771. The branches of the emergency system for hospitals shall be installed and connected to the alternate power source so that all functions for the emergency system shall be automatically restored to operation within _____ sec. after interruption of the normal source.

    A. 5   B. 10   C. 20   D. 30

Article 518
Places of assembly

772. A place of assembly is any room or space in an occupancy where _____ persons or more assemble.

    A. 50   B. 75   C. 100   D. 150

773. Nonmetallic raceways installed in places of assembly shall be encased with _____ in. of concrete.

    A. 2   B. 4   C. 5   D. 6

774. Flexible cords installed in cable tray for temporary wiring only shall have a permanent sign attached to the cable tray at intervals not to exceed _____ ft.

    A. 10   B. 25
    C. 50   D. 100

775. Electrical nonmetallic tubing shall be permitted to be installed above suspended ceilings where the suspended ceilings provide a thermal barrier of material that has at least a _____ minute flash rating.

    A. 5   B. 10   C. 15   D. 30

Article 520
Theaters, audience areas of motion picture and television studios, and similar locations

776. Dimmers installed in underground conductors shall be protected by OCPD's not exceeding _____ percent of their rating.

    A. 100   B. 115   C. 125   D. 150

777. Border lights shall be installed around stages in theaters on circuits rated at _____ amps or less.

    A. 20   B. 25   C. 30   D. 35

778. Receptacles used for electrical equipment or fixtures on stages shall be rated in _____.

    A. HP   B. Impedance   C. Watts   D. Amps

Article 525
Carnivals, circuses, fairs, and similar events

779. Service equipment for carnivals shall be installed in a location that is accessible to unqualified persons.

    A. True   B. False

780. Flexible cords or cables installed at circuses shall be continuous without splice or tap between boxes or fittings.

    A. True   B. False

781. Approved nonconductive mats shall be used to cover flexible cords or cables run on the ground, where accessible to the public at carnivals.

    A. True   B. False

782. All equipment requiring grounding at circuses shall be grounded by grounding electrode conductor.

    A. True   B. False

783. Amusement attractions (not supplying equipment) shall be maintained not less than _____ ft. in any direction from overhead conductors operating at 600 volts or less.

   A. 10   B. 15   C. 20   D. 25

784. Amusement rides shall not be located under or within _____ ft. horizontally or conductors operating in excess of 600 volts.

   A. 10   B. 15   C. 20   D. 25

785. Single conductor cables installed at fairs shall be permitted only in sizes _____ AWG or larger.

   A. 8   B. 6   C. 4   D. 2

786. Termination boxes installed outdoors shall be mounted so that the bottom of the enclosure is not less than _____ in. above the ground.

   A. 3   B. 6   C. 12   D. 18

787. Each ride shall be provided with a fused disconnect switch or circuit breaker located within sight and within _____ ft. of the operator's station.

   A. 6   B. 20   C. 40   D. 50

Article 530
Motion picture and television studios and similar locations

788. A set in a motion picture studio can be wired with PVC Schedule 40 conduit.

   A. True   B. False

789. Receptacles wired with DC power and located in plugging boxes of television studios shall not exceed 50 amps.

   A. True   B. False

790. Splices or taps shall be permitted for portable wiring used for stage effects where made with listed devices and the circuit is protected at not more than 20 amperes.

   A. True   B. False

791. DC pendant and portable lamps used to illuminate stages in motion picture studios shall not be required to be grounded at 150 volts-to-ground.

   A. True   B. False

792. Cables and cords smaller than 6 AWG shall be attached to the plugging box.

   A. True   B. False

793. A single externally operable switch shall be permitted to simultaneously disconnect all the contactors on any one location board, where located at a distance of not more than 6 ft. from the location board.

   A. True   B. False

Article 545
Manufactured buildings

794. Manufactured buildings shall have their service-entrance conductors installed after they have been erected at the building site.

    A. True   B. False

795. Electrical conductors and equipment for manufactured buildings shall be protected where subjected to physical damage while in transit.

    A. True   B. False

Article 547
Agricultural buildings

796. Cables installed in agricultural buildings shall be secured within _____ in. of boxes.

797. Motors shall be totally _____ so as to minimize the entrance of dust, moisture or corrosive particles.

798. Luminaires (lighting fixtures) shall be installed to minimize the entrance of _____, foreign matter, and dust in agricultural buildings.

799. The size bonding jumper required to bond the elements of an equipotential plane in agricultural buildings shall be _____ AWG copper.

800. The site-isolating device shall simultaneously disconnect all _____ conductors from the premises wiring.

    A. Grounded   B. Ungrounded
    C. Equipment Grounding   D. All of the above

801. The operating handle of the site-isolating device shall not be more than _____ above grade.

    A. 5 ft.   B. 6 ft.
    C. 6 ft. 6 in.   D. 6 ft. 7 in.

Article 550
Mobile homes, manufactured homes, and mobile home parks

802. Bathroom receptacle outlets shall be supplied by at least one _____ ampere circuit.

803. The power supply to a mobile home shall be a feeder assembly consisting of not more than one listed _____ ampere mobile home power supply cord. (General rule)

    A. 30   B. 40   C. 50   D. 100

804. Ground-fault circuit interrupters in mobile homes shall be provided for receptacle outlets located within _____ ft. of any sink.

    A. 5   B. 6   C. 10   D. 12

805. Receptacle outlets in mobile homes shall be installed so that no point along the floor line is more than _____ ft. measured horizontally from an outlet in that space.

    A. 2   B. 5   C. 6   D. 12

Test Questions

Section   Answer

806. The small appliance load is calculated at _____ VA for each 20 amp small appliance receptacle circuit.

      A. 1200   B. 1500   C. 1800   D. 2000

807. The service equipment shall be located in sight from and not more than _____ ft. from the exterior wall of the mobile home it serves.

      A. 10   B. 20   C. 25   D. 30

808. Nonmetallic cable located _____ in. or less above the floor, if exposed, shall be protected from physical damage.

      A. 6   B. 10   C. 15   D. 18

809. There shall be at least _____ small appliance circuits provided in mobile homes.

      A. Two   B. Three   C. Four   D. None of the above

Article 551
Recreational vehicles and recreational vehicle parks

810. Receptacle outlets shall be installed in recreational vehicles so that no point along the wall is more than _____ ft. from an outlet. (General rule)

      A. 3   B. 5   C. 6   D. 12

811. Recreational vehicles with a maximum of five 15 or 20 ampere circuits shall use a listed _____ ampere or larger main power-supply assembly.

      A. 20   B. 25   C. 30   D. 50

Article 553
Floating buildings

812. One set of service-entrance conductors shall be permitted to supply more than _____ set(s) of service equipment.

      A. One   B. Two   C. Three   D. None of the above

813. Floating buildings shall be supplied by _____ set(s) of feeder conductors from their service equipment. (General rule)

      A. One   B. Two   C. Both A and B   D. Neither A or B

Article 600
Electric signs and outline lighting

814. Wiring methods used to wire sign and outline lighting shall be required to terminate in the enclosure of the sign or transformer. (General rule)

      A. True   B. False

815. The sign enclosure shall be used to route a branch-circuit to supply floodlights to illuminate parking lots.

      A. True   B. False

816. Illuminated, portable outdoor signs shall be provided with a GFCI in the attachment cap of the supply cord.

    A. True   B. False

817. A 15 amp branch-circuit shall be brought to the front of each commercial store building to serve a sign with grade level access.

    A. True   B. False

818. A 20 amp branch-circuit for a sign shall be permitted to feed combination loads such as ballasts, lamps, and transformers.

    A. True   B. False

819. Metal poles supporting signs shall not be permitted to enclose supply conductors.

    A. True   B. False

820. Each sign shall be controlled by an circuit breaker that will open all ungrounded conductors.

    A. True   B. False

821. The disconnecting means for a sign shall be within sight of the sign that is controls.

    A. True   B. False

822. Portable signs shall not be required to be provided with factory-installed ground-fault circuit-interrupter protection for personnel.

    A. True   B. False

823. Circuits that only supply neon tubing installations shall not be rated in excess of _____ amperes.

    A. 15   B. 20   C. 30   D. 40

824. The branch-circuit load for a sign shall be calculated at a minimum of _____ volt-amperes.

    A. 1000   B. 1200   C. 1800   D. 2400

825. Where flexible nonmetallic conduit is used to install the secondary wiring of a transformer and a bonding jumper is required to bond metal electrode receptacles, the bonding conductor shall not be smaller _____ AWG copper.

    A. 14   B. 10   C. 12   D. 6

826. A sign shall be at least _____ ft. above areas accessible to vehicles.

    A. 10   B. 12   C. 14   D. 18

827. Each ballast shall be provided with a working space at least _____ ft. wide.

    A. 2   B. 3   C. 4   D. 5

Test Questions

Section   Answer

Article 604
Manufactured wiring systems

828. Locking receptacles and connectors shall be used in manufactured wiring systems.

    A. True   B. False

829. Component parts used in a manufactured wiring system shall be listed as an assembly for that system.

    A. True   B. False

Article 605
Office furnishings

830. Receptacle outlets for office furnishings shall be permitted to be installed in lighting accessories.

831. Individual partitions for office furnishings shall be permitted to have multiwired circuits.

    A. True   False

Article 610
Cranes and hoists

832. Contact conductors shall be installed in raceways.

    A. True   B. False

833. Taps without _____ overcurrent protection shall be permitted to brake coils.

834. Secondary conductors for resistors separated from controllers having a time of 5 seconds on and 75 seconds off have a demand of _____ percent for cranes.

835. Conductors external to motors and controls on cranes shall be at least _____ AWG. (General rule)

836. Main contact conductors carried along runways shall be supported on insulating supports placed at intervals not exceeding _____ ft.

    A. 10   B. 18   C. 20   D. 22

837. The continuous ampere rating of the circuit breaker shall not be less than _____ percent of the combined short-circuit ampere rating of the motors used for cranes.

    A. 20   B. 40   C. 50   D. 75

838. When servicing cranes that are energized, a working space in the direction of access to live parts shall be a minimum of _____ ft.

    A. 2   B. 2 1/2   C. 3   D. 3 1/2

Article 620
Elevators, dumbwaiters, escalators, moving walks, wheelchair lifts, and stairway chair lifts

839. Secondary conductors for resistors separated from controllers having a time of 10 seconds on and 70 seconds off shall have a demand of _____ percent for cranes.

840. Feeder conductors without _____ can be installed in a hoistway of an existing elevator.

841. The sum of the cross-sectional area of the individual conductors in raceways shall not exceed _____ percent of the interior cross-sectional area of the raceway.

842. The disconnecting means for escalators and moving sidewalks shall be located so that it is readily accessible to _____ persons.

843. Traveling cables for lighting circuits shall be permitted in parallel for conductors in sizes _____ AWG or larger, provided the ampacity is equivalent to at least that of 14 AWG copper.

    A. 22   B. 20   C. 18   D. 16

844. Flexible metal conduit of 3/8 in. nominal trade size shall be permitted between control panels and machine motors in elevators when not exceeding _____ ft. in length.

    A. 6   B. 10   C. 12   D. 18

845. Vertical runs of wireways in elevators shall be securely supported at intervals not exceeding _____ ft.

    A. 10   B. 12   C. 15   D. 20

846. Traveling cable shall be permitted to be run without the use of a raceway for a distance not exceeding _____ ft. in length as measured from the first point of support on the elevator car.

    A. 3   B. 5   C. 6   D. 10

847. Which of the following receptacles are required to be installed with ground-fault circuit-interrupter protection:

    A. Where installed in pits
    B. Where installed in machinery spaces
    C. Where installed in escalators
    D. All of the above

848. Traveling cables shall be permitted to be supported by looping the cables around supports for unsupported lengths less than _____ ft.

    A. 25   B. 50   C. 100   D. 150

Article 625
Electric vehicle charging system equipment

849. Where cord-and-plug connected electric vehicle supply equipment is used, the ground-fault circuit-interrupter protection for personnel shall be an integral part of the attachment plug or shall be located in the power supply cable not more than _____ in. from the attachment plug.

    A. 3   B. 6   C. 12   D. 18

850. For vehicle supply equipment rated more than _____ amps, the disconnecting means shall be provided and installed in a readily accessible location.

    A. 30   B. 40   C. 50   D. 60

851. Unless specifically listed for the purpose and location, the coupling means of the electric vehicle supply equipment shall be stored or located at a height of not less than _____ in. and not more than _____ ft. above the floor level.

    A. 12; 2   B. 12; 4   C. 18; 2   D. 18; 4

852. Unless specifically listed for the purpose and location, the coupling means of the electric vehicle supply equipment shall be stored or located at a height of not less than _____ in. and not more than _____ ft. above the parking surface.

    A. 12; 4   B. 18; 4   C. 24; 4   D. 30; 4

Article 630
Electric welders

853. Nonmotor generator arc welders without motors that have a duty cycle of 80 percent shall have an multiplier of _____ percent.

    A. 78   B. 84   C. 89   D. 95

854. Motor-generator arc welders with motors shall have an overcurrent protection rated or set at not more than _____ percent of the rated primary current of the welder.

    A. 150   B. 200   C. 300   D. 400

855. Resistance welders that have a duty cycle of 30 percent shall have a multiplier of _____ percent applied for sizing conductors.

    A. 50   B. 55   C. 63   D. 71

856. Resistance welders shall have an overcurrent device rated or set at not more than _____ percent of the ampacity of the supply conductors serving the welder.

    A. 150   B. 200   C. 300   D. 400

857. Welding cables supported by cable trays shall provide supports at not greater than _____ in. intervals to ensure proper support.

    A. 3   B. 4   C. 5   D. 6

Article 645
Information technology equipment

858. Branch-circuit conductors to data processing equipment shall be calculated at _____ percent of the connected load. (Based on the OCPD)

    A. 100   B. 115   C. 125   D. 150

859. Unless raised floors, Type DP data processing cables used to connect computer units shall be _____ as part of the system.

    A. Approved   B. Listed   C. Labeled   D. None of the above

Section	Answer

860. Branch-circuits to receptacles under raised floors in computer rooms shall be wired with _____.

    A. EMT   B. IMC   C. AC cable   D. All of the above

Article 650
Pipe organs

861. Self-excited generators supplying power to organs shall have a potential of not more than _____ volts.

    A. 15   B. 24   C. 30   D. 60

862. Circuit conductors to organs shall be protected with OCPD's rated at _____ amps or less.

    A. 6   B. 10   C. 15   D. 20

863. The smallest conductor used for organ circuits shall be _____ AWG for electromagnetic valve supply and the like.

    A. 14   B. 16   C. 18   D. 26

Article 660
X-ray equipment

864. The momentary rating used to size branch-circuit conductors to X-ray equipment shall be _____ percent.

    A. 25   B. 50   C. 75   D. 80

865. The long-time rating used to select OCPD's to protect circuits to X-ray equipment shall be _____ percent.

    A. 25   B. 50   C. 75   D. 100

866. The long-time rating for X-ray equipment shall be based on an operating time of _____ minutes or longer.

    A. 5   B. 7   C. 8   D. 9

Article 665
Induction and dielectric heating equipment

867. The generator output circuit shall be permitted to be _____ from ground when supplying components of induction heating equipment.

868. The supply circuit disconnecting means shall be permitted to serve as the _____ for an individual piece of dielectric heating equipment.

869. Circuit conductors supplying power to equipment other than a motor/generator for induction heating shall be sized at not less than _____ percent of the nameplate rating.

Test Questions

Section   Answer

Article 668
Electrolytic cells

870. Cell line conductors for electrolytic cells can be either covered, bare, or _____.

871. Removable links or removable conductors shall be permitted to serve as the _____ means for electrolytic cells.

Article 669
Electroplating

872. Circuit conductors supplying components of electroplating equipment shall be sized at _____ percent of the load.

873. Insulated conductors shall be permitted to be run without insulated supports for systems not exceeding _____ volts direct current.

874. Bare copper conductors can be used to connect conversion equipment to an electrolytic tank if supported on _____.

Article 670
Industrial machinery

875. Taps from _____ circuits can be made to supply power to industrial machinery.

876. All machines in industrial machinery spaces shall have individual _____ means.

Article 675
Electrically driven or controlled irrigation machines

877. Tap conductors to motors for irrigation machines shall not be less than _____ AWG.

878. Branch-circuit conductors and OCPD's shall be selected at _____ percent of the largest motor plus _____ percent of the FLC rating of the remaining motors.

879. The LRC in amps of irrigation machinery shall equal _____ times the LRC of the largest motor plus _____ percent of the FLC rating of the remaining motors.

880. Irrigation cables shall be supported at intervals not exceeding _____ ft.

Article 680
Swimming pools, fountains, and similar installations

881. Switching devices shall be located at least 5 ft. from the inside walls of the pool.

    A. True   B. False

882. Existing luminaires (lighting fixtures) located less than 5 ft. horizontally and 5 ft. vertically from the maximum water level of the pool shall be permitted.

    A. True   B. False

12-81

Section   **Answer**

_____ _____ 883. Receptacles located behind walls having hinged or sliding doors shall be permitted to be located within 10 ft. of the pool.

    A. True   B. False

_____ _____ 884. Receptacles used to supply power for a recirculating pump motor shall be permitted to be installed within 5 ft. to 10 ft. of the inside walls of the pool.

    A. True   B. False

_____ _____ 885. Flexible cords supplying fixed swimming pool equipment shall not exceed 3 ft. in length.

    A. True   B. False

_____ _____ 886. GFCI-protection shall be provided for all existing receptacles within 5 ft. of a hydromassage tub installed in an existing bathroom.

    A. True   B. False

_____ _____ 887. Motor circuits for electrical pool covers shall be GFCI-protected.

    A. True   B. False

_____ _____ 888. Receptacles shall be installed at least _____ ft. from the inside walls of a swimming pool.

    A. 5   B. 10   C. 15   D. 20

_____ _____ 889. Receptacles shall be installed at least a minimum of _____ ft. from and not more than _____ ft. from the inside wall of a permanently installed swimming pool for a dwelling unit.

    A. 5; 10   B. 5; 20   C. 10; 15   D. 10; 20

_____ _____ 890. Receptacles shall be located not more than _____ ft. above the grade level that serves the pool for a dwelling unit.

    A. 6   B. 6 1/2   C. 7   D. 7 1/2

_____ _____ 891. Receptacles located within _____ ft. of the inside walls of a swimming pool shall be protected by a ground-fault circuit-interrupter.

    A. 5   B. 10   C. 15   D. 20

_____ _____ 892. Ceiling fans shall not be installed over the pool or over the area extending _____ ft. horizontally from the inside walls of a pool.

    A. 3   B. 5   C. 6   D. 7 1/2

_____ _____ 893. Luminaires (lighting fixtures) can be installed over an outside pool if no part of the lighting unit is less than _____ ft. above the maximum water level.

    A. 6   B. 7 1/2   C. 10   D. 12

_____ _____ 894. Lighting units can be installed over an indoor pool area if the distance from the bottom of the fixture to the maximum water level is not less than _____ ft.

    A. 6   B. 7 1/2   C. 10   D. 12

Test Questions

Section   Answer

895. Switching devices shall be installed at least _____ ft. horizontally from the inside walls of a pool.

    A. 3   B. 4   C. 5   D. 6

896. Underground wiring shall not be installed within _____ ft. horizontally from the inside wall of the pool. (General rule)

    A. 5   B. 10   C. 12   D. 18

897. No luminaires (lighting fixtures) shall be installed for operation on supply circuits over _____ volts between conductors.

    A. 150   B. 240   C. 277   D. 480

898. Lighting units mounted in the walls of swimming pools shall be installed with the top of the fixture lens at least _____ in. below the normal level of the pool. (general rule)

    A. 4   B. 6   C. 12   D. 18

899. A _____ AWG insulated copper conductor shall be routed in PVC conduit to ground to the metal forming shell of a wet-niche fixture in the pool.

    A. 12   B. 10   C. 8   D. 6

900. The deck box shall have a height of _____ in. measured from the deck or _____ in. measured from the water, whichever is greater.

    A. 2; 6   B. 4; 6   C. 4; 8   D. 6; 10

901. The deck box located behind a solid permanent barrier shall be permitted to be located less than _____ ft. from the inside walls of the pool.

    A. 3   B. 4   C. 5   D. 6

902. All metallic parts (equipotential bonding grid) of a swimming pool shall be bonded with a _____ AWG solid copper bonding conductor.

    A. 12   B. 10   C. 8   D. 6

903. Wet-niche luminaires (lighting fixtures) for swimming pools shall be connected to an _____ AWG insulated equipment grounding conductor.

    A. 16   B. 14   C. 12   D. 10

904. At least one receptacle shall be installed a minimum of _____ ft. from and not more than _____ ft. from the inside wall of a spa or hot tub.

    A. 5; 10   B. 5; 20   C. 10; 15   D. 10; 20

905. Receptacles located within _____ ft. of the inside walls of a spa or hot tub shall be protected by a ground-fault circuit-interrupter.

    A. 5   B. 6   C. 10   D. 20

906. Switches shall be installed at least _____ ft. from the inside walls of a spa or hot tub.

    A. 5   B. 6   C. 10   D. 20

Section **Answer**

907. All receptacles within _____ ft. of the inside walls of a hydromassage tub shall be protected by a ground-fault circuit-interrupter.

   A. 3   B. 4   C. 5   D. 10

Article 690
Solar photovoltaic systems

908. A photovoltaic power source having one conductor of a two-wire system over _____ volts shall be solidly grounded.

   A. 50   B. 125   C. 150   D. 277

909. Storage batteries in solar photovoltaic systems for dwellings shall have cells operating at less than _____ volts.

   A. 12   B. 20   C. 30   D. 50

910. Solar photovoltaic systems in dwelling units shall have OCPDs protecting circuits to busbars sized at _____ percent less than the busbars.

   A. 100   B. 110   C. 120   D. 125

911. A _____ phase power conditioning unit shall automatically disconnect all ungrounded conductors of a solar photovoltaic system.

   A. 1   B. 2   C. 3   D. None of the above

912. Solar photovoltaic systems in a one-family dwelling unit with circuits rated over _____ volts-to-ground while energized shall not be accessible to other than qualified persons.

   A. 50   B. 150   C. 175   D. 200

Article 700
Emergency systems

913. A sign shall be placed at the service-entrance equipment indicating type and location of on-site emergency power sources.

   A. True   B. False

914. Emergency equipment shall be rated to operate all loads simultaneously.

   A. True   B. False

915. Ground-fault indicating means shall not be required on emergency systems.

   A. True   B. False

916. Emergency systems shall be tested periodically to verify proper operation.

   A. True   B. False

917. Inadvertent paralleling shall be permitted for isolation switches used to isolate transfer switch equipment.

   A. True   B. False

Test Questions

Section   Answer

918. The general rule requires all emergency wiring to be routed independently of the normal wiring of the electrical system.

    A. True   B. False

919. A separate service shall not be used for emergency power.

    A. True   B. False

920. A tap ahead of the main to serve an emergency system shall not be made to supply electrical power to such components.

    A. True   B. False

921. Under certain conditions, a generator shall be permitted to take more than 10 seconds to develop power to serve the emergency system loads.

    A. True   B. False

922. Emergency wiring and _____ wiring shall be permitted to enter the same enclosure to supply power to transfer equipment.

923. An emergency generator shall be permitted to be used for peak load _____.

924. Unit equipment shall be permitted to provide emergency power to _____ offices should the normal power fail.

925. Switches used to control emergency circuits shall be located so only _____ persons can have control.

926. Switches connected in _____ shall not be used to control emergency lighting circuits.

927. OCPD's for branch-circuits supplying emergency power equipment shall be _____ to authorized personnel only.

928. Feeder-circuit wiring for emergency systems shall be installed either in spaces fully protected by approved automatic fire suppression systems or shall be a listed electrical circuit protective system with a _____ hour fire rating.

    A. 1   B. 2   C. 3   D. 4

929. Storage batteries used as a source of power for emergency systems shall be of suitable rating and capacity to supply and maintain the total load for a period of _____ hours minimum.

    A. 1   B. 1 1/2   C. 2   D. 2 1/2

930. Battery pack units shall be permitted to supply emergency power if connected _____.

    A. On the lighting circuit of the area
    B. On any receptacle circuit
    C. On any branch-circuit
    D. Ahead of the main

931. An emergency system can be used for _____ load shaving in an industrial complex.

    A. Peak   B. Optional   C. Both A and B   D. Neither A or B

12-85

Section   Answer

932. All _____ containing emergency circuits shall be marked so they are readily identifiable from other circuits.

    A. Boxes   B. Enclosures   C. Both A and B   D. Neither A or B

933. A separate branch-circuit for unit equipment with a lock-on device can be routed to a(n) _____ area with three additional lighting circuits.

    A. Uninterrupted   B. Hazardous
    C. Both A and B   D. Neither A or B

934. Power of the emergency system shall be maintained to ensure the _____ of high-intensity discharge lighting until normal power is restored.

    A. Operation   B. Restriking   C. Both A and B   D. Neither A or B

935. Emergency power conductors supplying a building can be a _____.

    A. Storage battery   B. Generator set
    C. Separate service   D. All of the above

936. Where internal combustion engines are used as the prime mover, an on-site fuel supply shall be provided with an on-premise fuel supply sufficient for not less than _____ hours full-demand operation of the system.

    A. 1   B. 1 1/2
    C. 2   D. 2 1/2

Article 701
Legally required standby systems

937. Legally required standby systems shall be required by all governmental agencies having jurisdiction.

    A. True   B. False

938. The authority having jurisdiction shall not be required to witness a test on the complete installation of a legally required standby system.

    A. True   B. False

939. A written record shall be kept on test results of legally required standby systems.

    A. True   B. False

940. The wiring of legally required standby systems shall not occupy the same raceways and junction boxes as general wiring.

    A. True   B. False

941. Prime movers shall be solely dependent upon public utility gas companies for fuel supply for legally required standby systems.

    A. True   B. False

942. A tap ahead of the main supplying a legally required standby system can be made if the service main has sufficient separation.

    A. True   B. False

Test Questions

Section   Answer

943. Generators supplying power to legally required standby systems shall have an on-site fuel supply that provides fuel for at least 1 1/2 hours.

    A. True   B. False

944. Legally required standby systems shall be located to minimize hazards, floods, fire, etc.

    A. True   B. False

Article 705
Interconnected electric power production sources

945. The outputs of electric power production systems shall be _____ at the premises service disconnecting means.

946. Equipment intended to be operated and maintained as an internal part of a power production source exceeding _____ volts shall not be required to have disconnecting means.

947. An electric power production source shall be disconnected if the _____ power source is lost.

Article 710 is now Article 490
Over 600 volts, nominal, general

948. Where fuses can be energized by _____, a sign shall be placed on the enclosure door identifying this hazard.

949. Electrode-type boilers shall be supplied from an individual branch-circuit rated not less than _____ percent of the total load.

950. Control circuit voltages for electrode-type boilers shall not exceed _____ volts.

951. Switches with fused interrupters shall have the terminals for supply conductors located at the _____ of the enclosure.

952. Cutouts shall be located so that they are readily and safely accessible for re-fusing, with the top of the cutout not over _____ ft. above the floor or platform.

    A. 3   B. 5   C. 10   D. 15

953. Control and instrument transfer switch handles shall be in a readily accessible location at an elevation of not over _____ in.

    A. 48   B. 60   C. 67   D. 78

954. Oil-filled cutouts shall be so located that they will be readily and safely accessible for re-fusing, with the top of the cutout not over _____ ft. above the floor or platform.

    A. 3   B. 5   C. 6   D. 10

955. Operating handles requiring more than _____ lb. of torque shall be located and higher than 66 in. in either the open or closed position.

    A. 10   B. 15   C. 25   D. 50

Section	Answer

Article 725
Class 1, Class 2, and Class 3 remote-control, signaling, and power-limited circuits

956. Class 2 cables shall have a voltage rating of not less than _____ volts.

   A. 120   B. 150   C. 240   D. 300

957. A Class 1 nonpower limited remote-control circuit shall not exceed _____ volts.

   A. 50   B. 120   C. 240   D. 600

958. Class 1 circuit overcurrent protection shall not exceed _____ amps for 16 AWG conductors.

   A. 5   B. 7   C. 10   D. 12

959. Class 1 circuits and power supply circuits shall be permitted to occupy the same raceway only where _____.

   A. Equipment powered is functionally associated
   B. There is not a mixture of alternating or direct current circuits
   C. Class 1 circuits and power supply circuits are not of the same voltage
   D. None of the above

960. Conductors of Class 2 and Class 3 circuits shall be separated by at least _____ in. from conductors of any Class 1 circuits.

   A. 1   B. 2   C. 3   D. 4

961. Class 3 cables shall have a voltage rating of not less than _____ volts.

   A. 120   B. 150   C. 240   D. 300

Article 760
Fire alarm systems

962. Nonpower-limited fire alarm circuits shall have overcurrent protection not to exceed _____ amps for 18 AWG conductors.

   A. 5   B. 7   C. 10   D. 12

963. Multiconductor nonpower-limited fire alarm circuit cables located within 7 ft. of the floor cables shall be securely fastened at intervals of not more than _____ in.

   A. 3   B. 6   C. 12   D. 18

964. Power-limited fire alarm circuit conductors can be installed in the same raceway with which of the following wiring methods?

   A. Power conductors   B. Class 1 conductors
   C. Nonpower-limited circuit conductors   D. None of the above

	Section	Answer

965. Power-limited fire alarm circuit conductors shall be separated at least _____ in. from conductors of any electric light.

    A. 1   B. 2   C. 3   D. 4

966. Nonpower-limited fire alarm circuits shall have overcurrent protection not to exceed _____ amps for 16 AWG conductors.

    A. 5   B. 7   C. 10   D. 12

967. Multiconductor nonpower-limited fire alarm circuit cables installed within 7 ft. of the floor shall be securely fastened in an approved manner at intervals of not more than _____ in.

    A. 18   B. 24   C. 30   D. 36

Article 800
Communications circuits

968. Communication circuit conductors shall be routed independently of other circuit wiring. (General rule)

    A. True   B. False

969. The general rule requires communication circuits routed in shafts to be separated at least 1 1/2 in. from other power circuits.

    A. True   B. False

970. Communication circuits installed in a building can be Type CM cables.

    A. True   B. False

971. Communication cables routed vertically in a shaft shall be CMR cables.

    A True   B. False

972. Listed type CMCU communication cables can be placed under carpets.

    A. True   B. False

973. Supply service drops of less than 750 volts running above and parallel to communications service drops shall be permitted to have a minimum separation of 12 in. at any point in the span.

    A. True   B. False

974. Communication conductors passing over buildings shall have a clearance of 5 ft. if the roof can be easily walked upon.

    A. True   B. False

975. Open communication conductors shall be separated from open power conductors on buildings by at least 4 in.

    A. True   B. False

Section   Answer

976. Communication wires and cables shall have a vertical clearance of not less than _____ ft. from all points of roofs above which they pass.

    A. 8   B. 10   C. 12   D. 18

977. Communication wires and cables shall be separated at least _____ in. from electric power and light conductors if not in a raceway or cable.

    A. 2   B. 4   C. 6   D. 10

978. Where practicable, a separation of at least _____ ft. shall be maintained between open conductors of communication systems on buildings and lightning conductors.

    A. 6   B. 8   C. 10   D. 12

979. A bonding jumper not smaller than _____ AWG cu. shall be connected between the communications grounding electrode and power grounding electrode system at the building.

    A. 10   B. 8   C. 6   D. 4

980. A grounding electrode shall be bonded to the metal frame or available grounding terminal of the mobile home with a copper grounding conductor not smaller than _____ AWG.

    A. 12   B. 10   C. 8   D. 6

Article 810
Radio and television equipment

981. Television antennas and indoor lead-ins shall not be routed nearer than _____ in. from the general wiring.

982. Television antennas and indoor lead-ins can occupy the same box with other general wiring provided there is a(n) _____ between the wiring.

983. The minimum size copper grounding electrode conductor for grounding radio and television equipment is _____ AWG.

984. All inside conductors for transmitting stations shall be separated at least _____ in. from electric power circuits. (General rule)

985. The operating grounding electrode conductor for transmitting stations shall not be smaller than _____ AWG copper.

986. All external metal handles used by personnel to control a transmitter shall be _____ effectively.

987. Antenna conductors for television equipment shall be installed so they will not _____ under open electric light or power conductors where possible.

    A. Parallel   B. Cross   C. Both A and B   D. Neither A or B

988. Outdoor antennas and lead-in connectors shall not be attached to poles of over _____ volts between conductors.

    A. 120   B. 150   C. 250   D. 300

Article 820
Community antenna television and radio distribution systems

989. Coaxial cable shall have a separation of at least _____ in. from Class I circuit conductors.

    A. 1   B. 2   C. 3   D. 4

990. A separation of at least _____ ft. shall be maintained between any coaxial cable and lightning conductors.

    A. 3   B. 5   C. 6   D. 10

991. Service equipment grounding electrodes shall be bonded to the antenna system electrode with a _____ AWG copper conductor.

    A. 6   B. 8   C. 12   D. 14

992. The grounding conductor shall not be smaller than _____ AWG.

    A. 18   B. 16   C. 14   D. 12

Chapter 9
Tables

993. A conduit system is any run over 18 in. in length.

    A. True   B. False

994. A conduit system with more than two conductors is allowed 40 percent fill area.

    A. True   B. False

995. A nipple is _____ in. or less in length.

996. Derating for four or more conductors shall not be required for _____.

997. Nipples are permitted to be filled to _____ percent of their total cross-sectional area.

Appendix C
Conduit and tubing fill Tables for conductors

998. What size (EMT) conduit is required to enclose 12 - 10 AWG THWN copper conductors?

    A. 1/2 in.   B. 3/4 in.   C. 1 in.   D. 1 1/2 in.

999. What size (RMC) conduit is required to enclose 18 - 10 AWG THWN copper conductors?

    A. 1 1/4 in.   B. 1 1/4 in.   C. 2 in.   D. 3 in.

1000. What size (FMC) conduit is required to enclose 8 - 10 AWG THWN copper conductors?

    A. 1/2 in.   B. 3/4 in.   C. 1 in.   D. 1 1/4 in.

Test Problems

13

Chapter 13 is a compilation of problems based on the 2005 NEC. These problems are designed to reinforce as well as test the skills of each individual on problems that may appear on the Master's examination.

All problems are either illustrated and discussed in each chapter of this book, or they can be found in the 2005 NEC. For example, to solve a problem properly, the user may need to refer to the NEC instead of the *Master Electrician's Study Guide* to obtain the requirements for determining his or her answer. This procedure will prevent the user from depending completely on the information in this book and help the user learn to use the NEC as well. Each test in Chapter 13 begins with a listing of NEC references that will substantiate the answer determined for that particular problem.

Chapter 13 contains two-hundred and fifty (250) problems based on the NEC. Students can use these tests after studying each chapter to verify how much of the material they have retained during each session.

CHAPTER 13

ARTICLE 110	13-3
ARTICLE 210	13-3
ARTICLE 215	13-3
ARTICLE 220	13-4
ARTICLE 230	13-26
ARTICLE 240	13-26
ARTICLE 250	13-27
ARTICLE 310	13-28
ARTICLE 318	13-28
ARTICLE 370	13-28
ARTICLE 374	13-30
ARTICLE 422	13-30
ARTICLE 424	13-30
ARTICLE 430	13-31
ARTICLE 440	13-32
ARTICLE 450	13-33
ARTICLE 455	13-34
ARTICLE 460	13-34
ARTICLE 550	13-34
ARTICLE 600	13-35
ARTICLE 630	13-35
CHAPTER 9 TABLES AND EXAMPLES	13-35

Test Problems

Section   Answer

Chapter 13
Test Problems

Article 110
Requirements for electrical installations

1. What is the ampacity of a 2 AWG THWN copper conductor connected to an overcurrent protection device supplying power to equipment where all terminals are rated 60°C?

2. What is the ampacity of a 4/0 AWG THWN copper conductor connected to an overcurrent protection device supplying power to equipment where all terminals are rated 75°C?

Article 210
Branch-circuits

3. What is the voltage drop for the branch-circuit conductors run 200 ft. with 3 AWG THWN copper conductors with a 100 amp load supplied by a 240 volt, single-phase circuit with a voltage drop of 3 percent applied?

4. What is the voltage drop rating for branch-circuit conductors run 200 ft. with 2 AWG THWN copper conductors with a 100 amp load supplied by a 240 volt, three-phase circuit with a voltage drop of 3 percent applied?

5. What size OCPD and THWN copper conductors are required for a branch-circuit lighting load with a 5 amp noncontinuous load and 12 amp continuous load?

6. What size OCPD and THWN copper conductors are required for the branch-circuit lighting load with a 10 amp noncontinuous load and 8 amp continuous load?

7. Can a single-phase, 120 volt, A/C window unit with a 8 amp load be connected to an existing 20 amp (OCPD) branch-circuit?

Article 215
Feeder-circuits

8. What is the circular mil rating for feeder-circuit conductors run 100 ft. with a 84 amp computed load supplied by 240/120 volt, three-wire system with a voltage drop of 3 percent applied? (CM is based on VD calculation)

9. What size THWN copper conductors are required to supply a feeder-circuit with a continuous load of 150 amps and a noncontinuous load of 40 amps?

10. What size THWN copper conductors are required to supply a feeder-circuit with a continuous load of 172 amps and a noncontinuous load of 80 amps?

11. What size THWN copper conductors are required to supply the secondary side of a transformer supplying a continuous load of 50 amps and noncontinuous load of 20 amps?

12. What size THWN copper conductors are required to supply the secondary side of a transformer supplying a continuous load of 68 amps and noncontinuous load of 38 amps?

13. What size THWN copper conductors are required to supply the secondary side of a transformer having a continuous load output of 194 amps?

Section   Answer

14. What size THWN copper conductors are required to supply the secondary side of a transformer having a continuous load output of 212 amps?

Article 220
Branch-circuit, feeder, and service calculations

15. What is the general lighting load and receptacle load and small appliance plus laundry loads for a 3000 sq. ft. dwelling unit? (Compute the load in VA)

16. What is the general lighting load and receptacle load and small appliance plus laundry loads for a 3200 sq. ft. dwelling unit? (Compute the load in VA)

17. What is the general lighting load and receptacle load and small appliance plus laundry load for 20 - 980 sq. ft. dwelling units in a complex? (Compute the load in VA)

18. What is the general lighting load and receptacle load and small appliance plus laundry load for 15 - 1280 sq. ft. dwelling units in a complex? (Compute the load in VA)

19. What is the general lighting load and receptacle load and small appliance plus laundry in a multifamily dwelling unit with the following area? (Compute the load in VA)

    • 10 - 875 sq. ft. dwelling units
    • 10 - 920 sq. ft. dwelling units
    • 10 - 1050 sq. ft. dwelling units

20. What is the general lighting load and receptacle load and small appliance plus laundry in a multifamily dwelling unit with the following area? (Compute the load in VA)

    • 5 - 1000 sq. ft. dwelling units
    • 5 - 1050 sq. ft. dwelling units
    • 5 - 1200 sq. ft. dwelling units
    • 5 - 1280 sq. ft. dwelling units

21. What is the load in VA for the general-purpose lighting load in a 7000 sq. ft. office facility used continuously?

22. What is the load in VA for the general-purpose lighting load in a 8400 sq. ft. church used continuously?

23. What is the load in VA for the general-purpose lighting load in a 6800 sq. ft. bank used continuously?

24. What is the load in VA for the general-purpose lighting load in a 4800 sq. ft. restaurant used continuously?

25. What is the load in VA for the general-purpose lighting load in a 2400 sq. ft. beauty salon used continuously?

26. What is the lighting load in VA for the patient bedrooms in a 52,000 sq. ft. hospital?

27. What is the load in VA for the general-purpose lighting load in a hotel or motel with 50 units of 420 sq. ft. each?

28. What is the load in VA for the general-purpose lighting load in a 40,000 sq. ft. store including 20,000 sq. ft. of warehouse space? (Used at continuous operation)

Test Problems

	Section	Answer

29. What is the load in VA for the general-purpose lighting load in an school building with the following area: (Compute at continuous operation)

    • 20,000 sq. ft. classroom area
    • 4000 sq. ft. auditorium area
    • 1000 sq. ft. assembly hall area

30. What is the load in VA for 50, 120 volt, lighting ballast rated at 1.5 amps each and used for 8 hours a day?

31. What is the load in VA for the following luminaires (lighting fixtures):

    • 10 - 120 VA (each) recess luminaires (lighting fixtures)
    • 12 - 150 VA (each) incandescent luminaires (lighting fixtures)
    • 22 - 180 VA (each) electric discharge luminaires (lighting fixtures)
    • Compute at continuous operation

32. What is the lighting load in VA for 140 ft. of show window used at noncontinuous operation?

33. What is the lighting load in VA for 140 ft. of show window used at continuous operation?

34. What is the lighting load in VA for 100 ft. of track lighting used at noncontinuous operation?

35. What is the lighting load in VA for 100 ft. of track lighting used at continuous operation?

36. What is the lighting load in amps for a low-voltage lighting system supplied by an isolation transformer with a FLA of 30 amps used at noncontinuous operation?

37. What is the lighting load in amps for a low-voltage lighting system supplied by an isolation transformer with a FLA of 30 amps used at continuous operation?

38. What is the lighting load in VA for 15 continuous operated luminaires (lighting fixtures) with a 75 VA ballast in each unit and 15 noncontinuous operated luminaires (lighting fixtures) with 75 VA in each unit?

39. What is the lighting load in VA for 50 continuous operated luminaires (lighting fixtures) with a 150 VA ballast in each unit?

40. What is the lighting load in VA for a sign rated 1200 VA operating for six hours?

41. What is the lighting load in VA for a sign rated 2800 VA operating for three hours?

42. What is the demand load in VA for a 11 kW range?

43. What is the demand load in VA for a 10 kW and 12 kW cooking units?

44. What is the demand load in VA for 28 - 12 kW ranges installed in an apartment complex?

45. What is the demand load in VA for 55 - 11 kW ranges installed in an apartment complex?

46. What is the demand load in VA for a piece of cooking equipment with a 3 kW rating?

13-5

Stallcup's Master Electrician's Study Guide

Section   Answer

_____ _____ 47. What is the demand load in VA for two pieces of cooking equipment with a 3 kW rating?

_____ _____ 48. What is the demand load in VA for 12 - 2.5 kW ovens and 12 - 3 kW cooktops installed in an apartment complex?

_____ _____ 49. What is the demand load in VA for a piece of cooking equipment with a 8 3/4 kW rating?

_____ _____ 50. What is the demand load in VA for two pieces of cooking equipment with each having a 8 3/4 kW rating?

_____ _____ 51. What is the demand load in VA for 15 - 8 3/4 kW ovens and 15 - 8 3/4 kW cooktops installed in an apartment complex?

_____ _____ 52. What is the demand load in VA for 10 - 3 kW cooktops and 10 - 8 3/4 kW ovens installed in an apartment complex?

_____ _____ 53. What is the demand load in VA for a range with a 16 kW rating?

_____ _____ 54. What is the demand load in VA for 25 - 14 kW ranges installed in an apartment complex?

_____ _____ 55. What is the demand load in VA for 5 - 12 kW ranges, 5 - 14 kW ranges, 5 - 16 kW ranges and 5 - 18 kW ranges installed in an apartment complex supplied by an 208 volt, single-phase system? (**Note:** Service voltage is 208 V, 3Ø, 4-wire system)

_____ _____ 56. What is the demand load in VA for three pieces of cooking equipment with a 10 kW, 14 kW and 16 kW rating installed in a dwelling unit?

_____ _____ 57. What is the branch-circuit demand load in VA for a 12 kW cooktop, 8 kW oven and 4 kW oven installed in a dwelling unit?

_____ _____ 58. What is the demand load in VA for a 4 kW dryer?

_____ _____ 59. What is the demand load in VA for a 5 kW dryer?

_____ _____ 60. What is the demand load in VA for 23 dryers rated at 6500 VA each?

_____ _____ 61. What is the fixed appliance load in VA for the ungrounded (phase) conductors with the following loads:

- 6000 VA water heater   240 volt, single-phase
- 1400 VA disposal   120 volt, single-phase
- 1050 VA compactor   120 volt, single-phase

_____ _____ 62. What is the fixed appliance load in VA for the ungrounded (phase conductors) and grounded (neutral) conductors with the following loads:

- 5000 VA water heater   240 volt, single-phase
- 2600 VA water pump   240 volt, single-phase
- 1050 VA compactor   120 volt, single-phase
- 1200 VA disposal   120 volt, single-phase
- 1600 VA dishwasher   120 volt, single-phase
- 1000 VA microwave   120 volt, single-phase
- 800 VA blower motor   120 volt, single-phase

Test Problems

	Section	Answer

63. What is the fixed appliance load in VA for the ungrounded (phase conductors) and grounded (neutral) conductors for the following:

- 6000 VA water heater   240 volt, single-phase
- 2200 VA water pump   240 volt, single-phase
- 1000 VA blower motor   240 volt, single-phase
- 1000 VA compactor   120 volt, single-phase
- 1400 VA disposal   120 volt, single-phase
- 1400 VA dishwasher   120 volt, single-phase
- 1200 VA microwave   120 volt, single-phase

64. What is the fixed appliance load in VA for the phases and neutral for the following:

- 20 - 6000 VA water heaters   240 volt, single-phase
- 20 - 1200 VA compactors   120 volt, single-phase
- 20 - 1400 VA disposals   120 volt, single-phase
- 20 - 1600 VA dishwashers   120 volt, single-phase

65. What is the load in VA for a 10 kW heating unit, 1.44 kW attic fan and 6.5 kW A/C unit installed in a dwelling unit?

66. What is the load in VA for a 10 - 20 kW heating unit and 10 - 6 kW A/C unit installed in an apartment complex?

67. What is the load in VA for 1 HP, 240 volt, single-phase motor that is to be used for the largest motor load (ungrounded (phase) conductor)?

68. What is the load in VA for 1 HP 120 volt, single-phase motor that is to be used for the largest motor load (grounded (neutral) conductor)?

69. What is the load in VA for 48 general-purpose receptacles used to serve noncontinuous operated loads?

70. What is the load in VA for 48 general-purpose receptacles used to serve continuous operated loads?

71. What is the load in VA for 128 general-purpose receptacle outlets to cord-and-plug connected loads used at noncontinuous operation?

72. What is the load in VA for 150 ft. of multioutlet assembly used to cord-and-plug connected loads that are not used simultaneously?

73. What is the load in VA for 150 ft. of multioutlet assembly used to cord-and-plug connected loads that are used simultaneously?

74. What is the load in VA for a 208 volt, three-phase, 68 amp special appliance load operating for six hours and supplied by an individual branch-circuit?

75. What is the load in VA for a 208 volt, three-phase, 68 amp special appliance load operating for two hours and supplied by an individual branch-circuit?

76. What is the demand load in VA and amps for 15, 208 volt, three-phase cooking units rated 8 kW each?

77. What is the load in VA for a group of 208 volt, three-phase motors rated at 30 HP, 20 HP and 10 HP?

78. What is the load in VA for 4 compressors rated at 26.5 amps each and supplied by a 480 volt, three-phase supply?

13-7

Stallcup's Master Electrician's Study Guide

Section   Answer

79. Compute the following loads of a 3500 sq. ft. dwelling unit using the standard calculation and size the service-entrance conductors required for Phases A and B and the grounded (neutral) conductors supplied by an 120/240 volt, single-phase system: (Compute with or without the use of the residential calculation form)

- 3500 sq. ft. dwelling unit
- 2 small appliance circuits
- 1 laundry circuit
- 6500 VA A/C unit   240 volt, single-phase
- 10,000 VA heating unit   240 volt, single-phase
- 5500 VA water heater   240 volt, single-phase
- 10,000 VA cooktop   240 volt, single-phase
- 11,000 VA oven   240 volt, single-phase
- 4500 VA dryer   240 volt, single-phase
- 1000 VA blower motor   240 volt, single-phase
- 2600 VA water pump   240 volt, single-phase
- 1050 VA disposal   120 volt, single-phase
- 1200 VA compactor   120 volt, single-phase
- 1600 VA dishwasher   120 volt, single-phase
- 1400 VA microwave   120 volt, single-phase

(A) What size THWN copper conductors are required for the service?

(B) What size OCPD is required for the service based on the computed load?

(C) What size panelboard is required for the service?

(D) What size RMC is required based on min. size conductors?

80. Compute the following loads of a 2800 sq. ft. dwelling unit using the standard calculation and size the service-entrance conductors required for Phases A and B and the grounded (neutral) conductors supplied by an 120/240 volt, single-phase system: (Compute with or without the use of the residential calculation form)

- 2800 sq. ft. dwelling unit
- 2 small appliance circuits
- 1 laundry circuit
- 6000 VA water heater   240 volt, single-phase
- 6000 VA A/C unit   240 volt, single-phase
- 20,000 VA heating unit   240 volt, single-phase
- 8500 VA cooktop   240 volt, single-phase
- 8750 VA oven   240 volt, single-phase
- 5500 VA dryer   240 volt, single-phase
- 1200 VA blower motor   240 volt, single-phase
- 2400 VA water pump   240 volt, single-phase
- 1200 VA compactor   120 volt, single-phase
- 1400 VA disposal   120 volt, single-phase
- 1600 VA dishwasher   120 volt, single-phase
- 1000 VA microwave   120 volt, single-phase

(A) What size THWN copper conductors are required for the service?

(B) What size OCPD is required for the service based on the computed load?

(C) What size panelboard is required for the service?

(D) What size RMC is required based on min. size conductors?

Test Problems

Section   Answer

81. Compute the following loads of a 3500 sq. ft. dwelling unit using the optional calculation and size the service-entrance conductors required for Phases A and B and the grounded (neutral) conductors supplied by an 120/240 volt, single-phase system: (Compute with or without the use of the residential calculation form)

- 3500 sq. ft. dwelling unit
- 2 small appliance circuits
- 1 laundry circuit
- 6500 VA A/C unit   240 volt, single-phase
- 10,000 VA heating unit   240 volt, single-phase
- 5500 VA water heater   240 volt, single-phase
- 10,000 VA cooktop   240 volt, single-phase
- 11,000 VA oven   240 volt, single-phase
- 4500 VA dryer   240 volt, single-phase
- 1000 VA blower motor   240 volt, single-phase
- 2600 VA water pump   240 volt, single-phase
- 1050 VA disposal   120 volt, single-phase
- 1200 VA compactor   120 volt, single-phase
- 1600 VA dishwasher   120 volt, single-phase
- 1400 VA microwave   120 volt, single-phase

Note: See the standard calculation in problem 79 for sizing the neutral.

(A) What size THWN copper conductors are required for the service? _____ _____

(B) What size OCPD is required for the service based on the computed load? _____ _____

(C) What size panelboard is required for the service? _____ _____

(D) What size RMC is required based on max. size conductors? _____ _____

82. Compute the following loads of a 2800 sq. ft. dwelling unit using the optional calculation and size the service-entrance conductors required for Phases A and B and the grounded (neutral) conductors supplied by an 120/240 volt, single-phase system: (Compute with or without the use of the residential calculation form)

- 2800 sq. ft. dwelling unit
- 2 small appliance circuits
- 1 laundry circuit
- 6000 VA A/C unit   240 volt, single-phase
- 20,000 VA heating unit   240 volt, single-phase
- 6000 VA water heater   240 volt, single-phase
- 8500 VA cooktop   240 volt, single-phase
- 8750 VA oven   240 volt, single-phase
- 5500 VA dryer   240 volt, single-phase
- 1200 VA blower motor   240 volt, single-phase
- 2400 VA water pump   240 volt, single-phase
- 1200 VA compactor   120 volt, single-phase
- 1400 VA disposal   120 volt, single-phase
- 1600 VA dishwasher   120 volt, single-phase
- 1000 VA microwave   120 volt, single-phase

Note: See the standard calculation in problem 80 for sizing the neutral.

(A) What size THWN copper conductors are required for the service? _____ _____

(B) What size OCPD is required for the service based on the computed load? _____ _____

(C) What size panelboard is required for the service? _____ _____

(D) What size RMC is required based on min. size conductors? _____ _____

Stallcup's Master Electrician's Study Guide

Section   Answer

83. Compute the following loads of a 2800 sq. ft. dwelling unit using the standard calculation and size the service-entrance conductors required for Phases A and B and the grounded (neutral) conductors supplied by an 120/240 volt, single-phase system: (Compute with or without the use of the residential calculation form)

- 2800 sq. ft. dwelling unit
- 2 small appliance circuits
- 1 laundry circuit
- 6000 VA A/C unit   240 volt, single-phase
- 10,000 VA heating unit   Gas related load
- 5500 VA water heater   240 volt, single-phase
- 10,000 VA oven   Gas related load
- 12,000 VA cooktop   Gas related load
- 4500 VA dryer   240 volt, single-phase
- 1000 VA blower motor   240 volt, single-phase
- 2400 VA water pump   240 volt, single-phase
- 1000 VA disposal   120 volt, single-phase
- 1200 VA compactor   120 volt, single-phase
- 1400 VA dishwasher   120 volt, single-phase
- 1000 VA microwave   120 volt, single-phase

(A) What size THWN copper conductors are required for the service?

(B) What size OCPD is required for the service based on the computed load?

(C) What size panelboard is required for the service?

(D) What size RMC is required based on max. size conductors?

84. Compute the following loads of a 2800 sq. ft. dwelling unit using the standard calculation and size the service-entrance conductors required for Phases A and B and the grounded (neutral) conductors supplied by an 120/240 volt, single-phase system: (Compute with or without the use of the residential calculation form)

- 2800 sq. ft. dwelling unit
- 2 small appliance circuits
- 1 laundry circuit
- 6500 VA A/C unit (heat pump)   240 volt, single-phase
- 10,000 VA heating unit   240 volt, single-phase
- 6500 VA water heater   240 volt, single-phase
- 14 kW cooktop   240 volt, single-phase
- 16 kW oven   240 volt, single-phase
- 5500 VA dryer   240 volt, single-phase
- 1000 VA blower motor   240 volt, single-phase
- 2200 VA water pump   240 volt, single-phase
- 1400 VA disposal   120 volt, single-phase
- 1200 VA compactor   120 volt, single-phase
- 1400 VA dishwasher   120 volt, single-phase
- 1000 VA microwave   120 volt, single-phase

(A) What size THWN copper conductors are required for the service?

(B) What size OCPD is required for the service based on the computed load?

(C) What size panelboard is required for the service?

(D) What size RMC is required based on min. size conductors?

Test Problems

Section   Answer

85. Compute the following loads of a 4200 sq. ft. dwelling unit using the standard calculation and size the service-entrance conductors required for Phases A and B and the grounded (neutral) conductors supplied by an 120/240 volt, single-phase system: (Compute with or without the use of the residential calculation form)

- 4200 sq. ft. dwelling unit
- 2 small appliance circuits
- 1 laundry circuit
- 14 A A/C central unit (downstairs)   240 volt, single-phase
- 14 A A/C central unit (upstairs)   240 volt, single-phase
- 10,000 VA heating unit (downstairs)   240 volt, single-phase
- 10,000 VA heating unit (upstairs)   240 volt, single-phase
- 6500 VA water heater   240 volt, single-phase
- 12,000 VA oven   240 volt, single-phase
- 14,000 VA cooktop   240 volt, single-phase
- 5500 VA dryer   240 volt, single-phase
- 1000 VA blower motor   240 volt, single-phase
- 2200 VA water pump   240 volt, single-phase
- 1000 VA disposal   120 volt, single-phase
- 1200 VA compactor   120 volt, single-phase
- 1400 VA dishwasher   120 volt, single-phase
- 1200 VA microwave   120 volt, single-phase

(A) What size THWN copper conductors are required for the service?   _____ _____

(B) What size OCPD is required for the service based on the computed load?   _____ _____

(C) What size panelboard is required for the service?   _____ _____

(D) What size RMC is required based on min. size conductors?   _____ _____

86. Can a 5000 VA electric water heater be added to the existing dwelling unit when applying the optional calculation method without upgrading the service elements? The dwelling has the following loads: (service is rated 100 amps)   _____ _____

- 2200 sq. ft. dwelling unit
- 2 small appliance circuits
- 1 laundry circuit
- 10,000 VA range   240 volt, single-phase
- 5000 VA dryer   240 volt, single-phase
- 1050 VA disposal   120 volt, single-phase
- 1000 VA compactor   120 volt, single-phase
- 1600 VA dishwasher   120 volt, single-phase
- 5000 VA water heater   240 volt, single-phase (added load)

87. Can a 5820 VA A/C unit be added to the existing dwelling unit when applying the optional calculation method without upgrading the service elements? The dwelling has the following loads: (service is rated 100 amps)   _____ _____

- 1800 sq. ft. dwelling unit
- 2 small appliance circuits
- 1 laundry circuit
- 10,000 VA range   240 volt, single-phase
- 5000 VA dryer   240 volt, single-phase
- 920 VA disposal   120 volt, single-phase
- 1050 VA compactor   120 volt, single-phase
- 1600 VA dishwasher   120 volt, single-phase
- 5820 VA A/C unit   240 volt, single-phase (added load)

Stallcup's Master Electrician's Study Guide

Section	Answer

_____ _____ 88. Can three A/C window units rated 1820 VA each be added to the existing dwelling unit when applying the optional calculation method without upgrading the service elements? The dwelling has the following loads: (Service is rated 100 amps)

- 3800 sq. ft. dwelling unit
- 2 small appliance circuits
- 1 laundry circuit
- 10,000 VA range   240 volt, single-phase
- 5000 VA dryer   240 volt, single-phase
- 1050 VA disposal   120 volt, single-phase
- 1400 VA compactor   120 volt, single-phase
- 1600 VA dishwasher   120 volt, single-phase
- 1820 VA A/C window unit   120 volt, single-phase (added load)

_____ _____ 89. Compute the following load for a 860 sq. ft. mobile home using the standard calculation and size the service-entrance conductors required for Phases A and B and the grounded (neutral) conductors supplied by an 120/240 volt, single-phase system: (Compute with or without the use of the residential calculation form - mobile home)

- 860 sq. ft. mobile home   Use THWN copper conductors
- 2 small appliance circuits   with 75°C terminals
- 1 laundry circuit
- 8500 VA range   240 volt, single-phase
- 5000 VA water heater   240 volt, single-phase
- 5000 VA heating unit   240 volt, single-phase
- 720 VA disposal   120 volt, single-phase
- 920 VA compactor   120 volt, single-phase

90. What is the load in VA and amps for 25 multifamily dwelling units using the standard calculation and the size service-entrance conductors required for Phases A and B and the grounded (neutral) conductors supplied by 120/240 volt, single-phase system with the following loads? (Compute with or without the use of the residential calculation form - same size multifamily)

- 25 - 920 sq. ft. dwelling units   Use THWN copper conductors
- 2 small appliance circuits per unit   with 75°C terminals
- 1 laundry circuit per unit
- 25 - 9000 VA ranges   240 volt, single-phase
- 25 - 6000 VA water heaters   240 volt, single-phase
- 25 - 10,000 VA heating units   240 volt, single-phase
- 25 - 1200 VA disposals   120 volt, single-phase
- 25 - 1400 VA dishwashers   120 volt, single-phase

- Dryer facilities furnished by apartment complex

_____ _____ (A) What size THWN copper conductors are required for the service where they are paralleled six times per phase?

_____ _____ (B) What size OCPD is required for the service based on the computed load?

_____ _____ (C) What size grounded (neutral) conductors are required for the service where they are paralleled six times per phase?

_____ _____ (D) What size rigid metal conduit is required for each run?

Test Problems

Section   Answer

91. What is the load in VA and amps for 5 - 720 sq. ft. dwelling units, 5 - 860 sq. ft. dwelling units, 5 - 940 sq. ft. dwelling units and 5 - 980 sq. ft. dwelling using the standard calculation and the size service-entrance conductors required for Phases A and B and the grounded (neutral) conductors supplied by 120/240 volt, single-phase system with the following loads? (Compute with or without the use of the residential calculation form- different size multifamily)

- 5 - 720 sq. ft. dwelling units
- 2 small appliance circuits per unit
- 1 laundry circuit per unit
- 5 - 9000 VA ranges   240 volt, single-phase
- 5 - 5500 VA water heaters   240 volt, single-phase
- 5 - 10,000 VA heating units   240 volt, single-phase
- 5 - 1050 VA disposals   120 volt, single-phase
- 5 - 1000 VA dishwashers   120 volt, single-phase

- 5 - 860 sq. ft. dwelling units
- 2 small appliance circuits per unit
- 1 laundry circuit per unit
- 5 - 10,000 VA ranges   240 volt, single-phase
- 5 - 6000 VA water heaters   240 volt, single-phase
- 5 - 10,000 VA heating units   240 volt, single-phase
- 5 - 1200 VA disposals   120 volt, single-phase
- 5 - 1200 VA dishwashers   120 volt, single-phase

- 5 - 940 sq. ft. dwelling units
- 2 small appliance circuit per unit
- 1 laundry circuit per unit
- 5 - 11,000 VA ranges   240 volt, single-phase
- 5 - 6000 VA water heaters   240 volt, single-phase
- 5 - 10,000 VA heating units   240 volt, single-phase
- 5 - 1000 VA disposals   120 volt, single-phase
- 5 - 1400 VA dishwashers   120 volt, single-phase

- 5 - 980 sq. ft. dwelling units
- 2 small appliance circuits per unit
- 1 laundry circuit per unit
- 5 - 12,000 VA ranges   240 volt, single-phase
- 5 - 6500 VA water heaters   240 volt, single-phase
- 5 - 12,000 VA heating units   240 volt, single-phase
- 5 - 1200 VA disposals   120 volt, single-phase
- 5 - 1600 VA dishwashers   120 volt, single-phase

- Dryer facilities furnished by apartment complex

(A) What size THWN copper conductors are required for the service where they are paralleled six times per phase?   _____ _____

(B) What size OCPD is required for the service based on the computed load?   _____ _____

(C) What size grounded (neutral) conductors are required for the service where they are paralleled six times per phase?   _____ _____

(D) What size rigid metal conduit is required for each run?   _____ _____

Section **Answer**

_____ _____

92. What is the load in VA and amps for 25 multifamily dwelling units using the optional calculation and the size service-entrance conductors required for Phases A and B and the grounded (neutral) conductors when supplied by 120/240 volt, single-phase system with the following loads? (Compute with or without the use of the residential calculation form - same size multifamily)

- 25 - 920 sq. ft. dwelling units
- 2 small appliance circuits per unit
- 1 laundry circuit per unit
- 25 - 9000 VA ranges   240 volt, single-phase
- 25 - 6000 VA water heaters   240 volt, single-phase
- 25 - 10,000 VA heating units   240 volt, single-phase
- 25 - 1200 VA disposals   120 volt, single-phase
- 25 - 1400 VA dishwashers   120 volt, single-phase
- Dryer facilities are not furnished by the apartment complex

Note: See the standard calculation in problem 90 for sizing the grounded (neutral) conductors.

_____ _____

(A) What size THWN copper conductors are required for the service where they are paralleled six times per phase?

_____ _____

(B) What size OCPD is required for the service based on the computed load?

_____ _____

(C) What size rigid metal conduit is required for each run?

93. What is the load in VA and amps for 5 - 720 sq. ft. dwelling units, 5 - 860 sq. ft. dwelling units, 5 - 940 sq. ft. dwelling units and 5 - 980 sq. ft. dwelling using the optional calculation and the size service-entrance conductors required for Phases A and B and the grounded (neutral) conductors supplied by 120/240 volt, single-phase system with the following loads? (Compute with or without the use of the residential calculation form- different size multifamily)

- 5 - 720 sq. ft. dwelling units
- 2 small appliance circuits per unit
- 1 laundry circuit per unit
- 5 - 9000 VA ranges   240 volt, single-phase
- 5 - 5500 VA water heaters   240 volt, single-phase
- 5 - 10,000 VA heating units   240 volt, single-phase
- 5 - 1050 VA disposals   120 volt, single-phase
- 5 - 1000 VA dishwashers   120 volt, single-phase

- 5 - 860 sq. ft. dwelling units
- 2 small appliance circuits per unit
- 1 laundry circuit per unit
- 5 - 10,000 VA ranges   240 volt, single-phase
- 5 - 6000 VA water heaters   240 volt, single-phase
- 5 - 10,000 VA heating units   240 volt, single-phase
- 5 - 1200 VA disposals   120 volt, single-phase
- 5 - 1200 VA dishwashers   120 volt, single-phase

- 5 - 940 sq. ft. dwelling units
- 2 small appliance circuit per unit
- 1 laundry circuit per unit
- 5 - 11,000 VA ranges   240 volt, single-phase
- 5 - 6000 VA water heaters   240 volt, single-phase
- 5 - 10,000 VA heating units   240 volt, single-phase
- 5 - 1000 VA disposals   120 volt, single-phase
- 5 - 1400 VA dishwashers   120 volt, single-phase

Test Problems

                                                                                                     Section   Answer

- 5 - 980 sq. ft. dwelling units
- 2 small appliance circuits per unit
- 1 laundry circuit per unit
- 5 - 12,000 VA ranges   240 volt, single-phase
- 5 - 6500 VA water heaters   240 volt, single-phase
- 5 - 12,000 VA heating units   240 volt, single-phase
- 5 - 1200 VA disposals   120 volt, single-phase
- 5 - 1600 VA dishwashers   120 volt, single-phase

- Dryer facilities are not furnished by apartment complex

Note: See the standard calculation in problem 91 for sizing the grounded (neutral) conductors.

    (A) What size THWN copper conductors are required for the service where they are paralleled six times per phase?

    (B) What size OCPD is required for the service based on the computed load?

    (C) What size rigid metal conduit is required for each run?

94. What is the load in VA and amps for 25 multifamily dwelling units using the optional calculation and the size service-entrance conductors required for Phases A and B and the neutral supplied by 120/240 volt, single-phase system with the following loads? (Compute with or without the use of the residential calculation form - different size multifamily)

- 5 - 720 sq. ft. dwelling units
- 2 small appliance circuits per unit
- 1 laundry circuit per unit
- 5 - 9000 VA ranges   240 volt, single-phase
- 5 - 10,000 VA heating units   240 volt, single-phase
- 5 - 5500 VA water heaters   240 volt, single-phase
- 5 - 1050 VA disposals   120 volt, single-phase
- 5 - 1000 VA dishwashers   120 volt, single-phase

- 5 - 860 sq. ft. dwelling units
- 2 small appliance circuits per unit
- 1 laundry circuit per unit
- 5 - 10,000 VA ranges   240 volt, single-phase
- 5 - 6000 VA water heaters   240 volt, single-phase
- 5 - 10,000 VA heating units   240 volt, single-phase
- 5 - 1200 VA disposals   120 volt, single-phase
- 5 - 1200 VA dishwashers   120 volt, single-phase

- 5 - 940 sq. ft. dwelling units
- 2 small appliance circuit per unit
- 1 laundry circuit per unit
- 5 - 11,000 VA ranges   240 volt, single-phase
- 5 - 6000 VA water heaters   240 volt, single-phase
- 5 - 10,000 VA heating units   240 volt, single-phase
- 5 - 1000 VA disposals   120 volt, single-phase
- 5 - 1400 VA dishwashers   120 volt, single-phase

- 5 - 980 sq. ft. dwelling units
- 2 small appliance circuits per unit
- 1 laundry circuit per unit
- 5 - 12,000 VA ranges   240 volt, single-phase
- 5 - 6500 VA water heaters   240 volt, single-phase

13-15

Stallcup's Master Electrician's Study Guide

Section   Answer

- 5 - 12,000 VA heating units   240 volt, single-phase
- 5 - 1200 VA disposals   120 volt, single-phase
- 5 - 1600 VA dishwashers   120 volt, single-phase
- Dryer facilities furnished by apartment complex

House loads:

- Laundry loads   208 volt, single-phase
   38,675 VA   (Already computed)
- Outside lighting loads   120 volt, single-phase
   30 units, 180 VA each   (Electric discharge)
- Outside receptacle loads   120 volt, single-phase
   30 duplex receptacles   (Noncontinuous operation)
- Outside sign loads   120 volt, single-phase
   1800 VA   (Electric discharge)

Note: See the standard calculation in problem 91 for sizing the neutral.

     (A) What is the ampacity of the house loads?

     (B) What is the service ampacity for Phases A, B, and C and the neutral?

95. What is the load in VA and amps for 20 multifamily dwelling units using the standard calculation and the size service-entrance conductors required for Phases A and B and the grounded (neutral) conductors supplied by 120/240 volt, single-phase system with the following loads? (Compute with or without the use of the residential calculation form - same size multifamily)

    - 20 - 860 sq. ft. dwelling units
    - 2 small appliance circuits per unit
    - 1 laundry circuit per unit
    - 20 - heating units   Gas related load
    - 20 - ranges   Gas related load
    - 20 - 6500 VA water heaters   240 volt, single-phase
    - 20 - 1050 VA disposals   120 volt, single-phase
    - 20 - 1400 VA dishwashers   120 volt, single-phase

    - Dryer facilities furnished by apartment complex

    (A) What size THWN copper conductors are required for the service where they are paralleled three times per phase?

    (B) What size OCPD is required for the service based on the computed load?

    (C) What size grounded (neutral) conductors are required for the service where they are paralleled three times per phase?

    (D) What size EMT is required for each run?

96. What is the load in VA and amps for 15 multifamily dwelling units using the standard calculation and the size service-entrance conductors required for Phases A and B and the grounded (neutral) conductors supplied by 120/240 volt, single-phase system with the following loads? (Compute with or without the use of the residential calculation form - same size multifamily)

    - 15 - 920 sq. ft. dwelling units
    - 2 small appliance circuits per unit
    - 1 laundry circuit per unit

Test Problems

	Section	Answer

- 15 - 10,000 VA heating units   240 volt, single-phase
- 15 - 11,000 VA ranges   240 volt, single-phase
- 15 - 5500 VA water heaters   240 volt, single-phase
- 1440 VA A/C window unit   120 volt, single-phase
- 1800 VA A/C window unit   120 volt, single-phase
- 1920 VA A/C window unit   120 volt, single-phase
- 15 - 1000 VA disposals   120 volt, single-phase
- 15 - 1600 VA dishwashers   120 volt, single-phase
- Dryer facilities furnished by apartment complex

(A) What size THWN copper conductors are required for the service where they are paralleled four times per phase?

(B) What size OCPD is required for the service based on the computed load?

(C) What size grounded (neutral) conductors are required for the service where they are paralleled four times per phase?

(D) What size rigid PVC conduit (Schedule 80) is required for each run?

97. What is the load in VA and amps for 10 multifamily dwelling units using the standard calculation and the size service-entrance conductors required for Phases A and B and the neutral supplied by 120/240 volt, single-phase system with the following loads? (Compute with or without the use of the residential calculation form - same size multifamily)

- 10 - 1280 sq. ft. dwelling units
- 2 small appliance circuit per unit
- 1 laundry circuit per unit
- 10 - 20,000 VA heating units   240 volt, single-phase
- 10 - 6250 VA heat pumps   240 volt, single-phase
- 10 - 10,000 VA ranges   240 volt, single-phase
- 10 - 6000 VA water heaters   240 volt, single-phase
- 10 - 1000 VA disposals   120 volt, single-phase
- 10 - 1400 VA dishwashers   120 volt, single-phase
- Dryer facilities are furnished by apartment complex

(A) What size THWN copper conductors are required for the service where they are paralleled four times per phase?

(B) What size OCPD is required for the service based on the computed load?

(C) What size neutral conductor is required for the service where they are paralleled four times per phase?

(D) What size rigid PVC conduit (Schedule 40) is required for each run?

98. What is the load in VA and amps for 15 multifamily dwelling units using the standard calculation and the size service-entrance conductors required for Phases A and B and the neutral supplied by 120/208 volt, three-phase system with the following loads? (Compute with or without the use of the residential calculation form - same size multifamily - three-phase)

- 15 - 1020 sq. ft. dwelling units
- 2 small appliance circuits per unit
- 1 laundry circuit per unit
- 15 - 6500 VA A/C units   208 volt, three-phase
- 15 - 10,000 VA heating units   208 volt, three-phase

13-17

Stallcup's Master Electrician's Study Guide

Section   Answer

• 15 - 10,000 VA ranges	208 volt, three-phase
• 15 - 5000 VA dryers	208 volt, three-phase
• 15 - 6500 VA water heaters	208 volt, three-phase
• 15 - 1050 VA disposals	120 volt, three-phase
• 15 - 1400 VA dishwashers	120 volt, three-phase
• 15 - 1960 VA bathroom heaters & fans	120 volt, three-phase (ceiling mounted)

• Dryer facilities are not furnished by apartment complex

_____ _____ (A) What size THWN copper conductors are required for the service where they are paralleled four times per phase?

_____ _____ (B) What size OCPD is required for the service based on the computed load?

_____ _____ (C) What size grounded (neutral) conductors are required for the service where they are paralleled four times per phase?

_____ _____ (D) What size rigid metal conduit is required for each run?

99. What is the load in VA and amps for a store building including warehouse space using the standard calculation method for Phases A and B and the grounded (neutral) conductors supplied by a 120/240 volt, single-phase system with the following loads? (Compute with or without the use of the standard calculation form - office and warehouse space - 120/240 volt, single-phase)

   • 60,000 sq. ft. office space
   • 30,000 sq. ft. warehouse space

   120 volt, single-phase loads

   • 120 linear feet of show window (noncontinuous operation)
   • 120 ft. lighting track (noncontinuous operation)
   • 50 - 180 VA ballasts outside lighting (continuous operation)
   • 2400 VA sign lighting (continuous operation)
   • 82 receptacles (noncontinuous operation)
   • 30 receptacles (continuous operation)
   • 160 ft. multioutlet assembly (heavy-duty)

   240 volt, single-phase loads

   • 18,000 VA water heater
   • 50,000 VA heating unit
   • 29,400 VA A/C unit
   • 8640 VA freezer
   • 5820 VA ice cream box
   • 9680 VA walk-in cooler
   • 1 - 3/4 HP exhaust fan
   • 1 - 2 HP water pump

_____ _____ (A) What size THWN copper conductors are required for the service where they are paralleled six times per phase?

_____ _____ (B) What size OCPD is required for the service based on the computed load?

_____ _____ (C) What size neutral conductor is required for the service where they are paralleled six times per phase?

_____ _____ (D) What size rigid metal conduit is required for each run?

Test Problems

Section | Answer

100. What is the load in VA and amps for a store building including warehouse space using the standard calculation method for Phases A, B and C and the grounded (neutral) conductors supplied by a 120/208 volt, three-phase system with the following loads? (Compute with or without the use of the standard calculation form - office and warehouse space - 120/208 volt, three-phase)

- 60,000 sq. ft. office space
- 30,000 sq. ft. warehouse space

120 volt, single-phase loads

- 120 linear feet of show window (noncontinuous operation)
- 120 ft. lighting track (noncontinuous operation)
- 50 - 180 VA ballast outside lighting (continuous operation)
- 2400 VA sign lighting (continuous operation)
- 82 receptacles (noncontinuous operation)
- 30 receptacles (continuous operation)
- 160 ft. multioutlet assembly (heavy-duty)

208 volt, three-phase loads

- 18,000 VA water heater
- 50,000 VA heating unit
- 29,400 VA A/C unit
- 8640 VA freezer
- 5820 VA ice cream box
- 9680 VA walk-in cooler
- 1 - 3/4 HP exhaust fan
- 1 - 2 HP water pump

(A) What size THWN copper conductors are required for the service where they are paralleled six times per phase?

(B) What size OCPD is required for the service based on the computed load?

(C) What size grounded (neutral) conductors are required for the service where they are paralleled six times per phase?

(D) What size rigid metal conduit is required for each run?

101. What is the load in VA and amps for a store building including warehouse space using the standard calculation method for Phases A, B and C and the grounded (neutral) conductors supplied by a 277/480 volt, three-phase system with the following loads? (Compute with or without the use of the standard calculation form - office and warehouse space - 277/480 volt, three-phase)

- 60,000 sq. ft. office space - 277 volt lighting
- 30,000 sq. ft. warehouse space - 277 volt lighting

120 volt, single-phase loads

- 120 linear feet of show window (noncontinuous operation)
- 120 ft. lighting track (noncontinuous operation)
- 50 - 180 VA ballasts outside lighting (continuous operation)
- 2400 VA sign lighting (continuous operation)
- 82 receptacles (noncontinuous operation)
- 30 receptacles (continuous operation)
- 160 ft. multioutlet assembly (heavy-duty)

Stallcup's Master Electrician's Study Guide

Section   Answer

480 volt, three-phase loads

- 18,000 VA water heater
- 50,000 VA heating unit
- 29,400 VA A/C unit
- 8640 VA freezer
- 5820 VA ice cream box
- 9680 VA walk-in cooler
- 1 - 3/4 HP exhaust fan
- 1 - 2 HP water pump

(A) What size THWN copper conductors are required for the service where they are paralleled two times per phase?

(B) What size OCPD is required for the service based on the computed load?

(C) What size grounded (neutral) conductors are required for the service where they are paralleled two times per phase?

(D) What size rigid metal conduit is required for each run?

102. What is the load in VA and amps for a store building including warehouse space using the standard calculation method for Phases A, B and C and the grounded (neutral) conductors supplied by a 120/240 volt, three-phase system with the following loads? (Compute with or without the use of the standard calculation form - office and warehouse space - 120/240 volt, three-phase)

- 60,000 sq. ft. office space
- 30,000 sq. ft. warehouse space

120 volt, single-phase loads

- 120 linear feet of show window (noncontinuous operation)
- 120 ft. lighting track (noncontinuous operation)
- 50 - 180 VA ballast outside lighting (continuous operation)
- 2400 VA sign lighting (continuous operation)
- 82 receptacles (noncontinuous operation)
- 30 receptacles (continuous operation)
- 160 ft. multioutlet assembly (heavy-duty)

240 volt, three-phase loads

- 18,000 VA water heater
- 50,000 VA heating unit
- 29,400 VA A/C unit
- 8640 VA freezer
- 5820 VA ice cream box
- 9680 VA walk-in cooler
- 1 - 3/4 HP exhaust fan
- 1 - 2 HP water pump

103. What is the load in VA and amps for a office building including meeting hall using the standard calculation method for Phases A, B and C and the grounded (neutral) conductors supplied by a 277/480 volt, three-phase system with the following loads? (Compute with or without the use of the standard calculation form - office and meeting hall - 277/480 volt, three-phase)

- 130,000 sq. ft. office building - 277 volt lighting
- 5000 sq. ft. meeting hall - 277 volt lighting

13-20

Test Problems

Section   Answer

277 volt lighting

- 100 ft. lighting track (noncontinuous operation)
- 30 - 180 VA ballast outside lighting (continuous operation)
- 3800 VA sign lighting (continuous operation)
- 8000 VA isolation transformer for LVLS (continuous operation)

120 volt, single-phase loads

- 152 receptacles (noncontinuous operation)
- 138 receptacles (continuous operation)
- 200 ft. multioutlet assembly (heavy-duty)

208 volt, three-phase loads

- 2 - 1480 VA copying machines
- 10,000 VA water heater
- 34 - 225 VA data processors
- 10 - 175 VA word processors
- 4 - 1200 VA printers

480 volt, three-phase loads

- 50 HP elevator (15 minute intermittent duty)
- 30 kW heating unit

(A) What size THWN copper conductors are required for the service where they are paralleled four times per phase?

(B) What size OCPD is required for the service based on the computed load?

(C) What size grounded (neutral) conductors are required for the service where they are paralleled four times per phase?

(D) What size rigid metal conduit is required for each run?

104. What is the load in VA and amps for a school building using the standard calculation method for Phases A, B and C and the grounded (neutral) conductors supplied by a 277/480 volt, three-phase system with the following loads? (Compute with or without the use of the standard calculation form - school buildings - 277/480 volt, three-phase)

- 40,000 sq. ft. classroom area - 277 volt lighting
- 8000 sq. ft. auditorium area - 277 volt lighting
- 1500 sq. ft. assembly hall area - 277 volt lighting

120 volt, single-phase loads

- 190 receptacles (noncontinuous operation)
- 60 receptacles (continuous operation)
- 200 ft. multioutlet assembly (heavy-duty)

120 volt, single-phase cooking equipment loads

- 2 - 1.2 kW toasters
- 6 - 1.4 kW refrigerators
- 4 - 1.6 kW freezers

13-21

Stallcup's Master Electrician's Study Guide

Section   Answer

208 volt, single-phase cooking equipment loads

- 6 - 12 kW ranges
- 4 - 9 kW ovens
- 4 - 4 kW fryers

208 volt, single-phase motor loads

- 4 - 2 HP vent hood fans ⎫ Don not figure as
- 4 - 1/2 HP grill vent fans ⎭ cooking equipment.

480 volt, three-phase motor loads

- 20 - 3/4 HP exhaust fans

_____ _____ (A) What size THWN copper conductors are required for the service where they are paralleled two times per phase?

_____ _____ (B) What size OCPD is required for the service based on the computed load?

_____ _____ (C) What size grounded (neutral) conductors are required for the service where they are paralleled two times per phase?

_____ _____ (D) What size rigid metal conduit is required for each run?

_____ _____ 105. What is the load in VA and amps for a restaurant using the standard calculation method for Phases A, B and C and the grounded (neutral) conductors supplied by a 120/208 volt, three-phase system with the following loads? (Compute with or without the use of the standard calculation form - restaurant - 120/208 volt, three-phase)

- 8000 sq. ft. restaurant

120 volt, single-phase loads

- 40 ft. lighting track (continuous operation)
- 14 - 180 VA outside lighting (continuous operation)
- 2400 VA sign lighting (continuous operation)
- 50 receptacles (noncontinuous operation)
- 20 receptacles (continuous operation)
- 60 ft. multioutlet assembly (heavy-duty)

208 volt, single-phase loads

- 6450 VA hood fan ⎫ Do not figure as
- 5245 VA grill vent fan ⎭ cooking equipment.
- 15 A freezer
- 12,000 VA cooktop
- 2 - 10,000 VA ovens
- 12,000 VA range
- 18 A refrigerator
- 4250 VA ice cream box

208 volt, three-phase loads

- 2 - 30 kW heating units
- 2 - 6.5 kW A/C units
- 4400 VA broiler
- 2 - 3200 VA deep fat fryers
- 20 A walk-in cooler
- 6000 VA water heater

13-22

Test Problems

	Section	Answer

(A) What size THWN copper conductors are required for the service where they are paralleled two times per phase?   _____   _____

(B) What size OCPD is required for the service based on the computed load?   _____   _____

(C) What size grounded (neutral) conductors are required for the service where they are paralleled two times per phase?   _____   _____

(D) What size rigid metal conduit is required for each run?   _____   _____

106. What is the load in VA and amps for a hospital using the standard calculation method for Phases A, B and C and the grounded (neutral) conductors supplied by a 277/480 volt, three-phase system with the following loads? (Compute with or without the use of the standard calculation form - hospital - 277/480 volt, three-phase)

120 volt, single-phase loads

- 60 linear feet of show window (noncontinuous operation)
- 100 ft. of lighting track (noncontinuous operation)
- 28 - 225 VA ballast outside lighting (continuous operation)
- 4 - 3600 VA sign lighting (continuous operation)
- 228 receptacles (noncontinuous operation)
- 96 receptacles (continuous operation)
- 200 ft. multioutlet assembly (heavy-duty)
- 7 - 1000 VA copying machines
- 4 - 1400 VA soft drink machines

480 volt, three-phase loads

- 6 - 35,000 VA heating units
- 6 - 9840 VA A/C units
- 4 - 5420 VA water pumps
- 8 - 3 HP exhaust fans
- 1 - 7 1/2 HP sprinkler pump

277 volt, three-phase loads

- 52,000 sq. ft. hospital (patient rooms)

Emergency system

- 50,000 VA lift support branch load (already computed at continuous operation)
- 50,000 VA critical branch load (already computed at continuous operation)

X-ray equipment

- Short-time rating
two each at 3/4 seconds
rating = 200 MA
primary = 208 volt, single-phase
secondary = 100,000 volt, single-phase

- Long-term rating
two each at 15 minutes
rating = 20 MA
primary = 208 volt, single-phase
secondary = 200,000 volt, single-phase

Stallcup's Master Electrician's Study Guide

Section	Answer	
_____	_____	(A) What is the demand load in VA for the general lighting load?
_____	_____	(B) What is the load in amps for the branch-circuit to the X-ray equipment with the short-time rating?
_____	_____	(C) What size THWN copper conductors are required for the X-ray equipment with the short-time rating?
_____	_____	(D) What is the load in amps for the branch-circuit to the X-ray equipment with the long-time rating?
_____	_____	(E) What size THWN copper conductors are required for the X-ray equipment with the long-time rating?
_____	_____	(F) What is the total VA and amp rating for the feeder and service conductors for the X-ray equipment?
_____	_____	(G) What size THWN copper conductors are required to supply power for the feeder to the X-ray equipment?
_____	_____	107. What is the load in VA and amps for a hotel using the standard calculation method for Phases A, B and C and the grounded (neutral) conductors supplied by a 120/208 volt, three-phase system with the following loads? (Compute with or without the use of the standard calculation form - hotel - 120/208 volt, three-phase)

120 volt, single-phase loads

- 40 units - 410 sq. ft. each
- 40 linear feet of show window (noncontinuous operation)
- 50 ft. of lighting track (noncontinuous operation)
- 22 - 275 VA ballast outside lighting (continuous operation)
- 4 - 1800 VA sign lighting (continuous operation)
- 74 receptacles (continuous operation)
- 4 - 1000 VA copying machines
- 8 - 1400 VA soft drink machines
- 7200 VA house loads (continuous operation)
- 40 - 1/4 HP exhaust fans (noncontinuous operation)

208 volt, three-phase loads

- 2 - 20,000 VA heating units (office and lobby)
- 40 - 7 1/2 HP A/C and heat pump units
- 6 - 8.5 kW water heaters
- 28,200 VA house loads (continuous operation)
- 18,420 VA laundry facilities (continuous operation)

_____	_____	108. What is the load in VA and amps for a bank using the standard calculation method for Phases A, B and C and the grounded (neutral) conductors supplied by a 120/208 volt, three-phase system with the following loads? (Compute with or without the use of the standard calculation form - bank - 120/208 volt, three-phase)

120 volt, single-phase loads

- 7400 sq. ft. bank
- 60 linear feet of show window (noncontinuous operation)
- 80 ft. of lighting track (noncontinuous operation)
- 30 - 180 VA ballast outside lighting (continuous operation)
- 1800 VA sign lighting (continuous operation)

Test Problems

Section   Answer

- 88 receptacles (noncontinuous operation)
- 45 receptacles (continuous operation)
- 40 ft. multioutlet assembly (heavy-duty)
- 18 - 1/4 HP fans
- 6420 miscellaneous loads (continuous operation)
- 22 - 1,600 VA personal computers (continuous operation)

208 volt, three-phase loads

- 25,000 VA heating unit
- 7 1/2 HP A/C unit
- 8 kW water heater
- 3 - 5 HP pump motors
- 10 - 2 HP exhaust fans
- 14 kW range (noncontinuous operation)
- 18,420 VA miscellaneous loads (continuous operation)

109. What is the load in VA and amps for a welding shop using the standard calculation method for Phases A, B, and C and the grounded (neutral) conductors supplied by a 120/208 volt, three-phase system with the following loads? (Compute with or without the use of the standard calculation form - welding shop - 120/208 volt, three-phase)

120 volt, single-phase loads

- 8400 VA inside lighting loads (continuous operation)
- 4 - 180 VA outside lighting loads (continuous operation)
- 1200 VA sign lighting loads (continuous operation)
- 50 receptacles (continuous operation)

208 volt, three-phase loads

- 2 - 10,000 kW heating units
- 6500 VA A/C unit
- 5 HP air-compressor
- 2 - 2 HP grinders

Welders - resistance (40% duty cycle)

- 11 kW
- 9 kW

Welders - arc welder with motor (80% duty cycle)

- 16 kW
- 13 kW

Welders - arc welder without motor (70% duty cycle)

- 14 kW
- 8 kW

110. What is the load in VA and amps for a school using the optional calculation method for Phases A, B and C and the grounded (neutral) conductors supplied by a 277/480 volt, three-phase system with the following loads:

• Classroom area	=	40,000 sq. ft.
• Auditorium area	=	5000 sq. ft.
• Cafeteria area	=	5000 sq. ft.
• Hall area	=	1,200 sq. ft.
Total VA of school	=	845,794 VA

13-25

Stallcup's Master Electrician's Study Guide

Section   Answer

111. What is the load in VA and amps for a restaurant using the optional calculation method for Phases A, B and C and the grounded (neutral) conductors supplied by a 120/208 volt, three-phase system with the following loads:

     • Lighting loads   =   18,240 VA
     • Receptacle loads   =   16,400 VA
     • Special loads   =   98,215 VA
     • Heating or A/C loads   =   40,000 VA
     • Motor loads   =   7200 VA

112. Can a load of 125 amps be added to an existing service with 500 KCMIL THWN copper conductors having a maximum demand of 204 amps?

Article 230
Services

113. What size THWN copper conductors are required to supply a service with a continuous load of 92 amps and noncontinuous load of 84 amps?

114. What is the maximum setting for a 200 amp CB (over 600 volt) used as a main OCPD?

115. What is the maximum setting for a 200 amp fuse (over 600 volt) used as a main OCPD?

Article 240
Overcurrent protection

116. What size THWN copper conductors are required for a 10 ft. tap when supplied by a feeder-circuit with a 300 amp OCPD?

117. What size THWN copper conductors are required for a 10 ft. tap when supplied by a feeder-circuit with a 400 amp OCPD?

118. What size THWN copper conductors are required for a 25 ft. tap when supplied by a feeder-circuit with a 300 amp OCPD?

119. What size THWN copper conductors are required for a 25 ft. tap when supplied by a feeder-circuit with a 400 amp OCPD?

120. What size THWN copper conductors for the primary and secondary are required for a 25 ft. tap when supplied by a 300 amp OCPD with a 480 volt primary and 208 volt secondary? (**Note:** Tap includes primary plus XTMR and secondary)

121. What size THWN copper conductors for the primary and secondary are required for a 25 ft. tap when supplied by a 400 amp OCPD with a 480 volt primary and 208 volt secondary? (**Note:** Tap includes primary plus XTMR and secondary)

122. What size THWN copper conductors are required for the primary, when using a 100 ft. tap supplied by a 400 amp OCPD?

123. What size THWN copper conductors are required for the primary, when using a 100 ft. tap supplied by a 600 amp OCPD?

124. Can a 150 amp busway be tapped to a 450 amp busway without OCP at the tap?

125. What size THWN copper conductors are required for a 10 ft. tap motor circuit tap when supplied by a 400 amp OCPD? (Terminals are 75°C)

13-26

Test Problems

	Section	Answer

126. What size OCPD is required for a 10 ft. tap from the secondary to protect 4 - 1/0 AWG THWN copper conductors paralleled 3 times when supplying a 150 kVA separately derived transformer with a 480 volt primary and 208 volt secondary? (Protect secondary conductors)

127. What size OCPD is required for a 25 ft. tap from the secondary to protect 4 - 1/0 AWG THWN copper conductors paralleled 3 times when supplying a 150 kVA separately derived transformer with a 480 volt primary and 208 volt secondary? (Protect secondary conductors)

Article 250
Grounding

128. What size copper grounding electrode conductor is required to a metal water pipe if the service conductors are rated 350 KCMIL copper?

129. What size copper grounding electrode conductor is required to building steel if the service conductors are rated 350 KCMIL copper?

130. What size copper grounding electrode conductor is required to concrete encased electrode if the service conductors are rated 350 KCMIL copper?

131. What size copper grounding electrode conductor is required to ground ring if the service conductors are rated 350 KCMIL copper?

132. What size copper grounded service conductor is required to carry the maximum fault current if the service conductors are rated 350 KCMIL copper?

133. What size copper grounded service conductor is required to carry the maximum fault current if the service conductors are paralleled three times and rated 600 KCMIL copper?

134. What size copper grounding electrode conductor is required to building steel for a separately derived system when feeder-circuit conductors are rated 3/0 copper?

135. What size copper grounding electrode conductor is required to a metal water pipe for a separately derived system when feeder-circuit conductors are rated 3/0 copper?

136. What size copper grounding electrode conductor is required to driven rod for a separately derived system when feeder-circuit conductors are rated 3/0 copper?

137. What size copper bonding jumper is required for a separately derived system when feeder-circuit conductors are rated 3/0 copper?

138. What size copper equipment grounding conductor is required to ground a piece of equipment when supplied by a 35 amp OCPD?

139. What size copper equipment grounding conductor is required to ground a panelboard when supplied by a 200 amp OCPD?

140. What size copper equipment grounding conductor is required for three parallel runs in each conduit when supplied by a 600 amp OCPD?

13-27

Section	Answer

Article 310
Conductors for general wiring

141. What is the current-carrying capacity of 14 - 12 AWG THHN copper conductors pulled through 3/4 in EMT? (all current-carrying)

142. What is the current-carrying capacity of 6 - 10 AWG THWN copper conductors pulled through 3/4 in. EMT? (all current-carrying)

143. What is the current-carrying capacity of 12 - 10 AWG THWN copper conductors pulled through 1 in. EMT? (all current-carrying)

144. What is the ampacity of 4 - 8 AWG THHN copper conductors routed through an ambient temperature of 122°F? (only three are current-carrying)

145. What is the ampacity of 4 - 10 AWG THHN copper conductors routed through an ambient temperature of 102°F? (only three are current-carrying)

146. What is the ampacity of 12 - 12 AWG THHN copper conductors routed through an ambient temperature of 128°F? (all are current-carrying)

147. What is the ampacity of 8 - 10 AWG THHN copper conductors routed through an ambient temperature of 112°F? (all current-carrying)

148. What is the ampacity of 6 - 12 AWG THHN copper conductors routed through an ambient temperature of 100°F? (all current-carrying)

149. What is the ampacity of 12 - 10 AWG THHN copper conductors routed through an ambient temperature of 125°F? (all current-carrying)

150. What is the ampacity of 12 current-carrying 10 AWG THHN copper conductors, with diversity, located in the same raceway?

Article 318
Cable trays

151. What size ventilated cable tray is required for 12 - 300 KCMIL and 10 - 1000 KCMIL THWN copper conductors?

152. What size ventilated cable tray is required for 14 - 250 KCMIL and 8 - 1000 KCMIL THWN copper conductors?

Article 370
Outlet, device, pull and junction boxes, conduit bodies and fittings

153. What size octagon box is required to support a luminaire (lighting fixture) with the following:

- 2 - 12-2 AWG w/ground nonmetallic-sheathed cables
- 2 - 16 AWG fixture wires with fixture canopy
- 2 pigtails
- Fixture stud
- Hickey
- 2 romex connectors

Test Problems

                                                                                                      Section   Answer

154. What size octagon box is required to support a lighting fixture with the following:

    • 2 - 12-2 AWG w/ground nonmetallic sheathed cables
    • 2 - 14 AWG fixture wires with fixture canopy
    • 2 pigtails
    • Fixture stud
    • Hickey
    • 2 romex connectors

155. What size device box is required for a receptacle outlet with the following:

    • 2 - 12 AWG w/ground nonmetallic sheathed cables
    • 2 romex connectors
    • 1 duplex receptacle outlet

156. What size device box is required for a receptacle outlet with the following:

    • 2 - 12 AWG passing through
    • 1 - 14 AWG hot
    • 1 - 14 AWG neutral
    • 2 - 12 AWG EGC's
    • 1 duplex receptacle outlet with 14 AWG
    • 1 - 12 AWG pigtail
    • 1 - 12 AWG bonding jumper

157. What size square box is required to enclose the following:

    • 4 - 10 AWG hots
    • 4 - 10 AWG neutrals
    • 4 - 10 AWG EGC's
    • 2 cable clamps

158. What size square box is required to enclose the following:

    • 2 - 10 AWG hots
    • 2 - 10 AWG neutrals
    • 2 - 10 AWG EGC's
    • 2 - 12 AWG hots
    • 2 - 12 AWG neutrals
    • 2 - 12 AWG EGC's
    • 2 cable clamps

159. What size junction box is required to enclose the following:

    • 18 - 12 AWG conductors (9 x 2)
    • 18 - 12 AWG conductors (9 x 2)
    • 6 - 12 AWG conductors (3 x 2)
    • All conductors are spliced

160. What size junction box is required to enclose the following:

    • 12 - 8 AWG conductors (6 x 2)
    • 18 - 10 AWG conductors (9 x 2)
    • 22 - 12 AWG conductors (11 x 2)
    • 28 - 14 AWG conductors (14 x 2)
    • All conductors are spliced

161. What is the cross-sectional area of an LB (conduit body) that is connected to a 1 1/2 in. (41) EMT raceway?

Stallcup's Master Electrician's Study Guide

Section	Answer	
_____	_____	162. What is the minimum length for a junction box with a straight pull consisting of one run of 3 in. (78) raceways?
_____	_____	163. What is the minimum length for a junction box with a straight pull consisting of one run of 4 in. (103) raceways?
_____	_____	164. What is the minimum length for a junction box with a straight pull consisting of 3 in. (78), 2 in. (53), and 1 in. (27) raceways connected to its sides?
_____	_____	165. What is the minimum length for a junction box with a straight pull consisting of 4 in. (103), 3 in. (78), and 2 in. (53) raceways connected to its sides?
_____	_____	166. What is the minimum length for a junction box with an angle pull consisting of one run of 3 in. (78) raceways?
_____	_____	167. What is the minimum length for a junction box with an angle pull consisting of one run of 4 in. (103) raceways?
_____	_____	168. What is the minimum length for a junction box with an angle pull consisting of 4 in. (103), 3 in. (78), 2 in. (53), and 1 in. (27) raceways that are connected to the right wall and bottom wall?
_____	_____	169. What is the minimum length for a junction box with an angle pull consisting of 4 in., 3 in., and 1 in. raceways that are connected to the right wall and the bottom wall?

Article 374
Auxiliary gutters

Section	Answer	
_____	_____	170. What size metal auxiliary gutter and splicing space is required to enclose 3 - 500 KCMIL THWN copper conductors (feeder-circuit) that have 3 - 4/0, 3 - 2, and 3 - 4 AWG THWN copper conductors spliced to them?
_____	_____	171. What size metal auxiliary gutter and splicing space is required to enclose 3 - 350 KCMIL THWN copper conductors (feeder-circuit) that have 3 - 1/0, 3 - 4, and 3 - 8 AWG THWN copper conductors spliced to them?

Article 422
Appliances

Section	Answer	
_____	_____	172. What size THWN copper conductors and OCPD are required for a 12.5 amp continuous operated appliance?
_____	_____	173. What size THWN copper conductors and OCPD are required for a 15 amp continuous operated appliance?
_____	_____	174. What size THWN copper conductors and OCPD are required for a water heater with a 240 volt, single-phase supply and rated 5500 watts?
_____	_____	175. What is the minimum and maximum size OCPD allowed for a 15.5 amp continuous operated nonmotor appliance?

Article 424
Fixed electric space-heating equipment

Section	Answer	
_____	_____	176. What size THWN copper conductors and OCPD are required for a 40 amp baseboard heater supplied by a 240 volt, single-phase system?

Test Problems

Section   Answer

177. What size THWN copper conductors and OCPD are required for 2 - 12 amp baseboard heaters supplied by a 240 volt, single-phase system?

178. What size THWN copper conductors and OCPD are required for a 20 kW heating unit with a 3 amp blower motor supplied by a 240 volt, single-phase system?

Article 430
Motors, motor circuits, and controllers

179. What size THWN copper conductors are required to supply power to a 208 volt, single-phase, 3 HP, Design B motor that is installed in a commercial building?

180. What size THWN copper conductors are required to supply power to a 230 volt, single-phase, 5 HP, Design B motor that is installed in a commercial building?

181. What size THWN copper conductors are required to supply power to a 115 volt, single-phase, 1/2 HP, Design B motor that is installed in a commercial building?

182. What size THWN copper conductor are required to supply power to a 230 volt, three-phase, 10 HP, Design B motor having 75°C terminals?

183. What size THWN copper conductors are required to supply power to a 208 volt, three-phase, 30 HP, Design B motor having 75°C terminals?

184. What size THWN copper conductors are required to supply power to a 480 volt, three-phase, 50 HP, Design B motor having 75°C terminals?

185. What size THWN copper conductors are required to supply a group of 230 volt, three-phase, Design B motors that are rated 5 HP, 10 HP and 20 HP?

186. What size THWN copper conductors are required to supply a group of 460 volt, three-phase, Design B motors that are rated 30 HP, 40 HP and 50 HP?

187. What size THHN copper conductors are allowed in a controller where the upstream OCPD protecting the branch-circuit is rated 80 amps?

188. What size THWN copper conductors are allowed to be routed remotely from the magnetic starter where the upstream OCPD is rated 80 amps?

189. What is the minimum and next size nontime-delay fuses required to start and run a 30 HP, three-phase, 230 volt, Design B motor?

190. What is the minimum and next size nontime-delay fuses required to start and run a 50 HP, three-phase, 460 volt, Design B motor?

191. What is the minimum and next size time-delay fuses required to start and run a 25 HP, three-phase, 460 volt, Design B motor?

192. What is the minimum and next size time-delay fuses required to start and run a 20 HP, three-phase, 208 volt, Design B motor?

193. What is the minimum and maximum setting instantaneous trip circuit breaker required to start and run a 30 HP, three-phase, 460 volt, Design E motor?

194. What is the minimum and next size inverse time circuit breaker required to start and run a 30 HP, three-phase, 208 volt, Design B motor?

13-31

Section   Answer

195. What is the minimum and next size inverse time circuit breaker required to start and run a 200 HP, three-phase, 575 volt, Design B motor?

196. What is the maximum size nontime-delay fuses required to start and run a 30 HP, three-phase, 230 volt, Design B motor?

197. What is the maximum size nontime-delay fuses required to start and run a 40 HP, three-phase, 460 volt, Design B motor?

198. What is the maximum size time-delay fuses required to start and run a 30 HP, three-phase, 460 volt, Design B motor?

199. What is the maximum size time-delay fuses required to start and run a 20 HP, three-phase, 230 volt, Design B motor?

200. What is the maximum size inverse time circuit breaker required to start and run a 30 HP, three-phase, 208 volt, Design B motor?

201. What is the maximum size inverse time circuit breaker required to start and run a 50 HP, three-phase, 460 volt, Design B motor?

202. What size OCPD (CB) is required for a feeder-circuit supplying a 5 HP, 10 HP, 15 HP and 20 HP, three-phase, 230 volt, Design B motors?

203. What size OCPD (CB) is required for a feeder-circuit supplying a 20 HP, 30 HP and 40 HP, three-phase, 460 volt, Design B motors?

204. What is the minimum overload protection (OLR's) for a 40 HP, three-phase, 460 volt, Design B motor with a nameplate rating of 45 amps, a temperature rise of 40°C and a service factor of 1.15?

205. What is the maximum overload protection (OLR's) for a 40 HP, three-phase, 460 volt, Design B motor with a nameplate rating of 45 amps, a temperature rise of 40°C and a service factor of 1.15?

206. What size branch-circuit OCPD (maximum) is required for 14 AWG control circuit conductors located in the controller?

207. What size branch-circuit OCPD (maximum) is required for 14 AWG control circuit conductors that are run remote?

208. What size OCPD is required for a two-wire, 480 volt, 2800 VA control transformer?

209. What size OCPD is required for 14 AWG control circuit conductors tapped from the secondary side of a control transformer with 480 volt primary and 240 volt secondary?

Article 440
Air-conditioning and refrigerating equipment

210. What size THWN copper conductors are required to supply power to an A/C unit with a compressor having a full-load current of 22 amps and a condenser motor having a nameplate of 2.5 amps with terminals at 75°C?

211. What size THWN copper conductors are required to supply power to an A/C unit with a compressor having a full-load current of 18 amps and a condenser motor having a nameplate of 2 amps with terminals of 75°C?

Test Problems

	Section	Answer

212. What size THWN copper conductors are required to supply power to an A/C unit with a compressor having a full-load current of 37 amps and a condenser motor having a nameplate of 3 amps with terminals of 75°C?

213. What size THWN copper conductors are required to supply power to an A/C unit with the following:

   A/C unit #1
   • Compressor with full-load current of 24 amps
   • Condenser with nameplate current of 2.5 amps

   A/C unit #2
   • Compressor with full-load current of 24 amps
   • Condenser with nameplate current of 2.5 amps

   A/C unit #3
   • Compressor with full-load current of 24 amps
   • Condenser with nameplate current of 2.5 amps

214. What size THWN copper conductors are required to supply power to an A/C unit with the following:

   A/C unit #1
   • Compressor with full-load current of 32 amps
   • Condenser with nameplate current of 3 amps

   A/C unit #2
   • Compressor with full-load current of 26 amps
   • Condenser with nameplate current of 2.5 amps

   A/C unit #3
   • Compressor with full-load current of 24 amps
   • Condenser with nameplate current of 2.5 amps

215. What is the minimum size OCPD required for an A/C unit with a compressor load of 24 amps and a condenser load of 2.5 amps?

216. What is the minimum size OCPD required for an A/C unit with a compressor load of 32 amps and a condenser load of 3 amps?

217. What is the maximum size OCPD required for an A/C unit with a compressor load of 24 amps and a condenser load of 2.5 amps?

218. What is the maximum size OCPD required for an A/C unit with a compressor load of 32 amps and a condenser load of 3 amps?

Article 450
Transformers and transformer vaults

219. What size OCPD is required for a single-phase, 480 volt primary and 240 volt secondary, two-wire to two-wire system with 1/0 AWG THWN copper conductors?

220. What size OCPD is required for a three-phase, 480 volt primary and 240 volt secondary, three-wire to three-wire system with 400 KCMIL copper conductors?

221. What is the maximum (round up) size individual OCPD (circuit breaker) required for the primary side of a 1500 kVA transformer with a three-phase supply voltage of 12,470 volts? (Supervised Location)

Stallcup's Master Electrician's Study Guide

Section   Answer

_____ _____ 222. What is the maximum (round up) size OCPD (circuit breaker) required for a three-phase, 4160 volt primary and 480 volt secondary for a 450 kVA transformer with a 5.9 percent impedance in a nonsupervised location?

_____ _____ 223. What size OCPD (circuit breaker) is required for a three-phase, 4160 volt primary and 480 volt secondary for a 450 kVA transformer with a 5.9 percent impedance in a supervised location?

_____ _____ 224. What is the maximum (round up) size OCPD (circuit breaker) required for a single-phase, 240 volt primary and 120 volt secondary for a 20 kVA transformer? (Primary protection only)

_____ _____ 225. What size OCPD is required for a 480 volt, single-phase, two-wire to two-wire 3 kVA transformer?

_____ _____ 226. What is the minimum and maximum size OCPD required for a 480 volt, single-phase, two-wire to two-wire .6 kVA transformer?

_____ _____ 227. What is the maximum (round up) size OCPD required for a 480 volt primary and 240 volt secondary, three-phase, 30 kVA transformer? (Using the maximum size in the primary side)

_____ _____ 228. What is the maximum (round up) size OCPD required for a 480 volt primary and 240 volt secondary, three-phase, 50 kVA transformer? (Using the maximum size in the primary side)

Article 455
Phase converters

_____ _____ 229. What size THWN copper conductors are required to supply a phase converter that supplies a 230 volt, three-phase, 20 HP, Design B motor?

_____ _____ 230. What size OCPD is required for the phase converter when supplying a 25 HP, three-phase, 230 volt, Design B motor?

Article 460
Capacitors

_____ _____ 231. What size THHN copper conductors are required to supply power to a 460 volt, three-phase, 20 kVA capacitor serving a 460 volt, three-phase, 60 HP motor? (Motor supplied by 3 AWG THHN copper conductors)

_____ _____ 232. What size disconnect is required to supply power to a 208 volt, three-phase, 25 kVA capacitor serving a 208 volt, three-phase, 60 HP motor?

Article 550
Mobile homes

_____ _____ 233. How many 15 amp lighting circuits are required for a 900 sq. ft. mobile home supplied by a 120/240 volt, single-phase system?

	Test Problems
Section	Answer

234. What is the range, fixed appliance and heating load for a mobile home with the following:

- 8.5 kVA range
- 6000 VA water heater
- 450 VA disposal
- 700 VA dishwasher
- 5500 VA heating

Article 600
Electric signs and outline lighting

235. What size THWN copper conductors and OCPD are required for a branch-circuit feeding 10 neon signs with each ballast rated 1.5 amps each?

236. What is the lighting load in VA for a sign rated 1800 VA operating for two hours?

237. What is the lighting load in VA for a sign rated 2400 VA operating for six hours?

Article 630
Electric welders

238. What size THWN copper conductors are required to supply an arc welder without a motor and rated at 58 amps having a 70 percent duty cycle?

239. What size THWN copper conductors are required to supply an arc welder without a motor and rated at 42 amps having a 50 percent duty cycle?

240. What size THWN copper conductors are required to supply an arc welder with a motor and rated at 68 amps having a 80 percent duty cycle?

241. What size THWN copper conductors are required to supply an resistance welder rated at 84 amps having a 50 percent duty cycle?

242. What size OCPD and disconnect is required to protect the conductors for a welder without a motor and having a 62 amp load and 90 percent duty cycle? (Size to protect conductors)

243. What size OCPD is required for an motor arc welder with a 58 amp load and 80 percent duty cycle? (Size as a disconnect for welder)

244. What size OCPD and disconnect is required for an resistance welder with a 78 amp load and 40 percent duty cycle? (Size to protect conductors)

Chapter 9
Tables and examples

245. What size (EMT) conduit is required to enclose the following THWN copper conductors?

- 2 - 6 AWG THWN copper
- 2 - 8 AWG THWN copper
- 4 - 10 AWG THWN copper

Section   Answer

_____ _____ 246. What size (RMC) conduit is required to enclose the following THWN copper conductors?

- 4 - 8 AWG THWN copper
- 8 - 10 AWG THWN copper
- 12 - 12 AWG THWN copper

_____ _____ 247. What size PVC (Schedule 40) conduit is required to enclose the following XHHW copper conductors?

- 4 - 1/0 AWG XHHW copper
- 4 - 2 AWG XHHW copper
- 4 - 6 AWG XHHW copper

_____ _____ 248. What size EMT nipple is required to enclose 28 - 12 THWN copper conductors that are installed between a panelboard and junction box?

_____ _____ 249. What size EMT nipple is required to enclose the following THWN copper conductors?

- 28 - 10 AWG THWN copper
- 16 - 12 AWG THWN copper
- 12 - 14 AWG THWN copper

_____ _____ 250. What size EMT nipple is required to enclose the following THWN copper conductors?

- 14 - 12 AWG THWN copper
- 10 - 10 AWG THWN copper
- 4 - 6 AWG THWN copper

Topic Index

Symbols

120 VOLTS BETWEEN CONDUCTORS	6-25
15 AND 20 AMP BRANCH-CIRCUITS	6-16
277 VOLTS-TO-GROUND	6-25
30 AMP BRANCH-CIRCUIT	6-18
40 AND 50 AMP BRANCH-CIRCUITS	6-19
60°C CONDUCTORS	6-20
600 VOLTS BETWEEN CONDUCTORS	6-27
75°C CONDUCTORS	6-21
9 AMPS OR MORE	7-19, 7-21
90°C CONDUCTORS	6-22

A

AC CIRCUITS AND SYSTEMS	3-2
AC CIRCUITS OF 50 TO 1,000 VOLTS	3-2
AC CIRCUITS OF LESS THAN 50 VOLTS	3-2
AC grounded service	3-7
ACCESS OF OCCUPANT TO OCPD'S	4-23
ACTUAL POWER SINGLE-PHASE	6-3
ACTUAL POWER THREE-PHASE	6-4
AIR-CONDITIONING LOADS	6-42
AMPACITY OF CONDUCTORS FROM GENERATORS	7-6
AMPACITY RATING	8-56
ANGLE PULLS	5-18
APPLICATION AND SELECTION	8-59, 8-68
APPLICATIONS	4-30
APPLYING 50 PERCENT LOAD DIVERSITY FACTOR FOR CONDUCTORS	4-36
APPLYING ADJUSTMENT FACTORS FOR CONDUCTORS	4-34
APPLYING BOTH ADJUSTMENT AND CORRECTION FACTORS FOR CONDUCTORS	4-35
APPLYING CORRECTION FACTORS FOR CONDUCTORS	4-35
APPLYING DEMAND FACTORS	11-8, 11-11
APPLYING DEMAND FACTORS FOR RECEPTACLE LOADS	6-6
APPLYING Ex. 1	8-14
APPLYING Ex. 2	8-14
APPLYING Ex. TO SEC. 220-87	11-30
APPLYING STANDARD CALCULATION DIVIDING AMPS ON LINE	9-28
APPLYING STANDARD CALCULATION DIVIDING VA ON LINE A	9-16
APPLYING STANDARD CALCULATION SERVICE WITH 120 VOLT	9-14
APPLYING STANDARD CALCULATION SERVICE WITH ELECTRIC HEATING AND HEAT PUMP	9-13
APPLYING THE EXCEPTIONS	8-14
APPLYING THE OPTIONAL CALCULATION	9-28, 10-23, 11-27
APPLYING THE OPTIONAL CALCULATION FOR EXISTING UNITS	9-30
APPLYING THE OPTIONAL CALCULATION MULTIFAMILY - DIFFERENT SQUARE FOOTAGE	10-24
APPLYING THE OPTIONAL CALCULATION MULTIFAMILY - DIVIDING VOLT-AMPS ON LINE AND NEUTRAL	10-25
APPLYING THE OPTIONAL CALCULATION MULTIFAMILY - SAME SQUARE FOOTAGE	10-24
APPLYING THE OPTIONAL CALCULATION MULTIFAMILY - SERVICE WITH UNITS AND HOUSE LOADS	10-24
APPLYING THE STANDARD CALCULATION	9-1, 9-10, 10-1, 11-1
APPLYING THE STANDARD CALCULATION MULTIFAMILY	10-15
APPLYING THE STANDARD CALCULATION MULTIFAMILY - DIFFERENT SQUARE FOOTAGE	10-16
APPLYING THE STANDARD CALCULATION MULTIFAMILY - DIVIDING VOLT-AMPS ON LINE AND NEUTRAL	10-21
APPLYING THE STANDARD CALCULATION MULTIFAMILY - SAME SQUARE FOOTAGE	10-15
APPLYING THE STANDARD CALCULATION MULTIFAMILY - SERVICE WITH UNITS HAVING GAS HEATING	10-17
APPLYING THE STANDARD CALCULATION	

SERVICE WITH ELECTRIC HEATING 9-12
APPLYING THE STANDARD CALCULATION
SERVICE WITH GAS HEAT 9-10
ATTACHMENT 3-32
AUXILIARY GUTTERS 5-22

B

BACK-FED DEVICES 2-12
BALANCER SETS 7-5
BANKS SUPPLIED BY 120/208 VOLT,
THREE-PHASE POWER 11-26
BONDING 3-26, 3-32, 3-34
BONDING JUMPER 3-10
BONDING JUMPERS 3-28
BONDING OTHER ENCLOSURES 3-28
BOXES 5-6
BRANCH-CIRCUIT AND FEEDER-CIRCUIT
CONDUCTORS 8-2
BRANCH-CIRCUIT CONDUCTORS 4-38, 8-63
BRANCH-CIRCUIT REQUIREMENTS 8-70
BRANCH-CIRCUIT TAPS 4-7
BRANCH-CIRCUITS LARGER THAN 50 AMPS 6-19
BUSHINGS 7-9

C

CALCULATING AMPS 6-2
CALCULATING LOAD FOR CONDUCTORS 4-34
CALCULATING PRI. AND SEC.
CURRENTS OF XTMR'S 7-10
CANNOT BE LOCKED IN THE
OPEN POSITION 8-43
CAPACITOR 8-51
CIRCUIT AND SYSTEM GROUNDING 3-1
CIRCUIT BREAKERS 4-28
circuits 3-8, 3-9
CLAMP FILL 5-4
CLASS 1 CIRCUITS 8-54
CLASS 2 AND 3 CIRCUITS 8-54
CLEARANCE FROM BUILDING OPENINGS 1-5
CLEARANCES 1-7, 2-7
CODE LETTERS 8-16
COLUMN A 6-13
COLUMN A TABLE 220-19 6-13
COLUMN B 6-13
COLUMN B TABLE 220-19 6-13
COLUMN C 6-13
COLUMN C TABLE 220-19 6-13
COMBINATION LOAD 8-64
COMMERCIAL AND INDUSTRIAL 6-28
COMMERCIAL COOKING EQUIPMENT 6-38
Commercial Calculations 11-1
COMPRESSOR LOADS 11-11
Compressors 8-1

COMPUTED LOAD 6-35
CONCRETE-ENCASED ELECTRODE 3-19, 3-24
CONDUCTOR FILL 5-2, 5-3
CONDUCTOR GROUNDED FOR AC SYSTEM 3-9
CONDUCTORS 6-20, 6-40, 6-41, 6-43, 6-44
conductors 3-8, 3-9
CONDUCTORS FROM
GENERATOR TERMINALS 4-9
CONDUCTORS IN PARALLEL 4-54
CONDUIT BODIES 5-16
CONNECTIONS OF 10 FT. FROM
SECONDARY OF XTMR 4-46
CONNECTIONS OF 25 FT. FROM
SECONDARY OF XTMR 4-46
CONSTANT-VOLTAGE GENERATORS 7-2
CONTINUOUS AND NONCONTINUOUS
OPERATION 11-10
CONTINUOUS DUTY 6-29
CONTINUOUS LOAD 2-11
CONTINUOUS OPERATED LOADS 6-5
CONTROL TRANSFORMER
CIRCUIT CONDUCTORS 8-47
CONTROLLERS FOR MOTOR
COMPRESSORS 8-65
COOKING EQUIPMENT LOADS 10-4
COOKING EQUIPMENT LOADS COLUMN 2 9-2
CORD AND ATTACHMENT PLUG
CONNECTED MOTOR COMPRESSORS 8-69
CORD-AND-PLUG CONNECTED COMPUTERS 3-36
CORD-AND-PLUG CONNECTED MOTORS 8-40

D

DELTA BREAKERS 2-12
DEMAND FACTORS 6-5, 6-7, 9-2, 10-2
DEMAND LOAD 2 .... 9-2, 9-4, 10-4, 10-6, 10-7, 10-8
DEMAND LOAD 2 - THREE-PHASE SYSTEM 10-8
DEMAND LOAD 2 TABLE 220.55, COLUMN C 9-4
DEMAND LOAD 2 TABLE 220.55, NOTE 1 9-4
DEMAND LOAD 2 TABLE 220-55, NOTE 2 9-5
DEMAND LOAD 2 TABLE 220-55, NOTE 4 9-6
DERATING AMPACITY OF CONDUCTORS 4-34
DERATING BY TABLE 310-15(b)(2)(a)
AMBIENT TEMPERATURES 6-23
DESIGN E MOTORS 8-35, 8-38
DESIGNING MOTOR CURRENTS NOT
LISTED IN TABLES 430-147 THRU 430-150 8-55
DETERMINING AMPERAGE 6-27
DEVICE BOXES 5-11
DEVICE OR EQUIPMENT FILL 5-4
DIFFERENCE 2-8
DIFFERENT SIZE CONDUCTORS 5-8, 5-10, 5-12, 5-14
DISCONNECT REQUIRED FOR EACH 1-16
DISCONNECTING

MEANS 6-41, 6-42, 6-44, 6-45, 8-57, 8-70	GROUNDED SERVICE CONDUCTOR 3-27
disconnecting means 3-6, 3-7	GROUNDED SYSTEM 3-12
DISCONNECTING MEANS FOR	GROUNDED SYSTEMS 3-7
SERVICE EQUIPMENT 1-13	GROUNDING 3-34, 7-22
DOORWAYS 7-23	grounding 3-7, 3-8, 3-9
DRAINAGE 7-24	Grounding and Bonding 3-1
DRYER EQUIPMENT LOADS 6-14	grounding conductors 3-8, 3-9
DRYER LOAD 9-7, 10-11, 10-12	GROUNDING ELECTRODE 3-6, 3-11
	GROUNDING ELECTRODE CONDUCTOR 3-11
E	GROUNDING ELECTRODE SYSTEM 3-20
ELECTRIC DISCHARGE LOADS 6-30	grounding electrode system 3-8
ENCLOSING DIFFERENT SIZE CONDUCTORS 5-20	GROUNDING OF PANELBOARDS 2-13
ENCLOSING THE SAME SIZE CONDUCTORS 5-20	GROUNDING SEPARATELY
enclosure 3-7, 3-8	DERIVED AC SYSTEMS 3-9
Equipment 3-7, 3-8, 3-9	GROUNDING SERVICE SUPPLIED
EQUIPMENT BONDING JUMPER	BY AC SYSTEM 3-3
ON LOAD SIDE OF SERVICE 3-30	GROUNDING SUBPANELS
EQUIPMENT GROUNDING CONDUCTOR 3-7, 4-52	AND EQUIPMENT 3-16
EQUIPMENT GROUNDING CONDUCTOR FILL 5-5	GROUNDING USING SWITCHBOARD
EXCEPTION (a) THROUGH (e) 7-6	FRAMES 2-8
	GROUPING 1-15
F	GUARDING 7-21
FEEDER CIRCUIT PROTECTION -	GUARDS FOR ATTENDANTS 7-8
OVER 600 VOLTS 4-31	GUTTER SPACE 5-21
FEEDER TO MOBILE HOME -	
STANDARD CALCULATION 9-38	**H**
FEEDER-CIRCUIT CONDUCTORS 4-38	HEAT OR A/C LOADS 9-29, 11-13
Feeder-Circuits and Branch-Circuits 6-1	HEATING LOADS 6-41
FINDING AMPERAGE 7-11	HIGH-LEG MARKING 2-2
FINDING AMPERAGE SINGLE-PHASE CIRCUITS 6-28	HOSPITAL BUILDING SUPPLIED BY
FINDING AMPERAGE THREE-PHASE CIRCUITS 6-28	277/480 VOLT, THREE-PHASE POWER SOURCE 11-23
FIXED APPLIANCE LOAD 9-6, 10-11	HOT TUBS, SPAS, AND HYDROMASSAGE
FUSE MARKINGS 4-26	TUBS 3-34
FUSES 1-19, 4-25	HOTELS AND MOTELS SUPPLIED BY 120/208
	VOLT, THREE- PHASE POWER SOURCE 11-25
G	HUNG FROM WALL OR CEILING 7-12
GENERAL LIGHTING AND RECEPTACLE	HYDROMASSAGE BATHTUBS 3-35
LOADS AND SMALL APPLIANCE PLUS	
LAUNDRY LOADS 9-2, 10-2	**I**
GENERAL LIGHTING LOAD 10-2	IDENTIFYING HIGHER VOLTAGE-TO-GROUND 1-13
GENERAL LIGHTING LOADS 11-3	INDICATING 4-29
GENERAL PURPOSE 6-36	INDIVIDUAL 6-37
GENERAL-PURPOSE CIRCUITS 6-11	INDIVIDUAL CIRCUITS 6-12
Generators and Transformers 7-1	INDUSTRIAL AND COMMERCIAL BUILDINGS 3-21
GENERATORS RATED AT 65 VOLTS 7-4	INSTALLATION 2-6
GROUND RING 3-20, 3-25	INSTALLATION OF EQUIPMENT
GROUND-FAULT PROTECTION 1-18	BONDING JUMPER 3-31
GROUNDED AND UNGROUNDED SYSTEMS 3-12	INSTALLING TRANSFORMERS 7-11
GROUNDED CIRCUIT CONDUCTOR	INSTANTANEOUS TRIP CIRCUIT BREAKER 8-42
FOR GROUNDING EQUIPMENT 3-14	INSTANTANEOUS TRIP CIRCUIT BREAKERS 8-19
GROUNDED CONDUCTOR 3-8, 4-51	INVERSE TIME CIRCUIT BREAKERS 8-27
GROUNDED CONDUCTOR BROUGHT	INVERSE-TIME CIRCUIT BREAKERS 8-19, 8-36
TO SERVICE EQUIPMENT 3-5	ISOLATED EQUIPMENT GROUNDING
	CONDUCTOR 2-14

III

J

JUNCTION BOXES 5-17

K

KITCHEN EQUIPMENT 11-28

L

LARGEST LOAD BETWEEN HEAT AND A/C 9-9, 10-13
LARGEST MOTOR LOAD 9-9, 10-13, 11-13
LAUNDRY CIRCUIT 6-12
LIGHTING AND APPLIANCE PANELBOARD 2-9
LIGHTING LOADS 6-28, 11-2
LIGHTING TRACK LOADS 6-33
LISTED OCCUPANCIES 11-3
LOADING MORE THAN HALF
OF THE CONDUCTORS 4-36
LOADS 6-1
LOCATION 1-13, 7-12, 8-59
LOCATION OF GENERATORS 7-2
LOCATION OF THE DISCONNECTING
MEANS FOR THE CONTROL 8-42
LOCATION OF TRANSFORMER VAULTS 7-23
LOCKED IN THE OPEN POSITION 8-43
LOCKED-ROTOR CURRENT UTILIZING
CODE LETTER 8-16
LOCKED-ROTOR CURRENT UTILIZING HP 8-17
LOW-VOLTAGE LIGHTING LOAD 11-6

M

MADE AND OTHER ELECTRODES 3-25
MADE ELECTRODES 3-19
MAGNETIC STARTER CONTACTOR
AND ENCLOSURE 8-53
MAIN EQUIPMENT BONDING JUMPERS 3-29
MARKING 4-30
MAXIMUM NUMBER 1-14
MAXIMUM SIZE OVERLOAD PROTECTION 8-33
METAL FRAME OF BUILDING 3-23
METAL WATER PIPE 3-18, 3-21
METAL WATER PIPING 3-22
METHODS 3-27
METHODS OF COUNTING CONDUCTORS 5-1
MINIMUM SIZE OVERLOAD PROTECTION 8-33
MINIMUM SIZE SERVICE 9-9
MINIMUM WIRE BENDING SPACE 2-4
MORE THAN ONE BUILDING 1-16
MOTOR COMPRESSOR CONTROLLER RATING 8-65
MOTOR COMPRESSORS AND EQUIPMENT
ON 15 OR 20 AMP BR 8-68
MOTOR CONTROL AND MOTOR
POWER CIRCUIT CONDUCTORS 8-50
MOTOR LOADS 6-44, 11-12
MOTOR NOT OVER 1 HP 8-28

MOTORS OVER 2 HP THROUGH 100 HP 8-39
MOUNTED IN CEILING 7-12
MULTIOUTLET ASSEMBLIES 11-9
MULTIWIRE BRANCH-CIRCUITS 6-38

N

NAMEPLATE LISTING 8-55
NAMEPLATE MARKINGS 7-2
NEUTRAL 6-6
NONCONTINUOUS DUTY 6-29
NONCONTINUOUS OPERATED LOADS 6-4
NONSUPERVISED LOCATIONS 7-15
NONTIME-DELAY FUSES 8-18, 8-21
NOTE 1 6-13
NOTE 1 TABLE 220-19 6-13
NOTE 2 6-14
NOTE 2 TABLE 220-19 6-14
NOTE 4 6-14
NOTE 4 TABLE 220-19 6-14
NUMBER 1-10
NUMBER OF SERVICE DROPS OR LATERALS 1-1
NUMBER OF SERVICES PERMITTED TO A BUILDING 1-1

O

OCCUPYING THE SAME ENCLOSURE 8-52
OCPD'S FOR SYSTEMS OVER 600 VOLTS 4-31
OCPD'S PROVIDED 7-8
OCPD'S RATED 100 AMPS OR LESS 4-32
OCPD'S RATED OVER 100 AMPS 4-32
OCTAGON BOXES 5-7
OFFICE BUILDING SUPPLIED BY 277/480 VOLT,
THREE-PHASE POWER SOURCE 11-19
OPTIONAL CALCULATIONS FOR ADDITIONAL
LOADS TO EXISTING INSTALLATIONS 11-29
OTHER BOXES 5-12
OTHER DEVICES 3-28
OTHER GROUP INSTALLATIONS 8-32
OTHER LOADS 6-29, 9-30, 9-31
OTHER LOADS ADDED 120 VOLT,
A/C WINDOW UNITS 9-37
OTHER LOADS ADDED A/C LOAD 9-31
OTHER LOADS ADDED APPLIANCE LOAD 9-31
OTHER THAN HP RATED 8-35, 8-38
OUTLETS 6-31
OUTSIDE FEEDER TAPS 4-46
OUTSIDE LIGHTING LOAD 11-6
OUTSIDE OF THE BUILDING 1-5
OUTSIDE SIGN LIGHTING LOAD 11-6
OVER 250 VOLTS 3-29
OVERCURRENT
PROTECTION 6-40, 6-42, 6-43, 6-45, 7-14
Overcurrent Protection Devices

and Conductors 4-1
OVERCURRENT PROTECTION FOR
GENERATORS 7-2
OVERCURRENT PROTECTION FOR
SERVICE EQUIPMENT 1-16
OVERHEAD SERVICES 1-6

P

PANELBOARDS 2-8, 5-22
PERFORMANCE OF FAULT CURRENT PATH 3-14
PERMANENT-WIRED COMPUTERS 3-36
PERMISSIBLE LOADS 6-15
PHASE ARRANGEMENT 2-3
PLASTER RINGS AND EXTENSION RINGS 5-15
PLATE ELECTRODES 3-26
POINT OF ATTACHMENT 1-9
PORTABLE MOTOR OF 1/3 HORSEPOWER
OR LESS 8-34
POWER FACTOR 6-2
POWER PANEL 2-9
PREVENTING OVERLOAD CONDITIONS 7-6
PRIMARY AND SECONDARY -
600 VOLTS OR LESS 7-19
PRIMARY AND SECONDARY -
OVER 600 VOLTS 7-14
PRIMARY ONLY -
600 VOLTS OR LESS 7-17
PRIMARY ONLY -
OVER 600 VOLTS 7-14
PROTECTING FIXTURE CONDUCTORS 4-50
PROTECTION OF CIRCUIT CONDUCTORS 4-36
PROTECTION OF CONDUCTORS 8-48, 8-49
PROTECTION OF LIVE PARTS 7-8

R

RACEWAYS 8-52
Raceways, Gutters, Wireways, and Boxes 5-1
RATING 1-15, 6-34
RATING AND INTERRUPTING CAPACITY 8-57
RATING AND SETTING FOR INDIVIDUAL
MOTOR COMPRESSOR 8-60
RATING OR SETTING FOR EQUIPMENT 8-60
RATINGS 6-15
READILY ACCESSIBILITY OF OCPD'S 4-22
RECEPTACLE LOADS 6-36, 11-8
REPLACEMENT OF NONGROUNDING-
TYPE RECEPTACLES 3-12
REQUIRED 1-17
RESIDENTIAL 6-10
Residential Calculations Single-Family Dwellings 9-1
Residential Calculations Multifamily Dwellings 10-1
RESIDENTIAL COOKING EQUIPMENT 6-13
RESISTANCE OF MADE ELECTRODES 3-31
RESTAURANT SUPPLIED BY 120/208 VOLT,
THREE-PHASE POWER SOURCE 11-22
RESTAURANTS 11-29
ROD AND PIPE ELECTRODES 3-25
ROOM AIR-CONDITIONERS 8-70
ROUNDING UP OR DOWN OF OCPD 4-2
RUNNING OVERLOAD PROTECTION
FOR THE MOTOR 8-32

S

SAME SIZE CONDUCTORS 5-8, 5-10, 5-11, 5-13
SCHOOL BUILDING SUPPLIED BY 277/480 VOLT,
THREE-PHASE POWER SOURCE 11-21
SCHOOLS 11-29
SERVICE CONDUCTORS 4-38
service equipment 3-8
SERVICE EQUIPMENT BONDING 3-26
SERVICE MASTS 1-9
SERVICE MUST NOT PASS THROUGH
ONE BUILDING TO SUPPLY ANOTHER 1-5
SERVICE-ENTRANCE CONDUCTORS 1-10
Services 1-1
SETTING 1-19
SHOW WINDOW LIGHTING LOAD 11-4
SHOW WINDOW LOADS 6-32
SIGN LOADS 6-34
SINGLE MOTOR-COMPRESSORS 8-64
SINGLE-BRANCH CIRCUIT TO SUPPLY
TWO OR MORE MOTORS 8-28
SINGLE-PHASE CIRCUITS 6-8
SIZE AND RATING 1-7, 1-11
SIZE AND RATING - LATERAL 1-10
SIZE REQUIRED 6-34
SIZING 4-26, 4-28, 6-6
SIZING AND SELECTING OCPD'S 8-18
SIZING BUSWAY TAPS 4-44
SIZING CABLE TRAYS 5-23
SIZING CONDUCTORS FOR ADJUSTABLE
SPEED DRIVE SYSTE 8-5
SIZING CONDUCTORS FOR
CONTROL CIRCUIT 8-45
SIZING CONDUCTORS FOR DUTY
CYCLE MOTORS 8-5, 8-8
SIZING CONDUCTORS FOR
MOTOR CIRCUIT TAPS 4-44
SIZING CONDUCTORS FOR
MULTISPEED MOTORS 8-3
SIZING CONDUCTORS FOR
PART-WINDING MOTORS 8-7
SIZING CONDUCTORS FOR
SEVERAL MOTORS 8-8
SIZING CONDUCTORS FOR
SINGLE MOTORS 8-2
SIZING CONDUCTORS FOR
SINGLE-PHASE MOTORS 8-2

SIZING CONDUCTORS FOR TAPS 4-41	SECONDARY CONNECTIONS 4-11
SIZING CONDUCTORS FOR TAPS 10 FT. OR LESS IN LENGTH 4-42	SIZING OCPD FOR PHASE CONVERTERS 4-13
SIZING CONDUCTORS FOR TAPS AND CONNECTIONS INCLUDING TRANSFORMERS 4-43	SIZING OCPD FOR PROTECTION OF EQUIPMENT 4-2
SIZING CONDUCTORS FOR TAPS OVER 10 FT. TO 25 FT. IN LENGTH 4-43	SIZING OCPD FOR REMOTE-CONTROL, SIGNALING, AND POWER-LIMITED CIRCUIT CONDUCTOR 4-17
SIZING CONDUCTORS FOR TAPS OVER 25 FT. UP TO 100 FT. IN LENGTH 4-43	SIZING OCPD FOR TAPS 4-4
SIZING CONDUCTORS FOR THREE-PHASE MOTORS 8-2	SIZING OCPD FOR TAPS 10 FT. OR LESS IN LENGTH 4-4
SIZING CONDUCTORS FOR WELDER LOADS 4-47	SIZING OCPD FOR TAPS INCLUDING TRANSFORMER 4-5
SIZING CONDUCTORS FOR WYE-START AND DELTA RUN MOTOR 8-4	SIZING OCPD FOR TAPS OVER 10 FT. TO 25 FT. IN LENGTH 4-4
SIZING CONDUCTORS FROM GENERATOR TERMINALS 4-47	SIZING OCPD FOR TAPS OVER 25 FT. UP TO 100 FT. IN LENGTH 4-6
SIZING CONDUCTORS SUPPLYING MOTORS AND OTHER LOADS 8-9	SIZING OCPD FOR TRANSFORMER SECONDARY CONDUCTORS 4-16
SIZING CONDUCTORS TO AC/DC ARC WELDERS 4-47	SIZING OCPD FOR TWO OR MORE HERMETIC MOTORS 8-61
SIZING CONDUCTORS TO MOTOR OR NONMOTOR GENERATOR ARC WELDER 4-49	SIZING OCPD FOR TWO OR MORE MOTORS 8-27
SIZING CONDUCTORS TO RESISTANCE WELDERS 4-49	SIZING OCPD'S ABOVE THE AMPACITY OF THE CONDUCTORS 4-40
SIZING CONDUITS OR TUBING 5-20	SIZING OCPD'S BELOW THE AMPACITY OF THE CONDUCTORS 4-41
SIZING CONTROL CIRCUIT CONDUCTORS INSIDE ENCLOSURE 8-45	SIZING OCPD'S RATED 800 AMPS OR LESS 4-3
SIZING FIXTURE WIRE TO FLUORESCENT LAY-INS 4-50	SIZING OCPD'S RATED OVER 800 AMPS 4-3
SIZING GROUNDING ELECTRODE CONDUCTOR 3-18	SIZING OCPD'S TO ALLOW MOTORS TO START AND RUN 8-20
SIZING MAXIMUM OCPD 8-21	SIZING REMOTE CONTROL CIRCUIT CONDUCTORS 8-45
SIZING NIPPLES 5-21	SIZING TAPS 6-14
SIZING OCPD FOR AC AND REFRIGERATION EQUIPMENT CIRCUIT CONDUCTORS 4-13	SIZING THE BRANCH-CIRCUIT PROTECTIVE DEVICE 8-9
SIZING OCPD FOR CONDUCTORS OF 25 FT. FROM SECONDARY OF TRANSFORMER 4-10	SIZING THE CONTROLLER TO START AND STOP THE MOTOR 8-34
SIZING OCPD FOR CONTROL CIRCUIT 8-47	SIZING THE DISCONNECTING MEANS TO DISCONNECT BOTH 8-38
SIZING OCPD FOR ELECTRIC WELDER CIRCUIT CONDUCTORS 4-16	SMALL APPLIANCE AND LAUNDRY LOAD 10-2
SIZING OCPD FOR FIRE ALARM CIRCUIT CONDUCTORS 4-17	SMALL APPLIANCE CIRCUITS 6-11
SIZING OCPD FOR HERMETIC MOTOR AND OTHER LOADS WHEN A HERMETICALLY SEALED MOTOR IS THE LARGEST 8-61	SMALLEST RATED MOTOR PROTECTED 8-31
	SNAP SWITCHES RATED AT 30 AMPS OR LESS 2-10
SIZING OCPD FOR HERMETIC MOTORS AND OTHER LOADS WHEN A HERMETICALLY SEALED MOTOR IS THE LARGEST 8-62	SPECIAL APPLIANCE LOADS 11-10
	SPLIT-BUS PANEL 2-9
SIZING OCPD FOR MOTOR AND MOTOR CONTROL CIRCUIT CONDUCTOR 4-12	SQUARE BOXES 5-9
SIZING OCPD FOR MOTOR CIRCUIT TAPS 4-9	STANDARD CALCULATIONS 11-15
SIZING OCPD FOR MOTOR-OPERATED APPLIANCE CIRCUIT CONDUCTOR 4-12	STATIONARY AND PORTABLE MOTORS 8-37
	STATIONARY MOTOR 1/8 HORSEPOWER OR LESS 8-34
SIZING OCPD FOR OUTSIDE	STATIONARY MOTORS 8-36, 8-38, 8-39, 8-40
	STORAGE IN VAULTS 7-25
	STORE BUILDING SUPPLIED BY 120/208, THREE-PHASE VOLT POWER SOURCE 11-16
	STORE BUILDING SUPPLIED BY 120/240

VOLT, SINGLE-PHASE POWER SOURCE 11-15
STORE BUILDING SUPPLIED BY 120/240
VOLT, THREE-PHASE POWER SOURCE 11-18
STORE BUILDING SUPPLIED BY 277/480
VOLT, THREE-PHASE POWER SOURCE 11-17
STRAIGHT PULLS 5-17
STRUCTURAL STEEL 3-23
SUPERVISED LOCATIONS 7-16
SUPPLIED THROUGH A TRANSFORMER 2-12
SUPPLY FIXTURE WIRES FOR OUTSIDE
LIGHTING STANDARD 4-51
SUPPORT AND ARRANGEMENT OF
BUSBARS AND CONDUCTORS 2-1
SUPPORT FITTINGS FILL 5-4
SWIMMING POOLS 3-32
SWITCHBOARDS 2-6
Switchboards and Panelboards 2-1
SWITCHBOARDS AND PANELBOARDS ART 2-1

T

TABLE 450-3(a) 7-14
TERMINAL RATINGS 4-31
THREADED CONNECTIONS 3-27
THREADLESS COUPLINGS AND CONNECTORS ... 3-28
THREE-PHASE CIRCUITS 6-10
THREE-WIRE DC GENERATORS 7-5
TIME-DELAY FUSES 8-19, 8-21
TORQUE MOTOR 8-37, 8-41
TRACK LIGHTING LOAD 11-6
TRANSFORMER 8-50
TWO OR MORE BUILDINGS GROUNDED SYSTEMS 3-7
TWO-WIRE GENERATORS 7-3
TYPES OF EQUIPMENT GROUNDING CONDUCTORS . 3-17

U

UNDERGROUND SERVICES 1-10
UNGROUNDED CONDUCTORS 4-53
UNGROUNDED SYSTEM 3-12
UNGROUNDED SYSTEMS 3-8
UNLISTED OCCUPANCY 11-4
USED AS A SUBPANEL 2-7
USED AS SERVICE EQUIPMENT 2-2
USING 15 OR 20 AMP OCPD 8-62
USING A CORD-AND-PLUG CONNECTION
NOT OVER 250 VOLT 8-63
USING GROUNDED CONDUCTOR -
LOAD SIDE 3-15
USING GROUNDED CONDUCTOR -
SUPPLY SIDE 3-15
USING HANDLE TIES FOR CIRCUIT
BREAKERS 4-18
USING INSTANTANEOUS TRIP CB's 8-16

V

VENTILATION 7-22
VENTILATION OPENINGS 7-24
VERTICAL CLEARANCE FROM GROUND 1-8
VOLTAGE DROP 6-8
VOLTAGE LIMITATIONS 6-24

W

WALLS, ROOF, AND FLOOR 7-23
WATER HEATER LOADS 6-40
WATER PIPES AND ACCESSORIES 7-25
WELDING SHOPS SUPPLIED BY
120/208 VOLT POWER SOURCE 11-27
WITHIN SIGHT 8-42